TRANSPORT THEORY, INVARIANT IMBEDDING, AND INTEGRAL EQUATIONS

LECTURE NOTES

IN PURE AND APPLIED MATHEMATICS

Other Volumes in Preparation

G. Milton Wing

TRANSPORT THEORY, INVARIANT IMBEDDING, AND INTEGRAL EQUATIONS

Proceedings in Honor of G.M. Wing's 65th Birthday

Edited by

Paul Nelson
Texas A&M University
College Station, Texas

V. Faber
Thomas A. Manteuffel
Daniel L. Seth
Andrew B. White, Jr.
Los Alamos National Laboratory
Los Alamos, New Mexico

CRC Press
Taylor & Francis Group
Boca Raton London New York

CRC Press is an imprint of the
Taylor & Francis Group, an **informa** business

CRC Press
Taylor & Francis Group
6000 Broken Sound Parkway NW, Suite 300
Boca Raton, FL 33487-2742

© 1989 by Taylor & Francis Group, LLC
CRC Press is an imprint of Taylor & Francis Group, an Informa business

First issued in paperback 2017

No claim to original U.S. Government works

ISBN-13: 9780824781583 (pbk)
ISBN-13: 9781138441996 (hbk)

Visit the Taylor & Francis Web site at
http://www.taylorandfrancis.com

and the CRC Press Web site at
http://www.crcpress.com

Transport theory, invariant imbedding, and integral equations :
    proceedings in honor of G. M. Wing's 65th birthday / edited by Paul
Nelson ... [et al.] .
        p. cm. -- (Lecture notes in pure and applied mathematics ;
    115)
    Papers presented 1/20-22/88 at Eldorado Hotel, Santa Fe, New
Mexico.
    Includes bibliographies and index.
    ISBN 0-8247-8158-9 (alk. paper)
    1. Transport theory--Congresses. 2. Invariant imbedding-
--Congresses. 3. Integral equations--Congresses. 4. Wing, G. Milton
(George Milton), 1923- .   I. Wing, G. Milton (George Milton),
1923-   . II. Nelson, P. (Paul) III. Series : Lecture notes in pure
and applied mathematics ; v. 115.
QC175.A1T73   1989
530. 1'38--dc19   89-1086
                                                        CIP

Preface

On January 20-22, 1988, over seventy mathematicians, engineers and scientists gathered in the delightful surroundings of the Eldorado Hotel, near the plaza of Milt Wing's much beloved Santa Fe, to honor this truly outstanding individual in celebration of his 65th birthday anniversary. Following this preface Vance Faber has given a well-crafted biographical sketch of G. Milton Wing. The human dimension of Milt comes through in the after-dinner talk (this informal term seems more accurate than the formal "speech" or "lecture") that he gave on the occasion of the conference banquet, which is reprinted verbatim as the last piece of these proceedings. Here I wish primarily to attempt to provide some overall perspective of the conference and these proceedings, particularly the contributions of Milt Wing to the technical subjects represented in the title. These subjects are transport theory, invariant imbedding and integral equations. The technical contributions appearing herein are grouped in this manner, and are described thusly in this preface.

I cannot but open the discussion of transport theory by citing the widely known papers of Milt and Joseph Lehner [1,2]. These remarkably original and beautiful seminal papers on the spectral theory of the transport equation presented results that were so surprising that they provided the stimulus for literally hundreds of subsequent works, and laid the foundation for what remains, over thirty five years later, an active research field. (This conference provided me with my first opportunity to have the pleasure of meeting Joseph Lehner. Surely all present at the conference banquet will remember his admonition to Milt to the effect that "if the toothbrush is wet, then you have already brushed your teeth"—at least until we ourselves need that advice.) It also seems appropriate here to mention Milt's monograph on transport theory [3], from which many currently active researchers in the field—including the undersigned—learned to love (easy) and understand (not always so easy) the subject.

The first of the transport-theoretic papers is by G. C. Pomraning, C. D. Levermore and J. Wong (presented by Pomraning), and describes analytic results, obtained by a variety of methods, for (linear) particle transport in a random mixture of two immiscible fluids. Next V. Faber and Thomas A. Manteuffel, in a paper presented by Manteuffel, show that the conjugate gradient method provides substantial savings over the standard Neumann series solution (source iteration) of the transport equation, both in standard form and as preconditioned by the inverse of a diffusion operator (i.e. diffusion synthetic acceleration). Next follows a group of three contributions from Italian participants. The first of these, by M. Celaschi and B. Montagnini (presented by Montagnini) continues the diffusion approximation theme of the previous paper by presenting a quasi-diffusive approximation, in which Fick's law is generalized by replacing the diffusion coefficient by a suitably defined integral operator. Next L. Arlotti presents some existence and asymptotic behavior results for a time-dependent spatially homogeneous model of electron swarms. In the third of this group of papers, V. G. Molinari and P. Peerani (presented by Molinari) formally study dispersion and damping for an electron plasma in a (specified) electron field. The theme of linear transport theory is continued in the next paper, by A. K. Prinja, who uses a variational principle to develop backward (adjoint) transport equations and boundary conditions in the context of radiation damage studies. In the last of the contributions relating to linear microscopic transport, Y-W H. Liu presents a numerical study of a P-1 transverse leakage nodal spatial approximation.

Next there follows a group of three papers that are directed toward nonlinear microscopic transport. In the first of these, V. C. Boffi derives, from the nonlinear Boltzmann equation, conservation laws for a mixture of N different species of interacting particles, and discusses in some detail the solution of this system in the case $N = 2$. Next Peter Takáč discusses supersolutions and subsolutions of the full (time dependent spatially inhomogeneous force-free) Boltzmann equation, culminating with a weak L^1 convergence theorem for supersolutions. In the last of this group of papers, Hans G. Kaper and Man Kam Kwong, in a paper presented by Kaper, demonstrate the utility of classical comparison theorems in obtaining concavity and monotonicity results for nonlinear partial differential equations of the type arising in the study of reaction-diffusion phenomena. Following this there is a set of three contributions relating to macroscopic transport models (e.g. fluid dynamics). In the first of these, B. R. Wienke and B. A. Hills (presented by Wienke) consider models of the complex phenomenon of transport of gases (especially nitrogen) in biological systems, with particular reference to the application to decompression profiles for deep-water diving. Next, G. H. Pimbley exploits semigroup theory to study isentropic hydrodynamic flow in one spatial dimension. In the concluding paper of this group, H. W. Chiang presents an analytical study of mass and heat transfer and Brownian particle deposition from a fluid stream onto two spheres of unequal size that are in contact. In the concluding paper on transport theory, and the only one treating a discrete transport model *per se*, David C. Torney obtains an asymptotic (in N) estimate for the dominant eigenvalue of a random walk on an $N \times N$ square lattice torus having one trap site.

The subject of invariant imbedding, in its nascent form, stems from viewing transport phenomena primarily at the boundaries (i.e. "input/output"), as opposed to the focus on internal events that characterizes the classical approach via the transport equation. Elements of this idea can be found as far back as the work of Stokes [4] over a century ago, and certainly are present in the World War II era research of Ambarzumian [5] and of Chandrasekhar [6], but the method received much additional impetus—especially among applied mathematicians—from the collaborations of Milt Wing with Richard Bellman and Robert Kalaba, which were carried out in the fertile atmosphere of southern California during the late 1950's and early 60's. (See the list of publications accompanying the following biography of G. Milton Wing for precise literature citations to the many journal articles stemming from these collaborations.) Milt was one of the first to recognize that the basic ideas of invariant imbedding were inherently mathematical, and thus that the method could be carried over to a wide variety of applications. This fact is thoroughly illustrated in the beautifully written 1975 monograph [7] on the subject that was jointly authored by Milt and Richard Bellman. Again this field remains the subject of much currently active research.

In the first of the three papers contained herein that are devoted to this topic, T. W.

Mullikin establishes results regarding asymptotic stability (and instability), in the limit of large slab widths, for the solutions of the algebraic operator equations corresponding to formally neglecting the derivative term in the reflection equation for isotropically scattering homogeneous media. Next, James Corones outlines a concept of discrete invariant imbedding for multidimensional invariant imbedding. Finally, V. Faber, Daniel L. Seth and G. Milton Wing, in a paper presented by Vance Faber, describe some detailed computational algorithms, and associated computational experience, for implementation of the ideas of Corones in the two-dimensional case.

Although the fates kindly arranged it that Milt Wing would devote much of his energy to transport theory and related matters, there is ample evidence that an early and true mathematical love was the subject of integral equations. This first evidences itself in a set of unpublished Los Alamos reports written in the late 1950's. In recent years this has taken the form of active research in the subject of integral equations of the first kind, which persist in arising in diverse applications, notwithstanding the fact that we mathematicians know they cannot really be solved in the common case of imprecise experimental data. Much of the essence of this work is distilled in his widely read "primer" [8] on this subject, and many of the papers on integral equations that appear in these proceedings naturally concern either such equations or applications in which they arise.

In the first paper treating integral equations, James A. Cochran extends, to a wider class of kernels, an asymptotic estimate (in n) recently obtained by himself and M. A. Lukas for the nth eigenvalue of certain positive definite kernels. Next there follows a group of six contributions treating various applications of integral equations of the first kind. In the first of these, C. W. Groetsch provides a convergence analysis for an iterative method that has been widely applied to a model problem arising in Fourier optics. Next, Joseph F. McGrath, Stephen Wineberg, George Charatis and Robert J. Schroeder, in a paper presented by McGrath, show that the problem of determining internal properties of a plasma from its radiation profile gives rise to inversion of an Abel transform (cylindrical symmetry) or a Radon transform (general case), and they discuss aspects of accomplishing such inversions numerically. This is followed by a contribution from B. D. Ganapol, which is devoted to a certain scheme for numerically inverting Laplace transforms, and to experiences with this technique for problems arising in transport theory. Next in this group of six papers there is a subgroup of two papers devoted to some nonlinear problems that arise in certain applications. In the first of these, Kenneth M. Hanson describes the *maximum a posteriori* approach to inversion of the Abel transform arising in radiography of a cylindrically symmetric object, with consideration of the nonlinearity arising from background (film fogging, scattering). In the second paper of this subgroup, Katarzyna Jonca and Curtis R. Vogel, in a paper presented by Vogel, describe application of an algorithm based upon Tikhonov regularization and quasi-Newton/trust region procedures, for numerical solution of a model arising from the problem of determining, from airborne measurements of the local magnetic field, the location of the boundary between a layer of magnetized igneous rock and overlying unmagnetized sediments. In the final paper of this group of six, John D. Zahrt and Natalie D. Carter (presented by Zahrt) show how the problem of inverting Weierstrass transforms arises in analyzing small material samples by x-ray fluorescence, and describe an approach to such inversions that is based upon tabulation of transform pairs. Next, Kevin Cooper and P. M. Prenter, in a paper presented by Prenter, describe the theory and practice of the novel *multi-boundary alternating direction collocation* scheme for the numerical simulation of fluid flow around bodies of complex shapes. In the final technical paper of these proceedings, Yanping Lin uses a nonclassical H^1 projection to obtain optimal L^2 error estimates for various numerical approximations to the solution of a certain nonlinear integrodifferential equation of parabolic type.

The following papers that were presented at the conference were unavailable for inclusion in these proceedings (asterisks indicate speakers):

D. Allen Barnett, Jr. (Rensselaer Polytechnic Institute) and J. E. Morel* (Los Alamos National Laboratory), "A Multigrid Acceleration Method for the 1-D S_N Equations."

Paul Nelson (Texas A&M University), "A Survey of Rigorous Results in Numerical Transport Theory."

R. Sanchez (Centre d'Etudes Nucléaires de SACLAY), "Neumann Series Solutions of the Neutron Kinetic Equation."

H. B. Keller (California Institute of Technology), "Invariant Imbedding and Numerical Methods for Two Point Boundary Value Problems."

R. C. Allen, L. Baca and D. Womble* (Sandia National Laboratories, Albuquerque), "Invariant Imbedding and the Method of Lines for Multiprocessor Computers."

Jonq Juang (Wright State University), "The Abstract Riccati Equation Arising in Transport Theory."

S. Elaydi* and S. Sivasundaram (University of Colorado at Colorado Springs), "Stability of Volterra Equations with Infinite Delay."

M. Z. Nashed (University of Delaware), "Classes of Kernel Functions and Numerical Analysis of Integral Equations of the First Kind."

F. Hagin (Colorado School of Mines), "Some Observations about Fourier Integral Operators and Classical Asymptotics."

Many people and organizations contributed to the success of the conference and the preparation of these proceedings. The Organizing Committee consisted of V. Faber, T. Manteuffel, P. Nelson and A. B. White. Daniel L. Seth performed yeoman service, both as Conference Supervisor in seeing to the myriad of details attendant to any conference, and in similarly helping with many details necessary to the preparation of these proceedings. Gratitude is due the Los Alamos National Laboratory for financial support. Ms. Janet Hull of the laboratories' protocol office efficiently handled the many details of hotel accommodations, conference rooms, etc. Last, but certainly not least, Ms. Sheila Girard, of Group C-3 at Los Alamos, contributed most significantly in many ways as Conference Secretary.

The technical contributions to these proceedings were refereed.

Paul Nelson

References

[1] J. Lehner and G. M. Wing, "On the Spectrum of an Unsymmetric Operator Arising in the Transport Theory of Neutrons," *Communications on Pure and Applied Mathematics*, **8**: 217-234 (1955).

[2] J. Lehner and G. M. Wing, "Solution of the Linearized Boltzmann Equation for the Slab Geometry," *Duke Mathematical Journal*, **23**: 125-142 (1956).

[3] G. M. Wing, *Transport Theory*, John Wiley & Sons, New York-London, 1962.

[4] G. C. Stokes, "On the Intensity of Light Reflected or Transmitted through a Pile of Plates", *Mathematical and Physical Papers*, **2**, Cambridge University Press, 1883.

[5] V. A. Ambarzumian, "Diffuse Reflection of Light by a Foggy Medium," *C. R. Acad. Sci. SSR*, **38**: 229 (1943).

[6] S. Chandrasekhar, *Radiative Transfer*, Oxford University Press, Oxford, 1950.

[7] R. Bellman and G. M. Wing, *An Introduction to Invariant Imbedding*, John Wiley & Sons, New York-London-Sydney-Toronto, 1975.

[8] G. Milton Wing, "A Primer on Integral Equations of the First Kind: The Problems of Deconvolution and Unfolding," Report No. LA-UR-84-1234, Los Alamos National Laboratory, April 1984.

Contents

Contributors

L. ARLOTTI   Mathematical Institute, Faculty of Engineering, Ancona, Italy

V. C. BOFFI   Nuclear Engineering Laboratory of the University of Bologna, Bologna, Italy

NATALIE D. CARTER   Diagnostic Physics, Los Alamos National Laboratory, Los Alamos, New Mexico

M. CELASCHI   Department of Mechanics and Nuclear Construction, University of Pisa, Pisa, Italy

GEORGE CHARATIS   Experimental Physicist, KMS Fusion, Inc., Ann Arbor, Michigan

H. W. CHIANG   Whiteshell Nuclear Research Establishment, Atomic Energy of Canada Limited, Pinawa, Manitoba, Canada

JAMES A. COCHRAN   Department of Mathematics, Washington State University, Pullman, Washington

KEVIN COOPER   Mathematics Department, Colorado State University, Fort Collins, Colorado

JAMES CORONES   Applied Mathematical Sciences, Ames Laboratory—USDOE, and Department of Mathematics, Iowa State University, Ames, Iowa

V. FABER   Computer Research and Applications, Los Alamos National Laboratory, Los Alamos, New Mexico

B. D. GANAPOL   Department of Nuclear and Energy Engineering, University of Arizona, Tucson, Arizona

C. W. GROETSCH   Department of Mathematical Sciences, University of Cincinnati, Cincinnati, Ohio

KENNETH M. HANSON   Los Alamos National Laboratory, Los Alamos, New Mexico

B. A. HILLS   Hyperbaric Physiology, Marine Biomedical Institute, Galveston, Texas

KATARZYNA JONCA   Department of Mathematical Sciences, Montana State University, Bozeman, Montana

HANS G. KAPER   Mathematics and Computer Science Division, Argonne National Laboratory, Argonne, Illinois

MAN KAM KWONG*   Mathematics and Computer Science Division, Argonne National Laboratory, Argonne, Illinois

C. D. LEVERMORE   Department of Mathematics, University of Arizona, Tucson, Arizona

YANPING LIN   Department of Pure and Applied Mathematics, Washington State University, Pullman, Washington

Y-W H. LIU   National Tsing Hua University, Department of Nuclear Engineering, Hsinchu, Taiwan, Republic of China

THOMAS A. MANTEUFFEL   Computer Research and Applications, Los Alamos National Laboratory, Los Alamos, New Mexico

JOSEPH F. MCGRATH   Computational Mathematician, KMS Fusion, Inc., Ann Arbor, Michigan

V. G. MOLINARI   Nuclear Engineering Laboratory, University of Bologna, Bologna, Italy

B. MONTAGNINI   Department of Mechanical and Nuclear Construction, University of Pisa, Pisa, Italy

T. W. MULLIKIN   Department of Mathematics, Purdue University, West Lafayette, Indiana

P. PEERANI   Nuclear Engineering Laboratory, Laboratorio di Ingegneria Nucleare, University of Bologna, Bologna, Italy

G. H. PIMBLEY   Los Alamos National Laboratory, Los Alamos, New Mexico

G. C. POMRANING   School of Engineering and Applied Science, University of California at Los Angeles, Los Angeles, California

P. M. PRENTER   Mathematics Department, Colorado State University, Fort Collins, Colorado

A. K. PRINJA   Chemical and Nuclear Engineering Department, University of New Mexico, Albuquerque, New Mexico

ROBERT J. SCHROEDER   Experimental Physicist, KMS Fusion, Inc., Ann Arbor, Michigan

DANIEL L. SETH   Computer Research and Applications, Los Alamos National Laboratory, Los Alamos, New Mexico

PETER TAKÁČ   Department of Mathematics, Vanderbilt University, Nashville, Tennessee

DAVID C. TORNEY   Theoretical Biology and Biophysics, Los Alamos National Laboratory, Los Alamos, New Mexico

*Permanent address: Department of Mathematical Sciences, Northern Illinois University, DeKalb, Illinois

CURTIS R. VOGEL   Department of Mathematical Sciences, Montana State University, Bozeman, Montana

B. R. WIENKE   Applied Theoretical Physics Division, Los Alamos National Laboratory, Los Alamos, New Mexico

STEPHEN WINEBERG   Computational Mathematician, KMS Fusion, Inc., Ann Arbor, Michigan

G. MILTON WING   Computer Research and Applications, Los Alamos National Laboratory, Los Alamos, New Mexico

J. WONG   Computational Physics Division, Lawrence Livermore National Laboratory, Livermore, California

JOHN D. ZAHRT   Diagnostic Physics, Los Alamos National Laboratory, Los Alamos, New Mexico

G. Milton Wing: The Early Years

Milt Wing was born on January 21, 1923, in Rochester, New York. He recalls no exceptional mathematical abilities in grammar school and remembers high school geometry as being the first mathematics course which made him aware of the beauty of mathematics. Upon graduation from high school in 1939, he thought of aeronautical engineering as a possible college major but selected mathematics instead. He entered the University of Rochester on a scholarship. In his sophomore year (1941), a former high school classmate, Herb York (now a well-known physicist), told him that one of York's professors, Robert Marshak (Hans Bethe's student and later President of the American Physical Society), needed additional helpers. Late 1941 found Milt doing his very first applied mathematics numerical calculations on a mechanical calculator. Milt spent two years at 35¢ per hour doing hand calculations to supplement his scholarship (a paper based on some of this work appeared much later in 1950). Consequently, he spent a lot of time in the physics department where he met the noted physicist Victor Weisskopf and others. About 1943, both Marshak and Weisskopf disappeared. Marshak went to McGill in Montreal and Weisskopf went somewhere "out west." York went to Oak Ridge to work on an "important project" for the war effort. So many faculty members disappeared that the University employed undergraduates to teach courses in the V-12 Naval Training Program. In the summer of 1943, Milt began his teaching career at the age of 20 in this program. The two or three mathematics students who were left talked a faculty member (Professor Wladimir Seidel) into supervising a seminar—this was basically the advanced curriculum. Milt recalls that in early 1944 he got a letter from Marshak with the strange address, P. O. Box 1663, Santa Fe, New Mexico, asking for some of the calculations that they had done.

---

*This work was performed under the auspices of the U.S. Department of Energy.

Milt graduated in 1944 with a B.A. in mathematics and started graduate course work in the same seminar as before. He also continued teaching, but 15 hours now.

In December of 1944, Milt received a very official looking letter from P. O. Box 1663 stating that his name had been suggested as that of "someone who could make a significant effort in the defense of his country." He was requested to come to a site located within 50 miles of Santa Fe, New Mexico. So, shortly after his 22nd birthday, Milt joined the "missing" scientists at Los Alamos. Milt began by doing routine computing. It was here at Los Alamos that he first met Pfc. Richard Bellman and started a long relationship which led to many joint papers and a book on invariant imbedding. (Their first joint paper appeared in 1949.) After the war, Milt returned to Rochester to teach and received his Master's degree in mathematics in June 1947. He received a Ph.D. in mathematics from Cornell University in August 1949 under the direction of Harry Pollard and took a job as an instructor at UCLA. This was just the beginning of a long career which saw him return to Los Alamos from 1951-1958 and again in 1981 until the present. In between, he was affiliated with such places as the RAND Corporation and Sandia National Laboratories and taught at the University of New Mexico, University of California at Berkeley, University of Colorado, Texas Tech, and Southern Methodist University, where he was Chairman of the Department of Mathematics. His publication list follows.

V. Faber

1949

Summability with a governor of integral order, Bull. A. M. S., Vol. 55 (2) (1949), pp. 146-155.

Laplace transform solutions of a two-medium neutron aging problem (with R. Bellman, R. E. Marshak), Philos. Mag., Vol. 7 (40) (1949), pp. 297-308.

1950

Note on the effect of unequal molecular weights on the internal temperature density distribution of a star (with R. E. Marshak), Astrophys. J., Vol. 111 (2) (1950), pp. 447-448.

The mean convergence of orthogonal series, Am. J. Math., Vol. 72 (4) (1950), pp. 792-808.

1951

On the L^p theory of Hankel transforms, Pac. J. Math., Vol. 1 (2) (1951), pp. 313-319.

Averages of the coefficients of schlicht functions, Proc. A. M. S., Vol. 2 (4) (1951), pp. 658-662.

1955

On the spectrum of an unsymmetric operator arising in the transport theory of neturons (with J. Lehner), Commun. Pure Appl. Math., Vol. 8 (1) (1955), pp. 217-234.

1956

Solution of the linearized Boltzmann transport equation for the slab geometry (with J. Lehner), Duke Math. J., Vol. 23 (1) (1956), pp. 125-142.

Hydrodynamical stability and Poincare-Lyapunov theory I (with R. Bellman), Proc. Natl. Acad. Sci., Vol. 42 (11) (1956), pp. 867-870.

1957

On the principle of invariant imbedding and one-dimensional neutron multiplication (with R. Bellman, R. Kalaba), Proc. Natl. Acad. Sci., Vol. 43 (6) (1957), pp. 517-520.

1958

On the principle of invariant imbedding and neutron transport theory I: one-dimensional case (with R. Bellman, R. Kalaba), J. Math. Mech., Vol. 7 (2) (1958), pp. 149-162.

Invariant imbedding and neutron transport theory II: functional equations (with R. Bellman, R. Kalaba), J. Math. Mech., Vol. 7 (5) (1958), pp. 741-756.

Solution of time-dependent one-dimensional neutron transport problem, J. Math. Mech., Vol. 7 (5) (1958), pp. 757-766.

A note on the numerical integration of a class of nonlinear hyperbolic equations (with R. Bellman, I. Cherry), Q. Appl. Math. (16) (2) (1958), pp. 181-183.

Invariant imbedding and neutron transport theory III: neutron-neutron collision processes (with R. Bellman, R. Kalaba), J. Math. Mech., Vol. 8 (2) (1958), pp. 249-262.

1959

Invariant imbedding and neutron transport theory IV: generalized transport theory (with R. Bellman, R. Kalaba), J. Math. Mech., Vol. 8 (4) (1959), pp. 575-584.

1960

Invariant imbedding and neutron transport in a rod of changing length (with R. Bellman, R. Kalaba), Proc. Natl. Acad. Sci., Vol. 46 (1) (1960), pp. 128-130.

Invariant imbedding and neutron transport theory V: diffusion as a limiting case (with R. Bellman, R. Kalaba), J. Math. Mech., Vol. 9 (6) (1960) pp. 933-944.

Invariant imbedding and mathematical physics I: particle processes (review article, with R. Bellman, R. Kalaba), J. Math. Phys., Vol. 1 (4) (1960), pp. 280-308.

Dissipation functions and invariant imbedding I (with R. Bellman, R. Kalaba), Proc. Natl. Acad. Sci., Vol. 46 (8) (1960), pp. 1145-1147.

Invariant imbedding, conservation relations, and nonlinear equations with two-point boundary values (with R. Bellman, R. Kalaba), Proc. Natl. Acad. Sci., Vol. 46 (9) (1960), pp. 1258-1260.

Invariant imbedding and the reduction of two-point boundary value problems to initial value problems (with R. Bellman, R. Kalaba), Proc. Natl. Acad. Sci., Vol. 46 (12) (1960), pp. 1646-1649.

1961

Transport theory and spectral problems, Proc. Symp. Appl. Math., Vol. 11, Nuclear Reactor Theory (1961), pp. 140-150 (invited).

Asymptotic behavior of solutions of a system of functional equations (with M. J. Norris, J. R. Blum), Proc. A. M. S., Vol. 12 (3) (1961), pp. 463-469.

Invariant imbedding and transport theory: a unified approach, J. Math. Anal. Appl., Vol. 2 (2) (1961), pp. 277-292.

Invariant imbedding and variational principles in transport theory (with R. Bellman, R. Kalaba), Bull. A. M. S., Vol. 67 (4) (1961), pp. 396-399.

1962

Mathematical aspects of the problem of acoustic waves in stratified media, Q. Appl. Math., Vol. 19 (4) (1962), pp. 309-319.

Analysis of a problem of neutron transport in a changing medium, J. Math. Mech., Vol. 11 (1) (1962), pp. 21-34.

An Introduction to Transport Theory (John Wiley and Sons, New York, London, 1962).

1963

Transport theory, invariant imbedding, and two-point boundary value problems, Proc. Int. Symp. Non-Linear Differential Equations and Non-Linear Mechanics, Academic Press (1963), pp. 172-183 (invited).

1964

A correction to some invariant imbedding equations of transport theory obtained by "particle counting" (with P. B. Bailey), J. Math. Anal. Appl., Vol. 9 (1) (1964), pp. 170-174.

Invariant imbedding and the asymptotic behavior of solutions to initial value problems, J. Math. Anal. Appl., Vol. 9 (1) (1964), pp. 85-98.

On first order ordinary differential equations arising in diffusion problems (with I. I. Kolodner, J. Lehner), J. SIAM, Vol. 12 (2) (1964), pp. 249-269.

1965

On some problems in the optimal design of shields and reflectors in particle and wave physics (with M. Tierney, P. Waltman), SIAM Rev., Vol. 7 (1) (1965), pp. 100-113.

On a method for obtaining bounds on the eigenvalues of certain integral equations, J. Math. Anal. Appl., Vol. 11 (1-3) (1965), pp. 160-175.

Some recent developments in invariant imbedding with applications (with P. Bailey), J. Math. Phys., Vol. 6 (3) (1965), pp. 453-462.

A method for computing the eigenvalues of certain integral equations (with A. Roark), Numerische Mathematik, Vol. 7 (1965), pp. 159-170.

The method of invariant imbedding with applications to transport theory and other problems of mathematical physics, series of Colloquium Lectures in Pure and Applied Science, Socony Mobil Oil Co. (1965) (invited).

Existence and uniqueness theorems in invariant imbedding I: conservation principles (with R. Bellman, K. L. Cooke, R. Kalaba), J. Math. Anal. Appl., Vol. 10 (2) (1965), pp. 234-244.

1966

Transport Theory (Radiative Transfer), article in *Encyclopedia of Physics* (Reinhold Publishing Corporation, New York, 1966) (invited).

Invariant imbedding and transport problems with internal sources, J. Math. Anal. Appl., Vol. 13 (7) (1966), pp. 361-369.

A method for computing eigenvalues of certain Schrodinger-like equations (with R. Allen), J. Math. Anal. Appl., Vol. 15 (2) (1966), pp. 340-354.

1967

On certain Fredholm integral equations reducible to initial-value problems, SIAM Rev., Vol. 9 (4) (1967), pp. 655-670.

Convexity theorem for eigenvalues of certain integral equations, J. Math. Anal. Appl., Vol. 19 (2) (1967), pp. 330-336.

1968

On a generalization of a method of Bellman and Latter for obtaining eigenvalue bounds for integral operators, J. Math. Anal. Appl., Vol. 23 (2) (1968), pp. 384-396.

1969

Invariant imbedding and the calculation of eigenvalues for Sturm-Liouville systems (with M. Scott and L. Shampine), Computing, Vol. 4 (1969), pp. 10-23.

Solution of a certain class of non-linear two-point boundary value problems (with R. C. Allen, Jr., and M. Scott), J. Comput. Phys., Vol. 4 (2) (1969), pp. 250-257.

1970

Mathematical methods suggested by transport theory, Proc. Symp. Appl. Math. (1970), A.M.S.-SIAM, pp. 159-176 (invited).

Existence and uniqueness of solutions of a class of nonlinear elliptic boundary value problems (with L. Shampine), J. Math. Mech., Vol. 19 (1) (1970), pp. 971-979.

A numerical algorithm suggested by problems of transport in periodic media (with R. C. Allen Jr.), J. Math. Anal. Appl., Vol. 29 (1) (1970), pp. 141-157.

A numerical algorithm suggested by problems of transport in periodic media: the matrix case (with R. C. Allen, Jr., J. W. Burgmeier, and P. Mundorff), J. Math. Anal. Appl., Vol. 37 (3) (1970), pp. 725-740.

1971

Transport problems in media with periodic structure (with R. C. Allen, Jr.), Proc. Second Conf. Transp. Theory, LASL, available from National Technical Info. Service as CONF 710107 (1971) (invited).

1973

Generalized trigonometric identities and invariant imbedding (with R. C. Allen, Jr.), J. Math. Anal. Appl., Vol. 42 (2) (1973), pp. 397-408 (invited).

Problems for a Computer Oriented Calculus Course (with R. C. Allen, Jr.) (Prentice-Hall Inc., Englewood Cliffs, New Jersey, 1973).

1974

An invariant imbedding algorithm for the solution of inhomogeneous linear two-point boundary value problems (with R. C. Allen, Jr.), J. Comp. Phys., Vol. 14 (1) (1974), pp. 40-58.

Invariant imbedding and generalizations of the WKB method and the Bremmer series, J. Math. Anal. Appl., Vol. 48 (2) (1974), pp. 400-422.

1975

An Introduction to Invariant Imbedding (with R. Bellman) (John Wiley and Sons, Interscience Division, New York, London, 1975).

1978

A method for accelerating the iterative solution of a class of Fredholm integral equations (with R. C. Allen, Jr.) (by invitation), J. Optimization Theory Appl., Vol. 24 (1978), pp. 5-27.

On the effectiveness of the inverse Riccati transformation in the matrix case (with P. Nelson and A. K. Ray), J. Math. Anal. Appl., Vol. 65 (2) (1978), pp. 201-210.

1979

A theoretical result on the effectiveness of the addition formulae for two-point boundary systems (with P. Nelson and A. K. Ray), J. Math. Anal. Appl., Vol. 67 (2) (1979), pp. 329-399.

1981

Analysis of a nonlinear integral equation arising in the study of the magnetic field in the critical state model of superconductivity (with W. A. Beyer and A. Migliori), J. Math. Phys., Vol. 22 (4) (1981), pp. 908-913.

Reply to "Comment on 'The general critical state model in two dimensions and zero applied field: a uniqueness theorem and some consequences'" (with W. A. Beyer and A. Migliori), J. Appl. Phys., Vol. 52 (6) (1981), p. 4333.

1982

Analysis of Taylor's theory of toroidal plasma relaxation (with V. Faber and A. B. White, Jr.), J. Math. Phys., Vol. 23 (8) (1982), pp. 1524-1537.

On the advantages of invariant imbedding for the calculation of deep penetration problems (with G. Hughes, V. Faber, and A. B. White, Jr.), Trans. Am. Nucl. Soc., Vol. 41 (1982), pp. 486-487.

1983

Numerical experiments involving Galerkin and collocation methods for integral equations of the first kind (with R. C. Allen, Jr., and W. R. Boland), J. Comput. Phys., Vol. 49 (3) (1983), pp. 465-477.

Heat diffusion with time-dependent convective boundary conditions (with N. M. Becker, R. L. Bivins, Y. C. Hsu, H. D. Murphy, and A. B. White, Jr.), Int. J. Numerical Methods Eng., Vol. 19 (12) (1983), pp. 1871-1880.

1985

Analysis of a mathematical model of Pareto's Law of wealth distribution (with V. Faber and A. B. White, Jr), J. Math. Anal. Appl., Vol. 112 (2) (1985), pp. 579-594.

Singular values and condition numbers of Galerkin matrices arising from linear integral equations of the first kind (with R. C. Allen, Jr., W. R. Boland, and V. Faber), J. Math. Anal. Appl., Vol. 109 (2) (1985), pp. 564-590.

Helical and symmetric solutions of Taylor relaxation in a toroidal plasma (with V. Faber and A. B. White, Jr.), Plasma Phys. Controlled Fusion, Vol. 27 (4) (1985), pp. 509-515.

Condition numbers of matrices arising from the numerical solution of integral equations of the first kind, J. Integr. Equations, Vol. 9 (Suppl.) (1985), pp. 191-204.

Transport Theory (Radiative Transfer) (with P. Nelson), *Encyclopedia of Physics,* 3rd edition (Van Nostrand Reinhold Co., New York, 1985) (invited article).

1986

Asymptotic behavior of singular values and singular functions of certain convolution operators (with V. Faber, T. Manteuffel, A. B. White, Jr.), Comp. Appl. Math., Vol. 12A (6) (1986), pp. 733-747 (invited for Bellman memorial issue).

Singular values of fractional integral operators: a unification of theorems of Hille-Tamarkin and of Chang (with V. Faber), J. Math. Anal. Appl., Vol. 120 (2) (1986), pp. 745-760.

Asymptotic behavior of singular values of convolution operators (with V. Faber), Rocky Mountain J. Math., Vol. 16 (3) (1986), pp. 567-574.

1988

Effective bounds for the singular values of integral operators (with V. Faber), J. Integr. Equations Appl., Vol. 1 (1) (1988), pp. 54-64.

1989

Invariant imbedding in two-dimensions (with V. Faber and D. L. Seth), Proc. Conf. Trans. Theory, Invariant Imbedding, Integr. Equations.

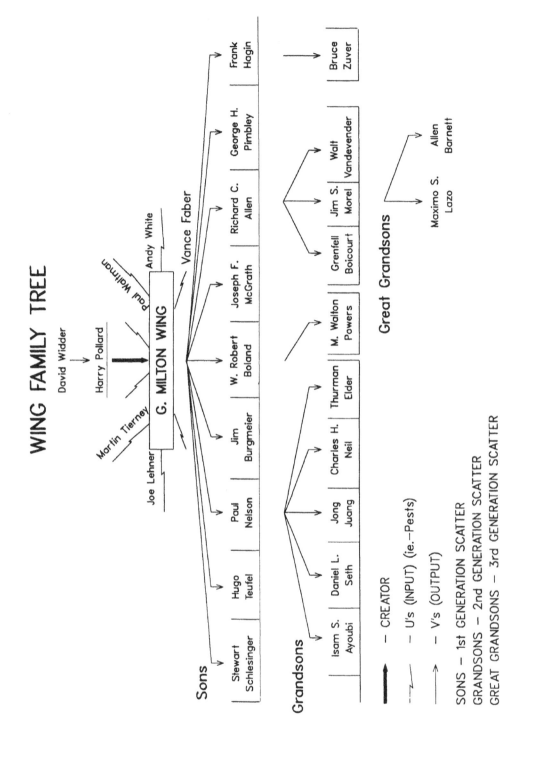

WING FAMILY TREE

Sons

Grandsons

Great Grandsons

➡ — CREATOR

⚡ — U's (INPUT) (ie.—Pests)

→ — V's (OUTPUT)

SONS — 1st GENERATION SCATTER
GRANDSONS — 2nd GENERATION SCATTER
GREAT GRANDSONS — 3rd GENERATION SCATTER

TRANSPORT THEORY, INVARIANT IMBEDDING, AND INTEGRAL EQUATIONS

Transport Theory in Binary Statistical Mixtures

G. C. POMRANING   School of Engineering and Applied Science, University of California at Los Angeles, Los Angeles, California

C. D. LEVERMORE   Department of Mathematics, University of Arizona, Tucson, Arizona

J. WONG   Computational Physics Division, Lawrence Livermore National Laboratory, Livermore, California

Presented as a tribute to
G. Milton Wing
on the occasion of his 65th birthday.

ABSTRACT

In this article, we discuss the formulation of a linear transport theory describing particle flow in a random mixture of two immiscible fluids. Topics to be discussed include: (1) a complete formulation in the case of Markov statistics as the description of the fluid mixing; (2) the use of asymptotics to derive various limiting descriptions of the Markov transport model; (3) the use of renewal theory to obtain results for non-Markov statistics in the absence of scattering; (4) a modification to the Markov description to (approximately) treat the non-Markov case when scattering is extant; and (5) a special case of the renewal analysis which leads to a singular eigenfunction problem, in the sense of "Caseology" from neutron transport theory. The paper concludes with a list of open questions in this area of investigation.

INTRODUCTION

A subject of recent interest has been the problem of describing linear particle transport in a statistical medium consisting of two randomly mixed immiscible fluids [1]-[5]. In this paper we present an overview of what is known in this area by attempting to weave the results of these five papers into a coherent tapestry. We also present the results of some recent work in which the effect of statistical fluctuations is to introduce a scattering-like term in the transport equation. If one seeks eigenfunctions of the resulting transport equation in the manner of Case [6],[7], one is led to a very interesting singular eigenfunction ("Caseology") problem.

The generic linear transport equation we shall treat is of the form [8]

$$\frac{1}{c}\frac{\partial \psi}{\partial t} + \vec{\Omega}\cdot\vec{\nabla}\psi + \sigma\psi = \frac{c}{4\pi}\sigma_s\psi + S \ , \tag{1}$$

where

$$E = \frac{1}{c}\int_{4\pi} d\vec{\Omega}\psi(\vec{\Omega}) \ . \tag{2}$$

In writing Eqs. (1) and (2) we are using the notation of radiative transfer, but our results are applicable to any linear transport description. The dependent variable in Eq. (1) is $\psi(\vec{r},\vec{\Omega},t)$, the specific intensity of radiation, with $\vec{r}$, $\vec{\Omega}$, and t denoting the spatial, angular (photon flight direction), and time variables, respectively. The quantity $E(\vec{r},t)$ in Eq. (2) is the radiative energy density, with c denoting the vacuum speed of light. The quantities $\sigma(\vec{r},t)$, $\sigma_S(\vec{r},t)$, and $S(\vec{r},t)$ denote the total cross section, scattering cross section, and (isotropic) external source, respectively. We have also assumed isotropic and coherent (no energy exchange) scattering in Eq. (1), but this simplification is not necessary for the essentials of the considerations to follow.

To treat the case of a binary statistical mixture, the quantities σ, σ_S, and S in Eq. (1) are considered as discrete random variables, each of which assumes one of two values at a phase space point $\vec{r}$, $\vec{\Omega}$, t. These two values are the values associated with each fluid component of the mixture. That is, as a particle traverses the mixture along any path, it encounters alternating packets of the two fluids (which we label by 0 and 1), and each fluid has deterministic values of σ, σ_S, and S. The statistical nature of the problem enters through the statistics of the fluid mixing. In partic-ular, we denote by $p_i(\vec{r}, t)$ the probability of finding material i at posi-tion $\vec{r}$ and time t. As we shall see, other statistical information of the fluid mixing is required, such as a correlation length $\lambda_c(\vec{r}, \vec{\Omega}, t)$. The precise definition of this correlation length will be given shortly.

Since σ, σ_s, and S in Eq. (1) are random variables, the solution of Eqs. (1) and (2) for ψ and E are also random variables, and we seek a transport description for $\overline{\psi}$ and $\overline{E}$, the ensemble averaged intensity and energy density. We let R denote any of the random variables σ, σ_s, S, ψ, and E, and write R as the sum of its ensemble average $\overline{R}$ and the statistical fluctuation $\tilde{R}$, i.e.,

$$R = \overline{R} + \tilde{R} . \tag{3}$$

We have

$$\overline{R} = p_0 R_0 + p_1 R_1 \quad ; \qquad \overline{\tilde{R}} = 0 . \tag{4}$$

If $R = \psi$ in Eq. (4), we have introduced ψ_i as the value of the intensity at $\vec{r}$, $\vec{\Omega}$, t if material i is extant at $\vec{r}$ and t. A similar definition holds for E_i. Now, use of Eq. (3) in Eq. (1) and a subsequent ensemble average of the result gives

$$\frac{1}{c} \frac{\partial \overline{\psi}}{\partial t} + \vec{\Omega} \cdot \vec{\nabla} \overline{\psi} + \overline{\sigma \psi} = \frac{c}{4\pi} \overline{\sigma}_s \overline{E} + \overline{S} + \overline{\left[\frac{c}{4\pi} \tilde{\sigma}_s \tilde{E} - \widetilde{\sigma \psi} \right]} . \tag{5}$$

Equation (5), while exact, is a formal result only. It does not represent a closed equation for the quantity of interest, namely $\overline{\psi}$. We need a formu-lation which gives the last term in Eq. (5), the cross correlation term, in terms of $\overline{\psi}$ to turn Eq. (5) into a useful result.

As a simple (and generally crude) approximation, one can ignore the cross correlation term in Eq. (5). This leads to the so-called "atomic mix" description, given by

$$\frac{1}{c} \frac{\partial \overline{\psi}}{\partial t} + \vec{\Omega} \cdot \vec{\nabla} \overline{\psi} + \overline{\sigma} \overline{\psi} = \frac{c}{4\pi} \overline{\sigma}_s \overline{E} + \overline{S} . \tag{6}$$

Equation (6) corresponds to a transport equation of the usual form, with the cross sections and external source simply being probability (volume) weighted averages of the corresponding quantities associated with the two fluids. On physical grounds, such a simple homogenization of σ, σ_s, and S is only justified when

$$\sigma_i \lambda_i \ll 1 \quad , \qquad i = 0,1 \ . \tag{7}$$

Here λ_i is a characteristic size of the fluid packets of fluid i, and Eq. (7) states that the characteristic packet sizes of both fluids must be small compared to the particle mean free path for Eq. (6) to be valid. The ultimate limit of the packet size is physically, of course, the size of an atom, thus accounting for the name "atomic mix" applied to Eq. (6).

If Eq. (7) is not true, the use of Eq. (6) as a transport description can be very wrong. For example, consider the simple problem of time-independent transmission through a source-free, no scattering mixture. Let material 0 be composed of optically thin packets ($\sigma_0 \lambda_0 \ll 1$), and let material 1 be composed of optically thick packets ($\sigma_1 \lambda_1 \gg 1$). Further, assume that material 1 is sparse ($p_1 \ll 1$). Physically, then, we have a near vacuum interspersed with a very few packets of essentially infinite optical thickness. Particles incident upon a piece of this mixture will, on the average, have a good chance of passing through the material. They will not often encounter the sparse absorbing packets, but will pass between these packets and hence will be transmitted through the mixture. On the other hand, the atomic mix cross section $\bar{\sigma}$ can be very large, and hence the homogenized material will have a large cross section everywhere. This in turn implies that the atomic mix transport description will lead to virtually no transmission through the mixture. This simple gedanken experiment indicates that the atomic mix model, Eq. (6), can be grossly in error under certain conditions. In particular, atomic mix will always underestimate the transmission for the simple problem we have just discussed.

In closing this introduction, we mention that the list of potential applications for a transport theory in statistical mixtures seems to be very large. The original motivation for this work was the need for an accurate transport description in the calculation of laser or beam driven inertially confined fusion pellets. At an interface between two materials, these pellets are susceptible to Rayleigh-Taylor instabilities which can lead to a two fluid turbulent mixture around the interface. Other areas of application also come to mind. In a boiling water nuclear reactor, the water, which acts as both coolant and moderator, is in a two fluid random state (liquid and vapor). A proper treatment of the neutron transport must take the statistical nature of the mixture into account. In shielding calculations through concrete, the random nature of the materials (e.g., gravel) in the concrete implies a need for a statistical transport treatment to obtain an accurate measure of the shield effectiveness. Still another area of application is the calculation of light transport through a two component random medium, such as sooty air or murky water. In general, there seems to be numerous areas of application for a transport theory for random media.

MARKOV STATISTICS

To compute the cross correlation terms which the atomic mix model neglects, we need to specify a statistical model for the fluid mixing. In this section, we assume a Markov model [1]-[3]. The Markov assumption is: "Given ordered points along a space-time ray $s_0 < s_1 < \ldots < s_{n-1} < s_n < s$, the probability that material i is at s, given the material information at all points s_0, s_1, ..., s_{n-1}, s_n, is the same as the probability that material i is at s, given the material information only at s_n." If we define $P_{ij}(s,r)$ as the conditional probability of finding material i at position s, given that material j is at position r, we can invoke the Markov assumption to obtain an explicit expression for this probability [1],[3]. A solution of the Chapman-Kolmogorov equations [9] gives this probability in terms of two arbitrary (with certain minimal constraints, such as positivity) functions. Specifically, we have [3]

$$P_{ij}(s,r) = p_i(s) + (-1)^{i+j} \frac{q(s,r)}{p_j(r)} e^{-\Gamma(s,r)} \quad , \qquad i,j = 0,1 \; , \qquad (8)$$

where

$$q(s,r) = \left[p_0(s)p_1(s)p_0(r)p_1(r) \right]^{1/2} \; , \qquad (9)$$

and

$$\Gamma(s,r) = \left| \int_r^s d\xi \, \frac{1}{\lambda_c(\xi)} \right| \; . \qquad (10)$$

The function $p_i(s)$ has already been identified as the probability of finding material i at position s. The function $\lambda_c(s)$, because of the way it occurs in Eqs. (8) and (10), is logically identified as a correlation length. The explicit functional dependences of these two functions p_i and λ_c depend upon the details of the Markovian mixture. Such details must come from the hydrodynamic mix model. Of course, such a mix model may not predict Markovian statistics, in which case Eq. (8) is not valid.

Under the Markov assumption, one can derive a 2 x 2 matrix transport equation. This is given by [3]

$$\left(\frac{1}{c}\frac{\partial}{\partial t} + \vec{\Omega}\cdot\vec{\nabla}\right)\begin{bmatrix}\bar{\psi}\\\chi\end{bmatrix} + \begin{bmatrix}\bar{\sigma} & \nu\\\nu & \hat{\sigma}\end{bmatrix}\begin{bmatrix}\bar{\psi}\\\chi\end{bmatrix} = \frac{c}{4\pi}\begin{bmatrix}\bar{\sigma}_s & \nu_s\\\nu_s & \hat{\sigma}_s\end{bmatrix}\begin{bmatrix}\bar{E}\\\eta\end{bmatrix} + \begin{bmatrix}\bar{S}\\T\end{bmatrix} , \qquad (11)$$

where

$$\chi = \overline{\left(\frac{\tilde{\sigma}\tilde{\psi}}{\nu}\right)} \quad ; \qquad \eta = \frac{1}{c}\int_{4\pi}d\vec{\Omega}\,\chi(\vec{\Omega}) = \overline{\left(\frac{\tilde{\sigma}_s\tilde{E}}{\nu_s}\right)} , \qquad (12)$$

and

$$\bar{S} = p_0 S_0 + p_1 S_1 \quad ; \qquad T = (p_0 p_1)^{1/2}(S_0 - S_1) , \qquad (13)$$

$$\bar{\sigma} = p_0\sigma_0 + p_1\sigma_1 \quad ; \qquad \hat{\sigma} = p_1\sigma_0 + p_0\sigma_1 + \frac{1}{\lambda_c} , \qquad (14)$$

$$\bar{\sigma}_s = p_0\sigma_{s0} + p_1\sigma_{s1} \quad ; \qquad \hat{\sigma}_s = p_1\sigma_{s0} + p_0\sigma_{s1} , \qquad (15)$$

$$\nu = (p_0 p_1)^{1/2}(\sigma_0 - \sigma_1) \quad ; \qquad \nu_s = (p_0 p_1)^{1/2}(\sigma_{s0} - \sigma_{s1}) . \qquad (16)$$

Thus to obtain a solution for $\bar{\psi}$, the ensemble averaged intensity, one must solve two coupled transport equations for the two unknowns $\bar{\psi}$ and χ. The variable χ is of little, if any, physical significance, but its solution must be found as part of the process of finding $\bar{\psi}$. We note that the statistics of the mixture is entirely described by the functions $p_i(\vec{r},t)$ and $\lambda_c(\vec{r},\vec{\Omega},t)$. From the transport point of view, these functions are assumed known, supplied by the hydrodynamic mixing model.

Equations (11) through (16) can be obtained in two different ways. The first derivation is entirely rigorous, but at this time we only know how to proceed in the absence of scattering [1],[3]. This is the so-called "method of smoothing," a projection operator technique as described by Keller [10]-[12] and Frisch [13]. In the absence of scattering, Eq. (1) is simply

$$\frac{d\psi}{ds} + \sigma\psi = S , \qquad (17)$$

where s denotes the space-time variable, i.e.,

$$\frac{d\psi}{ds} = \frac{1}{c}\frac{\partial\psi}{\partial t} + \vec{\Omega} \cdot \vec{\nabla}\psi \ . \tag{18}$$

Using the decomposition for ψ, σ, and S given by Eq. (3), an ensemble average of Eq. (17) gives

$$\frac{d\bar{\psi}}{ds} + \bar{\sigma}\bar{\psi} + \overline{(\tilde{\sigma}\tilde{\psi})} = \bar{S} \ , \tag{19}$$

which is just Eq. (5) in the absence of scattering ($\sigma_s = 0$). Subtracting Eq. (19) from Eq. (17) then gives

$$\frac{d\tilde{\psi}}{ds} + \bar{\sigma}\tilde{\psi} + \tilde{\sigma}\bar{\psi} + \tilde{\sigma}\tilde{\psi} - \overline{(\tilde{\sigma}\tilde{\psi})} = \tilde{S} \ . \tag{20}$$

The aim is to "solve" Eq. (20) for $\tilde{\psi}$ in terms of $\bar{\psi}$, $\bar{\sigma}$, $\tilde{\sigma}$, and $\tilde{S}$, and then evaluate the ensemble average of $(\tilde{\sigma}\tilde{\psi})$ to obtain a "closed" equation for $\bar{\psi}$, the ensemble averaged intensity.

To carry out this procedure, we introduce the projection operator P which performs the ensemble average. That is, for any random variable ϕ we have

$$P\phi = \bar{\phi} \ . \tag{21}$$

We also introduce the Green's operator G such that $\phi = Gf$ solves the equation

$$\frac{d\phi}{ds} + \bar{\sigma}\phi = f \ . \tag{22}$$

An explicit expression for G is given by

$$Gf = \int_0^s ds' e^{-\bar{\tau}(s,s')} f(s') \ , \tag{23}$$

where $\bar{\tau}$ is the optical depth between points s and s' given by

$$\bar{\tau}(s,s') = \left|\int_s^{s'} ds'' \ \bar{\sigma}(s'')\right| \ . \tag{24}$$

In terms of P and G, Eq. (20) can be written

$$\tilde{\psi} + \epsilon G(I - P)\overline{\sigma\tilde{\psi}} = G(\tilde{S} - \tilde{\sigma}\overline{\psi}) \ , \tag{25}$$

where I here is the identity operator. In writing Eq. (25) we have introduced a formal smallness parameter ϵ, so that we may solve for $\tilde{\psi}$ in powers of ϵ according to

$$\tilde{\psi} = \sum_{n=0}^{\infty} (-1)^n \, \epsilon^n [G(I - P)\tilde{\sigma}]^n G(\tilde{S} - \tilde{\sigma}\overline{\psi}) \ . \tag{26}$$

We then find, setting $\epsilon = 1$,

$$\overline{(\tilde{\sigma}\tilde{\psi})} = \sum_{n=0}^{\infty} (-1)^n \, P\{[\tilde{\sigma}G(I - P)]^{n+1}(\tilde{S} - \tilde{\sigma}\overline{\psi})\} \ . \tag{27}$$

Under the Markov assumption, the infinite series in Eq. (27) can be summed. Omitting the details [3], one finds the simple result

$$\overline{(\tilde{\sigma}\tilde{\psi})} = \nu\chi \ , \tag{28}$$

where the function χ satisfies the equation

$$\frac{d\chi}{ds} + \hat{\sigma}\chi + \nu\overline{\psi} = T \ . \tag{29}$$

Here ν, $\hat{\sigma}$, and T are defined by Eqs. (13), (14), and (16). Using Eq. (28) in Eq. (19) one finds that $\overline{\psi}$ satisfies

$$\frac{d\overline{\psi}}{ds} + \bar{\sigma}\overline{\psi} + \nu\chi = \overline{S} \ . \tag{30}$$

Recalling that d/ds is the convection operator given by Eq. (18), we see that Eqs. (29) and (30) agree with Eq. (11) in the absence of scattering ($\sigma_{si} = 0$).

The second derivation of Eq. (11) includes the scattering interaction, but is not entirely rigorous as we shall discuss after we sketch the details of the derivation. This treatment is based upon the master equation approach [13]-[15], as introduced in the present context by Vanderhaegen [2] and Levermore [3]. One writes an equation for the joint probability density, $P_i(\vec{r},\vec{\Omega},t,\psi)$, of finding material i and intensity ψ at point $\vec{r}$, $\vec{\Omega}$, and t. This is the so-called master equation, and is given by [3]

$$\frac{dP_i}{ds} + \frac{\partial}{\partial \psi}(S_i P_i - \psi_i \Sigma_i P_i) = \sum_{j=0}^{1} Q_{ij} P_j \quad , \qquad i = 0,1 \ , \tag{31}$$

where once again s denotes a space-time ray [see Eq. (18)], and Σ_i is the scattering operator, i.e.,

$$\Sigma_i = \sigma_i - \frac{\sigma_{si}}{4\pi} \int_{4\pi} d\vec{\Omega}'(\cdot) \ . \tag{32}$$

The matrix with elements Q_{ij} is an infinitesimal probability matrix for the Markov process, given explicitly by [3]

$$\underline{Q} = \begin{bmatrix} \left(-\dfrac{p_1}{\lambda_c} + \dfrac{dp_0/ds}{2p_0}\right) & \left(\dfrac{p_0}{\lambda_c} - \dfrac{dp_1/ds}{2p_1}\right) \\[4mm] \left(\dfrac{p_1}{\lambda_c} - \dfrac{dp_0/ds}{2p_0}\right) & \left(-\dfrac{p_0}{\lambda_c} + \dfrac{dp_1/ds}{2p_1}\right) \end{bmatrix} \ . \tag{33}$$

The probability of finding material i at position $\vec{r}$ and time t, $p_i(\vec{r},t)$, is related to the joint probability $P_i(\vec{r},\vec{\Omega},t,\psi)$ by a simple integral of P_i over ψ, i.e.,

$$p_i(\vec{r},t) = \int_0^\infty d\psi \ P_i(\vec{r},\vec{\Omega},t,\psi) \ . \tag{34}$$

The expected intensity in material i, $\psi_i(\vec{r},\vec{\Omega},t)$, is simply related to the first ψ moment of $P_i(\vec{r},\vec{\Omega},t,\psi)$ according to

$$p_i \psi_i = \int_0^\infty d\psi \ \psi P_i \ . \tag{35}$$

By multiplying Eq. (31) by ψ and integrating over $0 \le \psi < \infty$, we find that ψ_i satisfies the equation

$$\frac{d(p_i \psi_i)}{ds} - p_i S_i + p_i \Sigma_i \psi_i = \sum_{j=0}^{1} Q_{ij} p_j \psi_j \ . \tag{36}$$

From Eqs. (33) and (36), and writing out d/ds according to Eq. (18), one easily deduces

$$\left(\frac{1}{c}\frac{\partial}{\partial t} + \vec{\Omega} \cdot \vec{\nabla}\right)\begin{bmatrix} \psi_0 \\ \psi_1 \end{bmatrix} + \begin{bmatrix} \Sigma_0 & 0 \\ 0 & \Sigma_1 \end{bmatrix}\begin{bmatrix} \psi_0 \\ \psi_1 \end{bmatrix} - \begin{bmatrix} S_0 \\ S_1 \end{bmatrix}$$

$$= \begin{bmatrix} -\gamma_0 & \gamma_0 \\ \gamma_1 & -\gamma_1 \end{bmatrix}\begin{bmatrix} \psi_0 \\ \psi_1 \end{bmatrix} \ , \tag{37}$$

where

$$\gamma_i = \frac{(1 - p_i)}{\lambda_c} + \frac{1}{2p_i}\left(\frac{1}{c}\frac{\partial}{\partial t} + \vec{\Omega} \cdot \vec{\nabla}\right)p_i \ . \tag{38}$$

The γ_i are non-negative since Markov statistics demand that

$$\frac{1}{\lambda_c} \geq \frac{|dp_i/ds|}{2p_0 p_1} \tag{39}$$

for physical realizability [3]. This allows one to immediately deduce that the ψ_i as defined by Eq.(37) are non-negative for non-negative initial and boundary data. If one changes variables in Eq. (37) according to

$$\overline{\psi} = p_0 \psi_0 + p_1 \psi_1 \quad ; \quad \chi = (p_0 p_1)^{1/2}(\psi_0 - \psi_1) \ , \tag{40}$$

and uses Eq. (32) for the scattering operator Σ_i, one finds the result stated earlier, namely Eq. (11).

The master equation derivation just given is known to be rigorously correct in the case of time independent transport in general geometry with no scattering. In this instance, Eq. (11) reduces to an infinite, un-coupled set of ordinary differential equations for which the master equation approach is well understood [15]. In the more general time dependent multidimensional case including scattering, this transport description must at this time be considered a phenomenological model. This model, while not rigorously derived, is a reasonable model since it reduces to the proper limit in all known limiting cases, and is robust away from these limits.

We close this section by pointing out the simplification which occurs in the Markov mixing model if the statistics are homogeneous and time independent. By this we mean that the probabilities p_i and the correlation length λ_c are independent of space and time. In this case, the Markov assumption is equivalent to a classical Poisson process [1]. That is, as a particle streams along any ray it encounters alternating packets of fluids 0 and 1, with exponentially distributed chord lengths in each fluid. If we define $f_i(z)$ as the probability density function for a chord length z in the ith fluid, we have

$$f_i(z) = \frac{1}{\lambda_i} e^{-z/\lambda_i} , \tag{41}$$

where λ_i is the mean chord length in material i. In this case we have

$$p_i = \frac{\lambda_i}{\lambda_0 + \lambda_1} \quad ; \quad \frac{1}{\lambda_c} = \frac{1}{\lambda_0} + \frac{1}{\lambda_1} . \tag{42}$$

ASYMPTOTIC LIMITS

If one accepts Eq. (11) as the correct description for transport in a two component Markov mixture, one can use asymptotics to obtain various limits of this description. These limiting descriptions can be considered as simpler, but approximate, descriptions of particle transport in a two component Markov mixture. We consider three such asymptotic limits here, under completely general spatial and temporal variations of the statistical quantities $p_i(\vec{r},t)$ and $\lambda_c(\vec{r},\vec{\Omega},t)$. Of course, the physical realizability condition given by Eq. (39) is maintained at all times.

We first show how one can recover the atomic mix description given by Eq. (6). We recall that on physical grounds atomic mix should be a valid limit when the cross sections σ_i are small in the sense that the optical depths of the fluid packets are small, as expressed by Eq. (7). In the present context we express this limit of small cross sections by the inequality

$$\sigma_i \lambda_c \ll 1 . \tag{43}$$

This inequality can also be interpreted as expressing the fact that the correlation length is small compared to the particle mean free path in both fluid components. Equation (43) implies that $\hat{\sigma} \gg \bar{\sigma}$, ν, $\hat{\sigma}_s$, $\bar{\sigma}_s$, and ν_s. Accordingly, we introduce a formal smallness parameter ϵ into the Markov equation by replacing $\hat{\sigma}$ by $\hat{\sigma}/\epsilon$. That is, we write Eq. (11) as

$$\left(\frac{1}{c}\frac{\partial}{\partial t} + \vec{\Omega}\cdot\vec{\nabla}\right)\begin{bmatrix}\bar{\psi}\\\chi\end{bmatrix} + \begin{bmatrix}\bar{\sigma} & \nu\\\nu & \hat{\sigma}/\epsilon\end{bmatrix}\begin{bmatrix}\bar{\psi}\\\chi\end{bmatrix}$$

$$= \frac{c}{4\pi}\begin{bmatrix}\bar{\sigma}_s & \nu_s\\\nu_s & \hat{\sigma}_s\end{bmatrix}\begin{bmatrix}\bar{E}\\\eta\end{bmatrix} + \begin{bmatrix}\bar{S}\\T\end{bmatrix}. \qquad (44)$$

We seek a solution of Eq. (44) as a series in powers of ϵ according to

$$\bar{\psi} = \bar{\psi}^{(0)} + \epsilon\bar{\psi}^{(1)} + \dots \quad ; \quad \chi = \chi^{(0)} + \epsilon\chi^{(1)} + \dots . \qquad (45)$$

To leading order one finds

$$\bar{\psi} = \bar{\psi}^{(0)} \quad ; \quad \chi = \epsilon\chi^{(1)} , \qquad (46)$$

where $\bar{\psi}$ satisfies

$$\frac{1}{c}\frac{\partial\bar{\psi}}{\partial t} + \vec{\Omega}\cdot\vec{\nabla}\bar{\psi} + \bar{\sigma}\bar{\psi} = \frac{c}{4\pi}\bar{\sigma}_s\bar{E} + \bar{S} . \qquad (47)$$

This is just the atomic mix description [see Eq. (6)].

One can obtain a correction to atomic mix by considering the case of a small amount of one of the fluids, say fluid 0, with a large cross section and source, admixed with the second fluid. This is expressed by the replacements

$$p_0 \rightarrow \epsilon^2 p_0 \quad ; \quad \sigma_0 \rightarrow \sigma_0/\epsilon^2 \quad ;$$

$$\sigma_{s0} \rightarrow \sigma_{s0}/\epsilon^2 \quad ; \quad S_0 \rightarrow S_0/\epsilon^2 . \qquad (48)$$

Introducing these scalings into the parameters given by Eqs. (13) through (16), we find that the corresponding scaling of Eq. (11) is

$$\left(\frac{1}{c}\frac{\partial}{\partial t} + \vec{\Omega}\cdot\vec{\nabla}\right)\begin{bmatrix}\bar{\psi}\\[2mm]\chi\end{bmatrix} + \begin{bmatrix}\bar{\sigma} & \nu/\epsilon \\[2mm] \nu/\epsilon & \hat{\sigma}/\epsilon^2\end{bmatrix}\begin{bmatrix}\bar{\psi}\\[2mm]\chi\end{bmatrix}$$

$$= \frac{c}{4\pi}\begin{bmatrix}\bar{\sigma}_s & \nu_s/\epsilon \\[2mm] \nu_s/\epsilon & \hat{\sigma}_s/\epsilon^2\end{bmatrix}\begin{bmatrix}\bar{E}\\[2mm]\eta\end{bmatrix} + \begin{bmatrix}\bar{S}\\[2mm]T/\epsilon\end{bmatrix} . \tag{49}$$

Once again solving for $\bar{\psi}$ and χ as series in powers of ϵ according to Eq. (45), one finds to leading order

$$\bar{\psi} = \bar{\psi}^{(0)} \quad ; \quad \chi = \epsilon\chi^{(1)} , \tag{50}$$

where $\bar{\psi}$ and χ satisfy

$$\left(\frac{1}{c}\frac{\partial}{\partial t} + \vec{\Omega}\cdot\vec{\nabla}\right)\bar{\psi} + \bar{\sigma}\bar{\psi} + \nu\chi = \frac{c}{4\pi}(\bar{\sigma}_s\bar{E} + \nu_s\eta) + \bar{S} , \tag{51}$$

$$\nu\bar{\psi} + \hat{\sigma}\chi = \frac{c}{4\pi}(\nu_s\bar{E} + \hat{\sigma}_s\eta) + T . \tag{52}$$

Eliminating χ between these two equations, we find a renormalized transport equation of the usual form given by

$$\frac{1}{c}\frac{\partial\bar{\psi}}{\partial t} + \vec{\Omega}\cdot\vec{\nabla}\bar{\psi} + \sigma_{eff}\bar{\psi} = \frac{c}{4\pi}\sigma_{s,eff}\bar{E} + S_{eff} , \tag{53}$$

where the effective cross sections and source are given by

$$\sigma_{eff} = \bar{\sigma} - \frac{\nu^2}{\hat{\sigma}} \geq 0 \quad ; \quad \sigma_{s,eff} = \sigma_{eff} - \sigma_{a,eff} \geq 0 , \tag{54}$$

$$\sigma_{a,eff} = (\bar{\sigma} - \bar{\sigma}_s) - \frac{(\nu - \nu_s)^2}{(\hat{\sigma} - \hat{\sigma}_s)} \geq 0 \quad ; \quad S_{eff} = \bar{S} - \frac{(\nu - \nu_s)}{(\hat{\sigma} - \hat{\sigma}_s)}T \geq 0 . \tag{55}$$

This renormalized transport equation is clearly robust since all of the effective quantities are non-negative for all physical parameters. Equation (53) contains atomic mix as a limiting case corresponding to a correlation length small compared to the mean free paths, $1/\sigma_i$, of the two fluids.

That is, as $\hat\sigma \to \infty$ in Eqs. (53) through (55), Eq. (53) reduces to the atomic mix result, Eq. (47). Thus Eq. (53) can be considered as a generalization, in the usual transport equation form, of the atomic mix result. It should be emphasized however, that Eq. (53) should only be expected to be a quantitatively accurate description of the transport process when the scaling given by Eq. (48) corresponds to the physical problem of interest.

One can obtain an entirely different approximation to Eq. (11) by seeking a classic (isotropic) diffusion limit. Here we treat only the case of isotropic statistics; that is, the correlation length is independent of direction $\vec\Omega$. In this case, we introduce a scaling which suppresses the space and time derivatives, as well as the absorption and source terms. Specifically, we scale Eq. (11) as

$$
\left(\frac{\epsilon^2}{c}\frac{\partial}{\partial t} + \epsilon\vec\Omega \cdot \vec\nabla\right)\begin{bmatrix} \bar\psi \\ \chi \end{bmatrix} + \begin{bmatrix} \bar\sigma & \nu \\ \nu & \hat\sigma \end{bmatrix}\left\{\begin{bmatrix} \bar\psi \\ \chi \end{bmatrix} - \frac{c}{4\pi}\begin{bmatrix} \bar E \\ \eta \end{bmatrix}\right\}
$$

$$
= \epsilon^2\left\{\begin{bmatrix} \bar S \\ T \end{bmatrix} - \frac{c}{4\pi}\begin{bmatrix} (\bar\sigma - \bar\sigma_s) & (\nu - \nu_s) \\ (\nu - \nu_s) & (\hat\sigma - \hat\sigma_s) \end{bmatrix}\begin{bmatrix} \bar E \\ \eta \end{bmatrix}\right\}. \tag{56}
$$

This is a straightforward generalization to our matrix transport equation of the scaling introduced by Larsen and Keller [16] in their derivation of the diffusion limit corresponding to the usual scalar transport equation. Once again seeking a solution as a power series in ϵ, one finds (for isotropic statistics), with $O(\epsilon^2)$ error, the diffusion description given by

$$
\frac{1}{c}\frac{\partial}{\partial t}\begin{bmatrix} \bar E \\ \eta \end{bmatrix} - \vec\nabla \cdot \left\{\frac{1}{3(\bar\sigma\hat\sigma - \nu^2)}\begin{bmatrix} \hat\sigma & -\nu \\ -\nu & \bar\sigma \end{bmatrix}\vec\nabla\begin{bmatrix} \bar E \\ \eta \end{bmatrix}\right\}
$$

$$
+ \begin{bmatrix} (\bar\sigma - \bar\sigma_s) & (\nu - \nu_s) \\ (\nu - \nu_s) & (\hat\sigma - \hat\sigma_s) \end{bmatrix}\begin{bmatrix} \bar E \\ \eta \end{bmatrix} = \frac{4\pi}{c}\begin{bmatrix} \bar S \\ T \end{bmatrix}. \tag{57}
$$

Since $\bar{\sigma}\hat{\sigma} - \nu^2 \geq 0$ for all choices of the physical parameters, it is easily shown that the eigenvalues of the diffusion coefficient matrix are properly behaved.

To obtain boundary conditions for Eq. (57), one performs a boundary layer analysis by introducing an additional scaling in Eq. (56). Specifically, we replace z in Eq. (56) by z/ϵ, where z is a local, inward point- ing, normal spatial coordinate at any point on the system boundary. Solv- ing the resulting equation up to and including terms of $O(\epsilon)$, and carrying out the details of the asymptotic matching of this boundary layer solution to the interior solution, one finds a boundary condition at each surface point given by

$$\int_0^1 d\mu \left\{ \begin{bmatrix} \bar{E} \\ \eta \end{bmatrix} - \frac{\mu}{(\bar{\sigma}\hat{\sigma} - \nu^2)} \begin{bmatrix} \hat{\sigma} & -\nu \\ -\nu & \bar{\sigma} \end{bmatrix} \frac{\partial}{\partial z} \begin{bmatrix} \bar{E} \\ \eta \end{bmatrix} - \frac{2}{c} \begin{bmatrix} \Gamma_\psi \\ \Gamma_\eta \end{bmatrix} \right\} \mu H(\mu) = 0 \ . \quad (58)$$

Here μ is the cosine of the angle between $\vec{\Omega}$ and the z axis, $\underline{\Gamma}(\mu)$ is the incoming surface intensity (the transport boundary condition) integrated over the local azimuthal angle, and $H(\mu)$ is Chandrasekkar's H-function [17] for a purely scattering system with an isotropic phase function.

We have considered here three distinct scalings of the Markov transport equation which led to three distinct asymptotic limits. It seems likely that other asymptotic limits exist because of the richness of the Markov description. That is, this Markov description involves many parameters (i.e., $\bar{\sigma}$, $\bar{\sigma}_s$, $\hat{\sigma}$, $\hat{\sigma}_s$, ν, ν_s, and λ_c) and these paramters lead to many dimen- sionless variables which can be scaled in various ways. It also seems pos- sible to generalize the diffusion limit given by Eqs. (57) and (58) to include a direction dependent correlation length. (We recall that these equations were derived under the assumption of isotropic statistics.) This entire question of asymptotic limits of the Markov description, Eq. (11), seems ripe for additional investigation. Of particular interest would be the derivation of a so-called flux-limited diffusion approximation to this Markov description of the same general form as that available for the scalar transport equation [18]-[20].

RENEWAL STATISTICS

All of our discussion thus far has been concerned with a two component fluid mixture which obeys Markov statistics. In this section we review the progress reported to date in treating a fluid described by more general, non-Markov, statistics. In particular, we use the theory of alternating renewal processes [21] to treat the non-Markov problem, but only in the absence of scattering. How one includes the scattering interaction in the analysis is unknown at this time. The use of renewal theory in the present transport context was first suggested by Vanderhaegen, and subsequently generalized by Levermore et al. [5].

The equation we analyze is the simple one given by Eq. (17), which is the transport problem in the absence of scattering. We allow σ_i and S_i to have arbitrary s dependences. We imagine the entire line $-\infty < s < \infty$ populated statistically with alternating segments of the two fluids, which we again label by 0 and 1. The origin $s = 0$ is chosen at random on this infinite line. We assign a stochastic boundary condition to Eq. (17) of the form

$$\psi(0) = \left[\begin{array}{l} \psi_0 \text{ if } s = 0 \text{ is in material } 0 \\ \\ \psi_1 \text{ if } s = 0 \text{ is in material } 1 \end{array} \right. . \tag{59}$$

As before, we define $p_i(s)$ as the probability of finding material i at position s. We also define two conditional probabilities, namely

$$Q_i(r,s) = \text{Prob}\{[r,s] \text{ is in } i \mid s \text{ is in } i \text{ and } s + ds \text{ is not}\} , \tag{60}$$

and

$$R_i(r,s) = \text{Prob}\{[r,s] \text{ is in } i \mid s \text{ is in } i\} . \tag{61}$$

From these definitions, we see that $Q_i(r,s)$ is the probability that the entire interval [r,s] is in material i given that s is the right-hand boundary of material i. By contrast, $R_i(r,s)$ is the probability that the entire interval [r,s] is in material i, given that s is anywhere in material i. These probabilities describe in general inhomogeneous (spatially dependent) statistics. If the statistics are homogeneous, then p_i is independent of s, and the two conditional probabilities defined above depend only upon the single displacement argument, s - r.

As a prelude to writing the renewal equations whose solution leads to $\bar{\psi}(s)$, the ensemble averaged intensity, we define $G_i(r,s;b)$ as the solution of Eq. (17) given that the solution is b at r, and given that the entire interval [r,s] consists of material i. This is a nonstochastic solution which is easily found from Eq. (17) to be

$$G_i[r,s;b] = be^{-\tau_i(r,s)} + \int_r^s ds' \, S_i(s')e^{-\tau_i(s',s)} \, , \tag{62}$$

where

$$\tau_i(r,s) = \int_r^s ds'' \sigma_i(s'') \, . \tag{63}$$

Given this definition of $G_i(r,s;b)$, the ensemble averaged intensity $\bar{\psi}(s)$ is given by [5]

$$\bar{\psi}(s) = p_0(s)\psi_0(s) + p_1(s)\psi_1(s) \, , \tag{64}$$

where $\psi_0(s)$ and $\psi_1(s)$ satisfy the renewal equations

$$\psi_0(s) = G_0[0,s;\psi_0]R_0(0,s) + \int_0^s d_rR_0(r,s)G_0[r,s;\phi_1(r)] \, , \tag{65}$$

$$\phi_0(s) = G_0[0,s;\psi_0]Q_0(0,s) + \int_0^s d_rQ_0(r,s)G_0[r,s;\phi_1(r)] \, . \tag{66}$$

(The subscript r on the differentials d_rR_0 and d_rQ_0 means that this differential acts only on the r variable in R_0 and Q_0.) A second set of equations complementary to Eqs. (65) and (66) follows by interchanging the indices 0 and 1. The solution of these four coupled integral equations, when used in Eq. (64), gives an exact solution for the ensemble averaged intensity for the general statistics defined by Eqs. (60) and (61), and for general s dependences of $\sigma_i(s)$ and $S_i(s)$. It must be emphasized, however, that this treatment does not include the scattering interaction in the underlying transport equation.

The special case of homogeneous statistics is described entirely by the chord length distribution $f_i(z)$, such that $f_i(z)dz$ is the probability that material i has a chord length between z and z + dz. This distribution function is independent of position s on the line $-\infty < s < \infty$. That is, the length of any segment of material i on this infinite line is chosen from the same distribution $f_i(z)$. If we define

$$Q_i(z) = \int_z^\infty dz' \, f_i(z') \; , \tag{67}$$

then $Q_i(z)$ is interpreted as the probability of a given segment of material i exceeding a length z. The average chord length in material i, which we denote by λ_i, is given by

$$\lambda_i = \int_0^\infty dz \, z f_i(z) = \int_0^\infty dz \, Q_i(z) \; . \tag{68}$$

If we define

$$R_i(z) = \frac{1}{\lambda_i} \int_z^\infty dz' \, Q_i(z') \; , \tag{69}$$

it is easily argued that $R_i(z)$ is the probability that the right-hand boundary of a segment of material i is a distance greater than z from an arbitrary (random) point in the segment. For homogeneous statistics, the conditional probabilities defined for general inhomogeneous statistics by Eqs. (60) and (61) are given in terms of the single variable quantities defined by Eqs. (67) through (69) as

$$Q_i(r,s) = Q_i(s - r) \quad ; \quad R_i(r,s) = R_i(s - r) \; . \tag{70}$$

Additionally, for homogeneous statistics the probability p_i of finding material i at any point on the line $-\infty < s < \infty$ is given by

$$p_i = \frac{\lambda_i}{\lambda_0 + \lambda_1} \; . \tag{71}$$

A special case of homogeneous renewal statistics is Markov statistics which correspond to a simple exponential distribution, i.e.,

$$\lambda_i f_i(z) = Q_i(z) = R_i(z) = e^{-z/\lambda_i} \; . \tag{72}$$

If we specialize the general renewal equations given by Eqs. (65) and (66) to the case of homogeneous statistics, we find [5]

$$\psi_0(s) = \psi_0 e^{-\sigma_0 s} R_0(s) + \int_0^s d_r R_0(s - r) e^{-\sigma_0(s-r)} \phi_1(r)$$

$$+ \int_0^s dr \; e^{-\sigma_0(s-r)} S_0(r) R_0(s - r) \; , \qquad (73)$$

$$\phi_0(s) = \psi_0 e^{-\sigma_0 s} Q_0(s) + \int_0^s d_r Q_0(s - r) e^{-\sigma_0(s-r)} \phi_1(r)$$

$$+ \int_0^s dr \; e^{-\sigma_0(s-r)} S_0(r) Q_0(s - r) \; , \qquad (74)$$

with two compementary equations obtained by interchanging the indices in Eqs. (73) and (74). In writing Eq. (73) and (74) we have allowed the external source $S_0(s)$ to be spatially dependent, but have taken the cross section σ_0 to be constant, independent of s. It is clear that the integral terms in Eqs. (73) and (74) are of the convolution form, and hence these equations and their complements are easily solved by Laplace transformation. Hence in this case the problem of obtaining the ensemble average $\overline{\psi}(s)$ for renewal statistics reduces to the problem of performing a Laplace inversion.

The ability to obtain an exact analytic solution in the case just noted has been used to obtain a renormalized (scalar) transport equation for a fluid mixture obeying renewal statistics [5]. Specifically, for the case of spatially independent sources S_i, as well as homogeneous statistics and spatially independent cross sections σ_i, an exact solution was constructed for $\overline{\psi}(s)$ for arbitrary (but homogeneous) renewal statistics, i.e., for an arbitrary chord length distribution $f_i(z)$. From this solution, two quantities in particular were computed, namely: (1) the mean distance to collision for a particle entering the mixture at s = 0 in the case of a nonstochastic boundary condition [i.e., $\psi_0 = \psi_1$ in Eq. (59)]; and (2) the constant (independent of s) solution extant at s = ∞. A scalar transport equation of the form

$$\frac{1}{c} \frac{\partial \overline{\psi}}{\partial t} + \vec{\Omega} \cdot \vec{\nabla}\overline{\psi} + \sigma_{eff}\overline{\psi} = S_{eff} \; , \qquad (75)$$

was then proposed, with the σ_{eff} and S_{eff} chosen so that the solution of this scalar equation has the same two properties. This led to the definitions

$$\sigma_{eff} = \frac{p_0\sigma_0 + p_1\sigma_1 + \sigma_0\sigma_1 q\, \lambda_c}{1 + (p_0\sigma_1 + p_1\sigma_0)q\, \lambda_c} \;, \tag{76}$$

$$S_{eff} = \frac{p_0 S_0 + p_1 S_1 + (p_0\sigma_1 S_0 + p_1\sigma_0 S_1)q\, \lambda_c}{1 + (p_0\sigma_1 + p_1\sigma_0)q\, \lambda_c} \;, \tag{77}$$

with the quantity q given by

$$q = \frac{1}{\sigma_0}\left[\frac{1}{\tilde{Q}_0(\sigma_0)} - \frac{1}{\lambda_0}\right] + \frac{1}{\sigma_1}\left[\frac{1}{\tilde{Q}_1(\sigma_1)} - \frac{1}{\lambda_1}\right] - 1 \;. \tag{78}$$

Here λ_i is the mean chord length for fluid packets of material i, with

$$\frac{1}{\lambda_c} = \frac{1}{\lambda_0} + \frac{1}{\lambda_1} \;, \tag{79}$$

and $\tilde{Q}_i(\sigma_i)$ is the Laplace transform of $Q_i(z)$ evaluated at the transform variable σ_i. It is easily shown that $q \geq 0$ for all physical parameters, and hence Eq. (75) is a robust equation in the sense that $\sigma_{eff} \geq 0$ and $S_{eff} \geq 0$. In the special case of Markov statistics as given by Eq. (72), Eq. (78) predicts $q = 1$, and then Eqs. (76) and (77) reduce to the asymptotic Markov results given earlier, namely Eqs. (54) and (55) with $\nu_s = \hat{\sigma}_s = \bar{\sigma}_s = 0$.

   One major drawback of Eq. (75) and, more generally, of all of the renewal statistics results, is that the scattering interaction has not been included; we only have results for the transport problem with $\sigma_{si} = 0$. This supplies a motivation to consider a second independent procedure for deriving a standard transport equation involving effective quantities to take statistical effects into account for non-Markov statistics. This procedure includes scattering and reduces to Eq. (75) in the absence of scattering. Further, it yields, in addition to a standard (renormalized) scalar transport equation with effective quantities, a higher level approximation in the form of a two component vector description, as is Eq. (11) for Markov statistics. In fact, this scheme is based upon a modification of the Markov description given by Eq. (11)[5]. Specifically, one assumes that

Eq. (11) holds (approximately) in the non-Markov case by replacing λ_c in Eq. (14) by λ_{eff}, an effective correlation length. We define λ_{eff} by

$$\lambda_{eff} = q \, \lambda_c \, , \tag{80}$$

where q and λ_c are given by Eqs. (78) and (79). In both Eqs. (78) and (79), λ_i is the mean chord length for material i for whatever non-Markov (renewal) statistics is obeyed by the mixture.

The motivation behind the definition of λ_{eff} as given by Eq. (80) is found in an asymptotic analysis of Eq. (11) [with λ_c in $\hat{\sigma}$ in Eq. (14) replaced with λ_{eff}] with the scaling found in Eq. (49). Clearly one finds the renormalized (scalar) transport equation given by Eq. (53), but with the effective quantities, Eqs. (54) and (55), containing λ_{eff} in place of λ_c. In the absence of scattering, this modified Eq. (53) is identical to Eq. (75), and the expressions for σ_{eff} and S_{eff} given by Eqs. (54) and (55) with λ_{eff} in place of λ_c, are identical to Eqs. (76) and (77) if λ_{eff} is given by q λ_c. Hence by a proper choice of the single parameter λ_{eff} in the expression [see Eq. (14)]

$$\hat{\sigma} = p_1 \sigma_0 + p_0 \sigma_1 + \frac{1}{\lambda_{eff}} \, , \tag{81}$$

the Markov description has been modified to give both the correct σ_{eff} and S_{eff} for general non-Markov statistics. Accordingly, it has been suggested [5] that Eq. (53), with Eq. (81) used for $\hat{\sigma}$, is a reasonable low order approximate description of transport in a binary non-Markov mixture, including the effects of scattering. One would expect that a better, but still approximate, treatment would be to use Eq. (11), the vector description, with $\hat{\sigma}$ given by Eq. (81). In particular, Eq. (11) would be expected to be more accurate than Eq. (53) in a time independent problem in a source free ($S_i = 0$), purely absorbing ($\sigma_{si} = 0$), constant cross section mixture, since in this case Eq. (11) predicts a solution which is the sum of two exponentials whereas Eq. (53) gives a single exponential solution. A two exponential form clearly has a much better chance of predicting the true non-Markov solution than does a single exponential.

In summary, the treatment of particle transport in a two component non-Markov mixture is well founded in the absence of scattering. With the scattering interaction included, one must at present resort to reasonable, but approximate, models for non-Markov mixtures. An exact treatment, including scattering, in the sense of Eq. (11) for binary Markov mixtures, is not known at this time. As remarked earlier, even in the case of a binary Markov mixture, Eq. (11) is exact only insofar as the use of the master equation including scattering can be rigorously justified. An alternate approach to putting Eq. (11) on a firm footing would be an extension of the method of smoothing to include scattering.

A SINGULAR EIGENFUNCTION PROBLEM

The integral equations which arose from the renewal analysis given in the last section can be transformed into a very interesting coupled set of two transport-like equations under a certain, but rather general, assumed form for the statistics. These equations can be analyzed by a singular eigenfunction method as introduced by Case [6],[7] for the classic one speed, time independent, planar geometry transport equation. Here we consider the details of this treatment.

Specifically, we consider the case of homogeneous renewal statistics with no scattering, and cross sections σ_i which are independent of position. The renewal equations are then given by Eqs. (73) and (74) and their complements found by interchanging the indices. We assume that the function $Q_i(z)$ which describes the statistics [see Eq. (67)] can be represented as a Laplace transform according to

$$Q_i(z) = \int_0^\infty du\ q_i(u)e^{-z/u}\ ,\tag{82}$$

for some class of functions $q_i(u)$. From Eq. (67) it is clear that the function $Q_i(z)$ has certain properties, namely

$$Q_i(0) = 1\quad ;\quad Q_i(\infty) = 0\quad ;\quad dQ_i(z)/dz \le 0\ .\tag{83}$$

The first of these three conditions implies that

$$\int_0^\infty du\ q_1(u) = 1\ ,\tag{84}$$

as one condition on the function $q_i(u)$. From Eq. (68) we deduce that the mean chord length λ_i is given as the first moment of $q_i(u)$, i.e.,

$$\lambda_i = \int_0^\infty du \; u q_i(u) \; . \tag{85}$$

The function $R_i(z)$ corresponding to Eq. (82) is found from Eq. (69) to be

$$R_i(z) = \frac{1}{\lambda_i} \int_0^\infty du \; u q_i(u) e^{-z/u} \; . \tag{86}$$

We note that homogeneous Markov statistics is a special case of Eq. (82) given by

$$q_i(u) = \delta(u - \lambda_i) \; . \tag{87}$$

That is, with $q_i(u)$ given by this Dirac delta function, Eqs. (82) and (86) reduce to the exponential Markov description, Eq. (72).

Now, using Eq. (82) in the renewal result, Eq. (74), we find

$$\phi_0(s) = \psi_0 \int_0^\infty du \; q_0(u) \; \exp\left[-(\sigma_0 + \frac{1}{u})s\right]$$

$$+ \int_0^\infty du \; q_0(u) \int_0^s dr\left[S_0(r) + \frac{1}{u}\phi_1(r)\right] \exp\left[-(\sigma_0 + \frac{1}{u})(s - r)\right] \; . \tag{88}$$

If we define the function $\phi_0(s,u)$ by the equation

$$\phi_0(s,u) = \psi_0 \; \exp\left[-(\sigma_0 + \frac{1}{u})s\right]$$

$$+ \int_0^s dr\left[S_0(r) + \frac{1}{u}\phi_1(r)\right] \exp\left[-(\sigma_0 + \frac{1}{u})(s - r)\right] \; , \tag{89}$$

we then have

$$\phi_0(s) = \int_0^\infty du \; q_0(u) \; \phi_0(s,u) \; . \tag{90}$$

From Eq. (89) one deduces that $\phi_0(s,u)$ satisfies the equation

$$\frac{\partial \phi_0(s,u)}{\partial s} + (\sigma_0 + \frac{1}{u}) \; \phi_0(s,u) = S_0(s) + \frac{1}{u}\phi_1(s) \; , \tag{91}$$

with boundary condition

$$\phi_0(0,u) = \psi_0 \ . \tag{92}$$

Interchanging indices in Eqs. (90) through (92), we have the complementary equations

$$\phi_1(s) = \int_0^\infty du\ q_1(u)\ \phi_1(s,u)\ . \tag{93}$$

$$\frac{\partial\phi_1(s,u)}{\partial s} + (\sigma_1 + \frac{1}{u})\ \phi_1(s,u) = S_1(s) + \frac{1}{u}\ \phi_0(s)\ , \tag{94}$$

$$\phi_1(0,u) = \psi_1\ . \tag{95}$$

The boundary data ψ_0 and ψ_1 in Eqs. (92) and (95) are just the boundary data for the original transport problem [see Eq. (59)]. If we use Eq. (93) in Eq. (91), and Eq. (90) in Eq. (94), the structure of the equations for $\phi_i(s,u)$ becomes more transparent. We have, multiplying each equation by u,

$$u\ \frac{\partial\phi_0(s,u)}{\partial s} + (1 + \sigma_0 u)\phi_0(s,u) = \int_0^\infty du'\ q_1(u')\ \phi_1(s,u') + uS_0(u)\ , \tag{96}$$

$$u\ \frac{\partial\phi_1(s,u)}{\partial s} + (1 + \sigma_1 u)\phi_1(s,u) = \int_0^\infty du'\ q_0(u')\ \phi_0(s,u') + uS_1(u)\ , \tag{97}$$

These two equations are clearly coupled integro-differential equations of a transport-like form. Even though the underlying physics does not include the scattering interaction [i.e., the renewal equations given by Eq. (73) and (74), which were the starting point of our considerations, do not include scattering], the statistical fluctuations of the medium have introduced integral scattering-like terms in Eqs. (96) and (97). We note, however, several significant differences between Eqs. (96) and (97) and the classical one-speed, time independent, planar geometry transport equation. First, the range of the "angular" variable is $0 \leq u < \infty$, rather than the usual interval [-1,1]. Secondly, the "total cross section" is dependent upon u, a situation not normally encountered in classical transport theory. Lastly, the "scattering kernel" is independent of u, but depends upon u', again a situation which does not occur in any classical transport problem.

Proceeding, we analyze in a similar way the equation for $\psi_0(s)$ given by Eq. (73). Using Eq. (86) in Eq. (73), we find

$$\psi_0(s) = \psi_0 \int_0^\infty du \, \frac{u q_0(u)}{\lambda_0} \, \exp\left[-(\sigma_0 + \frac{1}{u})s\right]$$

$$+ \int_0^\infty du \, \frac{u q_0(u)}{\lambda_0} \int_0^s dr \left[S_0(r) + \frac{1}{u} \phi_1(r)\right] \exp\left[-(\sigma_0 + \frac{1}{u})(s - r)\right] \quad . \quad (98)$$

If we define $\psi_0(s,u)$ by the equation

$$\psi_0(s,u) = \psi_0 \, \exp\left[-(\sigma_0 + \frac{1}{u})s\right]$$

$$+ \int_0^\infty dr \left[S_0(r) + \frac{1}{u} \phi_1(r)\right] \exp\left[-(\sigma_0 + \frac{1}{u})(s - r)\right] \quad , \quad (99)$$

we then conclude

$$\psi_0(s) = \int_0^\infty du \, \frac{u q_0(u)}{\lambda_0} \, \psi_0(s,u) \quad . \quad (100)$$

Comparing Eqs. (89) and (99) we immediately conclude

$$\psi_0(s,u) = \phi_0(s,u) \quad . \quad (101)$$

By interchanging indices 0 and 1 in the above, we similarly have

$$\psi_1(s,u) = \phi_1(s,u) \quad . \quad (102)$$

Thus we conclude that the equations we previously derived for $\phi_0(s,u)$ and $\phi_1(s,u)$, namely Eqs. (96) and (97) with boundary conditions given by Eqs. (92) and (95), are, in fact, a closed set of equations for $\psi_0(s,u)$ and $\psi_1(s,u)$. It is the $\psi_i(s,u)$ that are needed to construct the emsemble averaged intensity $\bar{\psi}(s)$.

All of these algebraic manipulations can be summarized simply as follows. If the statistics of the mixture are homogeneous and defined by Eq. (82), then, in the absence of scattering and with constant cross sections σ_i, the ensemble averaged intensity $\bar{\psi}(s)$ is given by

$$\bar{\psi}(s) = p_0 \psi_0(s) + p_1 \psi_1(s) \ , \tag{103}$$

where $p_i = \lambda_i/(\lambda_0 + \lambda_1)$ is the probability of finding material i at any point in the mixture. The $\psi_i(s)$ are given by

$$\psi_0(s) = \int_0^\infty du \ \frac{u q_0(u)}{\lambda_0} \ \psi_0(s,u) \ , \tag{104}$$

$$\psi_1(s) = \int_0^\infty du \ \frac{u q_1(u)}{\lambda_1} \ \psi_1(s,u) \ , \tag{105}$$

where the $\psi_i(s,u)$ satisfy the coupled integro-differential equations

$$u \ \frac{\partial \psi_0(s,u)}{\partial s} + (1 + \sigma_0 u)\psi_0(s,u) = \int_0^\infty du' \ q_1(u') \ \psi_1(s,u') + u S_0(s) \ , \tag{106}$$

$$u \ \frac{\partial \psi_1(s,u)}{\partial s} + (1 + \sigma_1 u)\psi_1(s,u) = \int_0^\infty du' \ q_0(u') \ \psi_0(s,u') + u S_1(s) \ , \tag{107}$$

with boundary conditions given by the boundary data for the original transport problem [see Eq. (59)]

$$\psi_0(0,u) = \psi_0 \quad ; \quad \psi_1(0,u) = \psi_1 \quad , \quad u \geq 0 \ . \tag{108}$$

We remarked earlier that homogeneous Markov statistics is a special case of this analysis with $q_i(u)$ given by Eq. (87). In this case, Eqs. (106) and (107) reduce to the previous Markov result, Eq. (37) [or equivalently, Eq. (11)] in the absence of scattering and for homogeneous statistics. An interesting generalization of this Markov case involves writing the function $q_i(u)$ in Eq. (82) as a sum of delta functions according to

$$q_i(u) = \sum_{n=1}^{N} \alpha_{in} \ \delta(u - \beta_{in}) \ , \tag{109}$$

where, to satisfy the first constraint given by Eq. (83), we must have

$$\sum_{n=1}^{N} \alpha_{in} = 1 \ .$$ (110)

We also require $\beta_{in} > 0$ to satisfy the second constraint. In this case, the two coupled integro-differential equations given by Eqs. (106) and (107) simplify considerably and reduce to 2N coupled differential equations, which are easily solved explicitly by elementary means.

Returning now to the general equations given by Eqs. (106) and (107), one is led to an interesting singular eigenfunction problem. We consider the source free ($S_i = 0$), time independent (so that s now denotes a simple spatial variable) problem, and look for separable solutions of the form [6],[7]

$$\psi_i(s,u) = \phi_{i\nu}(u)e^{-s/\nu} \ .$$ (111)

[The eigenvalue ν here is not to be confused with the ν introduced in Eq. (16).] Use of Eq. (111) in Eqs. (106) and (107) yields

$$\left[1 + (\sigma_0 + \frac{1}{\nu})u\right]\phi_{0\nu}(u) = \int_0^\infty du' \ q_1(u')\phi_{1\nu}(u') \ ,$$ (112)

$$\left[1 + (\sigma_1 + \frac{1}{\nu})u\right]\phi_{1\nu}(u) = \int_0^\infty du' \ q_0(u')\phi_{0\nu}(u') \ .$$ (113)

These two coupled equations represent an eigenvalue problem for the eigenvalues ν and eigenfunctions $\phi_{i\nu}(u)$, in the sense of Case [6],[7]. The questions associated with this eigenvalue problem are the usual ones. One would first want to identify the (discrete and/or continuum) eigenvalues and construct the corresponding eigenfunctions. Secondly, for these eigenfunctions one would want to investigate the usual three associated items, namely orthogonality, normalization, and completeness [6],[7]. These investigations should take place for the general class of functions, or some subset, $q_i(u)$, defined by Eq. (82) which satisfy the constraints given by Eq. (83). At this time, we have only a few preliminary results which we now discuss.

Concerning the spectrum ν, one would expect on physical grounds that all of the eigenvalues would be real and non-negative. They will not occur in plus-minus pairs as in the usual transport problem [6],[7]. This is connected with the fact that the "angular" variable only assumes positive values, $0 \leq u < \infty$, as contrasted with the usual angular variable which assumes both positive and negative values over a symmetric range about the origin. In the special case when $q_i(u)$ is given by a sum of delta functions as in Eq. (109), Eqs. (112) and (113) reduce to 2N algebraic equations for $\phi_{i\nu}(\beta_n)$, and the spectrum consists entirely of 2N discrete eigenvalues. In this case there is no continuum.

One can also prove a result concerning the existance of discrete eigenvalues for a general $q_i(u) \geq 0$. Let us order the cross sections such that $\sigma_1 \geq \sigma_0$. Then, excluding the real line $0 \leq \nu \leq 1/\sigma_0$, the bracketed terms on the left-hand sides of Eq. (112) and (113) do not vanish. Accordingly, we can divide by these terms. A subsequent multiplication of Eq. (112) by $q_0(u)$, Eq. (113) by $q_1(u)$, integration of each result over u, and setting to zero the coefficient determinant for the unknowns consisting of the integral terms on the right-hand sides of Eqs. (112) and (113), we find the dispersion relationship given by

$$1 = \left[\int_0^\infty du \, \frac{q_0(u)}{\left[1 + (\sigma_0 - \frac{1}{\nu})u \right]} \right] \left[\int_0^\infty du \, \frac{q_1(u)}{\left[1 + (\sigma_1 - \frac{1}{\nu})u \right]} \right] . \qquad (114)$$

If we now specialize the discussion to non-negative $q_i(u)$ and invoke one of the conditions on the $q_i(u)$, namely Eq. (84), it is clear that the right-hand side of Eq. (114) is less than unity for any real value of ν excluding values on the cut $0 \leq \nu \leq 1/\sigma_0$. Thus we conclude in this case that, off the cut, there are no discrete eigenvalues (assuming that our physical intuition concerning the nonexistence of complex eigenvalues is correct).

In this regard, a few words are in order concerning the Markov model, in which case $q_i(u)$ is given by a Dirac delta function [see Eq. (87)]. This $q_i(u)$ is clearly non-negative, and the argument just given shows that there should be no discrete eigenvalues (off the cut). Yet in the Markov case we have also argued that the entire spectrum consists of just two discrete eigenvalues. This apparent contradiction is resolved by noting that these two discrete Markov eigenvalues lie on the cut; in fact, it is easily shown that these two eigenvalues lie in the range $1/\sigma_1 \leq \nu \leq 1/\sigma_0$, where $\sigma_1 \geq \sigma_0$. In this special case of $q_i(u)$ given by a delta function, the cut is not a cut in the usual sense of identifying the cut with a continuum of eigenvalues. In a similar vein, one would expect all of the 2N discrete eigenvalues associated with $q_i(u)$ given by Eq. (109) to lie in the range $0 \leq \nu \leq 1/\sigma_0$, at least when all of the α_{in} are positive.

For a smooth function $q_i(u)$ (i.e., excluding delta functions and the like), there is no mathematical reason at this time not to expect both continuum (on the cut) and discrete (off the cut) eigenvalues. It may also be the case, depending upon the details of $q_i(u)$, that discrete eigenvalues may exist imbedded in the continuous spectrum. However, one could argue physically that no eigenvalues should exist for $\nu > 1/\sigma_0$ (with $\sigma_0 < \sigma_1$) since this would correspond to exponential decay with a scale length greater than the largest mean free path in the underlying transport problem. A similar physical argument suggests that no eigenvalues should exist for $\nu < 1/\sigma_1$. Thus the best guess at this time is that all of the eigenvalues ν are real and lie in the range $1/\sigma_1 \leq \nu \leq 1/\sigma_0$, and these eigenvalues may be discrete and/or continuous depending upon the functions $q_i(u)$. Clearly, the entire question of the spectrum associated with Eqs. (112) and (113) is deserving of a rigorous and detailed mathematical investigation.

With regard to orthogonality, the only range of orthogonality which is pertinent here is halfrange orthogonality, $0 \leq u < \infty$. This is because the boundary data on the transport problem [see Eq. (108)] are given only for $u \geq 0$; the eigenvalue problem given by Eqs. (112) and (113) involves only positive values of u. This orthogonality relationship is easily derived. One multiplies Eq. (112) by $q_0(u)$ and Eq. (113) by $q_1(u)$, and writes the resulting equations for a second eigenvalue ν'. One then cross multiplies (in the matrix sense) these two sets of equations, integrates over u in the range $0 \leq u < \infty$, and subtracts one result from the other, in the usual way. The resulting orthogonality relationship is given by

$$\int_0^\infty duu \left[q_0(u)\phi_{0\nu}(u)\phi_{0\nu'}(u) + q_1(u)\phi_{1\nu}(u)\phi_{1\nu'}(u) \right] = 0 \quad , \qquad \nu \neq \nu' \quad . \quad (115)$$

The final item to be discussed is the completeness of the eigenfunctions $\phi_{i\nu}(u)$. For the eigenfunctions to be useful in solving the transport problem defined by Eqs. (106) through (108), it is necessary to be able to represent the boundary data given by Eq. (108) as a linear combination of these eigenfunctions. In this regard, we note that these boundary data are "isotropic," i.e., independent of u. This implies that the completeness of the eigenfunction $\phi_{1\nu}(u)$ only needs to be proven for isotropic functions.

[This is also true if one wishes to construct the Green's function for Eqs. (106) and (107) which is needed to treat, via eigenfunctions, the problem when external sources $S_i(s)$ are extant.] This is in contrast to the situation encountered in classical transport theory where completeness, both halfrange and fullrange, is needed for more or less arbitrary angular functions [6],[7]. Hence the completeness question here is very special; only halfrange completeness for isotropic functions is of interest.

We hope to have additional results in this singular eigenfunction context in the near future.

CONCLUDING REMARKS

In this article we have reviewed what is currently known concerning a linear transport description in a random mixture consisting of two immiscible fluids. With the exception of the material given in the last section, more detail can be found in the published literature [1]-[5]. In this concluding section, we wish to briefly discuss certain numerical results which have been obtained [5], as well as give a brief discussion of open questions in this area of investigation.

In the recent paper by Levermore et al [5], numerical calculations were summarized which compared the simple atomic mix description with both the Markov and renewal statistics descriptions. All comparisons were made for simple steady-state transmission problems. That is, these problems were time independent with no scattering ($\sigma_{si} = 0$) and no external sources ($S_i = 0$). A nonstochastic intensity was incident upon a halfspace, and the ensemble averaged intensity $\bar{\psi}$ was computed as a function of depth within the halfspace. The statistics were assumed homogeneous and the cross sections σ_i were taken as independent of space. Renewal statistics results were given for several different chord length distributions; however, all of these distributions were unimodal. Thus the conclusions which can be drawn from these numerical results only apply with confidence to the relatively simple transmission transport problem just described, and with relatively simple (i.e., unimodal) chord length distributions describing the homogeneous statistics.

With this caveat, the following four conclusions can tentatively be drawn [5].

1. The atomic mix model, Eq. (6), which completely neglects the statistical nature of the problem can be very much in error. The

magnitude of the error depends upon the physical parameters (σ_i, λ_i, p_i) involved. This error can be made arbitrarily large by certain choices for these parameters. Further, if the true statistics are not Markov, the Markov model given by Eq. (11) can, in certain cases, be just as much, or even more, in error than the atomic mix model.

2. As a practical matter, the information required from the fluid statistics is the mean and the variance of the chord length distributions; the mean alone is not sufficient. Different chord length distributions with the same mean (i.e., the same average fluid packet size) but different variances can give significantly different transport results. On the other hand, test calculations with different chord length distributions with the same mean and variance gave essentially the same transport results. It appears that the details of the chord length distributions as reflected in the moments higher than the variance have little effect on the transport description. We emphasize, however, that this conclusion is based upon relatively simple chord length distributions which involve a single mode. For complex, multimode distributions this conclusion may not be valid.

3. The modified Markov model, namely Eqs. (11) through (16) with λ_c in Eq. (14) replaced by λ_{eff} as defined by Eqs. (78) and (80), is a quite accurate transport model for non-Markov statistics. In this regard, we emphasize that this conclusion is based upon test calculations which did not involve scattering. As noted much earlier in this paper, even the Markov description (aside from its modification to approximately account for non-Markov statistics) is, at this time, not rigorously understood when scattering is present in the underlying transport problem.

4. The renormalized (scalar) transport equation given by Eqs. (53) through (55) with $\hat{\sigma}$ given by Eq. (81) is a simple, robust, and reasonably accurate transport model for both Markov and non-Markov statistics. [For Markov statistics, $q = 1$ and hence $\lambda_{eff} = \lambda_c$ in Eq. (81).] However, it is not as accurate (nor, however, is it as complex) as the modified Markov model discussed in item 3 above, from which it is derived as a certain asymptotic limit.

We conclude this paper by listing certain aspects of the development of a transport theory in a statistical medium which need further investigation. This list is undoubtedly incomplete, but includes those items which immediately come to mind. These are:

1. The entire question of physical realizability of all of our considerations could be studied. For example, can one envision an ensemble of partitioning of all space into two (or more) materials, such that the same prescribed homogeneous chord length distributions are extant along an arbitrary ray? It is known [22] that this is indeed possible in the special case of an exponential distribution (Markov statistics) in a plane. A second, but obviously related, question has to do with what statistics of mixing actually occur in nature. For example, the exponential distribution appears to be a fairly good description of rock fragment sizes [23]. This implies that for particle transport through such a system, the Markov description is appropriate. Is the Markov description equally applicable to other systems, such as the mixed region in the vicinity of a Rayleigh-Taylor unstable interface? Clearly, each physical system of interest needs to have its mixing statistics identified so that the proper transport model can be invoked.

2. The master equation derivation of Eq. (11), the Markov description, needs to be put on a firm foundation when scattering is extant. Alternately, the method-of-smoothing derivation of the Markov result needs to be generalized to include the scattering interaction. One or the other of these two approaches is needed to establish full confidence in Eq. (11) as the correct description of transport in a Markov mixture, including scattering.

3. In this paper, we have identified three asymptotic limits of the Markov description, Eq. (11). These are given by Eqs. (47), (53), and (57), respectively. An investigation is needed as to whether other asymptotic limits exist, and, if they do, whether or not they constitute reasonable approximations to Eq. (11) away from these limits.

4. A flux-limited diffusion approximation to Eq. (11) should be sought. This would be a generalization of the flux-limited diffusion treatment recently introduced [18]-[20] for the standard scalar transport equation.

5. A rigorous, or at least semi-rigorous in the sense of Eq. (11), treatment of scattering within the context of renewal theory is needed. At this time, the treatment of scattering for renewal

statistics is limited to the modification of the Markov description found by introducing an effective correlation length given by Eqs. (78) and (80) into the expression for $\hat{\sigma}$ [see Eq. (81)].

6. The details of the singular eigenfunction problem as outlined in the last section need to be carried out. The class of admissible functions $q_i(u)$ as defined by Eqs. (82) and (83) needs to be identified, and for this class of functions the eigenfunctions need be constructed and completeness proved.

7. All of the results reported to date have been for binary statistical mixtures. An extension to more than two materials should be made. This generalization seems relatively straightforward, using essentially the same methods which have been used to obtain the binary mixture results.

8. All of the work reported to date has concentrated on deriving equations for $\bar{\psi}$, the ensemble averaged intensity. Equations describing the higher moments of the random variable ψ, in particular the variance, would be of interest. The master equation approach seems ideally suited to generating results for the higher moments. In particular, multiplication of Eq. (31), the joint probability equation for $P_i(\vec{r},\vec{\Omega},t,\psi)$ by ψ^n prior to an integration over $0 \leq \psi < \infty$ should generate equations for the ensemble average of the nth moment in a straightforward manner.

9. More numerical tests are needed to validate the tentative conclusions we have listed at the beginning of this section. These tests should be for problems including scattering and external sources, and should involve multimodal chord length distributions.

In summary, while it appears that a good start has been made in developing a linear transport description applicable to transport in a statistical medium, much remains to be done before this area of inquiry can be considered to be unfruitful for further research.

Acknowledgement

This paper was prepared by the first author (GCP), who was partially supported in this task by the National Science Foundation.

REFERENCES

[1] C. D. Levermore, G. C. Pomraning, D. L. Sanzo, and J. Wong, *Linear transport theory in a random medium*, J. Math. Phys., 27 (1986) 2526.

[2] D. Vanderhaegen, *Radiative transfer in statistically heterogeneous mixtures*, J. Quant. Spectros. Radiat. Transfer, 36 (1986) 557.

[3] C. D. Levermore, *Transport in randomly mixed media with inhomogeneous anisotropic statistics*, in preparation.

[4] D. Vanderhaegen, *Impact of a mixing structure on radiative transfer in random media*, J. Quant. Spectros. Radiat. Transfer, in press.

[5] C. D. Levermore, G. C. Pomraning, and J. Wong, *Renewal theory for transport processes in binary statistical mixtures*, J. Math. Phys., in press.

[6] K. M. Case, *Elementary solutions of the transport equation*, Ann. Phys., 9 (1960) 1.

[7] K. M. Case and P. F. Zweifel, *Linear Transport Theory*, Addison-Wesley, Reading, MA, 1967.

[8] G. C. Pomraning, *The Equations of Radiation Hydrodynamics*, Pergamon, Oxford, 1973.

[9] E. Parzan, *Stochastic Processes*, Holden-Day, San Francisco, 1962.

[10] J. B. Keller, *Wave propagation in random media*, in Proc. Symposium on Applied Mathematics, vol. 13, Am. Math. Soc., Providence, RI, 1962, p. 227.

[11] J. B. Keller, *Stochastic equations and wave propagation in random media*, in Proc. Symposium on Applied Mathematics, vol. 16, Am. Math. Soc., Providence, RI, 1964, p. 145.

[12] J. B. Keller, *Effective Behavior of Heterogeneous Media*, in Statistical Mechanics and Statistical Methods in Theory and Application, U. Landman, ed., Plenum, New York, 1977, p. 632.

[13] U. Frisch, *Wave propagation in random media*, in Probabilistic Methods in Applied Mathematics, A. T. Bharucha-Reid, ed., Academic Press, New York, 1968, p. 75.

[14] J. A. Morrison, *Moments and correlation functions of solutions of some stochastic matrix differential equations*, J. Math. Phys., 13 (1972) 299.

[15] N. G. van Kampen, *Stochastic Processes in Physics and Chemistry*, North Holland, Amsterdam, 1981.

[16] E. W. Larsen and J. B. Keller, *Asymptotic solution of neutron transport problems for small mean free paths*, J. Math. Phys., 15 (1974) 75.

[17] S. Chandrasekhar, *Radiative Transfer*, Dover, New York, 1960.

[18] C. D. Levermore, *A Chapman-Enskog approach to flux-limited diffusion theory*, Lawrence Livermore National Laboratory Rpt. UCID-18229, Livermore, CA, 1979.

[19] C. D. Levermore and G. C. Pomraning, *A flux-limited diffusion theory*, Astrophys. J., 248 (1981) 321.

[20] G. C. Pomraning, *Initial and boundary conditions for flux-limited diffusion theory*, J. Comp. Phys., in press.

[21] D. R. Cox, *Renewal Theory*, Science Paperbacks, Chapman and Hall, London, 1982.

[22] P. Switzer, *A random set process in the plane with a Markovian property*, Ann. Math. Statis., 36 (1965) 1859.

[23] R. Englman, Z. Jaeger, and A. Levi, *Percolation theoretical treatment of two-dimensional fragmentation in solids*, Phil. Mag. B, 50 (1984) 307.

A Look at Transport Theory from the Point of View of Linear Algebra

V. FABER and THOMAS A. MANTEUFFEL   Computer Research and Applications, Los Alamos National Laboratory, Los Alamos, New Mexico

We show that the notion of "preconditioning" from linear algebra can provide a framework for the discussion of algorithms for the numerical solution of the transport equation. In this context we show that the conjugate gradient method yields substantial savings over the Neumann series solution of the standard integral formulation of the transport equation for optically thick regimes. Further, we show that the diffusion synthetic acceleration (DSA) algorithm is the Neumann series solution of the standard integral formulation of the transport equation preconditioned by the Green's function of a diffusion operator. The preconditioned conjugate gradient method, using DSA as a preconditioner, again yields substantial savings.

1 INTRODUCTION

In this paper, we propose to examine the integral equation formulation of the transport equation in slab geometry and the diffusion-synthetic acceleration (DSA) method [La1] for its solution from the point of view of linear algebra. We consider here only the one-dimensional isotropic transport equation, although these ideas can be expanded to the general transport equation.

2 PROBLEM STATEMENT

To fix ideas, the problem we wish to solve is to find $\psi(x,\mu)$ given by

This work was sponsored in part by the Air Force Office of Scientific Research contract No. AFOSR-86-0061 and NSF Grant DMS-8704169.

$$\begin{cases} \mu \dfrac{d}{dx}\,\psi + \psi = \dfrac{1}{2}\,\gamma \displaystyle\int_{-1}^{1} \psi\,d\mu + q(x,\mu) \ , \quad -a \le x \le a,\ -1 \le \mu \le 1 \\[2mm] \psi(-a,\mu) = g_-(\mu) \ , \quad \text{for } \mu > 0 \\[2mm] \psi(a,\mu) = g_+(\mu) \ , \quad \text{for } \mu < 0 \end{cases} \qquad (1)$$

where 2a represents a measure of the numer of mean free paths across the slab. We denote (1) as

$$\begin{cases} (\mu D + I)\psi = \gamma L\psi + q \\[2mm] B\psi = g(\mu) \end{cases} \quad , \qquad (2a)$$

where

$$D\psi = \frac{\partial \psi}{\partial x} \ , \qquad (2b)$$

$$L\psi = \frac{1}{2}\int_{-1}^{1}\psi(x,\mu')\,d\mu' \ . \qquad (2c)$$

3 INTEGRAL EQUATION FORMULATION

There exists an invertible operator H_μ^{-1} (the Green's function) such that $H_\mu^{-1}q$ is the solution to the differential equation

$$\begin{cases} (\mu D + I)\psi = q \\[2mm] \quad B\psi = 0 \end{cases} \cdot$$

It is given by

$$H_\mu^{-1}\psi = \begin{cases} \dfrac{1}{\mu}\,e^{-\frac{x}{\mu}}\displaystyle\int_{-a}^{x} e^{\frac{s}{\mu}}\psi(s,\mu)\,ds \ , \quad \mu > 0 \ , \\[5mm] -\dfrac{1}{\mu}\,e^{-\frac{x}{\mu}}\displaystyle\int_{x}^{a} e^{\frac{s}{\mu}}\psi(s,\mu)\,ds \ , \quad \mu < 0 \ . \end{cases} \qquad (3)$$

We have $(\mu D + I)H_\mu^{-1} = I$, and the solution to

$$\begin{cases} (\mu D + I)\psi = q \\[2mm] \quad B\psi = g \end{cases}$$

is given by $H_\mu^{-1}q + \psi_1$, where ψ_1 satisfies

$$\begin{cases} (\mu D + I)\psi_1 = 0 \\[2mm] \quad B\psi_1 = g \end{cases} \cdot$$

The function ψ_1 is given by

$$\psi_1 = \begin{cases} g_-(\mu)e^{-\frac{1}{\mu}(x+a)}, & \mu > 0, \\ g_+(\mu)e^{-\frac{1}{\mu}(x-a)}, & \mu < 0. \end{cases} \tag{4}$$

Thus, the solution to (2) satisfies

$$\psi = H_\mu^{-1}\gamma L\psi + H_\mu^{-1}q + \psi_1. \tag{5}$$

Operating on (5) by L gives

$$L\psi = LH_\mu^{-1}\gamma L\psi + q_1, \tag{6}$$

where the $q_1 = LH_\mu^{-1}q + L\psi_1$ depends only on q and g. We write this as

$$(I - K\gamma)\phi = q_1, \tag{7}$$

where $\phi = L\psi$ and $K = LH_\mu^{-1}$. Using (3) we see that

$$K\phi = \int_{-a}^{a} k(|x - s|)\phi(s)ds, \tag{8a}$$

where

$$k(t) = \int_0^1 \frac{1}{\mu} e^{-\frac{1}{\mu}t} d\mu. \tag{8b}$$

The properties of K are well known (cf. [Wi]);

i) K is self-adjoint, positive definite and compact on $L_2[-a,a]$.

ii) $\|K\|_{L_2} < 1$.

iii) $\|K\|_{L_2} = 1 - O(\frac{1}{a^2})$, for large a.

iv)   If σ_k, u_k are an eigenvalue/eigenvector pair of K, then

   a) $\sigma_k = O(\frac{1}{k})$ for large k,

   b) $u_k = f_k + O(\frac{1}{k})$, where f_k is the kth Fourier vector on $[-a,a]$ [FMWW].

4 SPLITTING AND PRECONDITIONING

Let T be any invertible operator, $TG = I$, and assume that $T - \gamma I$ is invertible. (We also need G to operate on the same space as K, but we avoid discussion of these mathematical niceties to get on with the story.) We can define a *splitting* of the operator $I - K\gamma$ by

$$I - K\gamma = (I - G\gamma) + (G - K)\gamma. \tag{9}$$

Equation (7) can now be rewritten as

$$[(I - G\gamma) + (G - K)\gamma]\phi = q_1 \ . \tag{10}$$

If we now apply $(I - G\gamma)^{-1}$, we get the *preconditioned* equation:

$$(I - G\gamma)^{-1}(I - K\gamma)\phi = [I + (I - G\gamma)^{-1}(G - K)\gamma]\phi = (I - G\gamma)^{-1}q_1 \ . \tag{11}$$

(Note that this is exactly the equation discussed in [AlWi].) If we had first multiplied (9) by T and then applied $(T - \gamma I)^{-1}$, we would have gotten the equivalent equation

$$[I + (T - \gamma I)^{-1}(I - TK)\gamma]\phi = (T - \gamma I)^{-1}Tq_1 = q_2 \ . \tag{12}$$

The term "preconditioned" arises because we hope to have chosen T so that the equation (12) is easy to solve, i.e., in some sense "well-conditioned." (Equation (12) is referred to in the DSA algorithm as the "accelerated" equation [La1].)

5 DISCRETIZATION AND ITERATION

Let us discuss for a moment how we would solve an equation like (7) or (11). We start, of course, with a *discretization*. For convenience, we use the same symbol for the discrete operator as the continuous one. The discretization yields a system of linear equations which we solve by using an *iteration* scheme. There are two different questions which arise here. One is the order of accuracy of the discretization (i.e., how close is the solution of the discrete problem to the solution of the continuous problem) and another is the rate of convergence of the iteration scheme (or, more properly, how much work does it take to get an approximate solution within a certain accuracy to the discrete equation). Of course, the two questions are intertwined since we really want an approximate solution within a certain accuracy to the continuous solution. For this analysis we choose to ignore the first question and concentrate on the second.

6 NEUMANN SERIES

With this discussion in mind, we note that a standard iteration method for solving the discrete analogue to (7) is

$$\phi_{n+1} = K\gamma\phi_n + q_1 \ . \tag{13}$$

Here K is a discrete operator and ϕ_n and q_1 are discrete vectors. Usually, one does not form the matrix K, but applies K in the form LH_μ^{-1}; i.e., for a set of μ_j $j = 1, \ldots, N$, one solves an equation with the operator H_{μ_j} ("invert" H_{μ_j}) and uses a quadrature rule to apply L. The intermediate quantity, $H_{\mu_j}^{-1}\gamma\phi_n$, is called ψ_n, the discrete approximation to ψ. The convergence rate of (13) is completely determined by the spectral radius of $K\gamma$.

Let ϕ be the solution to

$$\phi = K\gamma\phi + q_1 \ , \tag{14}$$

and let $e_n = \phi - \phi_n$. Combining (13) and (14) yields

$$e_{n+1} = K\gamma e_n = (K\gamma)^{n+1} e_0 \; . \tag{15}$$

If the spectral radius of $K\gamma$ is less than 1 (which we write as $\sigma(K\gamma) < 1$), then the iteration will converge. The iteration scheme (13) is perhaps the least sophisticated iterative scheme. Other schemes that have better convergence properties will be discussed later.

The equation (12) can be organized in the same fashion as (13). We write

$$\phi_{n+1} = (T - \gamma I)^{-1} (TK - I)\gamma\phi_n + q_2 \; . \tag{16}$$

Again, the error equation takes the form

$$e_{n+1} = (T - \gamma I)^{-1} (TK - I)\gamma e_n = [(T - \gamma I)^{-1} (TK - I)\gamma]^{n+1} e_0 \; . \tag{17}$$

When using (16), we must assume that we have an easy method for applying T and inverting $T - \gamma I$, just as we do for applying K. The success of this iteration hinges on the assumption that the spectral radius of $(T - \gamma I)^{-1} (TK - I)\gamma$ is sufficiently less than the spectral radius of $K\gamma$ to justify the extra work involved.

7 BOUNDARY CONDITIONS

There is another point to consider here. When T is a differential operator, for example, it is only invertible when the boundary conditions are considered, that is, when the problem

$$\begin{cases} Tu = f \\ B_T u = g \end{cases} \tag{18}$$

has a unique solution. Then, we can assume that the Green's function G satisfies

$$\begin{cases} TG = I \\ B_T Gf = 0 \end{cases} \quad \text{for all } f \; , \tag{19}$$

and the solution to (18) is given by

$$u = Gf + u_1 \; ,$$

where

$$\begin{cases} Tu_1 = 0 \\ B_T u_1 = g \end{cases} \; . \tag{20}$$

In this case, the partial sums for the Neumann series for ϕ in (12) are not given uniquely by

$$(T - \gamma I) \phi_{n+1} = (TK - I)\gamma\phi_n + Tq_1 \; , \tag{21}$$

but require boundary conditions to be given on ϕ_{n+1}. To obtain these boundary conditions, we recall that this scheme was derived from

$$(I - G\gamma)\phi_{n+1} = (K - G)\gamma\phi_n + q_1 \tag{22}$$

before we operated on both sides by T (see (10)). If we apply the boundary operator B_T to both sides of (22), we get (since $B_T G = 0$)

$$B_T \phi_{n+1} = B_T K \gamma \phi_n + B_T q_1 \ , \tag{23}$$

so the correct equation for the $(n + 1)st$ partial sum ϕ_{n+1} in the iteration (21) is

$$\begin{cases} (T - \gamma I) \phi_{n+1} = (TK - I)\gamma \phi_n + T q_1 \\ B_T \phi_{n+1} = B_T K \gamma \phi_n + B_T q_1 \end{cases} \ . \tag{24}$$

8 EXAMPLE

We include this example to show how DSA falls into this framework. Choose $T = -\frac{1}{3} D^2 + I$ with boundary equations of the type $\phi(\pm a) \overline{\mp} \beta \phi^1(\pm a) = 0$ (see (58)). Then G is the Green's function for this diffusion operator and the particular boundary conditions chosen. The partial sums of the Neumann series are found from the iteration (24) so that the boundary conditions on ϕ_{n+1} are completely determined by the chosen boundary conditions on T. This leads to the scheme

$$(-\frac{1}{3} D^2 + I - \gamma I)\phi_{n+1} = (TK - I)\gamma \phi_n + T q_1 \ . \tag{25}$$

Consider the quantity $(TK - I)\gamma \phi_n$. Define

$$\hat{\psi}_n = H_\mu^{-1} \gamma \phi_n \tag{26}$$

or equivalently

$$H_\mu \hat{\psi}_n = \gamma \phi_n \ . \tag{27}$$

Thus, since $L \phi_n = \phi_n$, we have

$$(TK - I)\gamma \phi_n = (TLH_\mu^{-1} - L)\gamma \phi_n = (TL - LH_\mu)\hat{\psi}_n \ . \tag{28}$$

Now, using the definition of T, we have

$$TL \hat{\psi}_n = (-\frac{1}{3} D^2 + I)L \hat{\psi}_n = -\frac{1}{3} D^2 L \hat{\psi}_n + L \hat{\psi}_n \ . \tag{29}$$

For LH_μ we get

$$LH_\mu \hat{\psi}_n = L(\mu D + I)\hat{\psi}_n = DL\mu \hat{\psi}_n + L \hat{\psi}_n \ . \tag{30}$$

(Note that $DL = LD$ from definitions of L and D in (2).) Thus,

$$(TL - LH_\mu)\hat{\psi}_n = -\frac{1}{3} D^2 L \hat{\psi}_n - DL\mu \hat{\psi}_n = D(-\frac{1}{3} DL \hat{\psi}_n - L\mu \hat{\psi}_n) \ . \tag{31}$$

Now operate on (27) by $L\mu$:

$$L\mu H_\mu \hat{\psi}_n = L\mu \gamma \phi_n \ . \tag{32}$$

From the definitions of L and ϕ_n we see that ϕ_n does not depend on μ. Thus,

$$L\mu \gamma \phi_n = \gamma \int_{-1}^{1} \mu \, \phi_n(x) \, d\mu = 0 \ . \tag{33}$$

Thus,

$$L\mu H_\mu \hat{\psi}_n = DL\mu^2 \hat{\psi}_n + L\mu \hat{\psi}_n = L\mu(\mu D + I)\hat{\psi}_n = 0 \ , \tag{34}$$

or

$$L\mu \hat{\psi}_n = -DL\mu^2 \hat{\psi}_n \ . \tag{35}$$

Substituting (35) into (31) yields

$$(TL - LH_\mu)\hat{\psi}_n = D \ (-\frac{1}{3} DL \hat{\psi}_n + DL\mu^2 \hat{\psi}_n) \tag{36}$$

$$= \frac{2}{3} D^2 L (\frac{3\mu^2 - 1}{2})\hat{\psi}_n$$

$$= \frac{2}{3} D^2 LP_2 \hat{\psi}_n = \frac{2}{3} D^2 LP_2 H_\mu^{-1} \gamma\phi_n \ ,$$

where

$$P_2(\mu) = \frac{3\mu^2 - 1}{2}$$

is the second Legendre polynomial. Thus, the quantity $(TK - I)\gamma\phi_n$ necessary for the computation of (25) can be found as the last term in Eq. (36). This agrees exactly with the DSA algorithm as described in [La1].

9 OTHER ITERATION SCHEMES

It is not necessary to solve Eq. (7) by the Neumann series iteration (13), nor is it necessary to solve the preconditioned Eq. (12) by the iteration (16). In this section, we will describe several iterative methods for solving (7) or (12). Consider a general linear system

$$Ax = b \ . \tag{37}$$

If x_0 is an initial guess at the solution, we let

$$r_0 = b - Ax_0 \tag{38}$$

be the initial residual. A class of iterative schemes called gradient or polynomial methods is characterized by choosing x_{i+1} to be x_i plus some linear combination of previous residuals, that is,

$$x_{i+1} = x_i + \sum_{j=0}^{i} \beta_{ij} r_j \ . \tag{39}$$

If $e_i = x - x_i$ is the error, it is easy to see that (cf. [FaMa])

$$e_{i+1} = p_{i+1}(A)e_0 \ , \tag{40}$$

where $p_{i+1}(\lambda)$ is a polynomial of degree $i + 1$ such that $p_{i+1}(0) = 1$. The convergence properties of such schemes depend upon choosing $p_{i+1}(\lambda)$ to be small on all of the eigenvalues of A.

Of course, it is not economical to keep all of the previous residuals. Three such algorithms that require only two auxiliary vectors are the nonstationary one-step method, or Richardson's iteration; a nonstationary two-step method known as the Chebychev iteration; and the conjugate gradient iteration. We will describe first the Richardson's iteration and show that the Neumann series is perhaps the simplest such algorithm. If we let

$$x_{i+1} = x_i + \alpha_i r_i \ , \tag{41}$$

then

$$e_{i+1} = e_i - \alpha_i r_i = (I - \alpha_i A)e_i \ ,$$

or

$$e_{i+1} = \prod_{j=0}^{i} (I - \alpha_j A)e_0 \ . \tag{42}$$

Thus, the polynomial is given by

$$p_{i+1}(\lambda) = \prod_{j=0}^{i} (1 - \alpha_j \lambda) \ , \tag{43}$$

and the roots of $p_{i+1}(\lambda)$ are $\dfrac{1}{\alpha_j}$, $j = 0, \ldots, i$. If the eigenvalues of A lie on some interval on the positive real axis, then the roots of $p_{i+1}(\lambda)$ might be chosen so that $p_{i+1}(\lambda)$ is the scaled and translated Chebychev polynomial on the interval. This polynomial has a minimax property that makes it "as-small as-possible" on the interval (cf. [GoVa]).

Now consider Eq. (13). Here $A = (I - K\gamma)$. If we rearrange the terms we have

$$\begin{aligned} \phi_{n+1} &= \phi_n + (q_1 - (I - K\gamma)\phi_n) \\ &= \phi_n + r_n \ . \end{aligned} \tag{44}$$

The corresponding error equation is

$$e_{n+1} = (I - A)e_n = (I - A)^{n+1}e_0 \ . \tag{45}$$

Thus, the iteration (13) is a Richardson iteration with $\alpha_i = 1$. The polynomial $p_i(\lambda)$ has all of its roots at 1.0. It is well known that $K\gamma$ is symmetric positive definite and $\|K\gamma\| < 1$. Thus, the eigenvalues of $A = (I - K\gamma)$ lie in some interval $[\varepsilon, 1.0]$, where $\varepsilon = 1 - \|K\gamma\|$. The *convergence factor* for the iteration (13) is

$$\|e_{i+1}\| \le \|K\gamma\|^{i+1} \|e_0\| = (1 - \varepsilon)^{i+1} \|e_0\| \ . \tag{46}$$

The convergence factor for choosing the scaled and translated Chebychev polynomial would be

$$\|e_{i+1}\| \le 2 \left[\frac{1 - \sqrt{\varepsilon}}{1 + \sqrt{\varepsilon}} \right]^{i+1} \|e_0\| \tag{47}$$

(cf. [GoVa]).

The iteration (41) with the α_i's chosen to be the reciprocals of the roots of the $i+1^{st}$ scaled and translated Chebychev polynomial will only yield the Chebychev polynomial in

(43) on the $i+1^{st}$ step. Because of certain recursion properties of the Chebychev polynomials, it is possible to design an algorithm of the form

$$r_i = b - Ax_i \tag{48a}$$

$$\Delta_i = \alpha_i r_i + \beta_i \Delta_{i-1} \tag{48b}$$

$$x_{i+1} = x_i + \Delta_i \quad, \tag{48c}$$

where the α_i's and β_i's are chosen so that *at every step $p_j(\lambda)$ is the jth* scaled and translated Chebychev polynomial. All that is needed is to know the interval on the real axis that contains the eigenvalues. This can be done dynamically using estimates made available during the iteration. We also mention that this iteration may be applied to certain linear systems with complex eigenvalues (cf. [Ma1],[Ma2]).

The conjugate gradient iteration (CG) (cf. [HeSt], [Re]) can be applied to (37) if, as in our case, the matrix A is symmetric positive definite. The iteration is of the form

$$x_{i+1} = x_i + \alpha_i p_i \quad, \quad \alpha_i = \frac{<r_i, r_i>}{<Ap_i, p_i>} \tag{49a}$$

$$r_{i+1} = r_i - \alpha_i Ap_i \tag{49b}$$

$$p_{i+1} = r_{i+1} + \beta_i p_i, \quad \beta_i = \frac{<r_{i+1}, r_{i+1}>}{<r_i, r_i>} \quad, \tag{49c}$$

where $< \cdot , \cdot >$ represents vector inner product. Like all of the iterations in this class, the error equation is of the form (40). Now, however, the polynomial $p_{i+1}(\lambda)$ is chosen so that e_{i+1} is optimal at each step in a certain norm. Therefore, the conjugate gradient iteration is at least as good as the Chebychev iteration if the error is measured in this norm.

Let us compare the convergence of the Neumann series with that of a Chebychev or conjugate gradient iteration for an optically thick slab, that is, for large a. Since $\|K\|_{L_2} = 1 - 0\,(\frac{1}{a^2})$, [Wi], we can replace ε in (46) and (47) by $(\frac{k_o}{a})^2$, where k_o is a constant with respect to a. We may ask how many steps of the Neumann series (46) yield the same error reduction as one step of a CG iteration (47). In other words, we seek m such that

$$\left[1 - \left[\frac{k_o}{a} \right]^2 \right]^m = \left[\frac{1 - \dfrac{k_o}{a}}{1 + \dfrac{k_o}{a}} \right]$$

for large a. Taking the log of both sides and expanding we find that

$$m = \frac{2a}{k_o} + 0\left[\frac{1}{a} \right] \quad. \tag{50}$$

For example, if $a = 128$, then $\varepsilon = 5.36 \times 10^{-5}$ and $m \doteq 273$. That is, each iteration of the Chebychev on conjugate gradient method is equivalent to approximately 273 iterations of the Neumann series. In practice, CG often performs even better. This is due to the fact that the CG iteration can take advantage of the distribution of eigenvalues in the interval containing the spectrum (cf. [HeSt]).

Another advantage of the CG iteration is that very good estimates of the extreme eigenvalues can be calculated at little cost from the iteration parameters ([GoVL]). Thus, $\|K\|$ can easily be approximated. Similar approximation can be calculated from the Chebychev iteration ([Ma2]).

10 PRECONDITIONED ITERATIVE SCHEMES

It is possible to combine a splitting or preconditioning with the iterative methods discussed in the previous section. The general idea is to multiply the equation (37) by a matrix C to yield the new system

$$CAx = Cb \tag{51}$$

with hopefully better properties than the original system. Equation (41) becomes

$$x_{i+1} = x_i + \alpha_i C r_i \ , \tag{52}$$

which yields

$$e_{i+1} = p_{i+1}(CA)e_0 \ . \tag{53}$$

Likewise, r_i can be replaced by Cr_i in (48b) to yield a preconditioned Chebychev iteration.

Now, it is the spectral properties of CA that become important. If the eigenvalues of CA lie in the right-half plane, parameters α_i and β_i can be chosen so that the preconditioned Chebychev iteration converges ([Ma1]).

In the context of this paper, we have $A = (I - K\gamma)$ and $C = (I - G\gamma)^{-1}$. As in (11) we have

$$CA = (I - G\gamma)^{-1}(I - K\gamma) = (I - (I - G\gamma)^{-1}(K - G)\gamma) \ . \tag{54}$$

Since both A and C are symmetric and positive definite, the spectrum of CA is positive and real. In this case, a preconditioned conjugate gradient iteration may be employed (cf. [CGO]):

$$x_{i+1} = x_i + \alpha_i p_i \ , \quad \alpha_i = \frac{<Cr_i, r_i>}{<Ap_i, p_i>} \ , \tag{55a}$$

$$r_{i+1} = r_i - \alpha_i A p_i \tag{55b}$$

$$P_{i+1} = Cr_{i+1} + \beta_i p_i \ , \quad \beta_i = \frac{<Cr_{i+1}, r_{i+1}>}{<Cr_i, r_i>} \ . \tag{55c}$$

Convergence rates for both Chebychev and preconditioned conjugate gradients are given by

$$\|e_{i+1}\|_A = 2\left[\frac{\sqrt{c}-1}{\sqrt{c}+1}\right]^{i+1} \|e_0\|_A \ , \tag{56}$$

where

$$c = Cond_A(CA) = \frac{\lambda_{max}}{\lambda_{min}} \ , \tag{57}$$

and λ_{max} and λ_{min} are the maximum and minimum eigenvalues of CA, respectively.

In particular, the conjugate gradient iteration can be used in conjunction with the DSA preconditioning. Before we examine this numerically, let us first derive a better understanding of the properties of the DSA preconditioning.

11 PROPERTIES OF THE DSA PRECONDITIONING

In this section, we illustrate the properties of the DSA splitting that make it a good preconditioning. Our point of take-off here is the example in Section 8. We focus on the case $\gamma = 1$, since it is the most difficult problem to solve. Let

$$\begin{cases} T_\beta \phi = (-\frac{1}{3} D^2 + I)\phi \\ B_\beta \phi = \phi(\pm a) \mp \beta \phi'(\pm a) = 0 \ . \end{cases} \tag{58}$$

The eigenvectors of T_β will be of the form

$$v_k(x) = a_k \cos\left[\frac{\eta_k}{a} x\right] + b_n \sin\left[\frac{\eta_k}{a} x\right] \ . \tag{59}$$

Putting (59) into (58) yields

$$\frac{\beta}{a} \eta_k = -\tan(\eta_k) \ , \tag{60a}$$

or

$$\frac{\beta}{a} \eta_k = \cot(\eta_k) \ . \tag{60b}$$

In Fig. 1 the graphs $y = \frac{\beta}{a} \eta$, $y = -\tan(\eta)$, and $y = \cot(\eta)$ are plotted. The intersections are values of η_k. Notice that for $\beta > 0$

$$\eta_k \in \left[(k-1)\frac{\pi}{2} \ , \ \frac{k\pi}{2}\right] \ , \tag{61}$$

and for $\frac{\beta k}{a} \ll 1$

$$\eta_k = \frac{k\pi}{2} - \frac{\beta k \pi}{2a} + O\left(\left[\frac{\beta k}{a}\right]^3\right) \ , \tag{62a}$$

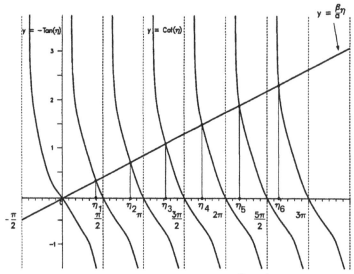

Figure 1. Values of η_k for $\dfrac{\beta}{a} = \dfrac{1}{3}$.

while for $\dfrac{\beta k}{a} \gg 1$

$$\eta_k = (k-1)\,\frac{\pi}{2} + \frac{2a}{\beta k \pi} - O\left[\left[\frac{a}{\beta k}\right]^3\right]\,. \tag{62b}$$

The values η_k migrate from the right end of $((k-1)\,\dfrac{\pi}{2},\dfrac{k\pi}{2})$ to the left. Putting this all together we see that the eigenvectors of T_β are

$$v_k = \begin{cases} \cos\left[\dfrac{\eta_k}{a}\,x\right]\,, & \text{for } k \text{ odd }, \\[3mm] \sin\left[\dfrac{\eta_k}{a}\,x\right]\,, & \text{for } k \text{ even }. \end{cases} \tag{63}$$

Finally, if G_β is the Green's function for T_β, the eigenvalues of G_β are given by

$$\lambda_k = \frac{1}{1 + \dfrac{1}{3}\left[\dfrac{\eta_k}{a}\right]^2}\,. \tag{64}$$

For $\dfrac{\beta k}{a} \ll 1$ we have

$$\lambda_k \doteq \frac{1}{1 + \dfrac{1}{3}\left[\dfrac{k\pi}{2a}\right]^2} \doteq 1 - \frac{1}{3}\left[\frac{k\pi}{2a}\right]^2\,. \tag{65}$$

The parameter β can be used to control the shape of the eigenvectors of G_β near the boundary of the slab. Figure 2 displays the condition of $CA = (I - G_\beta)^{-1}(I - K)$ as a function of β for $a = 1$. The β for which the condition is minimized will yield the best preconditioning (cf. (56)). Notice that the condition is approximately 1.25 over a wide range of β's near the value of $\beta = \dfrac{2}{3}$. This agrees with the results found in [La1]. In fact, the condition takes on the value 1.240276 for values of β in the interval $(.68, .76)$.

For very large a, the value of β in the range $[0,1]$ has little impact on the largest eigenvalues of G_β (cf. (62a)). The corresponding eigenvectors of G_β also closely match those of K. Let us now examine the operator K.

12 THE OPERATOR K

First consider H_μ. A singular value/vector decomposition of H_μ yields

$$
H_\mu^{-1}f = \sum_{b=1}^{\infty} \frac{\cos(\zeta_k)}{a\,\delta_0(\zeta_k)}
\begin{cases}
\displaystyle\int_{-a}^{a} \sin\left(\zeta_k\left(\frac{x+a}{2a}\right)\right) \sin\left(\zeta_k\left(\frac{s-a}{2a}\right)\right) f(s)\,ds, \quad \mu > 0 , \\[3mm]
\displaystyle\int_{-a}^{a} \sin\left(\zeta_k\left(\frac{x-a}{2a}\right)\right) \sin\left(\zeta_k\left(\frac{s+a}{2a}\right)\right) f(s)\,ds, \quad \mu < 0 ,
\end{cases}
\tag{66a}
$$

where

$$
a\,\delta_0(\zeta_k) = a\left[1 - \frac{\sin(2\zeta_k)}{2\zeta_k}\right] = \int_{-a}^{a} \sin^2\left(\zeta_k\left(\frac{x+a}{2a}\right)\right)dx
\tag{66b}
$$

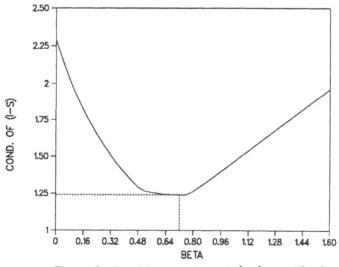

Figure 2. Condition number of (I–S) vs. $\beta(a=1)$.

is a normalization factor. The values ζ_k are the positive roots of the equation

$$\frac{|\mu|}{2a} \zeta = -\tan(\zeta) \quad . \tag{67}$$

The roots of (67) are very similar to the roots of (60a,b). Now, however, for $\frac{\mu k}{2a} \ll 1$

$$\zeta_k = k\pi - \frac{\mu k \pi}{2a} + 0\left[\left[\frac{\mu k}{2a}\right]^3\right] \quad , \tag{68a}$$

while for $\frac{\mu k}{2a} \gg 1$,

$$\zeta_k = \left[k - \frac{1}{2}\right]\pi + \frac{2a}{\mu k \pi} - 0\left[\left[\frac{2a}{\mu k}\right]^3\right] \quad . \tag{68b}$$

To find K we must apply L to (66a). Applying L to a single term of (66a) yields

$$\left[\frac{1}{2}\int_0^1 \frac{\cos(\zeta_k)}{a\,\delta_0(\zeta_k)} \sin\left[\zeta_k\left[\frac{x+a}{2a}\right]\right] \sin\left[\zeta_k\left[\frac{s-a}{2a}\right]\right] d\mu \right.$$

$$\left. + \frac{1}{2}\int_{-1}^0 \frac{\cos(\zeta_k)}{a\,\delta_0(\zeta_k)} \sin\left[\zeta_k\left[\frac{x-a}{2a}\right]\right] \sin\left[\zeta_k\left[\frac{s+a}{2a}\right]\right] d\mu \right] \quad . \tag{69}$$

Since $\zeta_k(\mu) = \zeta_k(-\mu)$ we can combine the two integrals above. Expanding each sin term and combining yields

$$\left[\int_0^1 \frac{\cos(\zeta_k)}{a\,\delta_0(\zeta_k)}\left[\cos^2\left[\frac{\zeta_k}{2}\right] \sin\left[\frac{\zeta_k}{2}\frac{x}{a}\right] \sin\left[\frac{\zeta_k}{2}\frac{s}{a}\right]\right.\right.$$

$$\left.\left. - \sin^2\left[\frac{\zeta_k}{2}\right] \cos\left[\frac{\zeta_k}{2}\frac{x}{a}\right] \cos\left[\frac{\zeta_k}{2}\frac{s}{a}\right]\right] d\mu \right] \quad . \tag{70}$$

While (70) does not constitute a singular value/vector decomposition of K, it does yield useful bounds for large a (since K is self-adjoint and positive definite, its singular vector/value decomposition is also its eigenvalue/vector decomposition). First, notice that for $\mu \neq 0$ as k increases, ζ_k migrates from $k\pi$ to $(k - \frac{1}{2})\pi$ and consequently $\cos(\zeta_k)$ migrates from 1 to 0. Using (68b), we see the terms decrease like

$$|\cos(k - \frac{1}{2})\pi + \frac{2a}{\mu k \pi})| = \sin(\frac{2a}{\mu k \pi}) \doteq \frac{2a}{\mu k \pi} \quad .$$

Now fix μ. Notice that $\sin(\frac{\zeta_k}{2}\frac{x}{a})$ is odd while $\cos(\frac{\zeta_k}{2}\frac{x}{a})$ is even, and thus they are orthogonal in $L_2[-a,a]$. Each term of (70) is a self-adjoint rank two operator, $R_k(\mu)$, already decomposed into its eigenvector expansion. For $\frac{k}{a} \ll 1$, one eigenvalue is close to 1 and the other is close to zero. Also, for $\frac{k}{a} \ll 1$, the operators $R_k(\mu)$ are nearly mutually orthogonal. Since the integration with respect to μ involves a very small range of ζ, the integrated terms retain these properties.

For $\dfrac{k}{a}$ not small, ζ_k is no longer close to $k\pi$ and thus, the $R_k(\mu)$ are no longer mutually orthogonal. Moreover, the integration with respect to μ involves a larger interval of ζ. The behavior is no longer clear. However, large k denotes high frequency. Notice that

$$\frac{1}{a}\int_{-a}^{a}\sin\left[\frac{\zeta_k}{2}\frac{x}{a}\right]\sin\left[\frac{\zeta_l}{2}\frac{x}{a}\right]dx = 2\left[\frac{\sin\left[\dfrac{\zeta_k-\zeta_l}{2}\right]}{(\zeta_k-\zeta_l)} - \frac{\sin\left[\dfrac{\zeta_k+\zeta_l}{2}\right]}{(\zeta_k+\zeta_l)}\right] \tag{71a}$$

$$= 0\left[\frac{1}{|\zeta_k-\zeta_l|}\right] \ .$$

Likewise,

$$\frac{1}{a}\int_{-a}^{a}\cos\left[\frac{\zeta_k x}{2a}\right]\cos\left[\frac{\zeta_l x}{2a}\right]dx = 2\left[\frac{\sin\left[\dfrac{\zeta_k-\zeta_l}{2}\right]}{(\zeta_k-\zeta_l)} + \frac{\sin\left[\dfrac{\zeta_k+\zeta_l}{2}\right]}{(\zeta_k+\zeta_l)}\right] \tag{71b}$$

is of the same order. While not mutually orthogonal the high-frequency terms are nearly orthogonal to the low-frequency terms. Often $\dfrac{k}{a} \ll 1$, for frequencies of interest. Now we will examine the behavior of the largest eigenvalues of K for $\dfrac{k\pi}{2a} \ll 1$. First let $\zeta_k = k\pi - \theta_k$. We have

$$0 \le \theta_k \le \frac{\mu k\pi}{2a} \ . \tag{72}$$

Notice that

$$\cos^2\left[\frac{\zeta_k}{2}\right] = \frac{1}{2}(1 + (-1)^k \cos(\theta_k)) \tag{73a}$$

$$\sin^2\left[\frac{\zeta_k}{2}\right] = \frac{1}{2}(1 - (-1)^k \cos(\theta_k)) \ . \tag{73b}$$

Thus, the odd terms of (70) are dominated by $\cos(\dfrac{\zeta_k x}{2a})\cos(\dfrac{\zeta_k s}{2a})$, while the even terms are dominated by $\sin(\dfrac{\zeta_k x}{2a})\sin(\dfrac{\zeta_k s}{2a})$. Compare this with the eigenvectors of G as given in (63).

Notice that $\cos(\zeta_k)$ is negative for k odd and positive for k even. Thus, each term of (70) includes a dominant positive part and a small negative part. Let us separate them and write

$$K = K_1 - K_2 \ , \tag{74}$$

where K_1 includes the dominant part and K_2 includes the small part of each term. Both K_1 and K_2 are self-adjoint and positive. The odd terms of K_1 are given by

$$\int_0^1 \frac{\cos(\theta_k)(1 + \cos(\theta_k))\delta_1(\zeta_k)}{2\delta_0(\zeta_k)} \frac{\cos\left[\frac{\zeta_k x}{2a}\right] \cos\left[\frac{\zeta_k s}{2a}\right]}{a\,\delta_1(\zeta_k)} d\mu \ , \tag{75a}$$

where

$$a\,\delta_1(\zeta_k) = a\left[1 + \frac{\sin(\zeta_k)}{\zeta_k}\right] = \int_{-a}^{a} \cos^2\left[\frac{\zeta_k s}{2a}\right] ds \tag{75b}$$

is a normalization factor. The even terms can likewise be written

$$\int_0^1 \frac{\cos(\theta_k)(1 + \cos(\theta_k))\delta_2(\zeta_k)}{2\delta_0(\zeta_k)} \frac{\sin\left[\frac{\zeta_k x}{2a}\right] \sin\left[\frac{\zeta_k s}{2a}\right]}{a\,\delta_2(\zeta_k)} d\mu \ , \tag{76a}$$

where

$$a\,\delta_2(\zeta_k) = a\left[1 - \frac{\sin(\zeta_k)}{\zeta_k}\right] = \int_{-a}^{a} \sin^2\left[\frac{\zeta_k s}{2a}\right] ds \tag{76b}$$

is another normalizing factor. Equation (72) implies that θ_k moves over a very short range as μ integrates from 0 to 1. The mean value theorem allows us to write (75a) and (76a) as

$$\frac{\cos\left[\frac{\hat{\zeta}_k x}{2a}\right] \cos\left[\frac{\hat{\zeta}_k s}{2a}\right]}{a\,\delta_1(\hat{\zeta}_k)} \int_0^1 \frac{\cos(\theta_k)(1 + \cos(\theta_k))\delta_1(\zeta_k)}{2\delta_0(\zeta_k)} d\mu \tag{77a}$$

for k odd and

$$\frac{\sin\left[\frac{\tilde{\zeta}_k x}{2a}\right] \sin\left[\frac{\tilde{\zeta}_k s}{2a}\right]}{a\,\delta_2(\tilde{\zeta}_k)} \int_0^1 \frac{\cos(\theta_k)(1 + \cos(\theta_k))\delta_2(\zeta_k)}{2\delta_0(\zeta_k)} d\mu \tag{77b}$$

for k even, where

$$\hat{\zeta}_k(x,s), \ \tilde{\zeta}_k(x,s) \in \left[k\pi - \frac{k\pi}{2a} \ , \ k\pi\right] \ . \tag{77c}$$

The terms that have been extracted from the integrals in (77a,b) represent the normalized outer product of the eigenvectors of K_1. The eigenvalues are approximated by evaluating the integrals. First, notice that from (67) we have

$$\frac{\mu}{2a}(k\pi - \theta_k) = \tan(\theta_k) \doteq \theta_k \ . \tag{78}$$

Solving (78) for θ_k yields

$$\theta_k \doteq \left[\frac{k\pi}{2a}\right]\left[\frac{\mu}{1+\frac{\mu}{2a}}\right] .$$

(79)

Next, notice that for k odd

$$\frac{\delta_1(\zeta_k)}{\delta_0(\zeta_k)} = \frac{\left[1+\dfrac{\sin(\zeta_k)}{\zeta_k}\right]}{\left[1-\dfrac{\sin(2\zeta_k)}{2\zeta_k}\right]} = \frac{\left[1+\dfrac{\sin(\theta_k)}{k\pi-\theta_k}\right]}{\left[1+\dfrac{\sin(2\theta_k)}{2(k\pi-\theta_k)}\right]} \doteq \left[1+\dfrac{1}{2}\dfrac{\theta_k^3}{k\pi}\right] .$$

Expanding the remaining terms to second powers of θ_k yields

$$\int_0^1 \frac{\cos(\theta_k)(1+\cos(\theta_k))\delta_1(\zeta_k)}{2\delta_0(\zeta_k)}\,d\mu \doteq \int_0^1\left[1-\frac{3}{4}\theta_k^2\right]d\mu$$

$$= 1-\frac{1}{4}\left[1+\frac{1}{a}\right]\left[\frac{k\pi}{2a}\right]^2 .$$

(80)

The term for k even yields the same result. We have established an asymptotic form for the eigenvectors $u_k^{(1)}$ of K_1, namely

$$u_k^{(1)} \doteq \begin{cases} \cos\left[\dfrac{\hat{\zeta}_k x}{2a}\right] , & k \text{ odd} , \\[3mm] \sin\left[\dfrac{\tilde{\zeta}_k x}{2a}\right] , & k \text{ even} , \end{cases}$$

(81)

where $\dfrac{\hat{\zeta}_k}{2}, \dfrac{\tilde{\zeta}_k}{2} \in (\dfrac{k\pi}{2}-\dfrac{k\pi}{4a}, \dfrac{k\pi}{2})$, and the eigenvalues $\sigma_k^{(1)}$ of K_1, namely

$$\sigma_k^{(1)} \doteq 1-\frac{1}{4}\left[\frac{k\pi}{2a}\right]^2 .$$

(82)

The odd terms of K_2 are given by

$$\int_0^1 \frac{\cos(\theta_k)(1-\cos(\theta_k))\delta_2(\zeta_k)}{2\delta_0(\zeta_k)}\ \frac{\sin(\frac{\zeta_k x}{2a})\sin(\frac{\zeta_k s}{2a})}{a\,\delta_2(\zeta_k)}\,d\mu ,$$

(83a)

and the even terms are given by

$$\int_0^1 \frac{\cos(\theta_k)(1-\cos(\theta_k))\delta_1(\zeta_k)}{2\delta_0(\zeta_k)}\ \frac{\cos(\frac{\zeta_k x}{2a})\cos(\frac{\zeta_k s}{2a})}{a\,\delta_1(\zeta_k)}\,d\mu .$$

(83b)

Again, using the mean value theorem, we write (83a) and (83b) as

$$\frac{\sin\left(\frac{\overline{\zeta}_k x}{2a}\right)\sin\left(\frac{\overline{\zeta}_k s}{2a}\right)}{a\,\delta_2(\overline{\zeta}_k)}\int_0^1 \frac{\cos(\theta_k)(1-\cos(\theta_k))\delta_2(\zeta_k)}{2\delta_0(\zeta_k)}\,d\mu \ , \tag{84a}$$

for k odd and

$$\frac{\cos\left(\frac{\dot{\zeta}_k x}{2a}\right)\cos\left(\frac{\dot{\zeta}_k s}{2a}\right)}{a\,\delta_1(\dot{\zeta}_k)}\int_0^1 \frac{\cos(\theta_k)(1-\cos(\theta_k))\delta_1(\zeta_k)}{2\delta_0(\zeta_k)}\,d\mu \ , \tag{84b}$$

for k even, where

$$\overline{\zeta}_k(x,s)\ ,\ \dot{\zeta}_k(x,s)\in\left[k\pi-\frac{k\pi}{2a},k\pi\right]\ . \tag{84c}$$

As before, the terms extracted from the integrals represent the normalized outer product of the eigenvectors of K_2. The eigenvalues are approximated by evaluating the integrals in (84a,b). This yields

$$\begin{aligned}\sigma_k^{(2)}=\int_0^1\frac{\cos(\theta_k)(1-\cos(\theta_k))\delta_1(\zeta_k)}{2\delta_0(\zeta_k)}\,d\mu&\doteq\int_0^1\frac{1}{4}\theta_k^2-\frac{1}{8}\theta_k^4\,d\mu\\&=\frac{1}{12}\left[\frac{k\pi}{2a}\right]^2-0\left[\frac{1}{a^4}\right]\ .\end{aligned} \tag{85}$$

To find approximations to the eigenvalues of K, we must consider both K_1 and K_2. One can show that $\hat{\zeta}_k$ and $\tilde{\zeta}_k$ in (81) are given approximately by setting $\mu=1/2\,(1-\frac{1}{8a})$ in (67), that is, by the roots of

$$\frac{1}{2}\,(1-\frac{1}{8a})\,\zeta_k=-Tan\,(\zeta_k)\ . \tag{86}$$

Using these values of ζ_k in (81) yields approximate eigenvectors u_k of K. The approximate eigenvalues are found from

$$\sigma_k=\frac{<Ku_k,u_k>}{<u_k,u_k>}=\frac{<K_1u_k,u_k>}{<u_k,u_k>}-\frac{<K_2u_k,u_k>}{<u_k,u_k>}\ . \tag{87}$$

The first term on the left of (87) is given by (82). The second term requires some work. Suppose k is odd. Then, $u_k=\cos\left(\frac{\zeta_k x}{2a}\right)$ is orthogonal to all the odd terms in the expansion of K_2. Consider an even term of K_2. The eigenvector of K_2 is $u_l^{(2)}\doteq\cos\left(\frac{\zeta_l x}{2a}\right)$. We have from (71b) and (72)

$$\frac{1}{a}\int_{-a}^a\cos\left(\frac{\zeta_k x}{2a}\right)\cos\left(\frac{\zeta_l x}{2a}\right)\,dx\doteq(\pm 1)\,\frac{2}{\pi}\left[\frac{2k}{k^2-l^2}\right]+0\,(\frac{1}{a})\ . \tag{88}$$

Thus, using (85) we have

$$\frac{<K_2 u_k,u_k>}{<u_k,u_k>} \doteq \sum_{l \text{ even}} \sigma_l^{(2)} \left[\frac{2}{\pi}\right]^2 \frac{(2k)^2}{(k^2-l^2)^2} = \frac{1}{3a^2} \sum_{l \text{ even}} \frac{k^2 l^2}{(k^2-l^2)^2} \, . \tag{89}$$

The sum in (89) can be bounded above and below by

$$\sum_{l \text{ even}} \frac{k^2 l^2}{(k^2-l^2)} \geq \frac{k}{2} \int_0^{\frac{k-1}{k}} \frac{w^2}{(1-w^2)^2} \, dw + \frac{k}{2} \int_{\frac{k+1}{k}}^{\infty} \frac{w^2}{(1-w^2)} \, dw \tag{90a}$$

$$\sum_{l \text{ even}} \frac{k^2 l^2}{(k^2-l^2)^2} \leq \frac{k}{2} \int_0^{\frac{k-1}{k}} \frac{w^2}{(1-w^2)^2} \, dw + \left[\frac{k^2(k-1)^2}{(k^2-(k-1)^2)^2}\right.$$
$$\left. + \frac{k^2(k+1)^2}{(k^2-(k+1)^2)^2}\right] + \frac{k}{2} \int_{\frac{k+1}{k}}^{\infty} \frac{w^2}{(1-w^2)^2} \, dw \, . \tag{90b}$$

The integrals in (90) become

$$\frac{k}{2} \int_0^{\frac{k-1}{k}} \frac{w^2}{(1-w^2)^2} \, dw = \frac{k^2}{4} \left[\frac{k-1}{2k-1} - \frac{\ln|2k-1|}{2k}\right] \tag{91a}$$

and

$$\frac{k}{2} \int_{\frac{k+1}{k}}^{\infty} \frac{w^2}{(1-w^2)^2} \, dw = \frac{k^2}{4} \left[\frac{k+1}{2k+1} + \frac{\ln|2k+1|}{2k}\right] \, . \tag{91b}$$

Putting (91a,b) into (90a,b) and rearranging into (89) yields

$$\frac{<K_2 u_k,u_k>}{<u_k,u_k>} \geq \frac{1}{3\pi^2} \left[\frac{k\pi}{2a}\right]^2 \left[\frac{(k^2-\frac{1}{2})}{(k^2-\frac{1}{4})} + \frac{\ln\left|\frac{2k+1}{2k-1}\right|}{2k}\right] \tag{92a}$$

$$\frac{<K_2 u_k,u_k>}{<u_k,u_k>} \leq \frac{1}{\pi^2} \left[\frac{k\pi}{2a}\right]^2 \left[\frac{1}{2} \frac{(k-1)(k-\frac{5}{6})}{(k-\frac{1}{2})^2} + \frac{1}{2} \frac{(k+1)(k+\frac{5}{6})}{(k+\frac{1}{2})^2}\right.$$
$$\left. + \frac{\ln|\frac{2k+1}{2k-1}|}{2k}\right] \, . \tag{92b}$$

Since the term in brackets in (92) is very nearly unity, we make the approximation

$$\frac{1}{3\pi^2}\left[\frac{k\pi}{2a}\right]^2 \le \frac{<K_2 u_k, u_k>}{<u_k, u_k>} \le \frac{1}{\pi^2}\left[\frac{k\pi}{2a}\right]^2 \, . \tag{93}$$

Combining (93) with (82) into (87) yields the approximate eigenvalue σ_k of K, namely

$$1 - \left[\frac{1}{4} + \frac{1}{\pi^2}\right]\left[\frac{k\pi}{2a}\right]^2 \le \sigma_k \le 1 - \left[\frac{1}{4} + \frac{1}{3\pi^2}\right]\left[\frac{k\pi}{2a}\right]^2 \, . \tag{94}$$

Now suppose k is odd in (87). We have $u_k = \sin\left(\frac{\zeta_k x}{2a}\right)$, which is orthogonal to all of the odd terms of $K_2(u_l^{(2)} = \cos\left(\frac{\zeta_l x}{2a}\right)$ for l odd). The even terms of K_2 yield an expansion identical to (88) above. Thus, (94) holds for k odd as well.

Notice that $\frac{1}{4} + \frac{1}{3\pi^2} \doteq .2838$ and $\frac{1}{4} + \frac{1}{\pi^2} \doteq .3513$. Compare this with the eigenvalues of G as given in (64) and (65) and the compare eigenvectors as given in (62) and (63) with the eigenvector approximations (81). Notice that the eigenvalues of G have asymptotic form

$$\lambda_k \doteq 1 - \frac{1}{3}\left[\frac{k\pi}{2a}\right]^2 \, . \tag{95}$$

Now, for $\frac{k\pi}{2a} \gg 1$, the roles of sin and cos in (70) are reversed. For k odd $\sin\left(\frac{\zeta_k x}{2a}\right)$ dominates and for k even $\cos\left(\frac{\zeta_k x}{2a}\right)$ dominates where ζ_k, $\tilde{\zeta}_k$ are near $(k - \frac{1}{2})\pi$. This is so in spite of the fact that the integration with respect to μ sweeps through almost all values in the interval $((k - \frac{1}{2})\pi, k\pi)$. The major portion of the integral involves the left end of the interval. Here, the terms decay like $\ln\left(\frac{k\pi}{2a}\right)/\left(\frac{k\pi}{2a}\right)$. The eigenvalues decay at a faster rate.

Suppose we denote the eigenvector decomposition of K and G by

$$K = U \Sigma U^* \, , \quad \Sigma = diag \, (\cdots \sigma_i \cdots) \, , \tag{96a}$$

$$G = V \Lambda V^* \, , \quad \Lambda = diag \, (\cdots \lambda_i \cdots) \, , \tag{96b}$$

where U and V are unitary operators. As the discussion above indicates, and numerical experiments support, U and V are very similar, especially for the large eigenvalues. We have

$$V^*U = I + \Delta \, , \tag{97}$$

where Δ is small. Now consider

$$(I - G)^{-1}(I - K) = I - (I - G)^{-1}(K - G) \tag{98}$$

from Eq. (11). Using (96a,b) we have

$$(I - G)^{-1}(K - G) = V(I - \Lambda)^{-1}V^*(U\Sigma U^* - V\Lambda V^*)$$
$$= V(I - \Lambda)^{-1}(V^*U)(\Sigma(U^*V) - (U^*V)\Lambda)V^* .$$

Substitution of (97) yields

$$(1 - G)^{-1}(K - G) = V(I - \Lambda)^{-1}(\Sigma - \Lambda)V^*$$
$$+ V(I - \Lambda)^{-1}(\Delta(\Sigma - \Lambda) + (I + \Delta)(\Sigma\Delta - \Delta\Lambda))V^* . \tag{99}$$

Notice that the second term is small because for large eigenvalues, $\Sigma - \Lambda$ and Δ are small. For small eigenvalues, Σ and Λ are both small. Thus, the first term is a good approximation. We have

$$(I - \Lambda)^{-1}(\Sigma - \Lambda) = diag(\cdots \rho_k \cdots) , \tag{100}$$

where

$$\rho_k = \frac{\sigma_k - \lambda_k}{1 - \lambda_k} . \tag{101}$$

Using the asymptotic form of both σ_k (94) and λ_k (95) yields

$$-.0539 = \left[\frac{1}{4} - \frac{3}{\pi^2}\right] \le \rho_k \le \left[\frac{1}{4} - \frac{1}{\pi^2}\right] = .1487 . \tag{102}$$

Of course, this is never achieved in practice due to the differences in the eigenvectors of G and K (which are reflected in Δ), the pollution due to intermediate eigenvalues, and perhaps most importantly, perturbations due to discretization.

The exact form of λ_k is given in (64). The intermediate eigenvalues of K can be approximated by

$$|\cos(\zeta_k)| \left[\frac{1 + |\cos(\zeta_k)|}{2}\right] , \quad \frac{\zeta_k}{4a} = -\tan(\zeta_k) , \tag{103}$$

as the equations (77a,b) suggest. A plot of numerically computed values σ_k, λ_k, and ρ_k against $\frac{k}{a}$ appears in Fig. 3. Here $a = 32$ was used. The curves for various values of a are similar, as are the values using analytic λ_k and approximate σ_k. Notice that the maximum value of ρ_k is approximately .23. This corresponds to the findings in [La1].

13 OTHER PRECONDITIONINGS

Also notice that in Fig. 3 the eigenvalues of K decrease as $\frac{1}{k}$ while those of G decrease as $\frac{1}{k^2}$. One would expect to do much better if one used as a preconditioner an operator whose spectrum more closely fit the spectrum of K. For example, one might consider using a preconditioning based upon the eigenvectors of G, which are given in (63), and the eigen-

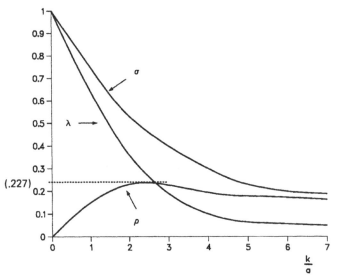

Figure 3. Values of σ_k, λ_k, and ρ_k versus $(\frac{k}{a})$ (a=32).

value approximations in (103). This leads to a preconditioning, $\hat{G}$, for which the spectral radius of

$$\hat{S} = (I - \hat{G})^{-1}(K - \hat{G}) \tag{104}$$

is on the order of .06. However, computation with such an operator is very expensive. Another preconditioning with attractive properties for problems with very large a is based upon the fast Fourier transform (FFT). Suppose, for example, that $k = 100$ frequencies are sufficient to resolve the solution. Further, suppose that $(100 \, \frac{\pi}{2a}) \ll 1$. The analysis of Sec. 12 indicates that the eigenvectors of K are given by (81), that is, are very nearly Fourier vectors on a somewhat larger region. To find that region consider the very first eigenvector of the diffusion operator. We have from (60a,b)

$$\frac{\beta}{a} \, \eta_1 = \cot (\eta_1) \ . \tag{105}$$

Suppose we let $\eta_1 = \frac{\pi}{2} - \theta_1$. Then (105) becomes

$$\frac{\beta}{a} \left[\frac{\pi}{2} - \theta_1 \right] \doteq \theta_1 \ , \tag{106}$$

which yields

$$\theta_1 = \frac{\dfrac{\pi\beta}{2a}}{1 + \dfrac{\beta}{a}} \ , \ \eta_1 = \frac{\dfrac{\pi}{2}}{1 + \dfrac{\beta}{a}} \ . \tag{107}$$

Now the end point of the larger region occurs where $\eta_1 \, \dfrac{x}{a} = \dfrac{\pi}{2}$, which yields

$$x = a + \beta \ . \tag{108}$$

Recall that $\beta \doteq .7$ gave the best results for $a = 1$ (see Fig. 2). This is very close to the extrapolated end point for the semi-infinite slab [Wi]. For a large, the significance of this extended region diminishes. We shall give numerical results using this preconditioning in a future report.

14 NUMERICAL RESULTS

The operators K and G_β were constructed using a finite element discretization on a mesh using 200 points. Eigenvector/value decompositions were performed to support the analysis of Sec. 11 and Sec. 12. Of course, only the first half of the eigenvectors and eigenvalues have any chance at accuracy. Over a large range of values of the boundary parameter β and the slab width a, the initial eigenvectors and eigenvalues behaved as predicted. The eigenvalues of K satisfy a relationship of the form

$$1 - \delta(a,k) \left[\frac{k\pi}{2a} \right]^2$$

for $\left[\frac{k\pi}{2a} \right] \ll 1$, where $\delta(a,k)$ varies slowly with k and a. Table 1 contains values of δ for various values of k and a. The numbers in bold print correspond to values of k and a with $\frac{k\pi}{2a} \doteq 1$. The values for $8 \le a \le 128$ were calculated on using 200 mesh points. The values for $a = 256$ were calculated using 500 mesh points. The values of $\delta(a,k)$ fit the bounds given in (94) fairly well for $\frac{k\pi}{2a} \le \frac{1}{2}$.

Also, the DSA preconditioning alone, that is using $(I - G)$ as a preconditioning for $(I - K)$ in the Neumann series, was compared with using DSA in a preconditioned conjugate gradient iteration (PCG). This produced surprising results. The conjugate gradient iteration performed better than predicted from bounds like (56). The bound (56) is only a bound and presumes even distribution of eigenvalues throughout the interval that contains the spectrum of $(I - G)^{-1}(I - K)$. The conjugate gradient algorithm has the capacity to take advantage of the distribution of the eigenvalues (cf. [HeSt]). For example, Table 2 contains a comparison of the relative residual for DSA alone, the expected relative residual using DSA with the PCG iteration and the actual relative residual using DSA with PCG. Here $a = 128$, $\beta = \frac{2}{3}$, and a mesh of 200 points was used.

15 SUMMARY

We have shown that standard notions of linear algebra can provide a framework for the discussion of algorithms for the numerical solution of the transport equation. In particular, we have seen that the DSA algorithm is the Neumann series applied to the standard integral

Table 1. VALUES OF δ VERSUS k

k \ a	8	16	32	64	128	256
1	0.2764	0.3047	0.3216	0.3365	0.3558	0.3672
2	0.2628	0.3002	0.3203	0.3366	0.3575	0.3568
3	0.2432	0.2931	0.3182	0.3358	0.3576	0.3548
4	0.2206	0.2837	0.3153	0.3350	0.3575	0.3541
5	**0.1976**	0.2725	0.3116	0.3540	0.3572	0.3537
10	0.1095	**0.2077**	0.2846	0.3256	0.3548	0.3526
20	—	0.1122	**0.2145**	0.2966	0.3452	0.3500
40	—	—	0.1449	**0.2223**	0.3131	0.3453
80	—	—	—	0.1173	**0.2996**	0.3258

Table 2. RELATIVE RESIDUAL VERSUS ITERATION COUNT

Iteration	DSA Alone	Expected PCG	Actual PCG
1	0.23	0.65 -1	0.20 -2
2	0.53 -1	0.43 -2	0.19 -4
3	0.12 -1	0.28 -3	0.92 -6
4	0.28 -2	0.18 -4	0.41 -7
5	0.64 -3	0.12 -5	0.11 -8
6	0.14 -3	0.76 -6	0.23 -10
7	0.34 -4		
8	0.78 -5		
9	0.18 -5		
10	0.41 -6		

equation formulation, Eq. (7), preconditioned by a certain diffusion operator $(T - \gamma I)$ given in Sec. 8. We have also seen that the boundary conditions can be handled straightforwardly.

The boundary conditions (cf. (23))

$$B\,\phi_{n+1} = BK\,\gamma\phi_n = BLH^{-1}\gamma\phi_n = BL\,\psi_n$$

can be thought of as expressing the fact that the prescribed boundary values of ϕ after the accelerated part of the scheme, ϕ_{n+1}, should remain the same as the values of ϕ before the acceleration, $L\psi_n$. Again this agrees completely with the boundary conditions used in [La1].

Using the spectral properties of the integral formulation of K and the DSA operator G, we have demonstrated why DSA works well for optically dense problems (i.e., large a). We have used this approach to suggest other preconditionings, $(I - \tilde{G})$, and iteration schemes whose convergence rates depend upon the spectrum of a fixed operator $(I - \tilde{G}\gamma)^{-1}(I - K\gamma)$.

Finally, we have demonstrated analytically in Sec. 9 that iterative acceleration of the basic Neumann iteration (13) can produce profound speedup. Also, we have demonstrated numerically that the PCG algorithm can produce significant acceleration of the DSA preconditioning.

ACKNOWLEDGMENT

The authors thank Milt Wing for many thoughtful discussions and Dan Seth for his help with numerical computations.

REFERENCES

[AlWi] R. C. Allen and G. M. Wing, "A method for accelerating the iterative solution of a class of Fredholm integral equations," J. Optimization Theory and Applications, 24 (1978), 5-27.

[CGO] P. Concus, G. H. Golub, and D. P. O'Leary, "A generalized conjugate gradient method for the numerical solution of elliptic partial differential equations," *Sparse Matrix Computations*, J. R. Bunch and D. J. Rose, eds. (Academic Press, New York, 1976).

[FaMa] V. Faber and Thomas A. Manteuffel, "Necessary and sufficient conditions for the existence of a conjugate gradient method," SIAM J. Numer. Anal., Vol. 21, No. 2, pp. 352-362 (1984).

[FMWW] V. Faber, Thomas A. Manteuffel, Andrew B. White, Jr., and G. Milton Wing, "Asymptotic behavior of singular values and singular functions of certain convolution operators," Comp. and Maths. with Appls., Vol. 12A, No. 6, pp. 733-747 (1986).

[GoVa] G. H. Golub and R. S. Varga, "Chebychev semi-iterative methods, successive overrelaxation iterative methods and second-order Richardson iterative methods," Numer. Math. **3**, 147 (1961).

[GoVL] G. H. Golub and C. VanLoon, *Matrix Computations* (John Hopkins University Press, Baltimore, 1983).

[HeSt] M. R. Hestenes and E. Stiefel, "Methods of conjugate gradients for solving linear systems," J. Res. Nat. Bur. Stand. **49**, 409-436 (1952).

[La1] E. W. Larsen, "Diffusion-synthetic acceleration methods for discrete-ordinates problems," Transport Theory and Statistical Physics **13** (1984), 107-126.

[Ma1] Thomas A. Manteuffel, "The Tchebychev iteration for nonsymmetric linear systems," Numer. Math. **28**, 307-327 (1977).

[Ma2] Thomas A. Manteuffel, "Adaptive procedures for estimating parameters for the nonsymmetric Tchebychev iteration," Numer. Math. **31**, 183-208 (1978).

[Re] J. K. Reid, "On the method of conjugate gradients for the solution of large sparse linear equations," in *Large Sparse Sets of Linear Equations*, J. K. Reid, ed. (Academic Press, New York, 1971), pp. 231-254.

[Wi] G. Milton Wing, *Transport Theory* (Wiley and Sons, New York, 1962).

A Quasi-Diffusive Approach to Linear Transport

M. CELASCHI and B. MONTAGNINI   Department of Mechanical and Nuclear Construction, University of Pisa, Pisa, Italy

A new method for approximately (but accurately) solving the linear transport equation, which resorts, at least in the simplest cases, to a diffusion-like equation, is presented. The method is based on a suitable generalization of the Fick's law, in which the diffusion coefficient is replaced by an integral operator. It turns out that this operator is self-adjoint and its kernel is similar, in many aspects, to a Green function, so that its inverse can be given approximately the form of a second order linear differential operator. This allows replacing the integrodifferential or integral transport equation by a set of differential equations and even, in some cases, by a single diffusion-like equation which, contrary to the ordinary diffusion equation, allows for an accurate evaluation of the particle density in

Work performed under the auspices of the Consiglio Nazionale delle Ricerche, Gruppo Nazionale per la Fisica Matematica, with a financial support by the Ministero della Pubblica Istruzione,Italy.

the neighborhood of the interfaces between different material regions or near localized sources.

1. INTRODUCTION

The solution of the (one speed) integrodifferential transport equation, when integrated over the angles specifying the direction of motion of the particles, tends, as well known, far apart from the physical discontinuities of the medium and the localized sources, to a solution of the (much simpler) diffusion equation. In the vicinity of a discontinuity or a localized source, however, the two solutions differ considerably, which means that the diffusion equation ceases to be a reliable model. Then, one may ask if a suitable modification of the underlying definitions (e.g. the definition of the diffusion coefficient) may lead to a significant improvement of its reliability and accuracy.

A new method for approximately (but accurately) solving the transport equation which gives at least a partial answer to the above question is here presented. It is based on a generalization of Fick's law, in which the diffusion coefficient is replaced by an integral operator. It turns out that this operator is self-adjoint and its kernel closely resembles a Green function, so that its inverse can be given, approximately, the form of a second order linear differential operator. This allows replacing the integrodifferential or integral transport equation by a set of differential equations and even, in some cases, by a single diffusion-like equation the validity of which, contrary to the ordinary diffusion equation, is not limited to the asymptotic regions and allows therefore for an accurate evaluation of the particle density also in the neighbourhood of the interfaces between different material regions or near localized sources (the procedure is reminiscent of Cadilhac's "secondary model" in thermalization theory [1]).

In order to contain the length of the exposition we shall be concerned here only with the principle of the method. The treatment will be purely formal. Moreover, we shall limit ourselves to plane geometry problems. Further examples, as well as some extensions, will appear in a forthcoming, more detailed paper.

2. ELEMENTARY GENERALIZATIONS OF THE FICK'S LAW

Let us consider the one-speed, isotropic scattering integral transport equation in plane geometry :

$$\rho(x) = \frac{1}{2}\int_{-\infty}^{\infty} E_1(|x-y|)\,[c(y)\rho(y) + s(y)]dy \tag{1}$$

where, in general,

$$E_n(t) = \int_1^{\infty} u^{-n}\exp(-tu)du, \qquad n\geq 1$$

while $c(x)$ is the mean number of secondaries per collision at the optical distance x from the origin, with $0\leq c(x)\leq c_1<1$ (but sometimes we allow $c_1=1$), $\rho(x)$ is the particle density and $s(x)$ the source.

Eq. (1) is equivalent [2] to the following system

$$j(x) = \frac{1}{2}\int_{-\infty}^{\infty} \text{sgn}(x-y)E_2(|x-y|)\,[c(y)\rho(y) + s(y)]\,dy \tag{2a}$$

$$\frac{d}{dx}j(x) = -\,[1-c(x)]\rho(x) + s(x) \tag{2b}$$

where Eq. (2a) is the definition of the current, $j(x)$, and Eq. (2b) is the equation of continuity.

The elementary approach consists in expanding $\rho(y)$, Eq.(2a), up to the second order term,

$$\rho(y) \approx \rho(x) + (y-x)\rho'(x) + \frac{1}{2}(y-x)^2\rho''(x) \tag{3}$$

Thus, Eq. (2a) can be given the following (approximate) form :

$$j(x) = j_s(x) + \beta_0(x)\rho(x) + \beta_1(x)\rho'(x) + \beta_2(x)\rho''(x) \tag{4}$$

where $j_s(x)$ denotes the current due to the particles coming directly from the source ("uncollided current") and is given by the following formula

$$j_s(x) = \frac{1}{2}\int_{-\infty}^{\infty} \text{sgn}(x-y)E_2(|x-y|)\,s(y)\,dy \tag{5}$$

while the coefficients $\beta_n(x)$ are defined as follows :

$$\beta_n(x) = \frac{1}{2}\int_{-\infty}^{\infty} \text{sgn}(x-y)E_2(|x-y|)\, c(y)\frac{(y-x)^n}{n!}\, dy \qquad (n \geq 0)$$

If $c(x)=c=$constant, we have $\beta_0(x) = \beta_2(x) = 0$, $\beta_1(x) = -c/3$ and, having introduced the "collided current"

$$j_c(x) = j(x) - j_s(x) \tag{6}$$

we see that Eq.(4) reduces to the classical Fick's law

$$j_c(x) = -D\frac{d\rho(x)}{dx}$$

with $D = -\beta_1 = c/3$.

If c is not constant, however, one should make use of the general formula, Eq.(4), i.e., in terms of $j_c(x)$,

$$j_c(x) = \beta_0(x)\rho(x) + \beta_1(x)\rho'(x) + \beta_2(x)\rho''(x) \tag{7a}$$

The continuity equation can also be written in terms of $j_c(x)$:

$$\frac{dj_c(x)}{dx} = -[1-c(x)]\rho(x) + q(x) \tag{7b}$$

where

$$q(x) = \frac{1}{2}\int_{-\infty}^{\infty} E_1(|x-y|)s(y)dy \tag{8}$$

After substituting Eq.(7a) into Eq.(7b) we get the following (mildly generalized) diffusion equation:

$$\gamma_2(x)\rho''(x) + \gamma_1(x)\rho'(x) + [\gamma_0(x)-1]\rho(x) + q(x) = 0 \tag{9}$$

(a term containing a third derivative has been discarded).
Here

$$\gamma_n(x) = \frac{1}{2}\int_{-\infty}^{\infty} E_1(|x-y|)c(y)\frac{(y-x)^n}{n!}dy \qquad (n \geq 0) ; \tag{10}$$

one has, moreover, $\gamma_0(x)=c(x)-\beta_0(x)$, $\gamma_1(x)=-\beta_0(x)\dot{=}\beta_1'(x)$, $\gamma_2(x)=-\beta_1(x)-\beta_2'(x)$. Eq.(9) (a particular case of that obtained by Boffi and Trombetti [3]) is the analogue of the "primary model" equation in neutron thermalization theory, according to Cadilhac's terminology [1]. In the case $c(x)=c=$constant it reduces obviously to the elementary diffusion equation

$$\frac{c}{3}\rho''(x) - (1-c)\rho(x) + q(x) = 0$$

where, if $s(x)$ is also constant, $q(x)$ may be simply replaced by s.

The weak point of this approach is that $\rho'(x)$ has a logarithmic singularity [2] at any point where $c(x)$ has a jump, which makes expansion (3) meaningless in the vicinity of such points. As these material discontinuities are very commonly encountered when dealing with neutron transport problems (even the simplest ones), the domain of applicability of the "primary" method is apparently restricted to some radiation problems involving a weak variation of the parameter c.

3. THE SYMMETRIC CONSERVATIVE FORM OF THE TRANSPORT EQUATION

Let $c(x)=c=$constant. Eq.(2a), written in terms of the collided current,

$$j_c(x) = \frac{c}{2}\int_{-\infty}^{x}E_2(x-y)\rho(y)dy - \frac{c}{2}\int_{x}^{\infty}E_2(y-x)\rho(y)dy$$

(see Eqs. (5) and (6)) gives, after an integration by parts,

$$j_c(x) = -\frac{c}{2}\int_{-\infty}^{\infty}E_3(|\,x-y\,|)\frac{d\rho(y)}{dy}dy \tag{11}$$

(recall that $E_n(x)=-E'_{n+1}(x)$; we assume, moreover, that $r(x)$ is at most $O(\exp(\alpha|x|))$, $\alpha<1$, as $|x|\to\infty$). Eq.(11) is a relationship between the (collided) current and the density gradient which can be considered as a generalization of the Fick's law, the usual diffusion coefficient D being replaced by an integral operator. We observe that the kernel of this operator, $(c/2)E_3(|x-y|)$, has the following properties:

(i)   it is bounded on R^2, positive and symmetric

(ii)   it is o(exp(-|x|)) as |x| $\to\infty$, for fixed y

(iii)   its partial derivative with respect to y has, for y=x, the finite jump

$$\frac{\partial}{\partial y}(\frac{c}{2}E_3)|_{y=x^-}^{y=x^+} = -c$$

Thus, not only this kernel is rather well behaved (it is actually much more well behaved than $(c/2)E_1(|x-y|)$, the kernel of Eq. (1)), but has properties closely resembling those of a Green function of a second order linear differential equation, in particular a diffusion equation.

Let us now try to generalize our argument to any c(x). We start again with the definition of current, Eq.(2a). Upon differentiation we have

$$\frac{dj(x)}{dx} = c(x)\rho(x) + s(x) - \frac{1}{2}\int_{-\infty}^{x}E_1(x-y)[c(y)\rho(y)+s(y)]dy - \frac{1}{2}\int_{x}^{\infty}E_1(y-x)[c(y)\rho(y)+s(y)]dy$$

After an integration by parts, both sides are multiplied by c(x) and integrated from -∞ to x. By reversing the order of the integrations we get

$$\int_{-\infty}^{x}c(x')\frac{dj(x')}{dx'}dx' = -\frac{1}{2}\int_{-\infty}^{x}dy\frac{d\rho(y)}{dy}\int_{x}^{\infty}dx'c(x')\int_{-\infty}^{y}c(y')E_1(x'-y')dy'$$

$$-\frac{1}{2}\int_{x}^{\infty}dy\frac{d\rho(y)}{dy}\int_{-\infty}^{x}dx'c(x')\int_{y}^{\infty}c(y')E_1(y'-x')dy'$$

$$-\int_{-\infty}^{x}c(x')\{[\gamma_0(x')-c(x')]\rho(x')-s(x')+q(x')\}dx' \qquad (12)$$

where use has been made of the definitions of q(x) and $\gamma_0(x)$, Eqs. (8) and (10). Now we set

$$D(x,y) = \left\{\begin{array}{ll} \frac{1}{2}\int_{x}^{\infty}dx'\int_{-\infty}^{y}c(x')c(y')E_1(x'-y')dy', & x\geq y \\ \frac{1}{2}\int_{-\infty}^{x}dx'\int_{y}^{\infty}c(x')c(y')E_1(y'-x')dy', & x<y \end{array}\right. \qquad (13)$$

and introduce a new unknown function,

$$i(x) = \int_{-\infty}^{x} c(x')\{\frac{dj(x')}{dx'} + [\gamma_0(x')-c(x')]\rho(x')-s(x')+q(x')\}dx' \tag{14}$$

or, since $q(x)-s(x)=-j'_s(x)$ (see Eqs. (6) and (8)),

$$i(x) = \int_{-\infty}^{x} c(x')\{\frac{d}{dx'}[j(x')-j_s(x')] + [\gamma_0(x')-c(x')]\rho(x')\}dx' =$$

$$= \int_{-\infty}^{x} c(x')\{\frac{dj_c(x')}{dx'} + [\gamma_0(x')-c(x')]\rho(x')\}dx' \tag{14'}$$

(an alternative definition is

$$i(x) = -\int_{x}^{\infty} c(x')\{\frac{dj_c(x')}{dx'} + [\gamma_0(x')-c(x')]\rho(x')\}dx \tag{14''}$$

With these definitions we see that Eq. (12) assumes the following form:

$$i(x) = -\int_{-\infty}^{\infty} D(x,y)\frac{d\rho(y)}{dy}dy \quad , \tag{15a}$$

again a "generalized" Fick's law. The continuity equation can also be written in terms of $i(x)$: since by Eq.(14'), $di/dx=c(x)dj_c(x)/dx + c(x)[\gamma_0(x)-c(x)]\, \rho(x)$, Eq. (7b) is equivalent to the following one

$$\frac{di(x)}{dx} = -c(x)[1-\gamma_0(x)]\rho(x) + c(x)q(x) \tag{15b}$$

Eqs. (15a,b) constitute an integrodifferential system which will be called "symmetric conservative form of the transport equation". If $c(x)=c=$constant, then $\gamma_0(x)=c$ and $i(x)=cj_c(x)$: this justifies our use of the name "pseudocurrent", when speaking of $i(x)$.

Note that the kernel $D(x,y)$ retains, even in the general case $c=c(x)$, the main properties of a Green function; in particular

$$\frac{\partial D(x,y)}{\partial y}\bigg|_{y=x^-}^{y=x^+} = -c(x)\gamma_0(x) \tag{16}$$

4. THE APPROXIMATE INVERSION OF THE DIFFUSION OPERATOR. THE "G_N METHOD".

We have seen that the kernel $D(x,y)$ of the integral operator

$$D = \int_{-\infty}^{\infty} D(x,y)dy$$

henceforth called "diffusion operator", is very similar to a Green function. So we put

$$D = G + E \tag{17}$$

where G is a Green operator, i.e. the inverse (with a - sign) of a second order Sturm-Liouville differential operator

$$L = \frac{d}{dx}u(x)\frac{d}{dx} - v(x)$$

associated with the conditions of vanishing at infinity (or other conditions, to be specified later) and E=D-G, also an integral operator, represents the error. If the approximation by a single Green function is found not sufficiently accurate, we can try

$$D = G_1 + G_2 + E^{(2)}$$

or, more generally,

$$D = \sum_{n=1}^{N} G_n + E^{(N)} \tag{18}$$

where each G_n is the Green operator of a Sturm-Liouville operator L_n. Then, assuming that the norm of $E^{(N)}$ is so small (in some underlying space), for some N, that we can take $E^{(N)}=0$ without practically affecting the solution, the (exact) symmetric conservative form of the transport equation

$$i = -D\frac{d\rho}{dx}$$

$$\frac{di}{dx} = -c(1-\gamma_0)\rho + cq$$

can be written (in an approximate way) as follows

$$i_n = -G_n\frac{d\rho}{dx} \qquad (n=1,...,N)$$

$$\sum_{n=1}^{N}\frac{di_n}{dx} = -c(1-\gamma_0)\rho + cq$$

or, using the relationships $G_n = -L^{-1}_n$ $(n=1,...,N)$

$$L_n i_n = \frac{d\rho}{dx} \qquad (n=1,...,N) \tag{19a}$$

$$\sum_{n=1}^{N}\frac{di_n}{dx} = -c(1-\gamma_0)\rho + cq \tag{19b}$$

The transport equation is thus converted into a differential system; it is remarkable that the first N equations of the system, i.e. $L_n i_n = \rho'$ $(n=1,...,N)$, can be solved independently.

The case $N=1$, i.e.

$$Li = \frac{d}{dx}u\frac{d}{dx}i - vi = \frac{d\rho}{dx} \tag{20a}$$

$$\frac{di}{dx} = -c(1-\gamma_0)\rho + cq \tag{20b}$$

has a distinct advantage, besides that of being the simplest approximation. Substituting Eq.(20b) into Eq.(20a) we get

$$-vi = \frac{d\rho}{dx} - \frac{d}{dx}u[-c(1-\gamma_0)\rho + cq] = \frac{d}{dx}\{[1+uc(1-\gamma_0)]\rho - ucq\}$$

Thus, if we introduce the new unknown function

$$\psi(x) = \{1+u(x)c(x)[1-\gamma_0(x)]\}\rho(x) - u(x)c(x)q(x)$$

system (20a,b) becomes

$$i = -\frac{1}{v}\frac{d\psi}{dx} \tag{21a}$$

$$\frac{di}{dx} = -\frac{c(1-\gamma_0)}{1+uc(1-\gamma_0)}\psi + \frac{cq}{1+uc(1-\gamma_0)} \tag{21b}$$

and, substituting i, as given by the first equation, into the second equation, we arrive finally at the following "quasi-diffusion" or "super-diffusion" equation

$$\frac{d}{dx}p(x)\frac{d\psi}{dx} - r(x)\psi + z(x) = 0 \tag{22}$$

with

$$p = \frac{1}{v}, \qquad r = \frac{c(1-\gamma_0)}{1+uc(1-\gamma_0)}, \qquad z = \frac{cq}{1+uc(1-\gamma_0)}$$

just in the style of Cadilhac's "secondary model".

As an example, we consider the classical Milne's problem [2,4] :

$$\rho(x) = \frac{c}{2}\int_0^{\infty}E_1(|x-y|)\rho(y)dy \qquad (c \leq 1, \ 0 \leq x < \infty) \tag{23}$$

with

$$\rho(x) \sim \exp(\kappa x) \qquad (x \to +\infty) \tag{24}$$

$\kappa = \kappa(c)$ being the positive root of the transcendental equation $(1/2\kappa)\ln[(1+\kappa)/(1-\kappa)] = 1/c$. In this problem the rôle of the source, which we can imagine as relegated to infinity, is played by an inhomogeneous condition, Eq.(24).

It is a simple matter to write down the symmetric conservative version of Milne's problem. As $c(x)=0$ for $x<0$, $c(x) = c$ (constant) for $x>0$, after some calculations we get

$$D(x,y) = \frac{c^2}{2}E_3(|x-y|) - \frac{c^2}{2}E_3(\max\{x,y\}) \qquad (0<x,y<\infty) \tag{25}$$

$$\gamma_0(x) = c[1 - \frac{1}{2}E_2(x)] \qquad (0<x<\infty) \tag{26}$$

Having in mind to approximate $D(x,y)$ by a sum of Green functions, we observe, following an idea by Coppa, Ravetto and Sumini [5,6,7,8], that $E_2(x)$ can be easily (and accurately) approximated by sums of exponentials. For this, it will be sufficient to approximate the integral which represents $E_2(x)$ by means of a quadrature formula :

$$E_2(x) = \int_1^\infty \exp(-xt)t^{-2}dt = \int_0^1 \exp(-x/\mu)d\mu \approx \sum_{\mu_n>0} w_n\exp(-x/\mu_n)$$

where μ_n, w_n are the points and the weights of the formula.

We also have

$$E_3(x) = \int_x^\infty E_2(y)dy \approx \sum_{\mu_n>0} w_n\mu_n\exp(-x/\mu_n)$$

If we choose the Gauss-Legendre scheme we have, in particular, at the lowest degree of approximation,

$$E_3(x) \approx (1/\sqrt{3})\exp(-\sqrt{3}x)$$

and, consequently,

$$D(x,y) \approx \begin{cases} (c^2/2\sqrt{3})\exp(-\sqrt{3}x)[\exp(-\sqrt{3}y)-1] & , \quad x>y \\ (c^2/2\sqrt{3})\exp(-\sqrt{3}y)[\exp(-\sqrt{3}x)-1] & , \quad x<y \end{cases} \tag{27}$$

$$\gamma_0(x) \approx c[1 - \tfrac{1}{2}\exp(-\sqrt{3}x)]$$

(to preserve Eq.(16), $\gamma_0(x)$ has been approximated consistently with $D(x,y)$).

Thus, the approximate $D(x,y)$ is already in the form of a Green function and we easily see that the coefficients $u(x)$, $v(x)$ of the corresponding differential operator are

$$u(x) = c^{-2}[1 - \tfrac{1}{2}\exp(-\sqrt{3}x)]^{-1}$$
$$v(x) = 3c^{-2}[1 - \tfrac{1}{2}\exp(-\sqrt{3}x)]^{-2}$$

The super-diffusion equation is, therefore,

$$\frac{1}{3}\frac{d}{dx}[1 - \tfrac{1}{2}\exp(-\sqrt{3}x)]^2\frac{d\psi}{dx} - [1 - \tfrac{1}{2}\exp(-\sqrt{3}x)]\{1-c[1 - \tfrac{1}{2}\exp(-\sqrt{3}x)]\}\psi = 0 \tag{28}$$

where

$$\psi(x) = \frac{\rho(x)}{c[1 - \tfrac{1}{2}\exp(-\sqrt{3}x)]}$$

The conditions are

$$\psi'(0) = 0 , \quad \psi(x) \sim (1/c)\exp(\kappa x) \qquad (x \rightarrow +\infty)$$

Of these, the first one requires an explaination: as the approximate kernel, Eq.(27), vanishes as $x \rightarrow 0$, (as well as the exact one, Eq.(25)), then $i(0) = -(1/v(0)) \psi'(0) = 0$.

Eq.(28) looks rather awkward. However, if we substitute the expression for $\psi(x)$ into it, after a few calculations we obtain

$$\frac{1}{3}\frac{d^2\rho}{dx^2} - (1-c)\rho = 0$$

i.e. the classical diffusion equation in the P_1 approximation, with

$$\rho'(0) - \sqrt{3}\rho(0) = 0 , \quad \rho(x) \sim \exp(\kappa x) \qquad (x \text{-> } +\infty)$$

The first condition (which can be immediately deduced from the condition $\psi'(0)=0$) is of the Mark's type [2]. The second one, being a condition on the asymptotic behaviour of the solution (and therefore much stronger than an ordinary boundary condition) requires an adjustment, namely that $\kappa = \sqrt{3(1-c)}$, consistently with the P_1 approximation.

Similar results are obtained in the general (N>1) case. They parallel those by Coppa et al. [5,6,7,8], who have applied the quadrature formula directly to the integral transport equation. A part from the nice fact that the Mark conditions now appear as a consequence of our choice of the Gauss-Legendre formula and are therefore natural conditions, our result is, however, rather unsatisfactory.

It means that, if we wish to minimize the error operator in the norm, i.e. if we aim at a uniform approximation, a single Green function cannot be expected to give exceptional results. To attain a good accuracy, the order N of our "G_N method", i.e. the number of Green functions involved, must be rather high, just as it should be with the P_N method, to which the G_N method is equivalent (at least in the case of a Gauss-Legendre formula).

The case N=1 is, however, so appealing that we try to improve our method as much as possible, by using a different approach. As well known, the coefficient $u(x)$ is the reciprocal, with a - sign, of the jump of $\partial_y G(x,y)$ at $y=x$, $G(x,y)$ being the Green function associated with L. It is natural therefore to choose $u(x)$ in such a way that this jump is equal to the jump of $\partial_y D(x,y)$, which amounts to take, by Eq.(16),

$$u(x) = \frac{1}{c(x)\gamma_0(x)} \tag{29}$$

Let now $\tilde{\rho}(x)$ be some function which is expected to be a reasonable approximation of the solution, e.g. the elementary diffusion solution of the problem at hand. Then we decide that a good way of determining the remaining coefficient, $v(x)$, consists in requiring that $E\tilde{\rho}' = 0$, that is $-LD\tilde{\rho}' = \tilde{\rho}'$ or, more explicitly, after having introduced the following approximate pseudocurrent

$$\tilde{i}(x) = - \int_{-\infty}^{\infty} D(x,y)\frac{d\tilde{\rho}(y)}{dy}dy \quad , \tag{30}$$

by requiring that

$$\frac{d}{dx}u(x)\frac{d\tilde{i}}{dx} - v(x)\tilde{i} = \frac{d\tilde{\rho}}{dx}$$

which immediately gives

$$v(x) = - \frac{\dfrac{d\tilde{\rho}}{dx} - \dfrac{d}{dx}u(x)\dfrac{d\tilde{i}}{dx}}{\tilde{i}} \tag{31}$$

With such $u(x)$, $v(x)$ the operator L acts on $-D\tilde{\rho}'$ as if it were the inverse of $-D$; it is hoped that it will be "nearly so" if $\tilde{\rho}$ is replaced by the true solution ρ (although this is likely to appear as a matter of faith, in lack of a rigorous proof). Of course, if the approximate $\rho(x)$ obtained by solving Eqs.(20a,b) or Eq.(22) with such $u(x)$,$v(x)$ turns out to be more accurate than the initial guess, $\tilde{\rho}(x)$ (as it is, to our experience), nothing prevents us from iterating the whole procedure . Again, a convergence proof of this iterative, or perhaps "bootstrap", method is lacking.

If a material discontinuity is present, however, this method suffers (just as it was for the "primary model") of a bad singularity affecting one of the coefficients, namely $v(x)$. But we can overcome this difficulty by means of the technique described in the following section.

5. THE METHOD OF SPLITTING

We return to Milne's problem and put (see Eq.(25))

$$D(x,y) = D_1(x,y) + D_2(x,y)$$

where

$$D_1(x,y) = \frac{c^2}{2}E_3(|x-y|) \qquad , \qquad D_2(x,y) = -\frac{c^2}{2}E_3(\max\{x,y\})$$

Then we write the symmetric form of the transport equation as follows

$$i_1(x) = -\int_0^\infty D_1(x,y)\frac{d\rho(y)}{dy}dy \tag{32a}$$

$$i_2(x) = -\int_0^\infty D_2(x,y)\frac{d\rho(y)}{dy}dy \tag{32b}$$

$$\frac{d}{dx}[i_1(x)+i_2(x)] = -c[1-\gamma_0(x)]\rho(x) \tag{32c}$$

Due to the particular form of $D_2(x,y)$, Eq.(32b) is equivalent to the following one

$$i_2(x) = \frac{c^2}{2}E_3(x)[\rho(x)-\rho(0)] + \frac{c^2}{2}\int_x^\infty E_3(y)\frac{d\rho(y)}{dy}dy$$

Then, upon differentiating, substituting into Eq.(32c) and recalling Eq.(26), we succeed in eliminating $i_2(x)$ and obtain

$$i_1(x) = -\int_0^\infty D_1(x,y)\frac{d\rho(y)}{dy}dy \tag{33a}$$

$$\frac{di_1(x)}{dx} = -c(1-c)\rho(x) - \frac{c^2}{2}\rho(0)E_2(x) \tag{33b}$$

If the additional term, $-(c^2/2)\rho(0)E_2(x)$, is interpreted as a source term, this system being formally analogous to our initial "symmetric conservative system", Eqs. (15a,b), we can try to solve it by the last method of the preceding section. Assuming for a moment that we have already determined the coefficients $u_1(x)$, $v_1(x)$ of the differential operator related to $D_1(x,y)$, Eqs.(33a,b) are first replaced by the following ones

$$L_1 i_1(x) \approx \frac{d}{dx}u_1(x)\frac{di_1}{dx} - v_1(x)i_1 = \frac{d\rho}{dx} \tag{34a}$$

$$\frac{di_1(x)}{dx} = -c(1-c)\rho(x) - \frac{c^2}{2}\rho(0)E_2(x) \tag{34b}$$

then, proceeding exactly as in sect. 4, Eqs. (20a,b)-Eq.(22), we obtain the following system

$$i_1(x) = -\frac{1}{v_1(x)}\frac{d\psi_1(x)}{dx} \tag{35a}$$

$$\frac{di_1(x)}{dx} = -\frac{c(1-c)}{1+c(1-c)u_1(x)}\psi_1(x) - \frac{c^2}{2}\rho(0)\frac{E_2(x)}{1+c(1-c)u_1(x)} \tag{35b}$$

where

$$\psi_1(x) = [1+c(1-c)u_1(x)]\rho(x) + \frac{c^2}{2}u_1(x)\rho(0)E_2(x) \tag{36}$$

and, finally, after substituting Eq.(35a) into (35b), we arrive at the following 2nd order differential equation

$$\frac{d}{dx}\frac{1}{v_1(x)}\frac{d\psi_1(x)}{dx} - \frac{c(1-c)}{1+c(1-c)u_1(x)}\psi_1(x) - \frac{c^2}{2}\rho(0)\frac{E_2(x)}{1+c(1-c)u_1(x)} = 0 \tag{37}$$

The conditions to be associated with Eq.(37) are also easily obtained. We have, by Eq.(36),

$$\psi_1(0) = [1+c(1-\frac{c}{2})u_1(0)]\rho(0) \tag{38}$$

On the other hand, by a classical result of transport theory ([2],p.77, form.(6.46)[*]) :

$$j(0) = -\frac{\sqrt{1-c}}{\kappa}\rho(0) \tag{39}$$

so that, by using Eqs.(32a), (2a) and integrating by parts,

$$i_1(0) = -\frac{c^2}{2}\int_0^\infty E_3(y)\frac{d\rho(y)}{dy}dy = \frac{c^2}{4}\rho(0) - \frac{c^2}{2}\int_0^\infty E_2(y)\rho(y)dy =$$

$$= \frac{c^2}{4}\rho(0) + cj(0) = [\frac{c^2}{4} - \frac{c\sqrt{1-c}}{\kappa}]\rho(0)$$

or, by Eq.(35a),

(*) - Eq.(39) can also be obtained (in approximate form) in the framework of the present method, without having recurse to general transport theory.

$$\psi'_1(0) = [\frac{c\sqrt{1-c}}{\kappa} - \frac{c^2}{4}] v_1(0) \rho(0) \tag{40}$$

Eqs.(38) and (40) are two initial conditions which determine completely the solution of the problem for any arbitrarily chosen $\rho(0)$ (which plays the rôle of a normalization factor). It remains only to determine $u_1(x)$ and $v_1(x)$. The calculations are particularly simple in the case c=1. According to the argument at the end of sect. 4, here applied to the kernel $D_1(x,y)$, we have, for c=1,

$$u_1(x) = [-\partial_y D_1(x,y)|_x^{x\dagger}] = 1$$

while $v_1(x)$ is obtained by taking, as a first approximation of the solution, the asymptotic behaviour $\tilde{\rho}(x)=C_1x+C_2$ (C_1,C_2 constants), or $\tilde{\rho}'(x)=C_1$. Then

$$\tilde{i}_1(x) = - \int_0^\infty D_1(x,y)C_1 dy = - C_1 [\frac{1}{3} - \frac{1}{2}E_4(x)]$$

and, by Eq.(31),

$$v_1(x) = \frac{6-3E_2(x)}{2-3E_4(x)}$$

Note that no singularities now appear. Eq.(37) can be integrated immediately and, using conditions (38) and (40) (recall that $\kappa/\sqrt{1-c} \to \sqrt{3}$ as $c\to 1$) and returning to the natural unknown $\rho(x)$ (Eq.(36)), we get

$$\rho(x) = \rho(0)\{\frac{3}{2} - \frac{1}{2}E_2(x) + \int_0^x [\frac{1}{\sqrt{3}} - \frac{1}{2}E_3(y)]v_1(y)dy\} \tag{41}$$

This approximate solution is then used as a new trial function and the whole procedure is repeated. Some values referring to the first (Eq.(41)) and the second iteration for the solution of Milne's problem with c=1 are reported in the table below. The treatment of Milne's problem for any c<1 is similar, but requires a numerical integration of Eq.(37).

An unpleasant aspect of our actual solution is the appearance of the factor $\rho(0)$ in Eq.(37) (and, consequently, in Eq.(41)), which gives a "local" character to the whole procedure. This can be avoided, however, by re-defining u(x), v(x) in such a way that they are smooth functions and, at the same time, reproduce the approximate solution obtained by the "local" method.

A "true" diffusion equation (but with a modified unknown function and modified coefficients) is therefore obtained, which can be dealt with just as the usual one.

Table. Some results for the Milne's problem with c=1

x	P1	1st iteration	2nd iteration	exact
0	1	1	1	1
0.1	1.1732	1.2526	1.2595	1.2608
0.2	1.3464	1.4580	1.4698	1.4714
0.4	1.6928	1.8406	1.8559	1.8587
0.6	2.0392	2.2082	2.2243	2.2271
0.8	2.3856	2.5690	2.5844	2.5868
1.0	2.7321	2.9256	2.9396	2.9419
2.0	4.4641	4.6819	4.6897	4.6902
extrap. length	0.5773	0.7092	0.7102	0.7104

Details are deferred to a further paper.

6. EXTENSIONS

Arguments very similar to those illustrated by our example concerning Milne's problem can be used in order to solve other classical transport problems (e.g. two adjacent half spaces). The method can also be extended, although at expense of a much heavier formalism, to anisotropic scattering and two- or three-dimensional problems.

REFERENCES

1. M. Cadilhac: Méthodes théoriques pour l'étude de la thermalisation des neutrons dans le milieux absorbants infinis et homogènes, CEA R 2368 (1964)

2. B. Davison: Neutron Transport Theory, Oxford Univ. Press (1958)

3. V.C. Boffi, T. Trombetti, Nucl. Sci. Eng. 50, 200 (1973)

4. K.M. Case, P.F. Zweifel : Linear Transport Theory, Add. Wesley, 1967

5. G. Coppa et al., Transp. Th. Stat. Phys., 10(4), 161 (1981)

6. G. Coppa et al., Ann. Nucl. Energy, 9, 169 (1982)

7. G. Coppa et al., Ann. Nucl. Energy, 9, 435 (1982)

8. G. Coppa et al., Transp. Th. Stat. Phys. 14(1), 83 (1985)

On the Asympotic Behaviour of Electrons in an Ionized Gas

L. ARLOTTI   Mathematical Institute, Faculty of Engineering, Ancona, Italy

The linear integrodifferential equation describing the behavior of electrons in a weakly ionized gas is studied. Sufficient conditions are given to assure that there exist solutions to the stationary problem and that each solution to the time-dependent problem decays to equilibrium.

1 INTRODUCTION

We consider a weakly ionized gas subject to a constant electric field; we suppose that recombination and ionization balance each other and that the electron-electron and ion-electron interactions are neglected. Therefore, only the collisions between the electrons and the neutral molecules of the gas are to be considered; nevertheless these collisions don't modify the distribution function of the molecules, which we suppose to be isotropic and independent of time and position. Moreover, we suppose that the cross-section describing the collisions is invariant with respect to a rigid rotation of all four velocities (before and after the collision) of the two colliding particles. When these conditions are realized, the behavior of the electrons in the given electric field can be described through their distribution function f in the velocity space which, as we know, evolves in time so as to satisfy the linear Maxwell-Boltzmann equation

$$\frac{\partial f}{\partial t} + \underline{a} \cdot \frac{\partial f}{\partial \underline{v}} = -\nu f + \int_{R^3} k(\underline{v}, \underline{v}') \nu(v') f(\underline{v}', t) \, d\underline{v}' \qquad (1)$$

In (1) vector $\underline{a}$ represents the acceleration of the generic electron due to the electric field; vector $\underline{v}$ denotes its velocity; ν is the collision frequency that depends, because of the assumptions already made, only on the electron speed $v = |\underline{v}|$; k is the collision kernel depending on the velocities before and after the collision.

The linear Maxwell-Boltzmann equation was used many years ago by numerous authors in the research on plasma physics, mainly to represent the stationary distribution of the electrons and therefore to obtain an estimate of the D.C. conductivity (see,e.g., [1,2], and the references contained in them). Nevertheless, Cavalleri and Paveri-Fontana [3] have only recently faced the study of the above mentioned equation from an analytical point of view, showing that a stationary distribution can exist only if the collision frequency is such that

$$\int_{V}^{+\infty} \nu \ (v) \ dv = +\infty \ \text{for any} \ V \geq 0;$$

we can remark that one author [4] has wrongly presented estimates of the D.C. conductivity even in cases when the stationary distribution doesn't exist.

An introduction to the physics of swarms, together with a brief historical sketch and up-to-date references can be found in the reports by Kumar [5,6].

Obviously both from the physical and mathematical points of view it is important to get not only the stationary solutions of the linear Maxwell- Boltzmann equation but also to prove the existence of solutions of the time-dependent problem and to study their asymptotic behavior, naturally by making assumptions that do not contrast with the experimental data. Therefore, so as not to exclude collision frequencies of the form $\nu(v) = \alpha v^n$ we cannot consider ν either bounded or (as we can for rigid spheres [7]) Lipschitz - continuous. Neither can we consider the condition

$$\lambda^* := \sup_{t > 0} \ \left\{ \ \text{ess} \inf_{\underline{v} \in R^3} \left[\ t^{-1} \int_0^t \nu \ (|\underline{v} - \underline{a} \ s|) \ ds \ \right] \right\} > 0$$

even if it allows us, in some cases, to face the study of the asymptotic problem by using perturbation theory [8,9]. Finally, since the medium is a pure scatterer, we must consider that, when ν is unbounded, the integral collision operator is also unbounded.

The aim of this paper is to study the linear Eq. (1), taking into account all the previous considerations suggested by the physical model. A similar study has recently been carried out by Frosali, van der Mee and Paveri-Fontana [10], who did not refer to Eq. (1), but to the more simple unidimensional model introduced in [11] by Kac to approximate the Boltzmann equation. But, as we know, in such a model the momentum is not preserved.

In this work in Section 2 we will introduce the necessary notations, and will make the exact assumptions on collision frequency ν and collision kernel k. In Section 3 we will prove (by using the semigroup theory in the complex Banach space $X = L^1(R^3)$) that, for suitable initial data, the Cauchy problem for the linear integrodifferential equation (1) has one and only one solution. In Section 4 we will study the stationary problem showing that, even in this abstract context, a necessary condition for the existence of stationary solutions to (1) is that ν is not integrable. Finally in Section 5 we will examine the case where the above condition is satisfied and furthermore the collision integral is represented by the BGK model. We will see that, in this case, Eq. (1) has stationary solutions and we will also show that each solution of the time dependent problem converges, as t goes to infinity, towards a stationary solution which is obviously uniquely determined by the initial datum.

2 NOTATIONS AND PRELIMINARY RESULTS

In this work we study Eq. (1) by supposing:

(i) that acceleration $\underline{a}$ is a constant vector;

(ii) that the collision frequency only depends on speed $v = |\underline{v}|$ and is a measurable non-negative function $\nu : R^+ \to R^+$;

(iii)   that

$J := \{ v \geq 0 : \forall h > 0 \text{ ess sup} \{\nu (v') : v' \in R_+ \cap [v - h , v + h]\} = + \infty\}$
is a finite subset of R^+;

(iv) that the collision kernel is a measurable function $k : R^6 \to R^+$ so that

$$\int_{R^3} k (\underline{v}, \underline{v}') \, d\underline{v} = 1 \quad \forall \; \underline{v}' \in R^3$$

Moreover $\forall f : R^3 \to C$ (set of all the complex numbers), $f \in L^1_{loc}(R^3)$, we put

$$A_0 f = - \underline{a} \cdot \frac{\partial f}{\partial \underline{v}}$$

where the right hand side makes sense as a distribution.

For every complex-valued and measurable function f defined on R^3, we then denote by Kf the complex-valued function

$$\underline{v} \; \to \; \int_{R^3} k (\underline{v}, \underline{v}') \, \nu (v') \; f (\underline{v}') \, d \underline{v}'$$

if the integral exists for a.e. $\underline{v} \in R^3$.

Obviously, if both f and νf belong to the Banach space $X = L^1 (R^3)$ (provided with the usual norm that we will indicate with $\| \cdot \|$), then Kf exists and belongs to X too.

Now we define in X the following operators:

Operator A

$D (A) = \{ f \in X : \nu f \in X , A_0 f \in X\}$

$A f = A_0 f - \nu f$

Operator K

$D (K) = \{ f \in X : \nu f \in X \}$

$$K f \quad = \quad K f.$$

These definitions allow us to write Eq.(1) in an abstract form as

$$\frac{df}{dt} \quad = \quad (A + K) f \qquad\qquad\qquad t > 0 \qquad (2)$$

and reduce the study of Eq.(1) to the study of operators A and $A + K$.

To reach this aim it is necessary to prove the validity of the following propositions, the proof of which is in the appendix.

PROPOSITION 1

If assumptions (i),(ii) and (iii) are verified, then the subset of R^3

$$E := \left\{ \underline{v} \in R^3 : \lim_{t \to 0+} \int_0^t v\, (\,|\,\underline{v} - \underline{a}\, s\,|\,)\, ds \neq 0 \right\} \qquad (3)$$

has measure $m(E) = 0$

Furthermore, if we put $\forall\, s,t \in R$, $\underline{v} \in R^3$

$$M\,(\underline{v},s,t\,) := \begin{cases} \exp\left[-|\int_s^t v\,(\,|\,\underline{v} - \underline{a}\,\sigma\,|\,)\, d\sigma\,|\,\right] & \text{if } |\int_s^t v\,(\,|\underline{v} - \underline{a}\,\sigma\,|\,)\, d\sigma\,| < +\infty \\ 0 & \\ & \text{if } |\int_s^t v\,(\,|\underline{v} - \underline{a}\,\sigma\,|\,)\, d\sigma\,| = +\infty \end{cases}$$

then for any $\underline{v} \in R^3$, $s \in R$, the function $t \to M(\underline{v},s,t)$ is differentiable a.e.; its derivative is equal to $-v(\,|\,\underline{v} - \underline{a}\,t\,|\,)M(\,\underline{v},s,t\,)$ on $(\,s, +\infty\,)$, it is equal to $v(\,|\,\underline{v} - \underline{a}\,t\,|\,)M(\,\underline{v},s,t)$ on $(-\infty,s\,)$.

PROPOSITION 2

If, for any $\varepsilon > 0$, $\int_\varepsilon^{+\infty} v\,(v)\, dv < +\infty$ then we have $\int_R v\,(\,|\,\underline{v} - \underline{a}\,t\,|\,)\, dt < +\infty$

for any $\underline{v} \in R^3$ so that $|\,\underline{v} \times \underline{e}\,| \notin J$ (where $\underline{e} := a^{-1}\,\underline{a}$); if

$$\int_{v_1}^{v_2} v\,(v)\, dv < +\infty \quad \forall\, 0 < v_1 < v_2 < +\infty \text{ then } \int_0^t v\,(\,|\,\underline{v} - \underline{a}\,s\,|\,)\, ds < +\infty$$

for any $0 < t < +\infty$, $\underline{v} \in R^3$ so that $|\,\underline{v} \times \underline{e}\,| \notin J$;

finally, if $\int_V^{+\infty} \nu(v)\,dv = +\infty$ for any $V \geq 0$, then we have for any $\underline{v} \in R^3$

$$\int_{R^+} \nu(|\underline{v} - \underline{a}t|)\,dt = +\infty.$$

3 TIME DEPENDENT PROBLEM

Here we want to see , by using the sole assumptions (i),(ii),(iii) and (iv), that for any initial datum $f_0 \in D(A)$ the Cauchy problem corresponding to Eq.(2) has one and only one continuously differentiable solution $t \to f(t)$; that $f(t) \geq 0$ if $f_0 \geq 0$; that $\| f(t) \| = \| f_0 \|$ if $f_0 \geq 0$. This result is an immediate consequence of the following theorems.

THEOREM 1

If vector $\underline{a}$ is constant and ν verifies the sole assumptions (ii) and (iii), then operator A is the infinitesimal generator of the strongly continuous semigroup { Z(t); $t \geq 0$ }given by

$$(Z(t) f)(\underline{v}) = m(\underline{v}, t)f(\underline{v} - \underline{a}t) \qquad (4)$$

(where $f \in X$; $m(\underline{v},t) := M(\underline{v}, 0, t)$)

THEOREM 2

If $\underline{a}$, ν and k verify assumptions (i) - (iv) then operator A + K is closable and its closure is the infinitesimal generator of the strongly continuous semigroup {T(t); $t \geq 0$} characterized by the equation

$$T(t) f = Z(t) f + \int_0^t T(t - s) K Z(s) f \, ds \qquad \forall f \in D(A) \quad (5)$$

This semigroup is stochastic, i.e. $\forall t \geq 0$ we have $T(t) \geq 0$, $\| T(t) \| = 1$ and $\| T(t) f \| = \| f \|$ $\forall f \in X^+$. Finally if $f \in D(A)$ then $T(t) f \in D(A)$ $\forall t \geq 0$. Therefore $\forall f \in D(A)$ the formula

$$T(t) f = Z(t) f + \int_0^t Z(t - s) K T(s) f \, ds \qquad (6)$$

is also true.

Proof of theorem 1

We can easily verify that formula (4) characterizes a contraction C_0 -
semigroup, if we take into account that the set E given by (3) is of measure zero. In
order to show that operator A is an extension of the infinitesimal generator A_1 of
$\{Z(t); t \geq 0\}$,suppose $f \in D(A_1)$ and fix $\lambda > 0$.

Then there exists $g \in X$ so that $f = (\lambda - A_1)^{-1} g = \int_0^{+\infty} e^{-\lambda s} Z(s) g\, ds.$

Obviously such an $f \in X$; we also have $vf \in X$, because

$$\| v f \| \leq \| v \int_0^{+\infty} Z(s)\, |g|\, ds \| = \int_{R^3 x R^+} - |g(-\underline{v})| \, \frac{\partial}{\partial s} m(\underline{v}, s)\, d\underline{v}\, ds \leq \| g \|$$

Finally we have

$$f = (\lambda + v) f - g \in X \qquad (7)$$

Indeed $\forall f \in C_0^1 (R^3)$ we obtain, by putting $\underline{v} = \underline{w} - \underline{a} t$ with $\underline{w} \cdot \underline{a} = 0$, $t \in R$

$$\int_{R^3} f\underline{a} \cdot \frac{d\phi}{d\underline{v}}\, d\underline{v} = - \int_{R^2} d\underline{w} \int_R \left[\frac{\partial \phi}{\partial t}(\underline{w} - \underline{a}t) \right] f(\underline{w} - \underline{a}t)\, dt \qquad (8)$$

Thanks to the formula giving f and to Proposition 1, some easy calculations
allow us to recognize that the right hand side of (8)

coincides with $\int_{R^3} \phi \left[(\lambda + v) f - g \right] d\underline{v}.$ This shows that (7) holds and

therefore that $f \in D(A)$ and $A f = A_1 f$.

We finally show that

$$\lim_{t \to 0+} t^{-1} \left[Z(t) f - f \right] = A f$$

It is known [12] that, if $f \in D(A)$, then the function $R^+ \ni t \to f(\underline{v} - \underline{a}t)$ is absolutely
continuous for a.e. $\underline{v} \in R^3$ and that

$$f(\underline{v} - \underline{a} t) - f(\underline{v}) = \int_0^t (A_0 f)(\underline{v} - \underline{a} s)\, ds$$

By virtue of this we easily see that for a.e. $\underline{v} \in R^3$ and $t > 0$

$$| t^{-1} [(Z (t) f) (\underline{v}) - f (\underline{v})] - (A f) (\underline{v}) | \leq \Sigma_{k=1} h_k (\underline{v} , t)$$

if

$$h_1 (\underline{v} , t) := | (m (\underline{v},t) - 1) (A_0 f) (\underline{v}) |$$

$$h_2 (\underline{v} , t) := | t^{-1} \int_0^t [(A_0 f) (\underline{v} - \underline{a} s) - (A_0 f) (\underline{v})] ds |$$

$$h_3 (\underline{v} , t) := \left| t^{-1} \int_0^t [(Z (s) v f) (\underline{v}) - (v f) (\underline{v})] ds \right|$$

$$h_4 (\underline{v} , t) := \left| t^{-1} \int_0^t [- \frac{\partial}{\partial s} m (\underline{v} , s)] [\int_0^s | A_0 f | (\underline{v} - \underline{a} s') ds'] ds \right|$$

To prove the thesis it is enough to show that

$$\lim_{t \to 0+} \| h_k (\underline{v} , t) \| = 0 \qquad k = 1,2,3,4$$

This statement is trivial for k = 1,2,3. In the case k = 4 it follows from

$$\| h_4 (\underline{v} , t) \| = - \| t^{-1} m (\underline{v} , t) \int_0^t | A_0 f | (\underline{v} - \underline{a} s) ds \| +$$

$$+ \| t^{-1} \int_0^t Z (s) | A_0 f | (\underline{v}) ds \| \to - \| A_0 f \| + \| A_0 f \| = 0$$

Proof of Theorem 2

For any $\lambda > 0$ and for any measurable function f defined on R^3 we put

$$(L_\lambda f) (\underline{v}) = \int_0^{+\infty} e^{-\lambda t} m (\underline{v} , t) f (\underline{v} - \underline{a} t) dt$$

if the integral exists for a.e. $\underline{v} \in R^3$.

It is easy to see that, if f is an extended real-valued and non-negative function such that $L_\lambda f \in X^+$, then $f \neq +\infty$ a.e. Indeed the above mentioned assumptions imply

$$\| L_\lambda f \| = \| hf \| \text{ if } h (\underline{v}) := \int_0^{+\infty} e^{-\lambda t} m (- \underline{v} , t) dt$$

By virtue of proposition 1, we are able to prove that $h(\underline{v}) > 0$ for a.e. $\underline{v} \in R^3$. Therefore $(hf)(\underline{v}) = +\infty$ for a.e. $\underline{v}$ such that $f(\underline{v}) = +\infty$. This fact and the condition $L_\lambda f \in X^+$ imply that f is finite a.e. Thanks to this result, by substituting the condition "locally integrable" with the condition "finite a.e.", we can repeat step by step the considerations contained in a previous paper [7] and show that, in this case too,

formula (5) defines a stochastic C_0 - semigroup having the closure of
A + K as its generator and that D(A) is invariant under the semigroup T(t).

4 STATIONARY SOLUTIONS

We now want to examine if Eq. (2) has stationary solutions, i. e. if there exist
functions $f \in X$ such that $T(t) f = f \; \forall \; t \geq 0$.

We will see that, in this abstract context too, the condition

$$(v) \quad \int_V^{+\infty} v\,(v)\,dv \; = \; +\infty \qquad \forall \; V \geq 0$$

is necessary for stationary solutions to exist, at least when the set J is either empty
or equal to $\{0\}$.

The results we will get can be seen more precisely in the following theorems:

THEOREM 3

Suppose that assumptions (i)-(iv) are satisfied and that Eq. (2) has a non-trivial
stationary solution; then v is not integrable on R$^+$. If we also suppose that set J is
contained in $\{0\}$, then v even verifies condition (v).

Proof

Suppose that function f is a non-trivial stationary solution to Eq.(2). From $f =$
$T(t)f$ we obtain $|f| = |T(t)f| \leq T(t)\,|f|$; this implies $\||f\|| \leq \|T(t)\,|f|\,\| = \||f\||$ and therefore
$T(t)|f| = |f|$. In other words $g:= |f|$ is also a stationary solution to (2). From (6) we
immediately deduce that $\forall \; \underline{v} \in R^3, \; t \geq 0$

$$g\,(\underline{v} + \underline{a}\,t) \geq \; (Z\,(t)\,g)\,(\underline{v} + \underline{a}\,t) \; = \; m\,(\underline{v} + \underline{a}\,t, t)\,g\,(\underline{v}) \; \geq$$

$$\geq \exp\left[\; -\int_R v\,(\,|\underline{v} - \underline{a}\,s\,|\,)\,ds\,\right]\,g(\underline{v})$$

Because of this inequality and of Prop. 2, if v were integrable on R$^+$ (or if
condition (v) were not satisfied in the case $J \subseteq \{0\}$), then we would have

$$\min_{t \to \infty} \lim g\,(\,\underline{v} + \underline{a}\,t\,) > 0 \quad \text{for a. e. } \underline{v} \in R^3 \text{ so that } g\,(\underline{v}) > 0.$$

Such a result is absurd because g is a non-trivial and non-negative integrable
function.

THEOREM 4

Suppose that, together with assumptions (i)-(iv), condition (v) is verified too. Then f belongs to D(A) and is a solution to the stationary problem if and only if f belongs to D(K) and is a solution to the integral equation

$$f = \int_0^{+\infty} Z(s) K f \, ds \qquad (9)$$

Proof

If $f \in D(A)$ is a stationary solution to (2), then because of (6) we have for any $t \geq 0$.

$$f = Z(t) f + \int_0^t Z(s) K f \, ds \qquad (10)$$

But (v) implies that $Z(t) \to 0$ as $t \to \infty$; from this and from (10) we easily see that such an f is a solution to (9) too.

Conversely, if (9) has a solution $f \in D(K)$, we can immediately see that such an f is also a solution to (10); that $f \in D(A)$, and finally that $(A + K) f = 0$. This implies $T(t) f = f \ \forall \, t \geq 0$ and therefore f is a solution to the stationary problem

5 ASYMPTOTIC BEHAVIOUR

In this section we will examine the BGK model: so from now on we will suppose that the integral operator K has rank one, i. e. that there exists $\psi \in X^+$ with $\|\psi\| = 1$ such that

(vi) $\qquad k(\underline{v}, \underline{v}') = \psi(\underline{v}) \qquad \forall \, \underline{v}' \in R^3$

We will also suppose that assumptions (i)-(iv) and condition (v) hold and finally that function ψ possesses the property

(vii) $\qquad \int_{R^3 \times R^+} \psi(\underline{v}) \, m(-\underline{v}, t) \, d\underline{v} \, dt < +\infty$

We will easily see that, in this case, Eq. (2) has stationary solutions. We will also be able to characterize the asymptotic behavior of the C_0 - semigroup $\{T(t); t \geq 0\}$ as time t becomes very large. More precisely, we will see that, if $\psi > 0$ a.e., then, for any $f_0 \in X$, function $t \to T(t) f_0$ strongly converges as $t \to \infty$. The limit is obviously a stationary solution to Eq. (2) and moreover it is uniquely determined by f_0. We can thus state that, at least when ψ is positive a.e., each solution to the time dependent problem decays to equilibrium.

First of all we shall examine the stationary problem. Regarding this, we have the following

PROPOSITION

If assumptions (i)- (vii) are verified, then

$$\chi := \int_0^{+\infty} Z\,(s)\,\psi\,ds \qquad\qquad (11)$$

is an element of $D(K) \cap X^+$ and $f \in D(A)$ is a solution to the stationary problem if and only if $f = c\,\chi$ $(c \in C)$.

PROOF

The first statement can be immediately verified, because of (vi) and (vii); the second one can be obtained by observing that, in this case, the right hand side of (9) coincides with $c\,\chi$, provided we put $c = \int vf$.

The asymptotic behavior of the solution, in the case in question, is finally described in the following:

THEOREM 5

We suppose that, together with the assumptions of the previous proposition, either the condition

(viii) $\psi\,(\underline{v}) > 0$ for a.e. $\underline{v} \in R^3$

or the condition

$$(ix) \qquad \int_{v_1}^{v_2} v\,(v)\,dv \;<\; +\infty \qquad \forall\; 0 < v_1 < v_2 < +\infty$$

holds. We then set

$N := \{\,\underline{v} \in R^3 : \chi\,(\underline{v}) = 0\,\}$

$Y_N := \{\,f \in X \;:\; f = 0 \text{ a.e. on } N\,\}$

Then, $\forall\; f_0 \in Y_N$, $f(t) := T(t)\,f_0$ also belongs to Y_N $\forall\; t \geq 0$; moreover function $t \to f(t)$ has a limit as $t \to +\infty$; more precisely we have:

$$\lim_{t \to +\infty} f\,(t) \;=\; c_0\,\chi\,, \text{ where } c_0 := \|\,\chi\,\|^{-1} \int_{R^3} f_0\,(\underline{v})\,d\underline{v}$$

If condition (viii) holds, then Y_N coincides with X and therefore, for any f belonging to X, function $t \to T\,(t)\,f$ has a limit as $t \to +\infty$.

Proof

The proof of the theorem is carried out in the following way:

In (j) we show that $R^3 - N$ cannot be a null set, on the contrary it is R^3 itself when $\psi > 0$ a.e. In (jj) we show that zero is the only purely imaginary number belonging to the point spectrum of $A + K$. In (jjj) we prove that $T(t)_{|Y_N}$ is an irreducible semigroup having $(A + K)_{|Y_N}$ as its generator. In (jk) we finally see that there exist $t > 0$, $\tau > 0$ so that

$$| T (t) - T (t + \tau) | \chi := \sup \{ | T (t) g - T (t + \tau) g | , |g| \leq \chi \} \neq 2 \chi \qquad (12)$$

The above mentioned results allow us to apply the "zero-two law" and its corollaries [13, p.346-349]; we can thus state for any $f_0 \in Y_N$ there exists

$$\lim_{t \to \infty} \quad T (t) f_0 \text{ and is equal } [\; 14 \; , \text{ p. 179 }] \text{ to } c_0 \chi.$$

(j) If assumption (viii) is satisfied, then for a.e. $\underline{v} \in R^3$ there exists $T > 0$ so that for a.e. $0 < s < T$

$$\int_0^T v (| \underline{v} - \underline{a} s |) \, ds < +\infty \text{ and } \psi (\underline{v} - \underline{a} s) > 0.$$

For these $\underline{v}$ we have, by virtue of (11)

$$\chi (\underline{v}) \geq \int_0^T m (\underline{v} , s) \psi (\underline{v} - \underline{a} s) \, ds > 0$$

This shows that in this case $m(N) = 0$

Suppose now that assumption (ix) holds and that there exists $\underline{v} \in R^3$ with $|\underline{v} \times \underline{e}| \notin J$ so that $\chi (\underline{v}) = 0$. From

$$0 \leq (Z (t) \chi) (\underline{v}) = \chi (\underline{v}) - \int_0^t (Z (s) K \chi) (\underline{v}) \, ds \leq \chi (\underline{v}) = 0$$

and from Prop.2 we then deduce that $\chi (\underline{v} - \underline{a} t) = 0 \; \forall \, t \geq 0$. On the set

$$F := \{ \underline{w} \in R^3 : \underline{w} \cdot \underline{a} = 0 , w \notin J \}$$

we can thus define an extended real - valued function $\underline{w} \to t_{\underline{w}}$ by putting $t_{\underline{w}} = \inf \{ s \in R : \chi (\underline{w} + \underline{a} s) > 0 \} \; \forall \, \underline{w} \in F$

Obviously $\forall \, \underline{w} \in F, s \in R$ we have $\chi(\underline{w} + \underline{a} s) = 0$ if $s \leq t_{\underline{w}}$, $\chi(\underline{w} + \underline{a} s) > 0$ if $s > t_{\underline{w}}$. But because of (11) we have $\forall \, \underline{w} \in F , t \in R$

$$\chi\,(\underline{w}+\underline{a}\,t)\;=\;\int_{-\infty}^{t}\,m\,(\underline{w}+\underline{a}\,t\,,\,t\text{-}s\,)\;\psi\,(\underline{w}+\underline{a}\,s\,)\;ds$$

Having arbitrarily fixed $\underline{w}\in F$, $t\in R$, the set $\{s<t:\psi\,(\underline{w}+\underline{a}\,s)>0\}$ is therefore of measure zero if $t\le t_{\underline{w}}$, positive if $t>t_{\underline{w}}$. In particular we have $\psi(\underline{v})=0$ for a.e. $\underline{v}\in N$; since ψ is non-trivial ($\|\,\psi\,\|=1$), the measure of R^3-N cannot be equal to zero.

(jj) As a consequence of assumption (vi), in the case we are examining the infinitesimal generator of the contraction C_0-semigroup $\{T(t);t\ge 0\}$ coincides with $A+K$ which is indeed a closed operator (see the proof of lemma 2 in [7]). Moreover the spectral bound of $A+K$ is equal to zero. Therefore, if there exist $\alpha\in R-\{0\}$ and a non-trivial $f\in X$ such that $(A+K)\,f=i\,\alpha\,f$, then we also have $(A+K)|f|=0$ (see [13] cor. 2.3 p. 297). From this, from the spectral mapping theorem for point spectrum ([13] th. 6.3 p. 85) and from (6) we immediately obtain

$$f\;=\;e^{-i\,\alpha\,t}\,Z\,(t)\,f\;+\;\int_{0}^{t}e^{-i\,\alpha\,s}\,Z\,(s)\,K\,f\,ds$$

and

$$|\,f\,|\;=\;Z\,(t)\,|\,f\,|\;+\;\int_{0}^{t}Z\,(s)\,K\,|\,f\,|\,ds$$

The comparison of these two formulas allows us to recognize that $|K\,f|=K\,|\,f\,|$ and that

$$\left|\,\int_{0}^{t}e^{-i\,\alpha\,s}\,Z\,(s)\,\psi\,ds\,\right|\;=\;\int_{0}^{t}Z\,(s)\,\psi\,ds\qquad\forall\,t\ge 0$$

But we have already seen (in the proof of the previous point (j)) that surely there exist $\underline{v}\in R^3$ and $T>0$ so that $g(s):=(Z(s)\,\psi\,)(\underline{v})>0$ for a.e. $0<s<T$. From the identity

$$\left|\,\int_{0}^{t}e^{-i\,\alpha\,s}\,g\,(s)\,ds\,\right|\;=\;\int_{0}^{t}g\,(s)\,ds$$

which is true for any $t\ge 0$, we easily obtain $\alpha=0$.

(jjj) We know that for any closed ideal $Y\subseteq X$, $Y\ne\{0\}$ there exists a measurable set $M\subseteq R^3$ so that

$$Y\;=\;\{g\in X:g=0\text{ a.e. on }M\}.\tag{13}$$

We want to prove that such an Y is T(t) - invariant for any $t \geq 0$ if and only if there exists an extended real-valued and measurable function defined on $F : \underline{w} \rightarrow \tau_{\underline{w}}$ (with $\tau_{\underline{w}} \leq t_{\underline{w}} \; \forall \; \underline{w} \in F$) so that

$$M = \{\underline{v} \in R^3 : \underline{v} = \underline{w} + \underline{a}\,t, \underline{w} \in F, t \leq \tau_{\underline{w}}\}. \tag{14}$$

Indeed if Y and M are respectively given by (13) and (14) then $\psi \in Y$ and moreover $Z(t)\,g \in Y \; \forall\, t \geq 0$ and $g \in Y$. From this and from formula (6), by virtue of (vi), we can see that $T(t)\,g \in Y \; \forall\, t \geq 0$, $g \in D(A) \cap Y$. But $D(A) \cap Y$ is dense in Y, because Y is Z(t)-invariant ([13] p.14); by continuity we therefore have $T(t)\,g \in Y \; \forall\, t \geq 0$, $g \in Y$ (i. e. Y is T(t) - invariant).

Conversely we now suppose that Y given by (13) is T(t)-invariant and show that the corresponding M is given by (14). In fact from $0 \leq Z(t) \leq T(t)$ it follows that Y is also Z(t)-invariant. All this implies ([13] p. 307 and p. 314) $A(D(A) \cap Y) \subset Y$ and $(A + K)\,(D(A) \cap Y) \subset Y$, and therefore $K(D(A) \cap Y) \subset Y$ too. We thus have $\psi \in Y$. Moreover, if $\underline{v} \in M$, then $\forall\, t > 0$, $\underline{v} - \underline{a}\,t \in M$ too, as a further consequence of the Z(t) - invariance of Y. These results show that M is given by (14). But each set M given by (14) is contained in N and therefore each Y given by (13) and T(t) - invariant contains Y_N. This proves that $T(t)_{|\, Y_N}$ is an irreducible semigroup having $(A + K)_{|\, Y_N}$ as its generator ([13] p. 14). In the particular case $m(N) = 0$ we have $Y_N = X$ and the semigroup T(t) is itself irreducible.

(jk) In order to prove that there exist $t > 0$ and $\tau > 0$ so that (12) is verified, we first observe that $|g| \leq \chi$ implies $|Z(t)\,g| \leq Z(t)\,\chi$ and moreover $Z(t)g \in D(K) \; \forall\, t \geq 0$. Therefore we can express T(t) g by using formula (5); we thus obtain (thanks to (vi))

$$T(t)\,g \; - \; T(t + \tau)\,g \; = \; Z(t)\,g \; - \; Z(t + \tau)\,g \; +$$

$$-\int_0^t T(t-s)\,[T(\tau) - I]\,\psi\,[\int_{R^3} v\,Z(s)\,g\,d\underline{v}]\,ds \; +$$

$$-\int_t^{t+\tau} [T(t + \tau - s)\,\psi]\,[\int_{R^3} v\,Z(s)\,g\,d\underline{v}]\,ds$$

and, from this,

$$|T(t)\,g - T(t + \tau)\,g| \; \leq \; Z(t)\,\chi \; + \; Z(t + \tau)\,\chi + \int_0^t |T(t-s)\,(T(\tau) - I)\,\psi|\,\|\,v\,Z(s)\,\chi\,\|\,ds$$

$$+\int_t^{t+\tau} [T(t + \tau - s)\,\psi]\,\|\,v\,Z(s)\,\chi\,\|\,ds$$

Because the right-hand side of this inequality is independent of g, it is also greater than $| T(t) - T(t + \tau) | \chi$ and therefore

$$\| | T(t) - T(t + \tau) | \chi \| \leq 2 \| Z(t) \chi \| + \| (T(\tau) - I) \psi \| \| \chi \|$$

But the right-hand side of the last inequality is surely less than $\| \chi \|$ if t is big enough and $\tau > 0$ is near to zero. For such t and τ (12) is obviously verified.

APPENDIX

All the statements contained in Prop. 1 and 2 can be easily obtained as consequences of the following considerations. Having arbitrarily fixed $\underline{v} \in R^3$ and t > 0 , only one of these three cases can occur:

$(\alpha) \; \underline{v}. \underline{a} < 0 ; \quad (\beta) \, \underline{v}. \underline{a} > a^2 t; \quad (\gamma) \, 0 \leq \underline{v} \cdot \underline{a} \leq a^2 t$

In the case (α) we have

$$\frac{1}{a} \int_{v}^{|\underline{v} - \underline{a} t|} \nu(v') \, dv' \leq \int_{0}^{t} \nu(|\underline{v} - \underline{a} s|) \, ds \leq \frac{v}{|\underline{v} \cdot \underline{a}|} \int_{v}^{|\underline{v} - \underline{a} t|} \nu(v') \, dv' \quad (15)$$

In the case (β) we obtain

$$\frac{1}{a} \int_{|\underline{v} - \underline{a} t|}^{v} \nu(v') \, dv' \leq \int_{0}^{t} \nu(|\underline{v} - \underline{a} s|) \, ds \leq \frac{|\underline{v} - \underline{a} t|}{|(\underline{v} - \underline{a} t) \cdot \underline{a}|} \int_{|\underline{v} - \underline{a} t|}^{v} \nu(v') \, dv' \quad ($$

and finally in the case (γ)

$$\int_{0}^{t} \nu(|\underline{v} - \underline{a} s|) \, ds = \int_{|\underline{v} \times \underline{e}|}^{v} \nu(v') \, v' \, [a^2 v'^2 - (\underline{v} \times \underline{a})^2]^{-1/2} \, dv' +$$

$$+ \int_{|\underline{v} \times \underline{e}|}^{|\underline{v} - \underline{a} t|} \nu(v') \, v' \, [a^2 \, v'^2 - (\underline{v} \times \underline{a})^2]^{-1/2} \, dv'$$

Therefore in this last case we have

$$\int_0^t v\,(\,|\,\underline{v}-\underline{a}\,s\,|\,)\,ds \geq \frac{1}{a}\,\left[\,\int_{|\underline{v}\times\underline{e}|}^v v\,(v')\,dv' + \int_{|\underline{v}\times\underline{e}|}^{|\underline{v}-\underline{a}\,t|} v\,(v')\,dv'\,\right] \qquad (17)$$

If we suppose that $w := |\,\underline{v}\times\underline{e}\,| \notin J$ and if we fix $h > 0$ so that $L := \operatorname{esssup}\{\,v(v') : w \leq v' \leq w + h\,\} < +\infty$, then we also have

$$\int_0^t v\,(\,|\,\underline{v}-\underline{a}\,s\,|\,)\,ds \leq 2\,\frac{L}{a}\,\sqrt{h^2 + 2\,hw}\ +$$

$$+\,\frac{w+h}{a\sqrt{h^2 + 2hw}}\,[\,H\,(\,v,\,w+h\,) + H\,(\,|\,\underline{v}-\underline{a}\,t\,|,\,w+h\,)\,] \qquad (18)$$

where $H\,(v\,,v') := \begin{cases} 0 \text{ if } v \leq v' \\ \int_{v'}^v v\,(v'')\,dv'' \text{ if } v > v' \end{cases}$

From the previous calculations we see that E is a null set, because

$$E \subseteq \{\,v \in R3 : \underline{v}\cdot\underline{a} = 0\,\} \cup \{\,\underline{v} \in R3 : \underline{v}\cdot\underline{a} \neq 0,\, v \in J\,\}$$

The first statement of Prop. 1 has thus been proved. The other ones can be similarly shown, by using the previous inequalities (15),(16),(17),and (18).

REFERENCES

1. S. Chapman and T.G. Cowling, The mathematical theory of non-uniform gases, Cambridge, Cambridge University Press (1970).

2. S. L. Paveri-Fontana, The moment method applied to the validation of Davydov's approximate results for weakly ionized gases, Ninth International Symposium on Rarefied gas dynamics, Göttingen (1974).

3. G. Cavalleri and S. L. Paveri-Fontana, Phys. Rev. A 6, 327-333 (1972).

4. M. C. Mackey, Biophys. J., 11, 75-95 (1971).

5. K. Kumar, H. R. Skullerud and R.E. Robson, Aust. J. Phys. 33, 343-448 (1980).

6. K. Kumar, Phys. Rep. 112, 319-375 (1985).

7. L. Arlotti, On the linear transport equation for rigid spheres , Transport Theory Stat. Phys., 17, 271-281 (1988)

8. T. Kato, Perturbation theory for linear operators, Berlin, Springer-Verlag (1966).

9. J. Voigt, J. Math. Anal. Appl. 106, 140-153 (1985).

10. G. Frosali, C. V. M. van der Mee and S. L. Paveri-Fontana, Conditions for runaway phenomena in the kinetic theory of particle swarms (manuscript).

11. M. Kac, Foundations of kinetic theory, in: J. Neyman (ed.), Proceedings of the third Berkeley Symposium on Mathematical Statistics and Probability, Vol.III, 171-197, University of California Press (1956).

12. J. Voigt, Functional Analytic Treatment of the Initial-Boundary-Value Problem for Collisionless Gases, Habilitationsschrift Universität München (1980).

13. R. Nagel (ed.), One-parameter semigroups of positive operators, Lecture Notes in Mathematics 1184, Berlin, Springer Verlag (1986).

14. E. B. Davies, One-Parameter Semigroups, London, Academic Press (1980).

Dispersion Characteristics and Landau Damping in a Plasma with an External Field

V. G. MOLINARI and P. PEERANI   Nuclear Engineering Laboratory, University of Bologna, Bologna, Italy

1. INTRODUCTION

The influence of external fields on plasma dynamics can produce additional instabilities and dampings through the interaction of waves that other-wise would propagate freely [1-3]. The aim of this paper is to analyse, in to the framework of kinetic theory, the change of the dispersion characte-ristics caused by a monochromatic external wave having phase velocity $\to\infty$, and to show how the coupling between electron plasma oscillations arises when an external field is present.

When the Laplace transform is inverted in order to obtain the time behavior of the electric field, an integral equation is obtained so that the poles of the integrals do not give directly the dispersion relation, but the characteristics of the wave propagation are determined by the coupling of waves as follows from the integral equation.

2. FORMULATION OF THE PROBLEM

The general Boltzmann equation for electron distribution function in a plasma,

$$\frac{\partial f}{\partial t} + \bar{v} \cdot \bar{\nabla} f + \frac{\vec{F}}{m} \cdot \frac{\partial f}{\partial \bar{v}} =$$

$$= \int_{R_3} K[\bar{v}' \to \bar{v}] \, f(\bar{v}') d\bar{v}' - \nu f, \tag{1}$$

can be put in a simplified integral operational form [4], if the following hypotheses are introduced:

- ions form a positive fixed background of neutralizing charges;
- interactions between charged particles are taken into account by the self-consistent electric field in the force term of Boltzmann equation (Vlasov theory);
- interactions of electrons with neutral atoms are on the contrary trea_ted in the collisional term by a suitable kernel (B.G.K. model);
- only electrostatic oscillations with long wavelength are considered.

With the previous assumptions eq(1) reduces to the form

$$f(\bar{r},\bar{v},t) = \nu \left[\int_0^\infty e^{-(\nu+\mathscr{D}+\bar{v}.\bar{\nabla})\tau} (\frac{\beta}{\pi})^{3/2} e^{-\beta(\bar{v}-v*(\tau))^2} d\tau \right] n \qquad (2)$$

where n= fdv is the electron density, $\beta = m_e/2kTe, \mathscr{D} \equiv \partial/\partial t$ and

$$\bar{v}*(\tau) = -\frac{e}{m} \int_0^\tau \bar{E}(\bar{r}-\bar{v}\tau',t-\tau')d\tau' \qquad (3)$$

results by the integration of the motion equation.

Eq. (2) is rather difficult to handle in its complete form; a useful approximation can be obtained by remarking that the term $\bar{v}*$ is proportional to the electric field $\bar{E}$, so if we restrict our analysis to the oscillations of small amplitude, a suitable expression to study weakly non-linear electrostatic phenomena in plasmas results by a siple expansion

$$f = \nu(\frac{\beta}{\pi})^{3/2} e^{-\beta v^2} \left[\int_0^\infty e^{-(\nu+\mathscr{D}+\bar{v}.\bar{\nabla})\tau} (1+2\beta\bar{v}.\bar{v}*(\tau))d\tau \right] n. \qquad (4)$$

The relation between the distribution function and the electric field is given by Maxwell equation in the electrostatic approximation

$$\mathscr{D}\bar{E} = -\frac{1}{\varepsilon_o} \bar{J} = \frac{e}{\varepsilon_o} \int_{R_3} f \bar{v} d\bar{v}$$

leading to the final form

$$\mathscr{D}\bar{E} = \frac{\nu e}{\varepsilon_o} \int_{R_3} d\bar{v}(\frac{\beta}{\pi})^{3/2} \bar{v} e^{-\beta v^2} \int_0^\infty e^{-(\nu+\mathscr{D}+\bar{v}.\bar{\nabla})\tau}(1+2\beta\bar{v}.\bar{v}*(\tau))d\tau \quad n. \qquad (5)$$

We will apply now this equation to the study of a plasma with an external electric field (for instance an electromagnetic wave propagating in a plasma) and of the effects arising from the interaction between the external wave and the intrinsic oscillating modes characteristic of plasma itself.

In this case the total electric field $\bar{E}$ is given by the sum of the external field $\bar{E}_o$ associated to the propagating wave and the self-consistent electric field $\bar{\varepsilon}$ connected with the spatial distribution of the charges of the plasma; assuming for the external field the form of a monochromatic wave, we have:

$$\bar{E}(\bar{r},t) = \bar{E}_o(\bar{r})\ e^{-i\omega_0 t} + \bar{\varepsilon}(\bar{r},t) \tag{6}$$

To allow the further Fourier transform, we assume for the space dependence of $\bar{E}_o(\bar{r})$ a properly simple form (f.i. $\bar{E}_o(r) = \bar{E}_o\ e^{ik_0 x}$); this arbitrary choice will be removed in the final analysis with the assumption of large phase vleocity ($v_\phi \to \infty$, $k_0 \to o$).

Likewise, according to eq.(3), also $\bar{v}*$ results by the sum of two terms

$$\bar{v}*(\tau) = \bar{v}_o + \bar{v}*_1 =$$

$$= -\frac{e}{m}\ \bar{E}_o(x,t) \left[\frac{1-e^{-i(k_0 v-\omega)\tau}}{i(k_0 v - \omega_0)}\right] - \frac{e}{m}\int_0^\tau \bar{\varepsilon}(\bar{r}-\bar{v}\tau',t-\tau')d\tau' \tag{7}$$

3. RESULTS

a) The linear case

As a first approach to the problem, we could consider the case of a constant density $n(\bar{r},t) = n_o$. With this approximation, eq.(5) with $\bar{E}$ and $\bar{v}*$ expressed respectively by relations (6) and (7) becomes linear and assumes the form

$$\mathscr{D}\bar{\varepsilon} - i\omega_o \bar{E}_o\ e^{i(k_0 x-\omega_0 t)} = I_o + I_1 \tag{8}$$

where

$$I_o = 2\beta\nu\ \frac{e}{\varepsilon_o}\int_{R_3} d\bar{v}(\frac{\beta}{\pi})^{3/2}\ v^2\ e^{-\beta v^2}\int_0^\infty e^{-\nu\tau}\ \bar{v}_o(\tau)d\tau \quad n_o =$$

$$= -2\beta\ w_p^2 \sqrt{\frac{\beta}{\pi}}\int_{-\infty}^{+\infty}\frac{u^2\ e^{-\beta u^2}}{\nu+i(k_0 u-\omega_0)}\ du\ \bar{E}_o\ e^{i(k_0 x-\omega_0 t)}$$

and

$$I_1 = 2\beta\nu \frac{e}{\varepsilon_o} \int_{R_3} d\bar{v} (\frac{\beta}{\pi})^{3/2} v^2 e^{-\beta v^2} \int_0^\infty e^{-\nu\tau} \bar{v}_1(\tau) d\tau \ n_o =$$

$$= -2\beta\nu \ \omega_p^2 \sqrt{\frac{\beta}{\pi}} \int_{-\infty}^{+\infty} du \ u^2 e^{-\beta u^2} \int_0^\infty d\tau \ e^{-\nu\tau} \int_0^\tau \bar{\varepsilon}(\bar{r}-\bar{v}\tau', t-\tau') d\tau'$$

with ω_p = plasma frequency = $\sqrt{n_o e^2/m\varepsilon_o}$ and u = component of the velocity parallel to the electric field.

Eq.(8) can be simply solved by taking the Fourier transform with respect to space and Laplace transform with respect to time; this leads to

$$s\overset{\approx}{\varepsilon} - \overset{\sim}{\varepsilon}_o - \frac{i\omega_s E_o}{(s+i\omega_o)(k-k_o)} = -2\beta \ \omega_p^2 \sqrt{\frac{\beta}{\pi}} \int_{-\infty}^{+\infty} \frac{u^2 e^{-\beta u^2}}{\nu+i(k_o u-\omega_o)} du \ \frac{E_o}{(s+i\omega_o)(k-k_o)}$$

$$-2\beta\omega_p^2 \sqrt{\frac{\beta}{\pi}} \int_{-\infty}^{+\infty} \frac{u^2 e^{-\beta u^2}}{\nu+s+iku} du \ \overset{\approx}{\varepsilon}$$

By putting

$$\Delta(k,s) \equiv s + 2\beta \ \omega_p^2 \sqrt{\frac{\beta}{\pi}} \int_{-\infty}^{+\infty} \frac{u^2 e^{-\beta u^2}}{\nu+s+iku} du$$

and

$$A_o(k_o,\omega_o) \equiv i\omega_o - 2\beta\omega_p^2 \sqrt{\frac{\beta}{\pi}} \int_{-\infty}^{+\infty} \frac{u^2 e^{-\beta u^2}}{\nu+i(k_o u-\omega_o)} du$$

we have

$$\overset{\approx}{\varepsilon}(k,s) = \frac{\overset{\sim}{\varepsilon}_o(k)}{\Delta(k,s)} + \frac{A_o E_o}{(s+i\omega_o)(k-k_o)\Delta(k,s)} \tag{9}$$

The time behavior of the electric field is then given by the inverse of Laplace transformation of eq.(9)

$$\overset{\sim}{\varepsilon}(k,t) = \frac{1}{2\pi i} \int \frac{\varepsilon_o(k)}{\Delta(k,s)} e^{st} ds + \frac{A_o E_o}{2\pi i} \int \frac{e^{st} ds}{(k-k_o)(s+i\omega_o)\Delta(k,s)}$$

and the normal modes are the poles of the integral given by solving the dispersion relation $\Delta(k,s) = 0$.

This relation is the same already obtained in reference [4], when electrostatic oscillations in a field-free plasma were analyzed; in that occasion we have shown some approximate solutions revealing the presence of Landau damping and the effects of short range collisions. In this case in addition to zeros of the dispersion relation, we have an additional pole $s_o = -i\omega_o$ in the second term of the r.h.s. of eq.(9).

So, calling s_j (with $j = 1, \ldots, n$) the zeros of Δ and c_j the residuals in the poles $c_j = \text{Res } 1/\Delta \big|_{s=s_j}$ the inverse transformation results

$$\tilde{\varepsilon}(k,t) = \sum_{1}^{n} {}_j \; c_j \left[\tilde{\varepsilon}_o + \frac{A_o E_o}{(k-k_o)(s_j+i\omega_o)} \right] e^{s_j t} + \frac{E_o e^{-i\omega_o t}}{(k-k_o)} \qquad (10)$$

We can remark how the oscillations at the forcing frequency ω_o and at the characteristic frequencies $\text{Im}(s_j)$ are completely independent; at this level of approximation there is no coupling between waves and no effect of the external field on dispersion characteristics of plasma and on Landau damping.

b) The non-linear (linearized) case

When we take for electron density the expression resulting from the Gauss equation

$$\bar{\nabla} \cdot \bar{\varepsilon} = - \frac{e}{\varepsilon_o} (n_o - n)$$

we introduce into eq. (5) a non-linear term; in fact we have

$$\mathscr{D}\bar{\varepsilon} - i\omega_o \bar{E}_o(x,t) = I_{00} + I_{01} + I_{11} \qquad (11)$$

where I_{00} and I_{10} are equal to I_0 and I_1 in eq.(8) and

$$I_{01} = \nu \int d\bar{v} \; (\frac{\beta}{\pi})^{3/2} \; \bar{v} \; e^{-\beta v^2} \int_{o}^{\infty} e^{-(\nu+\mathscr{D}+\bar{v}\cdot\bar{\nabla})\tau} \left[1+2\beta\bar{v}\cdot\bar{v}_o(\tau)\right] d\tau \; \bar{\nabla}\cdot\bar{\varepsilon}$$

$$I_{11} = \nu \int d\bar{v} \; (\frac{\beta}{\pi})^{3/2} \; \bar{v} \; e^{-\beta v^2} \int_{o}^{\infty} e^{-(\nu+\mathscr{D}+\bar{v}\cdot\bar{\nabla})\tau} \left[1+2\beta\bar{v}\cdot\bar{v}_1(\tau)\right] d\tau \; \bar{\nabla}\cdot\bar{\varepsilon}$$

The non-linearity comes from the term I_{11} where the operator applied to the unknown $\bar{\nabla}\cdot\bar{\varepsilon}$ depends on $\bar{\varepsilon}$ itself through $\bar{v}_1$. But if we suppose that the self-consistent electric field is small when compared to the external field, we could disregard this last term, and eq.(11) results so lineari-

Therefore it is possible to apply the same technique of the previous case, and get to the relation

$$\overset{\sim}{\varepsilon}(k,s) = \frac{\overset{\sim}{\varepsilon}_o(k)}{\Delta'(k,s)} + \frac{A_o E_o}{(k-k_o)(s+i\omega_o)\Delta'(k,s)} +$$

$$- 2\beta\nu \frac{e}{m}(k-k_o)E_o \frac{I(k,s;k_o,\omega_o)}{\Delta'(k,s)} \overset{\sim}{\varepsilon}(k-k_o,s+i\omega_o) \qquad (12)$$

where

$$\Delta'(k,s) \equiv s+2\beta w_p^2 \int_{-\infty}^{+\infty} \sqrt{\frac{\beta}{\pi}} \frac{u^2 e^{-\beta u^2}}{\nu+s+iku} \, du - ik\nu \int_{-\infty}^{+\infty} \sqrt{\frac{\beta}{\pi}} \frac{u \, e^{-\beta u^2}}{\nu+s+iku} \, du$$

and

$$I(k,s;k_o,\omega_o) \equiv \int_{-\infty}^{+\infty} \sqrt{\frac{\beta}{\pi}} \frac{u^2 e^{-\beta u^2} \, du}{(\nu+s+iku)[\nu+s+i\omega_o+i(k-k_o)u]}$$

We can see that the denominator Δ' shows an additional term with respect to the one of the previous case; nevertheless it is possible to verify that the approximated solutions of the relation $\Delta' = 0$, according to the methods outlined in reference [4], are the same of $\Delta=0$, being the correcti̲ve term comparable to the terms disregarded in the approximation. We can consider therefore that the poles of the r.h.s. of eq.(12) are about the same than the ones of eq.(9).

So the inverse transformation of eq.(12) differs from eq.(10) because of the presence of the last term, which gives rise to an integral term. In fact if we define

$$h(t) = L^{-1}[\frac{I}{\Delta'}] = \frac{1}{2\pi i} \int \frac{I}{\Delta'} e^{st} \, ds = \sum_{ij}^{n} b_j \, e^{s_j t}$$

where $\qquad b_j = \text{Res}[\frac{I}{\Delta'}]_{s=s_j} = \text{Res}[\frac{I}{\Delta}]_{s=s_j} = c_j I(k,s_j)$

the inverse transform of the last term results in a convolution product

$$L^{-1}[\frac{I}{\Delta'}\overset{\sim}{\varepsilon}(k-k_o,s+i\omega_o)] = L^{-1}[\frac{I}{\Delta'}] * L^{-1}[\overset{\sim}{\varepsilon}(k-k_o,s+i\omega_o)] =$$

$$= h(t) * [e^{-i\omega_o t}\overset{\sim}{\varepsilon}(k-k_o,t)] = \int_0^t h(t-\tau) e^{-i\omega_o\tau} \overset{\sim}{\varepsilon}(k-k_o,\tau)d\tau$$

and finally we have

$$\tilde{\varepsilon}(k,t) = \sum_{o}^{n} j \quad c_j [\tilde{\varepsilon}_o + \frac{A_o E_o}{(k-k_o)(s_j+i\omega_o)}] e^{s_j t} + \frac{E_o e^{-i\omega_o t}}{(k-k_o)} +$$

(13)

$$-2\beta\nu \frac{e}{m} (k-k_o)E_o \sum_{i}^{n} j \quad b_j \int_0^t e^{s_j(t-\tau)} e^{-i\omega_o \tau} \tilde{\varepsilon}(k-k_o,\tau)d\tau$$

We would point out that the last integral term is the one which introduces the coupling effects between waves. Moreover the dispersion equation is no longer given by $\Delta'=0$, but it can just be obtained by the solution of the integral equation (13).

As an example we report the solution when we consider only the contribution of the pole with smallest real part, which is the long-time persistent component, with the further simplification $k \gg k_o$ in the last term; eq.(13) reduces to

$$\tilde{\varepsilon}(k,t) = c_1 [\tilde{\varepsilon}_o + \frac{A_o E_o}{(k-k_o)(s_1+i\omega_o)}] e^{s_1 t} + \frac{E_o e^{-i\omega_o t}}{(k-k_o)} +$$

(14)

$$-2\beta\nu \frac{e}{m} k E_o b_1 e^{s_1 t} \int_0^t e^{-(s_1+i\omega_o)\tau} \tilde{\varepsilon}(k,\tau)d\tau$$

This integral equation can be simply turned to a differential one, by multiplying per $e^{-s_1 t}$ and derivating, obtaining so

$$\frac{d\tilde{\varepsilon}}{dt} - (s_1 - \alpha e^{-i\omega_o t})\tilde{\varepsilon} = -\frac{(s_1+i\omega_o)}{(k-k_o)} E_o e^{-i\omega_o t}$$

(15)

where $\alpha = 2\beta\nu \frac{e}{m} k E_o b_1$,

whose solution is of the kind

$$\tilde{\varepsilon} = [A + B e^{-\frac{i\alpha}{w_o} e^{-i\omega_o t}}] e^{s_1 t}$$

(16)

4. CONCLUSIONS

We have proved that an external electric field produces a coupling between electron waves, mathematically shown by the integral equation which arises when the Laplace transform is inverted.

In the example we have considered, the interaction with a single wave doesn't introduce new oscillating modes, but nevertheless results in a modulation of the amplitude of the intrinsic plasma oscillations caused by the external wave.

In a further development we intend to take into account the coupling effects between two or more electron waves with the purpose to point out the conditions for the instability decay.

REFERENCES

1. V.P. Silin, Parametric resonance in a plasma, Soviet Physics-JETP, 21: 1127-1134 (1965).
2. K. Nishikawa, Parametric excitation of coupled waves, I general formulation, J. Phys. Soc. Japan, 24: 916-922 (1068); II parametric plasmon-photon interaction, J. Phys. Soc. Japan, 24: 1152-1158 (1968).
3. V.N. Tsytovich, "Nonlinear effects in plasma", Plenum Press, New York (1970).
4. V.G. Molinari and P. Peerani, Electron oscillations in a plasma: effect of collisions on Landau damping, Nuovo Cimento, 5: 527-540 (1985).

Duality of Backward and Forward Transport Formulations in Radiation Damage

A. K. PRINJA   Chemical and Nuclear Engineering Department, University of New Mexico, Albuquerque, New Mexico

A variational principle is developed and used to establish a rigorous and systematic link between the forward and backward transport approaches in radiation damage work. Quite general boundary conditions are incorporated and the variational principle is shown to yield the adjoint system rather naturally and mechanically. Ad hoc physical interpretations of the adjoint flux, in the form of the importance function, are thus shown to be unnecessary.

1. INTRODUCTION

Radiation damage research involves the study of interaction of neutral and ionized atoms with the atoms of a target material (1) and the subsequent transport of these particles. To the extent that the medium can be represented as a random collection of scattering centres (not necessarily a good assumption in practice when single crystals are considered) the transport of incident and struck or recoiling atoms falls under the purview of linear transport theory (2). The pioneering work of Lindhard (3), greatly extended by Sanders (4), Winterbon (5) and Sigmund (6), formed the basis of the so−called "backward" transport approach to radiation damage problems − equations were written down directly for the quantity of interest, such as the reflection coefficient (or albedo), energy deposited in atomic motion, range profiles, etc. However, the approach was recognized as being limited to homogeneous infinite targets (7), i.e. free surfaces, interfaces and space−time variation of medium properties could not be incorporated. Despite these restrictions, considerable insight has been gained over the years using the backward equations, and under certain circustances, results of practical value have also been obtained.

More recently, Williams (8) has applied the Boltzmann equation for the angular flux to the transport of incident and recoiling atoms. As is well known from neutron transport, this approach, known as the "forward" transport approach, is very powerful and can incorporate quite general boundary conditions. It is beyond the scope of this paper to present a detailed exposition or critique of the two methods.

An extensive review may be found in the recent paper of Williams (9). Extensive applications of the forward equation have been made, to, for instance, range profile (10), energy deposition (11) and sputtering (12) calculations. However, the approach' is relatively new and has not found widespread use yet, although it has much to commend it.

Williams (9) has noted a link between the backward equation and the adjoint equation, but developed this only for infinite media, i.e. without regard to realistic boundary conditions. However, free surface boundary conditions are an important distinguishing and complicating feature of radiation damage problems. As will be discussed later, outgoing particles can experience internal reflection at free surfaces, and possibly at interfaces between different media. Whereas the formulation of these boundary conditions in forward form is quite straightforward, they are invariably glossed over in the backward approach. Thus, a systematic and rigorous link between the various transport theories is lacking.

In what follows, we develop such a link between the forward Boltzmann, "full backward" (i.e. adjoint) and "partial backward" (a'la Lindhard) formulations. At the core of this hierarchy is a Variational Principle, which is constructed to incorporate complex boundary conditions and which yields as the Euler–Lagrange equations the forward and full backward transport equations. The partial backward equations are subsequently derived. The virtues of a systematically constructed variational characterization over the more ad hoc prescriptions of the "Importance" function approach (13), popular in neutron transport, in generating adjoint formulations is briefly discussed.

2. THE VARIATIONAL PRINCIPLE AND ADJOINT SYSTEM

In this section we demonstrate through the agency of a Variational Principle that the adjoint system (i.e. adjoint equation, adjoint boundary conditions and sources) can be systematically and rather mechanically constructed for linear transport problems. We consider a general reflecting boundary condition on the angular flux at a free surface, which is representative of charged particle or neutral atom transport but may also arise in certain neutron transport modelling situations (e.g. for adjacent media with differing properties one of the media may be replaced by a reflecting condition at the interface if the albedo or reflection coefficient for that medium is known). Without loss of generality, we consider a one dimensional half space with phase space variable set $p = (x,E,\mu,t)$, where x measures distance into the half space, μ is the direction cosine, E is the particle energy and t is the time. The Boltzmann equation may be expressed compactly as:

$$\hat{B}\phi = \delta(x-x_0)\ \delta(E-E_0)\ \delta(\mu-\mu_0)\ \delta(t-t_0) = \delta(p-p_0) = Q(p), \tag{1}$$

where $\phi(x,E,\mu,t;\ x_0,E_0,\mu_0,t_0)\ (\equiv \phi(p;p_0))$ is the particle angular flux (in this case also the Greens function) and $\hat{B}$ is the Boltzmann operator, given by:

$$\hat{B} \equiv \frac{1}{v}\frac{\partial}{\partial t} + \mu\frac{\partial}{\partial x} - \frac{\partial}{\partial E}\ Se(E) - \hat{S} - \hat{R}. \tag{2}$$

Here $\hat{S}$ and $\hat{R}$ represent elastic scattering and recoil operators and, as usual, are integral operators with differential scattering cross–sections as kernels (9). We have considered only a single species, but the analysis may be readily generalized to multiple species. Also, $Se(E)$ represents the kernel for the continuous slowing down operator, which in charged particle transport represents inelastic interactions with the target atomic electrons (excitation and ionization). In general one may include any suitable Fokker–Planck operator, e.g. for small angle deflection, or energy straggling (diffusion).

A suitably general boundary condition may be expressed as:

$$\phi(0,E,\mu,t) = \int_0^{E_o} dE' \int_{-1}^{0} d\mu' \, \widehat{H} \, (E'-E_B) \, \widehat{H} \, (\mu'+\sqrt{E_B/E'}) \, \cdot$$

(3)

$$\cdot \, \Gamma(\mu' \to \mu; E' \to E) \, \phi(0,E',\mu',t), \qquad \mu > 0.$$

In radiation damage this boundary condition represents an anisotropic potential barrier at the surface (magnitude given by the surface binding energy, E_B) (6) that imposes a constraint on the outgoing particle's energy and angle (given by the step functions, $\widehat{H}(\mu)$), and which, if not satisfied, will result in internal reflection according to some general law Γ. Eliminating the step functions will yield a boundary condition suitable for neutrons perhaps. One could also add a term representing an incident source. In compact notation Eq. (3) may be written as:

$$\hat{H}\phi = 0; \quad x = 0, \quad \mu > 0.$$

(4)

We also prescribe an initial condition:

$$\phi(x,E,\mu,0) = g(x,E,\mu).$$

(5)

Eqs. (1)–(5) completely specify the transport problem in the forward formalism. Our objective is to generate from this the popular "partial" backward equations. In order to achieve this, we first develop the corresponding adjoint description, which we will refer to as the "full" backward description.

Given the complex boundary condition, Eq. (3), it does not seem to us very probable that the corresponding adjoint boundary condition can be written down easily using the physically motivated "Importance" arguments (13), which have worked quite well in the simpler neutron context of zero or prescribed incoming flux. We prefer to adopt a more systematic approach that does not rely on questionable physical interpretations of what is simply a mathematical artefact, the adjoint flux.

Thus, consider a functional of the angular flux:

$$F[\phi] = \int_0^T dt \int_0^{E_o} dE \int_{-1}^{0} d\mu \int_0^{\infty} dx \, f(x,E,\mu,t) \, \phi(\cdots) = (f,\phi)$$

(6)

where (a,b) is the inner product notation signifying integration over phase space, and f is an arbitrary known function. In particular, we will be interested in $f(p) = \delta(p-p_1)$. We proceed to develop a variational estimate for this functional.

Following Morse and Feshbach (14) and Pomraning (15) we introduce $\tilde{\phi}$, a trial function that is a first order accurate estimate of ϕ, and "conjugate variables" (14) $\tilde{\phi}^+$, $\tilde{\lambda}$, $\tilde{\gamma}$. Consider now the following functional of these variables:

$$G[\tilde{\phi},\tilde{\phi}^+,\tilde{\lambda},\tilde{\gamma}] = F[\tilde{\phi}] - (\tilde{\phi}^+,\hat{B}\tilde{\phi}-Q) - (\tilde{\lambda},\hat{H}\tilde{\phi})_x - (\tilde{\gamma},\tilde{\phi}-g)_t,$$

(7)

i.e. we have constrained the original functional by the defining equation and the boundary and initial conditions, but with all evaluated using the trial function $\tilde{\phi}$. The variables used to constrain the functional are the conjugate variables which in this sense are generalized Lagrange multipliers (or functions). The inner product $(a,b)_x$ is restricted to $x = 0$ and $\mu > 0$ (since it acts on the boundary term), while in $(a,b)_t$ time is fixed at $t = 0$ (the initial condition).

Thus,

$$\tilde{\lambda} \equiv \tilde{\lambda}(E,\mu,t), \;\; \mu > 0$$

$$\tilde{\gamma} \equiv \tilde{\gamma}(x,E,\mu)$$

and $\tilde{\phi}^+$ is defined over the whole phase space, p. We consider the conjugate variables $\tilde{\phi}^+$, $\tilde{\lambda}$, $\tilde{\gamma}$ to be first order estimates of corresponding exact functions ϕ^+, λ and γ respectively. Now, if $G[\cdots]$ is to be a valid variational principle for $F[\cdots]$, it must satisfy two conditions:

1) $G[\cdots]$, when evaluated with the exact function ϕ, must yield the desired functional $F[\phi]$.

2) First order errors in ϕ, ϕ^+, λ and γ must lead to second order errors in the functional $G[\cdots]$.

The first condition is clearly satisfied by virtue of Eqs. (1), (4) and (5):

$$G[\phi,\tilde{\phi}^+,\tilde{\lambda},\tilde{\gamma}] = F[\phi]. \tag{8}$$

The second is satisfied by requiring the first variation of G to vanish when evaluated with the exact solutions ϕ, ϕ^+, λ and γ. This procedure yields self–consistently the defining equations and boundary conditions for the conjugate variables. It is perhaps worth noting that this is standard in the calculus of variations (14) when seeking a variational formulation. The resulting Variational Principle, if valid, would yield as its Euler–Lagrange equations the defining equations. The two formulisms (the defining equations and the Variational Principle) are entirely complementary.

Taking the first variation of $G[\cdots]$ we obtain:

$$\delta G[\tilde{\phi},\tilde{\phi}^+,\tilde{\lambda},\tilde{\gamma}] = (f,\delta\tilde{\phi}) - (\delta\tilde{\phi}^+,\hat{B}\tilde{\phi}-Q) - (\tilde{\phi}^+,\hat{B}\delta\tilde{\phi})$$
$$- (\delta\tilde{\lambda},\hat{H}\tilde{\phi})_x - (\tilde{\lambda},\hat{H}\delta\tilde{\phi})_x - (\delta\tilde{\gamma},\tilde{\phi}-g)_t - (\tilde{\gamma},\delta\tilde{\phi})_t . \tag{9}$$

To proceed further, the third and fifth terms on the right hand side above are manipulated by integration by parts to interchange $\delta\tilde{\phi}$ with $\tilde{\phi}^+$ and $\delta\tilde{\phi}$ with $\tilde{\lambda}$. This results in operators which in general are different from $\hat{B}$ and $\hat{H}$. Also, additional boundary and initial time terms arise, which are then combined with similar terms in Eq. (9). We omit the algebraic details and quote below only the final result obtained after considerable rearrangement:

$$\delta G = (\delta\tilde{\phi},f-\hat{B}^+\tilde{\phi}^+) - (\delta\tilde{\phi}^+,\hat{B}\tilde{\phi}-Q) + (\delta\tilde{\phi},\mu\tilde{\phi}^++\tilde{\lambda})_x + (\delta\tilde{\lambda},\mu\tilde{\phi}^++\hat{H}^+\tilde{\phi}^+)_x$$
$$+ (\delta\tilde{\phi},1/\upsilon\ \tilde{\phi}^+-\gamma)_t - (\delta\tilde{\gamma},\tilde{\phi}-g)_t - (\delta\tilde{\phi},1/\upsilon\ \tilde{\phi}^+)_T. \tag{10}$$

In the last term above, the inner product is restricted to t = T, the final time. The new operators $\hat{B}^+$ and $\hat{H}^+$ are presented below. We now require that the first variation, δG, vanish for independent and arbitrary variations $\delta\tilde{\phi}$, $\delta\tilde{\phi}^+$, $\delta\tilde{\lambda}$ and $\delta\tilde{\gamma}$ around the exact functions ϕ, ϕ^+, λ and γ. This yields the following equations:

with

$$\hat{B}\phi = Q(p) \tag{11a}$$

$$\hat{H}\phi = 0; \ x = 0, \ \mu > 0 \tag{11b}$$

$$\phi = g; \ t = 0 \tag{11c}$$

with

$$\hat{B}^+\phi^+ = f(p) = Q^+(p) \tag{12a}$$

$$\mu\phi^+ + \hat{H}^+\phi^+ = 0; \ x = 0, \ \mu < 0 \tag{12b}$$

$$\phi^+ = 0; \ t = T \tag{12c}$$

Also,

$$\gamma = 1/\upsilon \ \phi^+; \ t = 0 \tag{13}$$

$$\lambda + \mu\phi^+ = 0; \ x = 0, \ \mu > 0 \tag{14}$$

The equation set (11) is clearly just the forward equation and associated boundary and initial conditions, as expected. Eq. (12a) is the corresponding adjoint equation with f, the adjoint source. $\hat{B}^+$ is the adjoint operator given by:

$$\hat{B}^+ \equiv -\frac{1}{\upsilon}\frac{\partial}{\partial t} - \mu\frac{\partial}{\partial x} + Se(E)\frac{\partial}{\partial E} - \hat{S}^+ - \hat{R}^+ \ . \tag{15}$$

where $\hat{S}^+$ and $\hat{R}^+$ are adjoint scattering and recoil integral operators, obtained in the usual way. The interesting result is the adjoint boundary condition given by Eq. (12b) which, explicitly stated, reads:

$$\mu\phi^+(0,E,\mu,t) + \text{(H)} \ (E-E_B) \ \text{(H)} \ (\mu + \sqrt{E_B/E} \ ) \int_0^{E_o} dE' \int_0^1 d\mu' \ \bullet$$

$$\tag{16}$$

$$\bullet \ \mu' \ \Gamma(\mu \to \mu'; \ E \to E') \ \phi^+(0,E',\mu',t) = 0, \quad \mu < 0 \ .$$

For no reflection, $\Gamma = 0$ and Equation (16) reduces to $\phi^+(0,E,\mu,t) = 0$, $\mu < 0$, interpreted in the traditional sense as zero "importance" for particles leaving the medium. Generalizing this concept to the case with reflection, i.e. Eq. (16), does not appear to be obvious. The fixed time condition for the adjoint flux becomes a final condition at t = T, emphasizing the backward nature of the adjoint equation. Finally, the remaining conjugate variables or Lagrange multipliers γ and λ are simply related to the adjoint flux as shown in Eqs. (13) and (14). Note that γ and λ are not required for the solution of ϕ and ϕ^+. However, a knowledge of γ and λ is necessary for obtaining second order accurate estimates of $F[\cdots]$, evident in Eq. (7), although this is again not the case if the trial function $\tilde{\phi}$ is chosen to satisfy the exact boundary and initial conditions.

3. THE BACKWARD FORMULATION

A useful reciprocity relation exists between the forward and adjoint equations (9). Subtracting the inner product of ϕ^+ with the forward equation, Eq. (11a), from the inner product of ϕ with the adjoint equation, Eq. (12a), and simplifying using boundary and initial conditions, we obtain:

$$(\phi^+, Q) = (\phi, Q^+). \tag{17}$$

With Q given by Eq. (1) and letting $Q^+ = \delta(p-p_1)$, Eq. (17) reduces to:

$$\phi^+(p;p_1) = \phi(p_1;p), \tag{18}$$

i.e. the adjoint flux in p due to a source in p_1 is equal to the forward flux in p_1 due to a source in p. This reciprocity relation can be used to replace the adjoint flux in the adjoint equation with the forward flux but with interchanged field and source variables:

$$\hat{B}^+(p) \; \phi(p_1;p) = \delta(p-p_1). \tag{19}$$

We now note that $\hat{B}^+$ operates on the source variable, p, leaving the field variables, p_1, as parameters. Thus, we may freely operate on the latter without affecting the adjoint operator. The advantage of this is that an equation for the desired physical quantity may be written down directly. For instance, such a quantity is typically a functional of the forward flux:

$$R(\bar{p}_1;p) = \int_{\Omega_1} dp_1 \; H(p_1) \; \phi(p_1;p), \tag{20}$$

where $\bar{p}_1$ is a subset of p_1, and Ω_1 is the volume in p_1 over which the response is desired. Thus operating on Eq. (19) gives:

$$\hat{B}^+(p) \; R(\bar{p}_1;p) = H(p) \; \Delta_{\Omega_1}(p) \tag{21}$$

where $\Delta_{\Omega_1}(p) = 1$, for $p \in \Omega_1$ and 0, otherwise. The equation for R becomes, explicitly:

$$\left[-\frac{1}{v}\frac{\partial}{\partial t} - \mu\frac{\partial}{\partial x} + Se(E)\frac{\partial}{\partial E} - \hat{S}^+ - \hat{R}^+ \right] R(\bar{p}_1; x,E,\mu, t) =$$

$$= H(x,E,\mu, t) \; \Delta_{\Omega_1}(x,E,\mu, t). \tag{22}$$

Corresponding boundary conditions on R may be obtained by similarly operating on the adjoint boundary conditions. We refer to Eq. (22) as the "full" backward equation for the desired quantity R, and is really just the adjoint equation. Depending on H, the function R may represent surface quantities (e.g. reflection coefficient) or internal quantities like energy deposition rates or range profiles. However, as it stands Eq. (22) is not of the form commonly encountered in the radiation damage literature. The difference is that in the equation above it is the source variables that are operational and not the field variables. This results in a non-canonical equation which is not appealing for physical problems, and the full backward equation is seldom used. To complete our hierarchy we develop the more popular form that we refer to as the "partial" backward equation. If we now restrict attention to infinite, space–time homogeneous media, then, as both Williams (9) and Winterbon (7) have noted translational invariance can be used to transform Eq. (22). This can be demonstrated as follows. Let R be desired at some field point $p_1 = (x_1, E_1, \mu_1, t_1)$, i.e.

$\Delta\Omega_1(p) = \delta(p-p_1)$. Then, it follows from homogeneity that the solution can depend only on $|x-x_1|$ and $|t_1-t|$, which is an expression of translational invariance in space and time. In this case we may use the transformation:

$$\frac{\partial}{\partial x} \equiv -\frac{\partial}{\partial x_1} \; ; \quad \frac{\partial}{\partial t} \equiv -\frac{\partial}{\partial t_1}$$

in Eq. (22).

The equation for R now takes the form popularly used in radiation damage work:

$$\left[\frac{1}{v}\frac{\partial}{\partial t} + \mu\frac{\partial}{\partial x_1} + Se(E)\frac{\partial}{\partial E} - \hat{S}^+ - \hat{R}^+ \right] R(p_1; p) = H(p)\,\delta(p-p_1). \quad (23)$$

valid only for an infinite medium, where the complex boundary conditions are not incorporated.

The "partial" backward nature of this equation is now clear. It is a curious mix of operational source and field variables and reflects the fact that in practical situations it is the initial source energy and angle variables that are the control variables while the information is desired at some final or field point. The price paid is the restriction to homogeneous media; for more realistic situations one must resort either to the "full" backward (i.e.adjoint) or forward transport formalism.

4. DISCUSSION AND CONCLUSIONS

A general result of our work has been to demonstrate the use of variational formulation of linear transport problems to generate self–consitently the adjoint system. Standard concepts in calculus of variations were used to incorporate generalized boundary conditions which, in rather a natural way, led to corresponding adjoint boundary conditions, without recourse to standard but not straightforward interpretations of the adjoint flux as an "Importance" function.

Specifically, we focussed on certain transport equations arising in radiation damage, the so–called (partial) backward equation, and using a Variational Principle, provided a rigorous link between the forward and backward approaches via the adjoint system. Well known limitations of the established approach were made apparent in a systematic fashion.

The variational approach is, of course, quite general and can be applied to all linear (and perhaps nonlinear) transport problems. Although its use for, and success in, constructing second order accurate estimates of functionals given only first order accurate estimates of the flux is widespread, it seems that apart from the early work of Pomraning (15) the Variational Principle has not been employed as extensively on more fundamental transport theoretic issues. We have demonstrated one such example here, where the variational technique has been used to reconcile in a rigorous manner two seemingly different formulations of the same problem.

REFERENCES

1. Chr. Lehmann, Interaction of Radiation with Solids and Elementary Defect Production, (North–Holland, Amsterdam, 1977).

2. M.M.R. Williams, Mathematical Methods in Particle Transport Theory, (Butterworth, London, 1971).

3. J. Lindhard, M. Scharff and H.E. Schiott, Notes on Atomic Collisions, II, *Mat. Fys. Medd. Dan. Vid. Selsk.*, 33, No. 14 (1963).

4. J.B. Sanders, On Penetration Depths and Collision Cascades in Solid Materials, (Thesis, Leiden, 1968).

5. K.B. Winterbon and J.B. Sanders, Analytical Calculations of Some Ion–Implantation Depth Distributions, *Rad. Effects*, 39:39–44 (1978).

6. P. Sigmund, Theory of Sputtering, I – Sputtering Yield of Amorphous and Polycrystalline Targets, *Phys. Rev.*, 184:383–416 (1969).

7. K.B. Winterbon, Ion Implantation Distributions in Inhomogeneous Materials, *App. Phys. Lett.*, 31:649–651 (1977).

8. M.M.R. Williams, *J. Phys. A*, 9:771 (1976).

9. M.M.R. Williams, The Role of the Boltzmann Transport Equation in Radiation Damage Calculations, *Prog. Nucl. Energy*, 3:1–65 (1979).

10. M.M.R. Williams, On the Path Length Distribution Function for Range Profiles, *Rad. Effects*, 31:233–239 (1977).

11. A.K. Prinja and M.M.R. Williams, The Energy Deposition of Slowing Down Particles in Heterogeneous Media, *Rad. Effects*, 46:235–248 (1980).

12. M.M.R. Williams, The Energy Spectrum of Sputtered Atoms, *Phil. Mag.*, 34:669–683 (1976).

13. J. Lewins, Importance, The Adjoint Function, (Pergamon Press, Oxford, 1965).

14. P.M. Morse and H. Feshbach, Methods of Theoretical Physics, (McGraw–Hill Book Co., New York, 1953).

15. G.C. Pomraning, A Derivation of Variational Principles for Inhomogeneous Equations, *Nucl. Sci. Eng.*, 29:220–236 (1967).

P-1 Transverse-Leakage Treatment of Interface-Current Nodal Transport Method

Y-W H. LIU   National Tsing Hua University, Department of Nuclear Engineering, Hsinchu, Taiwan, Republic of China

A P-1 transverse-leakage treatment is developed to improve the accuracy of the interface-current nodal transport method. In the three benchmark tests, errors of fluxes at 45 deg computed in x,y geometry using isotropic transverse-leakage treatment are reduced considerably when P-1 treatment is used. It requires, however, more than 50% of the CPU time compared to the isotropic treatment.

1. INTRODUCTION

Significant errors in nodal transport fluxes were found by Wagner and Lawrence in their nodal transport (NDOM and NTT) solutions of the International Atomic Energy Agency (IAEA) Stepanek benchmark problem [1],[2]. The errors were attributed to the isotropic transverse-leakage (ITL) approximation. Gelbard also concluded [3] on theoretical grounds that nodal transport fluxes, computed in x,y geometry using ITL, should be much more accurate on the coordinate axes than halfway between them. Later Lawrence developed an angle-dependent transverse-leakage treatment and showed that it gave excellent results in the Stepanek problem. It seems, however, that numerical methods in this improved NTT are still not wholly reliable and the formulations are very complicated. Here we developed a new approach, P-1 approximation, for the treatment of the angular dependence of the transverse-leakage term. It retains the important feature of the angle-dependent leakage treatment, but bears a much simpler formulation.

2. THEORY

2.1 Background

The interface-current nodal transport method developed by Lawrence is based on approximation of exact integral equations for the 1D scalar fluxes and for the outgoing angular fluxes on the surface of the node. Detailed derivation was given in the Appendix of Ref. 1 and is briefly repeated here for convenience. Using local coordinates, the two-dimensional transport equation with isotropic scattering for the k-th node is

$$\frac{1}{\Delta x} \mu \frac{\partial}{\partial x} \psi_g^k(x,y,\mu,\phi) + \frac{1}{\Delta y} \sqrt{1-\mu^2} \cos\phi \frac{\partial}{\partial y} \psi_g^k(x,y,\mu,\phi)$$

$$+ \Sigma_g^{t,k} \psi_g^k(x,y,\mu,\phi) = \frac{1}{4\pi} S_g^k(x,y) , \quad (x,y) \in V^k , \quad (1)$$

where μ and the azimuthal angle ϕ are defined such that

$$\Omega_x \equiv \mu , \quad \Omega_y \equiv \sqrt{1-\mu^2} \cos\phi .$$

The coordinates x and y are dimensionless in terms of the node mesh spacings Δx and Δy. The source term $S_g^k(x,y)$ involves contributions due to fission, in-scatter and external sources.

The one-dimensional transverse-integrated transport equation for the k-th node is obtained by integrating Eq. (1) over $y \in (-\frac{1}{2}, +\frac{1}{2})$. The result is

$$\frac{1}{\Delta x} \mu \frac{\partial}{\partial x} \psi_{gx}^k(x,\mu,\phi) + \Sigma_g^{t,k} \psi_{gx}^k(x,\mu,\phi)$$

$$= \frac{1}{4\pi} S_{gx}^k(x) - L_{gy}^k(x,\mu,\phi) , \quad x \in (-\frac{1}{2}, +\frac{1}{2}) , \quad (2)$$

where the one-dimensional angular flux is

$$\psi_{gx}^k(x,\mu,\phi) \equiv \int_{-\frac{1}{2}}^{\frac{1}{2}} dy \, \psi_g^k(x,y,\mu,\phi) , \quad (3)$$

and transverse-leakage term specifying the net loss of particles due to streaming across the y-directed faces is

$$L_{gy}^k(x,\mu,\phi) \equiv \frac{1}{\Delta y} \sqrt{1-\mu^2} \cos\phi \, [\psi_g^k(x,\frac{1}{2},\mu,\phi) - \psi_g^k(x,-\frac{1}{2},\mu,\phi)] . \quad (4)$$

The 1D scalar fluxes are approximated by polynomials of up to second orders with coefficients computed using moments weighting applied to the 1D integral equations. The surface-averaged angular fluxes are approximated by a double P_1 expansion which accounts for azimuthal dependence. For example, the outgoing flux on the plus-x-directed face of a two-dimensional node is given by

$$\psi_{gx+}^{out,k}(\mu,\phi) = \frac{1}{2\pi} \left[4\psi_{gx+}^{out,k} - 6J_{gx+}^{out,k} \right] + \frac{1}{2\pi} \left[12J_{gx+}^{out,k} - 6\psi_{gx+}^{out,k} \right]\mu$$

$$+ \frac{1}{2\pi} \left[3J_{gxT+}^{out,k} \right] \sqrt{1-\mu^2} \cos\phi \ . \tag{5}$$

where

$$\psi_{gx+}^{out,k} \equiv \int_0^{2\pi} d\phi \int_0^1 d\mu \ \psi_{gx+}^{out,k}(\mu,\phi) \tag{6}$$

$$J_{gx+}^{out,k} \equiv \int_0^{2\pi} d\phi \int_0^1 d\mu \ \mu \ \psi_{gx+}^{out,k}(\mu,\phi) \tag{7}$$

$$J_{gxT+}^{out,k} \equiv \int_0^{2\pi} d\phi \int_0^1 d\mu \ \sqrt{1-\mu^2} \cos\phi \ \psi_{gx+}^{out,k}(\mu,\phi) \ . \tag{8}$$

Several different approximations have been developed for the angular dependence of the transverse-leakage term introduced by the transverse integration procedure. The simplest approach is to assume that the transverse leakage is isotropic. However, as mentioned earlier, substantial errors due to isotropic transverse-leakage treatment were detected by Lawrence. He later developed an angle-dependent leakage formulation, obtained by using the y-direction analogs of Eq. (5) to evaluate the angular dependence of the transverse-leakage term appearing in the x-direction equations. Although the angle-dependent leakage formulation reduces the errors associated with the isotropic transverse-leakage approximation, it diverges for relatively fine-mesh calculations and the formulations are very complicated.

2.2 P-1 Transverse-Leakage Treatment

Here we developed a new approach for the treatment of the angular dependence of the transverse-leakage term, called P_1 transverse-leakage treatment. It retains the important feature of Lawrence's angle-dependent leakage treatment, but bears a much simpler formulation.

For clarity, the group index is dropped in the following derivations. The transverse-leakage term is expanded using P_1 approximation as

$$L_y(\mu,\phi) = \frac{1}{4\pi} \left(f_0 + 3f_{1y}\Omega_y + 3f_{1x}\Omega_x \right) \tag{9}$$

where

$$f_o \equiv \int d\Omega \; L_y(\mu,\phi) \tag{10}$$

$$f_{1y} \equiv \int d\Omega \; \Omega_y \; L_y(\mu,\phi) \tag{11}$$

$$f_{1x} \equiv \int d\Omega \; \Omega_x \; L_y(\mu,\phi) \; . \tag{12}$$

Using Eq. (4) it can be derived easily that

$$L_y(\mu,\phi) = \frac{1}{4\pi} \frac{1}{\Delta y} \left[\Delta J_y + (3\delta J_y - \frac{1}{2}\delta\psi_y) \; \Omega_y + (\frac{9}{8} \Delta J_{yT}) \; \Omega_x \right] \tag{13}$$

where

$$\Delta J_y \equiv (\; J_{y+}^{out} - J_{y+}^{in} \;) + (\; J_{y-}^{out} - J_{y-}^{in} \;) \tag{14}$$

$$\Delta J_{yT} \equiv (\; J_{yT+}^{out} - J_{yT+}^{in}) + (\; J_{yT-}^{out} - J_{yT-}^{in}) \tag{15}$$

$$\delta J_y \equiv (\; J_{y+}^{out} + J_{y+}^{in} \;) - (\; J_{y-}^{out} + J_{y-}^{in} \;) \tag{16}$$

$$\delta\psi_y \equiv (\; \psi_{y+}^{out} + \psi_{y+}^{in} \;) - (\; \psi_{y-}^{out} + \psi_{y-}^{in} \;) \; . \tag{17}$$

For simplicity, flat spatial approximation is used for the transverse-leakage treatment. The resultant interface-current equations in the x-direction are

$$\underline{\psi}_x^{out} = [P_x] \; \{\underline{Q}_x - \underline{L}_y^o\} + [R_x] \; \underline{\psi}_x^{in} + \underline{E}_x \; L_y^{1x} \tag{18}$$

$$\underline{J}_{xT}^{out} = \underline{F}_x \; L_y^{1y} + [G_x] \; \underline{J}_{xT}^{in} \; . \tag{19}$$

where

$$\underline{\psi}_x^{out} \equiv col \; [J_{x+}^{out}, \; J_{x-}^{out}, \; \psi_{x+}^{out}, \; \psi_{x-}^{out}]$$

$$\underline{J}_{xT}^{out} \equiv col \; [J_{xT+}^{out}, \; J_{xT-}^{out}]$$

$$\underline{E}_x \equiv col \; [-E_{1x}, \; E_{1x,} \; -E_{2x}, \; E_{2x}]$$

$$\underline{F}_x \equiv col \; [F_{1x}, \; F_{1x}]$$

$$[G_x] \equiv \begin{bmatrix} 0 & G_{1x} \\ G_{1x} & 0 \end{bmatrix} \, .$$

The entries of $\underline{E}_x$, $\underline{F}_x$ and $[G_x]$ are

$$E_{1x} \equiv \frac{1}{2} \Delta x \, \beta_{13} \tag{20}$$

$$E_{2x} \equiv \frac{1}{2} \Delta x \, \beta_{12} \tag{21}$$

$$F_{1x} \equiv - \frac{1}{4} \Delta x \, [\beta_{11} - \beta_{13}] \tag{22}$$

$$G_{1x} \equiv \frac{3}{2} \, [E_2(\Sigma) - E_4(\Sigma)] \, , \tag{23}$$

where

$$\beta_{1n} \equiv \int_{-\frac{1}{2}}^{\frac{1}{2}} dx \, E_n[\Sigma(\frac{1}{2} + x)] \, , \tag{24}$$

$$\Sigma \equiv \Sigma^t \Delta x$$

and the zero spatial moment of the transverse-leakage terms are

$$L_y^o \equiv \Delta J_y / \Delta y \tag{25}$$

$$L_y^{1y} \equiv (3 \delta J_y - \frac{1}{2} \delta \psi_y) / \Delta y \tag{26}$$

$$L_y^{1x} \equiv \frac{9}{8} \Delta J_{yT} / \Delta y \, . \tag{27}$$

The corresponding equation for the isotropic leakage approximation is

$$\underline{\psi}_x^{out} = [P_x] \, \{\underline{Q}_x - \underline{L}_y^o\} + [R_x] \, \underline{\psi}_x^{in} \, . \tag{28}$$

The matrices $[P_x]$ and $[R_x]$ are defined in the Appendix of Ref. 1.

The same as Lawrence's angle-dependent leakage approximation, the P_1 transverse-leakage approximation adds one additional unknown (the transverse component Eq. (8) of the outgoing current) per surface. Thus P_1 leakage formulation involves three unknowns per surface in two-dimensional calculations.

Using flat P_1 transverse-leakage approximation leads to the following equation for the 1D scalar flux:

$$\phi_x^k(x) = \Delta x \int_{-\frac{1}{2}}^{\frac{1}{2}} dx_0 \ E_1[\Sigma|x-x_0|] \ \frac{1}{2} \ [S_x^k(x_0) - \Delta J_y^k)/\Delta y]$$

$$+ \int_0^1 d\mu \ \psi_{x-}^{in,k}(\mu) \ \exp[-\Sigma(\tfrac{1}{2}+x)/\mu]$$

$$+ \int_{-1}^0 d\mu \ \psi_{x+}^{in,k}(\mu) \ \exp[-\Sigma(\tfrac{1}{2}-x)/|\mu|]$$

$$- \Delta x \int_{-\frac{1}{2}}^x dx_0 \ E_2(\Sigma(x-x_0)) \ (\frac{9}{16} \ \Delta J_{yT}^k)/\Delta y)$$

$$+ \Delta x \int_x^{\frac{1}{2}} dx_0 \ E_2(\Sigma|x-x_0|) \ (\frac{9}{16} \ \Delta J_{yT}^k)/\Delta y) \ . \tag{29}$$

The first three terms are identical to those in the equation of the 1D scalar flux when isotropic transverse-leakage approximation is used (i.e. Eq. (A-10) of Ref. 1). The remaining two terms account for the corrections due to P_1 transverse-leakage treatment.

Taking spatial moments of Eq. (29), the last two terms cancel each other for even moments. For odd moments, they combine to

$$- \Delta x \int_{-\frac{1}{2}}^{\frac{1}{2}} dx \ f_i(x) \int_{-\frac{1}{2}}^x dx_0 \ E_2(\Sigma(x-x_0)) \ (\frac{9}{8} \ \Delta J_{yT}^k/\Delta y)$$

where $f_i(x)$ are base functions as defined in Eq. (3) of Ref. 1.

Comparing with Lawrence's angle-dependent transverse-leakage formulation [1] which involves Ki_n, the Bickley-Nayler function and more terms, it is easily seen that when P_1 treatment is used for the transverse-leakage term the resultant flux moment equations and the interface-current equations are much simpler and easier to implement. Only minor changes on the existing NTT version are necessary.

3 BENCHMARK TESTS

The flat P_1 transverse-leakage treatment was implemented on DIF3D [4],[5] as a nodal transport option (NTT-P_1). Three test problems were run to see the improvements of accuracy by using this treatment.

The first test was a three-group calculation of the flux produced by a square, isotropic, group-1 source, 6 cm on a side, in a large sodium pool. This ("Green's function") computation was performed using Lawrence's NTT-P_0 option [1] and the new NTT-P_1 option in DIF3D. The S_{16} ONEDANT values were used as the standard. Results are found in Table 1. There are some indications here of the error pattern predicted in Ref. 3. The use of P_1 transverse-leakage treatment in NTT gives better results at 45 deg than the ITL approximation.

Table 1 Errors in Group-1 Fluxes in x,y Green's Function

Method	θ* $\diagdown$ r[†]	73.5 cm	88.5 cm	103.5 cm	142.5 cm
S_{16}		1.002E-3	1.438E-4	2.094E-5	1.457E-7
		% errors in fluxes			
NDT[a]	$0°$	-36%	-45%	-54%	-69%
NDT	$45°$	-34%	-43%	-53%	-69%
NTT-P_0[b]	$0°$	6%	2%	-3%	-13%
NTT-P_0	$45°$	-30%	-36%	-43%	-56%
NTT-P_1[c]	$0°$	-5%	-9%	-12%	-21%
NTT-P_1	$45°$	-11%	-15%	-21%	-32%

Flux errors in groups 2 and 3 are very small.

*Angle between x-axis and radius-vector to field-point.

†r is distance from origin to center of mesh box.

[a]Nodal diffusion option, DIF3D.

[b]Nodal transport option, DIF3D (quadratic ITL).

[c]Nodal transport option, DIF3D (flat P_1 transverse-leakage).

Mesh widths: 3 cm x 3 cm. Problems rerun with 6 cm x 6 cm mesh-boxes. No significant change in results. When necessary, fluxes were linearly interpolated between nearest-neighbor fluxes. S_{16} fluxes normalized to a source density of 10 neuts/sec/cm^2.

Cross sections:

$$\Sigma_t^1 = .17457, \quad \Sigma_r^1 = 3.4317E-2, \quad \Sigma_s^{1 \to 2} = 3.2581E-2, \quad \Sigma_s^{1 \to 3} = 1.2092E-4$$

$$\Sigma_t^2 = .33738, \quad \Sigma_r^2 = 5.8333E-3, \quad \Sigma_s^{2 \to 3} = 4.4281E-3,$$

$$\Sigma_t^3 = .86415, \quad \Sigma_r^3 = 5.5191E-3$$

The second test problem was the Stepanek IAEA problem, run with isotropic scattering as in Ref. 2. Only the zone-3 average flux, $\bar{\Phi}_3$, which seems, in Ref. 2, very difficult to compute will be discussed here. For 32 x 32, 48 x 48 and 64 x 64 mesh-boxes we get, via NTT-P_0, $\bar{\Phi}_3$= .0294 x 10^{-3}, .0302 x 10^{-3}, and .0304 x 10^{-3}, respectively. Via NTT-P_1, we get $\bar{\Phi}_3$= .0373 x 10^{-3} for 64 x 64. Taking, as standard, the S_8 TWODANT value of Ref. 2., i.e., $\bar{\Phi}_3$ = .0367 x 10^{-3}, the improvement of results using P_1 transverse-leakage treatment in NTT is quite obvious and satisfactory in this problem.

The last test problem is a grossly simplified x,y fast reactor with a 250- x 250-cm core, surrounded by a 25-cm-thick blanket, which, in turn, is surrounded by a 50-cm sodium reflector. It is seen in Table 2 that the NTT-P_0 group-1 corner fluxes are substantially in error. The use of NTT-P_1 removes half of the error.

Some CPU time and region requirements for the test problems are gives in Table 3. NTT-P_1 requires more than 50% of CPU time compared to NTT-P_0. This is understandable since it involves three unknowns per surface rather than two.

Table 2 Errors in Group-1 Fluxes in Blanket at Blanket/Reflector Boundary

Method	Blanket Corner Mesh-Box	Mesh-Box on axis
TWODANT (S_8, 80x80)	7.810×10^4	1.814×10^6
	% Errors	
NDT (40x40)	-20.0%	-6.1%
NTT-P_0[a] (40x40)	-19.7%	.8%
NTT-P_1 (56x56)	-9.0%	0%

Flux errors in groups 2 and 3 are very small.

Fission spectrum: χ_1 = 0.75781, χ_2 = 0.24179, χ_3 = 3.9749E-4.

Cross sections: Reflector - same as in Table 1.

Blanket: Σ_t^1 =.14358, Σ_r^1 =4.3300E-2, $\Sigma_s^{1\rightarrow2}$=3.6562E-2, $\Sigma_s^{1\rightarrow3}$=2.0360E-5,

Σ_t^2 =.27747, Σ_r^2 =6.3974E-3, $\Sigma_s^{2\rightarrow3}$=2.1452E-3, Σ_t^3 =.45030,

Σ_r^3 =1.1349E-2, $\nu\Sigma_f^1$=1.0830E-2, $\nu\Sigma_f^2$ =1.2360E-4, $\nu\Sigma_f^3$ =5.3218E-4,

Core: Σ_t^1 =.12819, Σ_r^1 =3.5350E-2, $\Sigma_s^{1\rightarrow2}$=2.6438E-2, $\Sigma_s^{1\rightarrow3}$=2.8500E-5,

Σ_t^2 =.22033, Σ_r^2 =7.2400E-3, $\Sigma_s^{2\rightarrow3}$=1.4161E-3, Σ_t^3 =.48002,

Σ_r^3 =1.8084E-2, $\nu\Sigma_f^1$=1.7022E-2, $\nu\Sigma_f^2$ =8.4460E-3, $\nu\Sigma_f^3$ =1.8978E-2.

[a]Calculation was repeated for a 56 x 56 mesh without significant changes in results.
When necessary, fluxes were linearly interpolated between nearest-neighbor fluxes.

Table 3 CPU Time and Region Requirements

Method	Green's Function		IAEA LWR	
	CPU(sec)	Region(k)	CPU(sec)	Region(k)
NDT	19	1800	–	–
NTT-P_0	37	2000	48	1700
NTT-P_1	88	2500	75	2260

4 SUMMARY AND CONCLUSIONS

The limitations of the ITL approximation are apparently quite serious. Errors of fluxes at 45 deg computed in x,y geometry using ITL are reduced considerably when P-1 transverse-leakage treatment is applied. Although NTT-P_1 gives improving results, it suffers from divergence as mesh gets smaller. Similar situation occurs in Lawrence's angle-dependent transverse-leakage treatment. The reason remains to be investigated.

ACKNOWLEDGMENTS

The author would like to express her appreciation to Dr. Ely Gelbard of Argonne National Laboratory for his suggestion of the P_1 form and helpful discussions. Work supported by the U.S. Department of Energy, Nuclear Energy Programs, under Contract W-31-109-Eng-38.

REFERENCES

1. R.D. LAWRENCE, "Three-Dimensional Nodal Diffusion and Transport Methods for the Analysis of Fast-Reactor Critical Experiments," *Proc. Topl. Mtg. Reactor Physics and Shielding*, Chicago, Illinois, September 17-19, 1984, p.814.
2. R.D. O'DELL, J. STEPANEK, M.R. WAGNER, "Intercomparison of the Finite-Difference and Nodal Discrete Ordinates and Surface Flux Transport Method for a LWR Pool Reactor Benchmark Problem in X-Y Geometry," *Proc. Topl. Mtg. Advances in Reactor Computations*, Salt Lake City, Utah, March 28-31, 1983, Vol. II, p. 941.
3   E.M. GELBARD, "Model Problem Study of Transverse-Leakage Treatment in Nodal Methods," *Trans. Am. Nucl. Soc.*, 49, 214(1985).
4. K.L. DERSTINE, "DIF3D: A Code to Solve One, Two, and Three-Dimesional Finite-Difference Diffusion Theory Problems," ANL-82-64, Argonne National Laboratory (Apr. 1984).
5. R.D. LAWRENCE, "The DIF3D Nodal Neutronics Option for Two- and Three-Dimensional Diffusion Theory Calculations in Hexagonal Geometry," ANL-83-1, Argonne National Lab. (Mar. 1983).

Systems of Conservation Equations in Nonlinear Particle Transport Theory

V. C. BOFFI   Nuclear Engineering Laboratory of the University of Bologna, Bologna, Italy

A system of conservation equations is derived by an appropriate physical modeling of the system of nonlinear integro-partial differential Boltzmann equations governing the distribution functions of a mixture of N different species of interacting particles. This system is discussed on both mathematical and physical grounds, and analytical solutions to it are constructed for the case of N=2.

INTRODUCTION

Spatially inhomogeneous evolution problems are today of increasing interest in both science and technology. Here we refer to that class of such problems falling within the frame of particle transport theory which are formulated via the integro-partial differential Boltzmann equation. It is well known that in particle transport the exact treatment of the space variable is a very arduous task even in the linear forceless case (compare, for instance, the relevant section in [1]). In the nonlinear case such a task gets clearly much more complicated; significant progress has been, however, made recently with respect to either the existence and uniqueness of the solution to the nonlinear Boltzmann equation and the construction of explicit results for such a solution in some special cases (see, for instance, [2]-[6], and [1], [7]-[12], respectively). The state of the art is still, however, far from being fully satisfactory, and, moreover, it covers only the case of a single species of particles, mostly in the absence of external forces. If the case of a mixture of N different species of particles (N≥2),

123

interacting between themselves and against the particles of a bath, in which they diffuse upon the action of external forces, is instead to be treated (a system of N nonlinear integro-partial differential Boltzmann equations is now to be faced!), the state of the art can be then regarded as practically empty. (It may be remarked that in the general mathematical literature, papers devoted to systems of integro-partial differential equations are very rare; we just quote the two of [13] and [14], going back to 1983, both which are not suitable for the purposes of the present investigation). Actually, the Boltzmann system for a mixture has been considered in [15], but for the spatially homogeneous case, the leading idea being rather simple. Indeed, in the limit of constant collision frequencies, and then integrating over the velocity domain, the Boltzmann system for the distribution functions $f_1, f_2, \ldots, f_N$ of the particles of the N species of the given mixture has been shown [15] to be directly convertible into a system of N nonlinear first-order ordinary differential equations for the N number densities $\rho_1, \rho_2, \ldots, \rho_N$, with $\rho_j \equiv \rho_j(t) = \int_{R3} f_j(\bar{v}, t) d\bar{v}$ (j=1,2,...,N). This latter has thus been investigated - on the basis of the theory of the dynamical systems - for various physical situations involving both removal and scattering as well as creation by collision. In this latter case, the mean number of secondary particles created by collision is taken as constant (compare also Refs.16-20). More recently, the idea of [15] has been resumed in [21] for a spatially inhomogeneous mixture of N gases of particles, and it has been shown that - also in the spatially inhomogeneous case - the Boltzmann system for the distribution functions $f_1, f_2, \ldots, f_N$ can be converted into a system of differential equations for the N number densities $\rho_1, \rho_2, \ldots, \rho_N$, where now $\rho_j \equiv \rho_j(\bar{x}, t) = \int_{R3} f_j(\bar{x}, \bar{v}, t) d\bar{v}$. Of course, in order to attain such conversion, the assumption of constant collision frequencies is no longer sufficient, as it was in the spatially homogeneous case. It must be combined, indeed, with an additional set of assumptions concerning both the scattering probability distributions and the initial data. A second feature is that the differential system for the number densities, which is the system of nonlinear conservation equations associated with the physical situation considered, exhibits a peculiar functional-hyperbolic structure. It becomes fully hyperbolic if scattering is ignored, and only removal between the particles of the mixture is accounted for. Here we resume again the idea of [15] with the aim at extending the results of [21] to include creation-by-collision effects. In the first part of the paper, the general theory is developed, and the physical model, according to which the original Boltzmann system for the distribution functions f_j (j=1,2,...,N) is converted into a nonlinear functional-hyperbolic conservation system for the number densities ρ_j, is illustrated in detail. The differences with respect to the analogous conservation system derived in [21] , where creation effects have been neglected, are then verified. The role played by the hyperbolicity of the conservation system is also underlined in connection with a new constitutive equation, which is formulated for the current densitiy $\bar{J}_j(\bar{x},t) = \int_{R3} \bar{v}\; f_j(\bar{x},\bar{v},t)\; d\bar{v}$. The problem of reconstructing f_j from the ρ_j's is also discussed, especially for the case when the full hyperbolicity of the conservation system is not achieved. In the second part, applications of the conservation system for the ρ_j's are instead carried out. In particular, we concentrate on seeking nontrivial solutions to such a system - of both explicit and implicit form - in the case N=2, thus enlarging the number of solutions already proposed in Refs. 21-23. Finally, we draw some conclusions illustrating the results obtained on both mathematical and physical grounds.

1 GENERAL THEORY

1.1 Outline of the problem

We consider a mixture of N different species of particles, the general species having mass m_j (j=1,2,...,N), and being released, at time t=0, by an external source with an intensity $S_j(\bar{x},\bar{v})$, function of both position $\bar{x}$ and velocity $\bar{v}$, in the interior of a bath consisting of certain other particles, filling up, with a fixed distribution, the whole 3-dimensional Euclidean space R_3 (for the bath j=N+1, m_{N+1} and ρ_{N+1} will denote the mass and the fixed number density, respectively, of the bath particles. Henceforth, we shall distinguish the particles of the mixture and those of the bath as test particles (t.p.) and field particles (f.p.), respectively). The t.p. of the species j, once emitted, will diffuse in the bath - upon the action of an externally applied constant force $\bar{F}_j$ - by binary collisions between themselves, the t.p. of each of the other species $\ell \neq j$ and the f.p. of the bath. Every collision can result in an event of either scattering (S), removal (R) or creation (C), this last effect having been neglected in the study of spatially inhomogeneous mixtures previously undertaken [21]. For this physical situation it is requested, in many applications, to determine at time t>0 the N distribution functions $f_1, f_2, ..., f_N$ of the N species of particles constituting the given mixture, such that the corresponding (positive) initial datum $f_j(\bar{x},\bar{v},0)=S_j(\bar{x},\bar{v})$ be correctly reproduced. The relevant Cauchy problem is then formulated via the system of nonlinear integro-partial differential Boltzmann equations governing the N f_j's. We write, following the so-called scattering kernel formulation [24], and supposing that all the physical parameters are constant in time and position, the system of equations

$$\frac{\partial f_j}{\partial t} + \bar{v} \cdot \frac{\partial f_j}{\partial \bar{x}} + \frac{\bar{F}_j}{m_j} \cdot \frac{\partial f_j(\bar{x},\bar{v},t)}{\partial \bar{v}} \equiv \frac{D f_j}{Dt} =$$

$$- \sum_{k=1}^{N} f_j(\bar{x},\bar{v},t) \int_{R_3} g_{jk}(|\bar{v}-\bar{w}|) f_k(\bar{x},\bar{w},t) d\bar{w} +$$

$$\sum_{\ell=1}^{N} \iint_{R_3} g_{j\ell}^{S}(|\bar{v}'-\bar{w}'|) \pi_{j\ell}^{S}(\bar{v}',\bar{w}'{\rightarrow}\bar{v}) f_j(\bar{x},\bar{v}',t) f_{\ell}(\bar{x},\bar{w}',t) d\bar{v}' d\bar{w}' +$$

$$\sum_{m=1}^{N} \sum_{n=m}^{N} \iint_{R_3} \chi_{mn}^{j}(\bar{v}',\bar{w}'{\rightarrow}\bar{v}) g_{mn,j}^{C}(|\bar{v}'-\bar{w}'|) f_m(\bar{x},\bar{v}',t) f_n(\bar{x},\bar{w}',t) d\bar{v}' d\bar{w}' -$$

$$\rho_{N+1} g_{j,N+1}(|\bar{v}|) f_j(\bar{x},\bar{v},t) +$$

$$\rho_{N+1} \int_{R_3} g_{j,N+1}^{S}(|\bar{v}'|) \pi_{j,N+1}^{S}(\bar{v}'{\rightarrow}\bar{v}) f_j(\bar{x},\bar{v}',t) d\bar{v}' +$$

$$\rho_{N+1} \int_{R_3} \chi_{j,N+1}^{j}(\bar{v}'{\rightarrow}\bar{v}) \ g_{(j,N+1),j}^{C}(|\bar{v}'|) f_j(\bar{x},\bar{v}',t) d\bar{v}' \qquad (1)$$

$$(\bar{x},\bar{v} \ \epsilon \ R_3; \ t \ \epsilon \ [0,\infty[).$$

Equations (1) are to be integrated upon the initial condition

$$f_j(\bar{x},\bar{v},0) = S_j(\bar{x},\bar{v}) > 0. \qquad (1a)$$

For the meaning of the symbols appearing in Eq.(1) we shortly recall what follows.

$\dfrac{D f_j}{Dt}$ → substantial derivative of f_j;

$g_{mn,j}^{C}=g_{nm,j}^{C}$→ microscopic frequency of the collisions between a particle m of velocity $\bar{v}'$ and a particle n of velocity $\bar{w}'$ producing (by any process but elastic scattering) one or more particle j (if both m and n are different from j, the interaction is a chemical reaction; if either m or n coincides with j, then we have a self-generation. Of course, $g_{jj,j}^{c}=0$, and if $g_{jn,j}^{c} {\neq} 0$, then $g_{jn,n}^{c}=0$);

$\chi_{mn}^{j} d\bar{v}$ → number of the created particles j that attain a final velocity in $d\bar{v}$ at $\bar{v}$;

$g_{j\ell}^{S}$, $g_{j\ell}^{R}$ → microscopic frequency of scattering and removal collisions, respectively, between species j and species ℓ.

If creation by collision produces only one participating species, then

$$g_{j\ell} = g^S_{j\ell} + g^R_{j\ell} + \sum_{m=1}^{N} g^C_{j\ell,m} = g^S_{j\ell} + g^A_{j\ell} \rightarrow \text{ total (microscopic)}$$
collision frequency;   (2a)

$$g_{j,N+1} = g^S_{j,N+1} + g^R_{j,N+1} + g^C_{(j,N+1),j} \text{ and } \chi^j_{j,N+1} \rightarrow \text{ analogous}$$

quantities for the interactions between particle j and the f.p. of the bath.   (2b)

As for the scattering probability distributions $\pi^S_{j\ell}$ and $\pi^s_{j,N+1}$, we recall that they must obey the normalization conditions

$$\int_{R_3} \pi^S_{j\ell}(\bar{v}',\bar{w}'\rightarrow\bar{v})\,d\bar{v} = 1; \qquad \int_{R_3} \pi^S_{j,N+1}(\bar{v}'\rightarrow\bar{v})\,d\bar{v} = 1. \quad (2c)$$

We observe then that in the limit of all $g^C \rightarrow 0$ (no creation), the system, Eq.(1), reduces to the one considered in [21]; if also $N\rightarrow1$, then [1] and [10] are in order. The corresponding system for the spatially homogeneous case has been formulated in [15].

1.2 Physical modeling of the system, Eq.1

We show next that the system, Eq.(1), can be solved for the general f_j, once an appropriate physical model is adopted. Let us indeed make the following assumptions.

-i) All the considered collision frequencies are nonnegative constants, namely ($\alpha = S,R$)

$$g^\alpha_{j\ell}(|\bar{v}-\bar{w}|) \rightarrow g^\alpha_{j\ell} \geq 0; \quad g^C_{j\ell,m}(|\bar{v}-\bar{w}|) \rightarrow g^C_{j\ell,m} \geq 0, \quad (3a)$$

and consequently

$$g_{j\ell}(|\bar{v}-\bar{w}|) \rightarrow g^S_{j\ell} + g^R_{j\ell} + \sum_{m=1}^{N} g^C_{j\ell,m} \geq 0. \quad (3b)$$

Analogously we set

$$\rho_{N+1}g^\alpha_{j,N+1}(|\bar{v}|) \rightarrow g^\alpha_{j,N+1} \geq 0;$$
$$\rho_{N+1}g^C_{(j,N+1),j}(|\bar{v}|) \rightarrow g^C_{(j,N+1),j} \geq 0 \quad (3c)$$

so that

$$\rho_{N+1}g_{j,N+1}(|\bar{v}|) \rightarrow g_{j,N+1} = g^S_{j,N+1} + g^R_{j,N+1} + g^C_{(j,N+1),j} \geq 0. \quad (3d)$$

Equations (3) amount to assuming that all the relevant cross sections follow the $1/|\bar{v}|$ - law, which is physically plausible.

-ii) As for the scattering probability distributions, we set, independent of the velocity before collision,

$$\pi_{j\ell}^{S}(\bar{v}',\bar{w}'\rightarrow\bar{v})=\delta(\bar{v}-\bar{v}_{j\ell}) \; ; \quad \pi_{j,N+1}^{S}(\bar{v}'\rightarrow\bar{v})=\delta(\bar{v}-\bar{v}_{j,N+1}) , \qquad (4)$$

with δ denoting the Dirac delta function. Equations (4), which satisfy the conditions of Eq.(2c), can be interpreted as a special B.G.K. model [25], according to which the general $\pi_{j\ell}^{S}$ is the limit of a Maxwellian with fixed drift velocity $\bar{v}_j$, when its variance goes to zero. This special B.G.K. model is of a stochastic nature, as momentum and energy are not conserved in a collision. However, the H-theorem holds, and the t.p. of the general species j are pushed towards the equilibrium distribution $\delta(\bar{v}-\bar{v}_{j\ell})$ in absence of any other physical event save collisions with the t.p. of the species ℓ.

-iii) Analogously, we set for the creation probability distributions

$$\chi_{mn}^{j}(\bar{v}',\bar{w}'\rightarrow\bar{v}) = \eta_{mn}^{j}\delta(\bar{v}-\bar{v}_{mn}^{j}) ;$$
$$\chi_{j,N+1}^{j}(\bar{v}'\rightarrow\bar{v}) = \eta_{j,N+1}^{j}\delta(\bar{v}-\bar{v}_{j,N+1}^{j}) , \qquad (5)$$

where

$$\eta_{mn}^{j}=\int_{R_3}\chi_{mn}^{j}(\bar{v}',\bar{w}'\rightarrow\bar{w})\,d\bar{v} \quad \text{and} \quad \eta_{j,N+1}^{j}=\int_{R_3}\chi_{j,N+1}^{j}(\bar{v}'\rightarrow\bar{v})\,d\bar{v},$$

which, in general, would depend on the velocities before collision. We denote the mean number of secondary particles created by collisions as η^{j}. It is assumed that both η_{mn}^{j} and $\eta_{j,N+1}^{j}$ are greater than unity.

-iv) Finally, we take the general initial datum $S_j(\bar{x},\bar{v})$ to be separable and monokinetic, namely

$$S_j(\bar{x},\bar{v}) = Q_j(\bar{x})\delta(\bar{v}-\bar{v}_{jo}) , \qquad (6)$$

where $Q_j(\bar{x})$ is positive and bounded, and $\bar{v}_{jo}$ is a certain velocity at which the particles j are injected in the bath at time t=0.

Accounting for all Eqs.(3),(4),(5) and (6), and introducing the (unknown) number density

$$\rho_j(\bar{x},t) = \int_{R_3} f_j(\bar{x},\bar{v},t)\,d\bar{v}, \qquad (7)$$

the system, Eq.(1), with $\rho_{N+1}(\bar{x},t)=1$, becomes

$$\frac{\partial f_j}{\partial t} + \bar{v} \cdot \frac{\partial f_j}{\partial \bar{x}} + \frac{\bar{F}_j}{m_j} \cdot \frac{\partial f_j(\bar{x},\bar{v},t)}{\partial \bar{v}} \equiv \frac{Df_j}{Dt} =$$

$$- \sum_{k=1}^{N+1} g_{jk}\rho_k(\bar{x},t) f_j(\bar{x},\bar{v},t) + \sum_{\ell=1}^{N+1} g_{j\ell}^S \delta(\bar{v}-\bar{v}_{j\ell}) \rho_\ell(\bar{x},t) \rho_j(\bar{x},t) +$$

$$(8)$$

$$\{ \sum_{m=1}^{N} \sum_{n=m}^{N} g_{mn,j}^C \; \eta_{mn}^j \; \delta(\bar{v}-\bar{v}_{mn}^j) \; \rho_m(\bar{x},t) \; \rho_n(\bar{x},t) +$$

$$+ g_{(j,N+1),j}^C \; \eta_{j,N+1}^j \; \delta(\bar{v}-\bar{v}_{j,N+1}^j) \; \rho_j(\bar{x},t) \}.$$

Equation (8) is to be integrated upon the initial condition

$$f_j(\bar{x},\bar{v},0) = Q_j(\bar{x}) \delta(\bar{v}-\bar{v}_{jo}). \tag{8a}$$

If the number densities $\rho_1, \rho_2, \ldots, \rho_N$ were supposed to be known, Equation (8) would thus represent, formally, a system of N linear ordinary differential equations for the N f_j's in which the coefficient of the term in f_j is a linear combination of all the ρ_j's and the known term is a quadratic form of the ρ_j's themselves, respectively. With respect to the velocity, equation (8) would be, instead, regarded as a multivelocity problem, with special velocities like $\bar{v}_j, \bar{v}_{mn}, \bar{v}_{j,N+1}$ and $\bar{v}_{jo}$, respectively. We observe that, altogether there are, for any j,

$$M = N + \frac{(N+1)!}{2(N-1)!} + 3 \tag{8b}$$

such special velocities.

1.3 Analytical solutions to Eq.(8)

If now we integrate Eq.(8) along the characteristics $\bar{x}-\bar{v}t+ (\bar{F}_j/2m_j)t^2$ = const.; $\bar{v}-(\bar{F}_j/m_j)t$ = const., we get for the distribution function f_j of the t.p. of the species j the following result, reproducing the initial datum, Eq.(8a), as $t \rightarrow 0$, namely

$$f_j(\bar{x},\bar{v},t) = Q_j(\bar{x}-\bar{v}_{j_o}t-\frac{\bar{F}_j}{2m_j}t^2)\ \delta(\bar{v}-\bar{v}_{j_O}-\frac{\bar{F}_j}{m_j}t)\ x$$

$$\exp\{-\sum_{k=1}^{N+1}g_{jk}\int_0^t \rho_k[\bar{x}-\bar{v}_{j_O}(t-u)-\frac{\bar{F}_j}{2m_j}(t^2-u^2),u]du\}\ +$$

$$\sum_{\ell=1}^{N+1}g_j^S\ell\int_0^t \delta[\bar{v}-\bar{v}_{j}\ell - \frac{\bar{F}_j}{m_j}(t-u)]\ x$$

$$\rho_\ell[\bar{x}-\bar{v}_j\ell(t-u)-\frac{\bar{F}_j}{2m_j}(t-u)^2,u]\ \rho_j[\bar{x}-\bar{v}_j\ell(t-u)-\frac{\bar{F}_j}{2m_j}(t-u)^2,u]\ x$$

$$\exp\{-\sum_{k=1}^{N+1}g_{jk}\int_u^t \rho_k[\bar{x}-\bar{v}_j\ell(t-u')-\frac{\bar{F}_j}{2m_j}(t-u')(t-2u+u'),u']\ du'\}du\ +$$

$$\sum_{m=1}^{N}\sum_{n=m}^{N}g_{mn,j}^C\ \eta_{mn}^j\ \int_0^t \delta[\bar{v}-\bar{v}_{mn}^j-\frac{\bar{F}_j}{m_j}(t-u)]\ x \qquad\qquad (9)$$

$$\rho_m[\bar{x}-\bar{v}_{mn}^j(t-u)-\frac{\bar{F}_j}{2m_j}(t-u)^2,u]\ \rho_n[\bar{x}-\bar{v}_{mn}^j(t-u)-\frac{\bar{F}_j}{2m_j}(t-u)^2,u]\ x$$

$$\exp\{-\sum_{k=1}^{N+1}g_{jk}\int_u^t \rho_k[\bar{x}-\bar{v}_{mn}^j(t-u')-\frac{\bar{F}_j}{2m_j}(t-u')(t-2u+u'),u']du'\}du\ +$$

$$g_{(j,N+1),j}^C\ \eta_{j,N+1}^j\ \int_0^t \delta[\bar{v}-\bar{v}_{j,N+1}^j-\frac{\bar{F}_j}{m_j}(t-u)]\ x$$

$$\rho_j[\bar{x}-\bar{v}_{j,N+1}^j(t-u)-\frac{\bar{F}_j}{2m_j}(t-u)^2,u]\ x$$

$$\exp\{-\sum_{k=1}^{N+1}g_{jk}\int_u^t \rho_k[\bar{x}-\bar{v}_{j,N+1}^j(t-u')-\frac{\bar{F}_j}{2m_j}(t-u')(t-2u+u'),u']du'\}du.$$

We realize thus that f_j is expressed as a function of all the ρ_j's, and will be fully explicit once the ρ_j's are known. Actually, we can exploit Eq.(9) to determine the ρ_j's and thus the higher moments of f_j, always in terms of the ρ_j's.

1.4 The autonomous system for the ρ_j's

By integrating both sides of Eq.(9) over the velocity domain, accounting for the definition of Eq.(7) and differentiating the resulting equation with respect to t, we get for the ρ_j's the system

$$\frac{\partial \rho_j}{\partial t} + \frac{\partial}{\partial \bar{x}} \cdot \bar{J}_j(\bar{x},t) + \sum_{\ell=1}^{N} g_j^A \; \rho_\ell(\bar{x},t) \; \rho_j(\bar{x},t) -$$

$$(10)$$

$$\sum_{m=1}^{N} \sum_{n=m}^{N} g_{mn,j}^C \; \eta_{mn}^j \; \rho_m(\bar{x},t) \; \rho_n(\bar{x},t) + \nu_j \rho_j(\bar{x},t) = 0,$$

whose initial condition is

$$\rho_j(\bar{x},0) = Q_j(\bar{x}) > 0, \tag{10a}$$

and where

$$\nu_j = g_{j,N+1}^R - (\eta_{j,N+1}^j - 1) \; g_{(j,N+1)}^C, j \gtrless 0. \tag{10b}$$

As for the current density

$$\bar{J}_j(\bar{x},t) = \int_{R_3} \bar{v} \; f_j(\bar{x},\bar{v},t) \; d\bar{v} \;, \tag{11}$$

we find accordingly that

$$\bar{J}_j(\bar{x},t) = (\bar{v}_{j_0} + \frac{\bar{F}_j}{m_j} t) \; \rho_j(\bar{x},t) + \bar{J}_j^{SC}(\bar{x},t) \tag{12}$$

with

$$\bar{J}_j^{SC}(\bar{x},t) = \sum_{\ell=1}^{N+1} g_{j\ell}^S \int_0^t (\bar{v}_{j\ell} - \bar{v}_{jo} - \frac{\bar{F}_j}{m_j}u)\, \rho_\ell[\bar{x} - \bar{v}_{j\ell}(t-u) - \frac{\bar{F}_j}{2m_j}(t-u)^2, u]\, x$$

$$\rho_j[\bar{x} - \bar{v}_{j\ell}(t-u) - \frac{\bar{F}_j}{2m_j}(t-u)^2, u]\, x$$

$$\exp\{-\sum_{k=1}^{N+1} g_{jk} \int_u^t \rho_k[\bar{x} - \bar{v}_{j\ell}(t-u') - \frac{\bar{F}_j}{2m_j}(t-u')(t-2u+u'), u']du'\}du +$$

$$\sum_{m=1}^{N} \sum_{n=m}^{N} g_{mn,j}^C \eta_{mn}^j \int_0^t (\bar{v}_{mn}^j - \bar{v}_{jo} - \frac{\bar{F}_j}{m_j}u)\, \rho_m[\bar{x} - \bar{v}_{mn}^j(t-u) - \frac{\bar{F}_j}{2m_j}(t-u)^2, u]\, x$$

$$\rho_n[\bar{x} - \bar{v}_{mn}^j(t-u) - \frac{\bar{F}_j}{2m_j}(t-u)^2, u]\quad x \tag{12a}$$

$$\exp\{-\sum_{k=1}^{N+1} g_{jk} \int_u^t \rho_k[\bar{x} - \bar{v}_{mn}^j(t-u') - \frac{\bar{F}_j}{2m_j}(t-u')^2, u']du'\}du +$$

$$g_{(j,N+1),j}^C\, \eta_{j,N+1}^j \int_0^t (\bar{v}_{j,N+1}^j - \bar{v}_{jo} - \frac{\bar{F}_j}{m_j}u)\quad x$$

$$\rho_j[\bar{x} - \bar{v}_{j,N+1}^j(t-u) - \frac{\bar{F}_j}{2m_j}(t-u)^2, u]\quad x$$

$$\exp\{-\sum_{k=1}^{N+1} g_{jk} \int_u^t \rho_k[\bar{x} - \bar{v}_{j,N+1}^j(t-u') - \frac{\bar{F}_j}{2m_j}(t-u')(t-2u+u'), u']du'\}du,$$

coinciding with the result derivable by inserting Eq.(9) in Eq.(11) and eliminating appropriately the source term.
The system, Eq.(10), of complicated functional-hyperbolic type, is thus an autonomous system for the ρ_j's; it is also the system of nonlinear conservation equations into which the original system of nonlinear integro-partial differential Boltzmann equations for the f_j's has been converted, on the basis of the rather simplifying assumptions, Eqs.(3)-(6). Systems of nonlinear conservation equations, having different structure, are encountered also in other fields of mathematical physics; they are, in general, hyperbolic and 1-dimensional in space (see, for instance, the review article of [26]). Some challenging problems should be faced now. The first would be the interpretation of Eq.(11), which is the closure condition of our procedure, as a constitutive equation for $\bar{J}_j$ in terms of all the ρ_j's. For the case N=1, in absence of creation, some

results in this direction have been reported in [27]. It has been thus recognized therein that the first addend, in the r.h.s. of Eq.(10), which is proportional to ρ_j through the drift velocity $(\bar{v}_{jo}+\bar{F}_j/m_j)t$, is the one to be expected in the frame of classical continuum mechanics, whereas the second addend - which is a highly nonlinear functional, depending on all the ρ_j's and their previous history - is quite new and would need a more articulated interpretation. In the sequel, however, we shall be concentrating on the problem of solving analytically the system, Eq.(10), for the ρ_j's and of reconstructing the distribution function f_j by plugging back in Eq.(9) the ρ_j's so found. Such a problem will depend strongly on whether $\bar{J}_j^{sc}(\bar{x},t)$, Eq.(12a), is different from zero or not.

1.5 The case $\bar{J}_j^{sc}(\bar{x},t) \neq 0$

In this case the system, Eq.(10), is quite impracticable, and thus Eq.(9) turns out to be useless. If no forces are present, however, f_j may be recovered by means of Eq.(9), but independently of solving the system, Eq.(10), for the ρ_j's. Indeed, when $\bar{F}_j=0$, Eq.(9) is first rewritten as

$$f_j(\bar{x},\bar{v},t) = Y_{jo}(\bar{x},t)\,\delta(\bar{v}-\bar{v}_{jo}) + \sum_{\ell=1}^{N+1} Y_{j\ell}^S(\bar{x},t)\,\delta(\bar{v}-\bar{v}_{j\ell}) +$$

$$\sum_{m=1}^{N}\sum_{n=m}^{N} Y_{mn,j}^C(\bar{x},t)\,\delta(\bar{v}-\bar{v}_{mn}^j) + Y_{j,N+1}^C(\bar{x},t)\,\delta(\bar{v}-\bar{v}_{j,N+1}^j), \qquad (13)$$

where the contributions $Y_{jo}, Y_j^S, Y_{mn,j}^C, Y_{j,N+1}^C$ due to the source, the scattering, the creation by collision and the self-generation, respectively, are immediately identified by comparison with Eq.(9), and for any j, are just as many as M, the number of the special velocities $\bar{v}_{jo}, \bar{v}_{j\ell}, \bar{v}_{mn}^j$ and $\bar{v}_{j,N+1}^j$ as given by Eq.(8b). By integrating both the sides of Eq.(13) over the velocity domain, we further get

$$\rho_j(\bar{x},t) = Y_{jo}(\bar{x},t) + \sum_{\ell=1}^{N+1} Y_{j\ell}^S(\bar{x},t) +$$

$$\sum_{m=1}^{N}\sum_{n=m}^{N} Y_{mn,j}^C(\bar{x},t) + Y_{j,N+1}^C(\bar{x},t). \qquad (14)$$

For a given j the M unknowns $Y_{jo}, Y_{j\ell}^S, Y_{mn,j}^C, Y_{j,N+1}^C$ are then determined by inserting both Eqs.(13) and (14) in Eq.(8). A canonical system of M nonlinear hyperbolic equations is then derived by equating the coefficients of the same delta-term at both sides. For simplicity's sake we restrict here our calculations to the case N=2, and refer to the realistic physical situation in which, beside scattering and removal, creation effects consist of self-generation of the t.p. of species 2 by collision against the t.p. of species 1, plus the self-generation of both species 1 and 2 by collision against the f.p. of the bath. The only surviving creation collision frequencies are thus $g_{12,2}^c = g_{21,2}^c$, $g_{13,1}^c$ and $g_{23,2}^c$, respectively. There are, consequently, altogether 11 unknowns that we relabel as $Z_{j\ell}$ ($j=1,2$; $\ell=0,1,2,3+j$), that is, we set $Z_{jo} \equiv Y_{jo}$; $Z_{j\ell} \equiv Y_{j\ell}^s$ ($\ell=1,2,3$); $Z_{j4} \equiv Y_{j3}^c$ and $Z_{25} \equiv Y_{12,2}^c$, respectively. Correspondingly, we set $\bar{v}_{j3}^j \equiv \bar{v}_{j4}$ and $\bar{v}_{12}^2 \equiv \bar{v}_{25}$. The resulting system of 11 nonlinear hyperbolic (conservation) equations can be written in a compact form as (δ_ℓ^j denoting the Kronecker index)

$$\frac{\partial Z_{j\ell}}{\partial t} + \bar{v}_{j\ell} \cdot \frac{\partial Z_{j\ell}}{\partial \bar{x}} + L_{j\ell}(Z_{1\ell}, Z_{2\ell}) Z_{j\ell} = (1-\delta_\ell^o)\Phi_{j\ell}(Z_{1\ell}, Z_{2\ell}), \quad (15)$$

to be integrated upon the initial conditions

$$Z_{jo}(\bar{x},0) = Q_j(\bar{x}) \; ; \quad Z_{j\ell}(\bar{x},0) = 0 \; , \quad (\ell>0) \qquad (15a)$$

as follows from the definition of the $Z_{j\ell}$'s. In Eq.(15), $L_{j\ell}$ is a linear combination of the $Z_{j\ell}$'s given explicitly by

$$L_{j\ell}(Z_{1\ell}, Z_{2\ell}) = (g_{j3} - g_{j3}^s \delta_\ell^3 - g_{14}^s \delta_\ell^4) +$$

$$\sum_{m=0}^{4+j-1} \{g_{jj} - (2-\delta_m^j)g_{jj}^s [\prod_{\substack{n=0 \\ n\neq j}}^{4} (1-\delta_\ell^n)](1-\delta_\ell^5 \delta_1^{j\pm 1})\} Z_{jm} +$$

$$(g_{j,j\pm 1} - g_{j,j\pm 1}^s \delta_\ell^{j\pm 1} - g_{j5}^s \delta_\ell^5 \delta_1^{j\pm 1}) \sum_{m=0}^{3+(j\pm 1)} Z_{j\pm 1,m} \; , \qquad (16a)$$

whereas $\Phi_{j\ell}$, which is essentially a quadratic form, is in turn given by

$$\Phi_{j\ell}(Z_{1\ell}, Z_{2\ell}) = g_{j\ell}^s (\sum_{m\neq\ell}^{4} Z_{1m} + \delta_2^j Z_{1\ell})^{\delta_\ell^1 + (2-j)\sum_{m=1}^{4}\delta_\ell^m + (j-1)\delta_\ell^5} x$$

$$(\sum_{m\neq\ell}^{5} Z_{2m} + \delta_1^j Z_{2\ell})^{\delta_\ell^2 + (j-1)\sum_{m=1}^{5}\delta_\ell^m} \; . \qquad (16b)$$

In Eqs.(16) the upper (lower) sign holds for $j=1$ ($j=2$). We recall further that $g_{14}^s \equiv g_{13,1}^c \eta_{13}^1$, $g_{24}^s \equiv g_{23,2}^c$, $g_{25}^s \equiv g_{12,2}^c \eta_{12}^2$. It is worthwhile remarking that the order, 11x11, of the system, Eq.(15), can be reduced if the number, 11, of the special velocities is correspondingly decreased. This can be attained, for instance, if the velocities admissible after scattering were equal, namely $\bar{v}_{j3} = \bar{v}_{j2} = \bar{v}_{j1}$ for $j=1$ and $j=2$, respectively . The system, Eq.(15), will be then 7x7, and will follow from the appropriate modifications of both Eqs.(15) and (16).

1.6 The case $\overline{J}_j^{SC}(\overline{x},t)=0$

It is easily verified that $\overline{J}_j^{SC}(\overline{x},t)$ vanishes in the following two circumstances.

-i) The first is when scattering and creation are negligible, and only removal, at the fixed velocity $\overline{v}_{jo}$, between the t.p., and the t.p. and the f.p. of the mixture considered is accounted for. The system, Eq.(10), for the ρ_j's becomes then

$$\frac{\partial \rho_j}{\partial t} + (\overline{v}_{jo}+ \frac{\overline{F}_j}{m_j}\, t)\cdot\frac{\partial \rho_j}{\partial \overline{x}} + \sum_{\ell=1}^{N+1} g_{j\ell}^{R}\ \rho_\ell(\overline{x},t)\rho_j(\overline{x},t) = 0 \ ;$$

$$\rho_j(\overline{x},0)=Q_j(\overline{x}), \tag{17}$$

which is hyperbolic, in the sense of the extended thermo-dynamics (see, for instance, [28]). Some analytical solutions to it have been obtained for N=2 in Refs. 21-23.

-ii) The second circumstance in which $\overline{J}_j^{SC}(\overline{x},t)$ vanishes, is when $\overline{v}_{j\ell} =\overline{v}_{mn}^{j}=\overline{v}_{jo}$ (ℓ,n=1,2,...,N+1) and $\overline{F}_j=0$. Now both scattering and creation events take place, all t.p. of species j maintaining their initial velocity $\overline{v}_{jo}$ until they are removed by absorption. The system, Eq.(10), is again hyperbolic and reads as

$$\frac{\partial \rho_j}{\partial t} + (\overline{v}_{jo} + \frac{\overline{F}_j}{m_j}\, t)\cdot\frac{\partial \rho_j}{\partial \overline{x}} + \sum_{\ell=1}^{N} g_{j\ell}^{A}\ \rho_\ell(\overline{x},t)\ \rho_j(\overline{x},t) -$$

$$\sum_{m=1}^{N}\sum_{n=m}^{N} g_{mn,j}^{C}\ \eta_{mn}^{j}\rho_m(\overline{x},t)\rho_n(\overline{x},t) + \nu_j\rho_j(\overline{x},t) = 0 \ ;$$

$$\rho_j(\overline{x},0) = Q_j(\overline{x}). \tag{18}$$

We shall next consider some new solutions, for N=2, to the systems Eqs.(17) and (18), restricting ourselves to nontrivial, nonnegative solutions.

2 APPLICATIONS

2.1 The case N=2

Both the systems, Eqs.(17) and (18), can be treated simultaneously by recasting them in the common form

$$\frac{\partial \rho_1}{\partial t} + (\bar{v}_{10} + \frac{\bar{F}_1}{m_1}t) \cdot \frac{\partial \rho_1}{\partial \bar{x}} + C_{11}\rho_1^2 + C_{12}\rho_1\rho_2 + C_{13}\rho_1 = 0 \; ; \quad \rho_1(\bar{x},0) = Q_1(\bar{x}) \; ;$$

$$\tag{19}$$

$$\frac{\partial \rho_2}{\partial t} + (\bar{v}_{20} + \frac{\bar{F}_2}{m_2}t) \cdot \frac{\partial \rho_2}{\partial \bar{x}} + C_{21}\rho_1\rho_2 + C_{22}\rho_2^2 + C_{23}\rho_2 = 0 \; ; \quad \rho_2(\bar{x},0) = Q_2(\bar{x}).$$

Observe that, according to the physical situation, the six coefficients $C_{j\ell}$ (j=1,2; ℓ=1,2,3) can exhibit different signs. In the case of only removal (Eq.(17) for N=2), $C_{j\ell} \geq 0$ for all j and ℓ. For the case of the system, Eq.(18), for N=2, referring to the physical situation when creation reduces to only self-generation (as in connection with the derivation of Eq.(15)), we have $\bar{F}_j = 0$ and

$$C_{jj} = g_{jj}^R \geq 0 \; ;$$

$$C_{12} = g_{12}^R + g_{12,2}^C > 0 \; ; \quad C_{21} = g_{21}^R - (\eta_{21}^2 - 1)g_{21,2}^C \gtrless 0 \; ; \quad (g_{12,2}^C = g_{21,2}^C)$$

$$\tag{20}$$

$$C_{j3} \equiv \nu_j = g_{j3}^R - (\eta_{j3}^j - 1) g_{j3,j}^C \gtrless 0.$$

In discussing the system, Eq.(19), it is crucial to consider the drift velocity $\bar{v}_{jo} + (\bar{F}_j/m_j)t$, acting as a coefficient, linear in time, of the gradient of ρ_j. Here we shall confine our attention to the case

$$\bar{v}_{20} = \bar{v}_{10} \; ; \quad \bar{F}_2/m_2 = \bar{F}_1/m_1 \; , \tag{21}$$

that is, when the velocities at which the t.p. of both species are injected and the forces per unit mass coincide. The general case, $\bar{v}_{20} \# \bar{v}_{10}$, $\bar{F}_2/m_2 \# \bar{F}_1/m_1$, is more difficult, and will be the object of a separate paper.

2.2 The system, Eq.(19), under the conditions, Eq.(21)

Upon the change to the new independent variables

$$\bar{\xi} = \bar{x} - \bar{v}_{jo}t - \frac{\bar{F}_j}{2m_j} t^2 \; ; \quad \eta = t \; , \tag{22}$$

the system, Eq.(19), becomes (j=1,2)

$$\frac{\partial \rho_j \left(\bar{\xi} + \bar{v}_{j_0}\eta + \frac{\bar{F}_j}{2m_j}\eta^2, \eta \right)}{\partial \eta} +$$

$$\sum_{\ell=1}^{3} C_{j\ell}\rho_\ell \left(\bar{\xi} + \bar{v}_{j_0}\eta + \frac{\bar{F}_j}{2m_j}\eta^2, \eta \right) \rho_j \left(\bar{\xi} + \bar{v}_{j_0}\eta + \frac{\bar{F}_j}{2m_j}\eta^2, \eta \right) = 0$$

(23)

with $\rho_j(\bar{\xi}, 0) = Q_j(\bar{\xi})$. Proceeding by substitution, we then solve the first of Eq.(23) for ρ_1 as a function of ρ_2, obtaining

$$\rho_1 \left(\bar{\xi} + \bar{v}_{10}\eta + \frac{\bar{F}_1}{2m_1}\eta^2, \eta \right) = \frac{Q_1(\bar{\xi})}{h(\rho_2)}$$

(24)

with

$$h(\rho_2) = \exp\left[C_{12} \int_0^\eta \rho_2 \left(\bar{\xi} + \bar{v}_{10}u + \frac{\bar{F}_1}{2m_1}u^2, u \right) du + C_{13}\eta \right] +$$

(24a)

$$C_{11}Q_1(\bar{\xi}) \int_0^\eta \exp\left[C_{12} \int_u^\eta \rho_2 \left(\bar{\xi} + \bar{v}_{10}u' + \frac{\bar{F}_1}{2m_1}u'^2, u' \right) du' + C_{13}(\eta-u) \right] du.$$

Equation (24) is then inserted in Eq.(23) with j=2 to yield for ρ_2 the autonomous nonlinear functional equation

$$\frac{\partial \rho_2 \left(\bar{\xi} + \bar{v}_{20}\eta + \frac{\bar{F}_2}{2m_2}\eta^2, \eta \right)}{\partial \eta} + C_{22}\rho_2^2 \left(\bar{\xi} + \bar{v}_{20}\eta + \frac{\bar{F}_2}{2m_2}\eta^2, \eta \right) +$$

$$C_{23}\rho_2 \left(\bar{\xi} + \bar{v}_{20}\eta + \frac{\bar{F}_2}{2m_2}\eta^2, \eta \right) +$$

$$C_{21}Q_1(\bar{\xi})\rho_2 \left(\bar{\xi} + \bar{v}_{20}\eta + \frac{\bar{F}_2}{2m_2}\eta^2, \eta \right) / h\left[\rho_2 \left(\bar{\xi} + \bar{v}_{10}\eta + \frac{\bar{F}_1}{2m_1}\eta^2, \eta \right) \right] = 0. \quad (25)$$

This equation is amenable to a standard form once we use the conditions Eq.(21). Indeed, regarding $\bar{\xi}$ as a parameter, for the new dependent variable

$$y(\bar{\xi}, \eta) = h(\rho_2)$$

(26)

we find the nonlinear second-order ordinary differential equation

$$y\ddot{y} + Ay\dot{y} + B\dot{y}^2 + Cy^2 + D\dot{y} + Ey + F = 0,$$

(27)

$$y(0) = 1 \; ; \qquad \dot{y}(0) = C_{13} + C_{11}Q_1 + C_{12}Q_2 . \qquad (27a)$$

In Eq.(27) the six coefficients A,B,C,D,E,F are expressed in terms of the original six coefficients $C_{j\ell}$ and of the initial datum Q_1 as

$$A = C_{23} - 2\,\frac{C_{22}}{C_{12}}\,C_{13}\; ; \qquad D = Q_1(C_{11} + C_{21} - 2\,\frac{C_{22}}{C_{12}}\,C_{11})\; ;$$

$$B = -(1 - \frac{C_{22}}{C_{12}})\; ; \qquad E = -Q_1(C_{11}C_{23} + C_{13}C_{21} - 2\,\frac{C_{22}}{C_{12}}\,C_{13}C_{11})\,;$$

$$C = -C_{13}(C_{23} - \frac{C_{22}}{C_{12}}\,C_{13})\,; \qquad F = -C_{11}Q_1^2(C_{21} - \frac{C_{22}}{C_{12}}\,C_{11}). \qquad (28)$$

Once y is determined from Eqs.(27)+(27a) we can recover ρ_2 by just inverting Eq.(26). It is thus found that

$$\rho_2(\bar{\xi} + \bar{v}_{10}\eta + \frac{\overline{F}_1}{2m_1}\,\eta^2, \eta) = (C_{12}y)^{-1}(\dot{y} - C_{13}y - C_{11}Q_1)\;, \qquad (29)$$

which together with Eq.(24) for ρ_1 constitutes the solution to the system, Eq.(23), under the conditions, Eq.(21).

2.3 Analytical solutions to Eqs.(27)+(27a)

We list henceforth a number of analytical solutions to Eqs.(27)+(27a) (compare also [29]). We do not pretend, of course, to exhaust the class of such solutions, even if, because of the physical meaning of the original coefficients $C_{j\ell}$ and, consequently, of the coefficients A,B,C,D,E,F in Eq.(27), this class seems to be less ample than expected.

-i) $C_{13} = C_{23} = 0$

In this case A = C = E = 0 so that Eq.(27) becomes

$$y\ddot{y} + B\dot{y}^2 + D\dot{y} + F = 0 \qquad (30)$$

with $y(0) = 1$; $\dot{y}(0) = C_{11}Q_1 + C_{12}Q_2$. If we further assume that $C_{22} = C_{21} = 0$, $C_{11}Q_1 = 2$, then Eq.(30) is also

$$y\ddot{y} = \dot{y}^2 - 2\dot{y} \qquad (31)$$

with $y(0)=1$; $\dot{y}(0)=2+C_{12}Q_2$. Omitting the trivial solution $y=$const., it is easily verified that Eq.(31) admits the analytical solution

$$y = (C_{12}Q_2)^{-1}[(2+C_{12}Q_2)\exp(C_{12}Q_2\eta)-2], \qquad (32)$$

satisfying the relevant initial conditions.

-ii) $C_{22} = C_{12}$; $C_{21} = C_{11}$

In this case $B=D=F=0$ so that Eq.(27) becomes linear and reads as

$$\ddot{y} + A\dot{y} + Cy + E = 0 \qquad (33)$$

with

$$y(0) = 1 ; \qquad \dot{y}(0) = C_{13} + C_{11}Q_1 + C_{12}Q_2$$
$$A = C_{23}-2C_{13}; \quad C = -C_{13}(C_{23}-C_{13}); \quad E = -C_{11}Q_1(C_{23}-C_{13}) \qquad (33a)$$

A particular solution to Eq.(33) is $y=-E/C$, to which we must add the solution of the associated homogeneous equation. This does not imply any difficulty, and leads to the analytical solution

$$y = -\frac{C_{11}Q_1}{C_{13}} + (1+\frac{C_{11}Q_1}{C_{13}}+\frac{C_{12}Q_2}{C_{23}})\exp(C_{13}\eta) - \frac{C_{12}Q_2}{C_{23}}\exp[(C_{13}-C_{23})\eta], (34)$$

satisfying the relevant initial conditions.

-iii) $C_{22} = C_{21} = C_{11} = 0$; $C_{13} = C_{23} = C_3$

Equation (27) becomes

$$y\ddot{y} + C_3y\dot{y} - \dot{y}^2 - C_3^2y^2 = 0 \qquad (35)$$

with $y(0)=1$; $\dot{y}(0)=C_3+C_{12}Q_2$. Dividing by y^2, a first quadrature is made possible yielding for the new dependent variable $z=\ln y$ the first-order first-degree linear ordinary differential equation

$$\dot{z} + C_3z = C_3 + C_{12}Q_2 + C_3^2\eta \qquad (36)$$

with $z(0) = 0$. Equation (36) is solved by

$$z = C_3^{-1}C_{12}Q_2[1-\exp(-C_3\eta)] + C_3\eta, \qquad (37)$$

and therefore

$$y = \exp\{C_3^{-1}C_{12}Q_2[1-\exp(-C_3\eta)] + C_3\eta\} , \qquad (38)$$

satisfying the relevant initial condition.

-iv) $C_{21} = C_{11} = 0$; $C_{13} = C_{23} = C_3$; $C_{12} = C_{22}$

In this case, B=C=D=E=F=0, and Eq.(27) reduces simply to

$$\dot{y} - C_3 y = C_{22} Q_2 \qquad\qquad (39)$$

which is solved by

$$y = C_3^{-1}[(C_3 + C_{22} Q_2) \exp(C_3 \eta) - C_{22} Q_2] \qquad\qquad (40)$$

satisfying the relevant initial conditions y(0)=1, and
$\dot{y}(0) = C_3 + C_{22} Q_2$, respectively.

-v) $C_{13} = C_{23} = C_{11} = 0$; $C_{12} = C_{22}$

In this case, A=B=C=E=F=0, and Eq.(27) is then

$$y\ddot{y} + C_{21} Q_1 \dot{y} = 0 \qquad\qquad (41)$$

with y(0)=1; $\dot{y}(0)=C_{22} Q_2$. A first quadrature leads to

$$\dot{y} = C_{22} Q_2 - C_{21} Q_1 \ln y , \qquad\qquad (42a)$$

which is an equation with separable variables, namely

$$\frac{dy}{C_{22} Q_2 - C_{21} Q_1 \ln y} = d\eta . \qquad\qquad (42b)$$

Introducing the function logarithmic integral [30]

$$\mathrm{li}\ u = \int_0^u \frac{du'}{\ln u'} , \qquad (u>1) \qquad\qquad (43)$$

and resorting to an integral listed in [31], we get then an
analytical solution of implicit form

$$\mathrm{li}[\exp(-\frac{C_{22} Q_2}{C_{21} Q_1})y] = \mathrm{li}[\exp(-\frac{C_{22} Q_2}{C_{21} Q_1})] - C_{21} Q_1 [\exp(-\frac{C_{22} Q_2}{C_{21} Q_1})]\eta . \qquad (44)$$

This solution is mathematically consistent with the
definition, Eq.(43), if with Q_1, Q_2, C_{22} positive, $C_{21}<0$, as
occurs in the case of Eqs.(20).

2.4 Reconstruction of ρ_2

With reference to case -iv) of the preceeding section, we invert now Eq.(40) for η_2 by means of Eq.(29). Going back to the original variables $(\bar{x},t)$ we find then

$$\rho_2(\bar{x},t) = \cfrac{C_3 Q_2(\bar{x}-\bar{v}_{10}t-\cfrac{\bar{F}_1}{2m_1}t^2)}{[C_{22}Q_2(\bar{x}-\bar{v}_{10}t-\cfrac{\bar{F}_1}{2m_1}t^2)+C_3]\exp(C_3 t)-C_{22}Q_2(\bar{x}-\bar{v}_{10}t-\cfrac{\bar{F}_1}{2m_1}t^2)} , \quad (45)$$

which is nonnegative, as physically due. According to the sign of C_3 we realize that

$$\lim_{t\to\infty} \rho_2(\bar{x},t) = 0 \qquad\qquad (46a)$$

for $C_3 \geq 0$, whereas for $C_3 < 0$, ρ_2 tends to a constant (positive) value, namely

$$\lim_{t\to\infty} \rho_2(\bar{x},t) = -\frac{C_3}{C_{22}} . \qquad\qquad (46b)$$

We further observe that, in the limit of $C_3 \to 0$, Eq.(45) degenerates as

$$\rho_2(\bar{x},t) = \cfrac{Q_2(\bar{x}-\bar{v}_{10}t-\cfrac{\bar{F}_1}{2m_1}t^2)}{1 + C_{22}Q_2(\bar{x}-\bar{v}_{10}t-\cfrac{\bar{F}_1}{2m_1}t^2)t} , \qquad\qquad (47)$$

and, when $C_3 \leq 0$, we must set $\bar{F}_1 = 0$. The other examples of reconstruction of ρ_2 from the relevant y, through Eq.(29), can be carried out similarly.

CONCLUSIONS

It has been shown that, given the set of hypotheses, Eqs.(3)-(6), the nonlinear Boltzmann system, Eq.(1), can be solved for the distribution function f_j of the t.p. of the general species j - in terms of all the number densities ρ_j, Eq.(7)- by means of Eq.(9). This latter equation has been then exploited for deriving the autonomous nonlinear conservation system, Eq.(10), for the ρ_j's themselves, and this system has been successively faced in the limiting case of the functional $\bar{J}_j^{sc}$, Eq.(12a), taken to be zero. Analytical results for non trivial nonnegative solutions to such systems have been obtained for N=2, accounting for only self-generation effects. The problems of reconstructing f_j via Eq.(9) for the case of $\bar{J}_j^{sc} \neq 0$ has also been sketched by introducing the new dependent variables $Z_{j\ell}(\bar{x},t)$, provided by the system, Eq.(15). Various problems are left open by the present treatment, and among them we list the following ones.

-i) More sophisticated models for collision frequencies, scattering and creation probability distributions should be adopted in order to produce, for assigned initial data, new constitutive equations for $\bar{J}_j$, Eq.(12), to be then interpreted in the frame of continuum mechanics.

-ii) Analytical solutions to the system, Eq.(19), should be constructed by relaxing the conditions, Eq.(21), which seem to be crucial, considering also values of N>2. Also the systems, Eq.(10) and Eq.(15), both concerning the case with $\bar{J}_j^{sc} \neq 0$, would deserve further investigation.

-iii) The connections between the present theory and the theory of the nonlinear hyperbolic conservation systems should be worked out, especially by trying to extend the results of the latter from the 1-dimensional scheme (in which it is mostly framed thus far) to a 3-dimensional one.

-iv) The boundary value problem associated with the system, Eq.(1) and Eq.(10) (or Eq.(19)), when $\bar{x} \in D_3 \subset R_3$, should be investigated upon an appropriate formulation of the relevant boundary conditions.

-v) The connections with the discrete velocity model of the Boltzmann equations should be identified (see ,for instance, [32] and [33])

ACKNOWLEDGMENTS

The research work dealt with in this paper has been carried out under a 40% - Project, supported by the Italian Board of Education. Also the support of the C.N.R.'s National Group for Mathematical Physics is gratefully acknowledged.

REFERENCES

[1] V.C.BOFFI and G.SPIGA, An analytical study of space-dependent evolution problems in particle transport theory, Il Nuovo Cimento, 7D: 327-338 (1986).

[2] R.ILLNER and M.SHINBROT, The Boltzmann equation: global existence for a rare gas in an infinite vacuum, Comm.Math.Phys., 95: 217-226 (1984).

[3] S.UKAI, Local solutions in Gevrey classes to the nonlinear Boltzmann equation without cutoff, Japan. J.Appl. Math., 1: 141-156 (1984).

[4] N.BELLOMO and G.TOSCANI, On the Cauchy problem for the nonlinear Boltzmann equation: global existence and uniqueness and asymptotic stability, J.Math.Phys., 26: 334-338 (1985).

[5] G.TOSCANI and V.PROTOPOPESCU, Global existence for a mixed problem associated with the nonlinear Boltzmann equation, C.R.Acad.Sci.Paris, Sér. I Math.,302: 255-258 (1986).

[6] G.TOSCANI, On the nonlinear Boltzmann equation in unbounded domains, Arch.Rat.Mech.Anal., 95: 37-49 (1986).

[7] W.RUIJGROK and T.T.WU, A completely solvable model of the nonlinear Boltzmann equation, Physica, 113A : 401-416 (1982).

[8] V.V.STRUMINSKIY, Analytical methods for solving the Boltzmann equation, Fluid Mech.-Soviet Res. 13: 55-69 (1984).

[9] H.CORNILLE, Oscillating Maxwellians, J.Phys.A, 18: L839-L844 (1985).

[10] H.CORNILLE, Exact solution for the spatially inhomogeneous nonlinear Kac model of the Boltzmann equation, J.Math.Phys., 26: 1203-1214 (1985).

[11] H.CORNILLE, Oscillating and absolute Maxwellians: exact solutions for (d>1)-dimensional Boltzmann equation, J.Math.Phys.,27: 1373-1386 (1986).

[12] V.C.BOFFI and G.SPIGA, Spatial effects in the study of
     nonlinear evolution problems of particle transport
     theory, in Proceedings of the X International
     Conference on Transport Theory (La Jolla, Ca.,23-26
     march 1987) G.C.Pomraning, ed., Trans.Th. Stat.
     Phys., to appear.

[13] A.BORZYMOWSKI and J.URBANOWICZ, Uniqueness of the
     solution of a tangential derivative problem for an
     infinite system of nonlinear-integrodifferential
     equations of parabolic type, Ann.Pol.Math., 41: 99-
     109 (1983).

[14] P.OLSZEWSKI, Cauchy's problem for a certain system of
     integro-differential equations of order 2p,
     Demonstratio Math., 16: 27-38 (1983).

[15] V.C.BOFFI,V.FRANCESCHINI and G.SPIGA, Dynamics of a gas
     mixture in an extended kinetic theory, Phys.Fluids,
     28: 3232-3236 (1985).

[16] V.C.BOFFI and G.SPIGA, Extended kinetic theory for gas
     mixtures in the presence of removal and
     regeneration effects, Zeit.Angew.Math.Phys., 37:
     27-42 (1986).

[17] V.C.BOFFI and G.SPIGA, Calculation of the number
     densities in an extended kinetic theory of gas
     mixtures, Trans.Th.Stat.Phys., 16: 175-188 (1987).

[18] G.SPIGA,P.VESTRUCCI and V.C.BOFFI, Dynamics of gas
     mixtures in an extended kinetic theory, in
     Proceedings of the Italian-Polish Meeting on
     Selected Problems of Modern Continuum Mechanics
     (Bologna, 03-06 june 1987), A.Kosinski,
     T.Manacorda, A.Morro, T.Ruggeri, eds.,157, Tecno-
     print, Bologna, 1987.

[19] G.SPIGA, Dynamical systems in nonlinear transport
     theory, Report 7/87,Mathematics Department of the
     University of Bari, 1987.

[20] V.C.BOFFI and G.SPIGA, A dynamical analysis of the
     breeding process in nuclear engineering, in Book of
     Abstracts FICIAM 87 (Paris, La Villette, 28 june-03
     july 1987), 125, INRIA, Paris, 1987.

[21] V.C.BOFFI and K.AOKI, A system of conservation
     equations arising in nonlinear dynamics of gas
     mixtures, Il Nuovo Cimento, 10D: 145-159 (1988).

[22] V.C.BOFFI and K.AOKI, Nonlinear hyperbolicity in
     kinetic theory of a gas mixture, Il Nuovo Cimento,
     to appear.

[23] V.C.BOFFI, Nonlinear spatially inhomogeneous evolution problems in gas kinetic theory, Atti dell'Accademia Gioenia delle Scienze, Catania, to appear.

[24] V.C.BOFFI and V.MOLINARI, Nonlinear transport problems by factorization of the scattering probability, Il Nuovo Cimento, 65B: 29-44 (1981).

[25] P.L.BHATNAGAR,E.P.GROSS and M.KROOK, A model for collision processes in gases .I., Phys.Rev., 94: 511-525 (1954).

[26] R.J.DI PERNA, Measure-valued solutions to conservation laws, Arch.Rat.Mech.Anal., 88: 223-270 (1985).

[27] V.C.BOFFI and T.RUGGERI, Spatially inhomogeneous evolution problems in kinetic theory, in Proceedings of the ISIMM Symposium on Kinetic Theory and Extended Thermodynamics (Bologna, 18-22 may 1987), I.Müller and T.Ruggeri, eds., 41, Tecnoprint, Bologna, 1987.

[28] T.RUGGERI, Mathematical structure of the extended thermodynamics, in Proceedings of the ISIMM Symposium on Kinetic Theory and Extended Thermodynamics (Bologna, 18-22 may 1987), I.Müller and T.Ruggeri, eds., 279, Tecnoprint, Bologna, 1987.

[29] G.M.MURPHY, Ordinary Differential Equations and Their solutions, Van Nostrand, Princeton, 1960.

[30] M.ABRAMOWITZ and I.STEGUN,eds., Handbook of Mathematical functions, Dover, New York, 1964.

[31] W.GRÖBNER and N.HOFREITER, Integraltafel, Vol.I, Springer-Verlag, Wien-New York, 1965.

[32] G.TOSCANI, Global existence and asymptotic behavior for the discrete velocity models of the Boltzmann equation, J.Math.Phys., 26: 2918-2921 (1985).

[33] H.CORNILLE, Exact solutions of the Broadwell model in 1+1 dimensions, J.Phys.A: Math.Gen. 20: 1973-1988 (1987).

The Nonlinear Boltzmann Equation:
Representation Theory, Super- and Subsolutions

PETER TAKÁČ   Department of Mathematics, Vanderbilt University, Nashville, Tennessee

1. INTRODUCTION

In this article we investigate the global existence and convergence properties of standard super- and subsolutions of the spatially inhomogeneous, nonlinear Boltzmann equation in the absence of exterior forces; cf. C. Cercignani [7] or C. Truesdell & R. G. Muncaster [17]:

$$(1) \qquad \frac{\partial f}{\partial t} + \vec{\xi} \cdot \nabla_{\vec{x}} f = J(f, f), \qquad f(0) = \phi,$$

where the single particle distribution $f = f(t, \vec{x}, \vec{\xi})$ is a nonnegative real-valued function of the time $t \in \mathbb{R}_+ = [0, \infty)$ and of the position and velocity $(\vec{x}, \vec{\xi}) \in \Omega \times \mathbb{R}^3 \subset \mathbb{R}^3 \times \mathbb{R}^3$. The collisional operator J, for interaction potentials with a "cut-off", is the difference of the "gain" and "loss" operators:

$$(2) \qquad J(f, f) = Q(f, f) - f \cdot R(f),$$

where Q is a positive bilinear operator defined by

$$(3) \qquad Q(f,g)(t,\vec{x},\vec{\xi}) = \iint_{\mathbb{R}^3 \times D} B(\theta,|\vec{v}|) f(t,\vec{x},\vec{\xi}') g(t,\vec{x},\vec{\eta}') d\vec{\eta} d\theta d\varsigma,$$

and R is a positive linear operator defined by

$$(4) \qquad R(g)(t,\vec{x},\vec{\xi}) = \int_{\mathbb{R}^3} \left(\int_D B(\theta,|\vec{v}|) d\theta d\varsigma \right) g(t,\vec{x},\vec{\eta}) d\vec{\eta}.$$

In these definitions $(\vec{\xi},\vec{\eta})$ are the pre-collisional velocities, $(\vec{\xi}',\vec{\eta}')$ are the post-collisional velocities, $\vec{v} = \vec{\eta} - \vec{\xi}$ is the relative velocity, and the angles θ and ς are respectively the polar and azimuthal angles of the impact parameter $\vec{\omega} \in \mathbb{R}^3$, $|\vec{\omega}| = 1$, defined by

$$\vec{\omega} = \vec{\omega}(\theta,\varsigma) = -(\cos\varsigma\,\vec{e}_1 + \sin\varsigma\,\vec{e}_2)\sin\theta + \frac{\vec{v}}{|\vec{v}|}\cos\theta,$$

where $\{\vec{e}_1, \vec{e}_2, \frac{\vec{v}}{|\vec{v}|}\}$ form an orthonormal system in $\mathbb{R}^3$, and $(\theta,\varsigma) \in D = [0, \frac{\pi}{2}] \times [0, 2\pi]$. The velocities $(\vec{\xi}',\vec{\eta}')$ are determined through the relations

$$(5) \qquad \vec{\xi}' = \vec{\xi} + (\vec{v}\cdot\vec{\omega})\vec{\omega}, \ \vec{\eta}' = \vec{\eta} - (\vec{v}\cdot\vec{\omega})\vec{\omega}$$

which can be derived from the momentum and energy conservation laws.

The kernel $B(\theta,v)$ is determined by the details of the collision. For the sake of simplicity we assume that Ω is a three-dimensional torus $\Omega = \mathbb{R}^3/\mathbb{Z}^3$ endowed with the Lebesgue measure. This choice of Ω corresponds to a space-periodic particle distribution with a fixed period. Consequently the nonlinear Boltzmann equation (1) is equivalent to

$$(6) \qquad \frac{\partial f^{\#}}{\partial t} + f^{\#} \cdot R^{\#}(f) = Q^{\#}(f,f), \qquad f(0) = \phi,$$

where $f^{\#}(t,\vec{x},\vec{\xi}) = f(t,\vec{x} + \vec{\xi}t,\vec{\xi})$ for all $t \in \mathbb{R}_+$ and $(\vec{x},\vec{\xi}) \in \Omega \times \mathbb{R}^3$.

As the main topic of this article we investigate the existence and convergence properties of a pair of nonnegative $L^1(\Omega \times \mathbb{R}^3)$-valued functions f and g of the time $t \in \mathbb{R}_+$ satisfying $f \leq g$ and

$$(7) \qquad \frac{\partial f^{\#}}{\partial t} + f^{\#} \cdot R^{\#}(g) = Q^{\#}(f,f), \qquad f(0) = \phi,$$

in a suitable weak sense defined below, and the function g has all properties expected from a solution of (1), i.e. g conserves mass, does not increase the initial energy and entropy, and $g^{\#}$ is uniformly L^1-continuous in time with $g(0) = \phi$. We refer to g and f respectively as *super-* and *subsolutions* of the nonlinear Boltzmann equation (cf. Definition 2.1 below). Our main result is a weak L^1-convergence theorem for the supersolutions g. More precisely, let $\{\phi_n\}_{n=1}^{\infty}$ and $\{B_n\}_{n=1}^{\infty}$ be nondecreasing sequences of nonnegative initial values and scattering kernels, respectively, such that $\phi_n \longrightarrow \phi$ and $B_n \longrightarrow B$ a.e. as $n \longrightarrow \infty$. Then, if the sequence $\{g_n^{\#}\}_{n=1}^{\infty}$ of the corresponding supersolutions of (7) is uniformly equicontinuous in time, it is also relatively weakly L^1-compact, and every cluster point $g^{\#}$ of this sequence is a supersolution of (7). We prove this result under very natural hypotheses of finite mass, energy and entropy for the initial particle distribution $\phi \geq 0$:

$$(8) \qquad \phi(1 + |\vec{\xi}|^2), \qquad \phi \log \phi \in L^1(\Omega \times \mathbb{R}^3),$$

cf. L. Arkeryd [1,3,4].

The proof of our main result is based upon the following convexity property: given a supersolution g, the smallest corresponding subsolution $f = F(g)$ calculated from the monotone approximation scheme of E. Wild [19] is represented by a convex mapping F of the variable g. This observation allows us to use a standard weak L^1-compactness argument for approximating equations (cf. L. Arkeryd [1] and D. Morgenstern [13]) in order to show the inequality $F(g) \leq g$, since the convexity and lower semicontinuity of F imply its weak lower L^1-semicontinuity. In this proof also the convexity of the $\mathcal{H}$-function (entropy) plays a similar role. As a consequence of the convexity of F and $\mathcal{H}$ we obtain that the set $\mathcal{G}$ of all supersolutions g of the Boltzmann equation is convex and sequentially weakly L^1-compact.

Several authors (cf. L. Arkeryd [2], W. Greenberg, J. Voigt & P. F. Zweifel [9] and D. Morgenstern [13]) raised the question about what equation holds for a weak L^1-limit of a subsequence of the solutions to suitable approximating equations. In our proof below the supersolution g is such a limit, and the equation (7) is a partial answer to this question. Partial because we do not know whether $f = g$ in (7) except for the spatially homogeneous Boltzmann equation for Maxwellian molecules with a "cut-off". Also the uniform (norm-) L^1-continuity in time of every subsolution $f^{\#}$ is of particular interest, cf. L. Arkeryd [3,4].

While the existence of solutions of the spatially homogeneous Boltzmann equation has been proved in most physically interesting situations, this problem remains open for the spatially inhomogeneous Boltzmann equation except for the cases near the vacuum (cf. N. Bellomo & G. Toscani [6], K. Hamdache [10], R. Illner & M. Shinbrot [11], S. Kaniel & M. Shinbrot [12], M. Shinbrot [14], G. Toscani [16] and others), near an equilibrium (cf. Y. Shizuta & R. Asano [15], S. Ukai [18] and others), and near a spatially homogeneous solution (cf. L. Arkeryd, R. Esposito & M. Pulvirenti [5]). A rather special exception are nonstandard (Loeb) solutions introduced by L. Arkeryd [3,4] which satisfy (1) in a rather weak sense, but with very general initial data ϕ. Their L^1-regularity is still an open problem.

As for the organization of this article, in Section 2 we first introduce the concepts of super- and subsolutions of the Boltzmann equation (Definition 2.1) and then state our main result Theorem 2.2 together with our temporal L^1-continuity result for every subsolution f (Proposition 2.3). In Section 3 we develop a representation theory for the smallest subsolution $f = F(g)$ of the nonlinear Boltzmann equation (7). We summarize the results of this theory in Theorem 3.1 in terms of path integrals. In Section 4 we prove our main results (Theorem 2.2 and Proposition 2.3). Finally in Section 5 we show some additional properties of super- and subsolutions, such as mass and energy inequalities, which seem to be relevant for future investigations of our onjecture that every supersolution g is an actual solution of the Boltzmann equation (1), i.e. that $f = g$ in (7).

2. MAIN RESULTS

We denote by $\mathcal{M}^*$ the set of all Lebesgue measurable functions $f : \mathbb{R}_+ \times \Omega \times \mathbb{R}^3 \longrightarrow [-\infty, \infty]$ and set $\mathcal{M}_+^* = \{f \in \mathcal{M}^* : f \geq 0 \text{ a.e.}\}$ with the Lebesgue integral naturally extended to the set $\mathcal{M}_+^*$ with values in $\mathbb{R}_+^* = [0, \infty]$. Given $\kappa \in \mathbb{R}$, we denote by $L_\kappa^1(\Omega \times \mathbb{R}^3)$ the Lebesgue space of all Lebesgue measurable functions $f : \Omega \times \mathbb{R}^3 \longrightarrow \mathbb{R}$ whose norm

$$\|f\|_\kappa = \int_\Omega \int_{\mathbb{R}^3} |f(\vec{x}, \vec{\xi})| (1 + |\vec{\xi}|^2)^{\kappa/2} d\vec{x} d\vec{\xi}$$

is finite, and by $L_\kappa^1(\Omega \times \mathbb{R}^3)_+$ its positive cone. Finally we denote by $C_b(\mathbb{R}_+ \longrightarrow L_\kappa^1(\Omega \times \mathbb{R}^3))$ the Banach space of all continuous vector-valued functions $f : \mathbb{R}_+ \longrightarrow L_\kappa^1(\Omega \times \mathbb{R}^3)$ whose norm

$$\|\|f\|\|_\kappa = \sup\{\|f(t)\|_\kappa : t \in \mathbb{R}_+\}$$

is finite, and by $C_b(\mathbb{R}_+ \longrightarrow L_\kappa^1(\Omega \times \mathbb{R}^3))_+$ its positive cone.

For mathematical reasons we impose the following standard cut-off hypothesis on the scattering kernel $B(\theta, v)$: We assume that $B : [0, \frac{\pi}{2}] \times \mathbb{R}_+ \longrightarrow \mathbb{R}_+$ is a Lebesgue measurable function such that there exist constants $\gamma \in \mathbb{R}_+$, $\lambda_1 \in [0, 3)$ and $\lambda_2 \in [0, 2)$ satisfying

$$(9) \qquad \int_0^{\pi/2} B(\theta, v) d\theta \leq \gamma(v^{-\lambda_1} + v^{\lambda_2}) \text{ for a.e. } v > 0.$$

Consequently the loss operator R defined by (4) is a positive linear operator from $L_\kappa^1(\mathbb{R}^3)$ into the linear sum of itself and its dual space $L_{-\kappa}^\infty(\mathbb{R}^3)$ endowed with the weighted supremum norm

$$\text{ess sup}\{|f(\vec{\xi})|(1 + |\vec{\xi}|^2)^{-\kappa/2} : \vec{\xi} \in \mathbb{R}^3\},$$

for every $\kappa \geq \lambda_2$. Note that R is independent from time and position. Furthermore, given g with $g^\# \in C_b(\mathbb{R}_+ \longrightarrow L_\kappa^1(\Omega \times \mathbb{R}^3))_+$ for some $\kappa \geq \lambda_2$, we have, for a.e. $(\vec{x}, \vec{\xi}) \in \Omega \times \mathbb{R}^3$,

$$(10) \qquad \int_0^t R^\#(g)(\tau, \vec{x}, \vec{\xi}) d\tau = \int_0^t R(g)(\tau, \vec{x} + \vec{\xi}\tau, \vec{\xi}) d\tau < \infty$$

for all $t \in \mathbb{R}_+$ (cf. Lemma 4.1 below).

Given such g, we say that a function f is a *mild solution* of (7) if it satisfies $f^{\#} \in C_b(\mathbb{R}_+ \longrightarrow L^1(\Omega \times \mathbb{R}^3))_+$ and

(11)
$$f^{\#}(t) = \phi \exp(-\int_0^t R^{\#}(g)(\tau)d\tau)$$
$$+ \int_0^t Q^{\#}(f,f)(s) \exp(-\int_s^t R^{\#}(g)(\tau)d\tau)ds \quad \text{for all } t \in \mathbb{R}_+,$$

where $\phi \in L^1(\Omega \times \mathbb{R}^3)_+$ is given. Note that (11) makes sense whenever $f, g \in \mathcal{M}_+^*$ and g satisfies (10).

DEFINITION 2.1. We say that a function g is a *supersolution* of (7) if it satisfies $g^{\#} \in C_b(\mathbb{R}_+ \longrightarrow L^1(\Omega \times \mathbb{R}^3))_+$ with the moments satisfying

(12)
$$\mathcal{M}g(t) \equiv \int_\Omega \int_{\mathbb{R}^3} g(t, \vec{x}, \vec{\xi})d\vec{x}d\vec{\xi} = \mathcal{M}\phi,$$

(13)
$$\vec{\mathcal{P}}g(t) \equiv \int_\Omega \int_{\mathbb{R}^3} \vec{\xi}g(t, \vec{x}, \vec{\xi})d\vec{x}d\vec{\xi} = \vec{\mathcal{P}}\phi,$$

(14)
$$\mathcal{E}g(t) \equiv \int_\Omega \int_{\mathbb{R}^3} |\vec{\xi}|^2 g(t, \vec{x}, \vec{\xi})d\vec{x}d\vec{\xi} \leq \mathcal{E}\phi$$

(mass, momentum and 2× energy, respectively) and the entropy satisfying

(15)
$$\mathcal{H}g(t) \equiv \int_\Omega \int_{\mathbb{R}^3} g(t) \log g(t)d\vec{x}d\vec{\xi} \leq \mathcal{H}\phi$$

for every $t \in \mathbb{R}_+$, and if there exists a mild solution f of (7) such that $\mathcal{M}f : \mathbb{R}_+ \longrightarrow \mathbb{R}_+$ is nonincreasing, and $0 \leq f(t) \leq g(t)$ for every $t \in \mathbb{R}_+$. Here we assume that the initial particle distribution $\phi \geq 0$ satisfies (8). We say that every such a function f is a *subsolution* of (7) corresponding to the supersolution g. Finally, if $f = g$, then f is called a *mild solution* of the Boltzmann equation (6). ∎

All integrals in this definition are absolutely convergent, cf. L. Arkeryd [1] or our proof of Lemma 4.2 below, since $\phi(1 + |\vec{\xi}|^2) \in L^1(\Omega \times \mathbb{R}^3)_+$ entails $\phi \log^- \phi \in L^1(\Omega \times \mathbb{R}^3)_+$ where $\log^- \alpha = \max\{-\log \alpha, 0\}$ and $\log^+ \alpha = \max\{\log \alpha, 0\}$ for all $\alpha > 0$. Also $f^{\#}, g^{\#} \in C_b(\mathbb{R}_+ \longrightarrow L^1(\Omega \times \mathbb{R}^3))$ combined with $0 \leq f \leq g$ and (14) implies $f^{\#}, g^{\#} \in C_b(\mathbb{R}_+ \longrightarrow L^1_\kappa(\Omega \times \mathbb{R}^3))_+$ whenever $\kappa < 2$, cf. our proof of Proposition 2.3 below. Now we are ready to state our main result:

THEOREM 2.2. *Let $M > 0$, $E > 0$ and $H \in \mathbb{R}$ be some constants, and denote by $\Phi(M, E, H)$ the set of all $\phi \geq 0$ satisfying (8) with*

(16)
$$\mathcal{M}\phi \leq M, \quad \mathcal{E}\phi \leq E \quad \text{and} \quad \mathcal{H}\phi \leq H.$$

Assume that $B \geq B_{n+1} \geq B_n \geq 0$ satisfy (9), and $0 \leq \phi_n \leq \phi_{n+1} \leq \phi \in \Phi(M, E, H)$ for $n = 1, 2, \cdots$. If g_n is a supersolution of (7) with B_n and ϕ_n in place of B and ϕ, *respectively, for $n = 1, 2, \cdots$, and if the functions $g_n^{\#} \in C_b(\mathbb{R}_+ \longrightarrow L^1(\Omega \times \mathbb{R}^3))$ are uniformly equicontinuous in time, then there exists a supersolution g of (7) with B and ϕ. Moreover, every subsolution f of (7) corresponding to the supersolution g satisfies also*

$$(17) \qquad \begin{aligned} &\|f^{\#}(t+h) - f^{\#}(t)\|_0 \le C_0(-\log|h|)^{-1} \\ &\text{for all } t, t+h \in \mathbb{R}_+ \text{ with } 0 < |h| < 1, \end{aligned}$$

where $C_0 = C_0(|\Omega|, \gamma, \lambda_1, \lambda_2, M, E, H) \ge 0$ is a constant depending only on the Lebesgue measure $|\Omega|$ of the torus Ω and the constants appearing in (9) and (16). ∎

The following result shows that f need not be a subsolution of (7) in order to satisfy (17):

PROPOSITION 2.3. *Let M, E, H and $\Phi(M, E, H)$ be as in Theorem 2.2, and let $B \ge 0$ satisfy (9). Assume that $f, g : \mathbb{R}_+ \longrightarrow L^1(\Omega \times \mathbb{R}^3)_+$ are two functions defined on all of $\mathbb{R}_+$ which satisfy $0 \le f(t) \le g(t) \in \Phi(M, E, H)$ for all $t \in \mathbb{R}_+$, $f, g \in \mathcal{M}_+^*$, $\mathcal{M}f : \mathbb{R}_+ \longrightarrow \mathbb{R}_+$ is nonincreasing and*

$$(18) \qquad f^{\#}(t+h) \ge f^{\#}(t) \exp\left(-\int_0^h R^{\#}(g)(t+\tau)d\tau\right) \text{ for all } t, h \in \mathbb{R}_+.$$

Then f satisfies (17) with the same constant C_0. Moreover, $f : \mathbb{R}_+ \longrightarrow L_\kappa^1(\Omega \times \mathbb{R}^3)$ is a uniformly continuous function for every $\kappa < 2$. ∎

Note that (18) follows from (11) by multiplying (11) by the factor $\exp(-\int_0^h R^{\#}(g)(t+\tau)d\tau)$ and comparing the product to (11) with $t + h$ in place of t. In particular, if f is a solution of the Boltzmann equation (6) in the sense that $f(t) \in \Phi(M, E, H)$ for all $t \in \mathbb{R}_+$, $f \in \mathcal{M}_+^*$, $\mathcal{M}f : \mathbb{R}_+ \longrightarrow \mathbb{R}_+$ is nonincreasing, and (11) holds with $g = f$, then f satisfies also (17), i.e. $f^{\#}$ is uniformly continuous from $\mathbb{R}_+$ into $L^1(\Omega \times \mathbb{R}^3)$. Thus, in Theorem 2.2, if $g_n = f_n$ is a mild solution of (6) with B_n and ϕ_n in place of B and ϕ, respectively, for $n = 1, 2, \cdots$, then the functions $g_n^{\#}$ are uniformly equicontinuous in time, and so supersolution g of (7) exists. The continuity of nonstandard (Loeb) solutions of the Boltzmann equation was proved in L. Arkeryd [3,4] by nonstandard methods.

An example of the scattering kernel B which satisfies (9) is $B(\theta, v) = \beta_s(\theta)v^{(s-4)/s}$ for inverse-power interaction potentials $U(r) = r^{-s}(r > 0)$ with a cut-off, where $s \in [2, \infty]$ is a constant.

3. REPRESENTATION THEORY FOR SUBSOLUTIONS

Throughout this entire section we assume that $g \in \mathcal{M}_+^*$ is a function which satisfies (10), and set, for a.e. $(\vec{x}, \vec{\xi}) \in \Omega \times \mathbb{R}^3$,

$$(19) \qquad p^{\#}(s,t;\vec{x},\vec{\xi}) = \exp(-\int_s^t R^{\#}(g)(\tau,\vec{x},\vec{\xi})d\tau)$$

for all $0 \le s \le t < \infty$, together with $p^{\#}(s,t;\vec{x},\vec{\xi}) = p(s,t;\vec{x}+\vec{\xi}t,\vec{\xi})$ in our #-notation. Then the smallest solution f of (11), in the sense that $f \in \mathcal{M}_+^*$ and (11) holds with $f(t)$ Lebesgue measurable on $\Omega \times \mathbb{R}^3$ for every $t \in \mathbb{R}_+$, can be calculated form the monotone approximation scheme of E. Wild [19]: $f(t) = \lim_{m \to \infty} f_m(t)$ pointwise a.e. in $\Omega \times \mathbb{R}^3$ for all $t \in \mathbb{R}_+$, where $f_m \in \mathcal{M}_+^*$, $m = -1, 0, 1, 2, \cdots$, is defined by

$$(20) \qquad f_{-1} = 0, \quad f_{m+1}^{\#}(t) = \phi p^{\#}(0,t) + \int_0^t Q^{\#}(f_m, f_m)(s)p^{\#}(s,t)ds$$

pointwise a.e. in $\Omega \times \mathbb{R}^3$ for all $t \in \mathbb{R}_+$. Clearly $h_m = f_m - f_{m-1} \ge 0$, $m \ge 0$, so the limit

$$(21) \qquad F(g)(t) = \lim_{m \to \infty} f_m(t) = \sum_{m=0}^{\infty} h_m(t)$$

exists pointwise a.e. in $\Omega \times \mathbb{R}^3$ for all $t \in \mathbb{R}_+$, and $F(g) \in \mathcal{M}_+^*$ is the smallest solution of (11) because, if $f \in \mathcal{M}_+^*$ is a solution of (11), then (20) implies $0 \le f_m \le f$ by induction on m. Note that $F(g)$ is a solution of (11) by (20), (21) and the Lebesgue monotone convergence theorem. The m-th iterate f_m has the following physical interpretation:

$f_m(t,\vec{x},\vec{\xi})$ is the density of particles at the time $t \in \mathbb{R}_+$ at the point $(\vec{x},\vec{\xi}) \in \Omega \times \mathbb{R}^3$ which collided at most m-times with other particles during the time interval $(0,t)$. The factor $p^{\#}(s,t;\vec{x},\vec{\xi})$ is the probability that a particle P which at the time $s \in [0,t]$ occupies the position $\vec{x} + s\vec{\xi}$ with the velocity $\vec{\xi}$ does not collide with other particles during the time interval (s,t). Consequently (21) gives the total density of particles if one assumes that *during any bounded time interval no particle undergoes infinitely many collisions with other particles.*

This interpretation suggests the following representation theory for $F(g)$ in terms of path integrals: Let us consider the following more general gain operator Q^{δ}, $\delta \in \mathbb{R}_+$, introduced by D. Morgenstern [13],

$$(22) \qquad Q^{\delta}(f,g) = Q(f, \Psi^{\delta}g) \quad \text{for all } f, g \in \mathcal{M}_+^*,$$

where $\Psi^{\delta}g = g$ if $\delta = 0$, and

$$\Psi^{\delta}g(t,\vec{x},\vec{\xi}) = c\delta^{-3}\int_{\mathbb{R}^3} \psi(|\vec{x}-\vec{y}|/\delta)g(t,\vec{y},\vec{\xi})d\vec{y}$$

if $\delta > 0$ where $\psi : \mathbb{R}_+ \longrightarrow \mathbb{R}_+$ is a nonincreasing function satisfying $\psi(s) = 1$ for $s \in [0, \frac{1}{2}]$ and $\psi(s) = 0$ for $s \in [1, \infty)$, $c = (\int_{\mathbb{R}^3} \psi(|\vec{x}|)d\vec{x})^{-1}$ and $g(t, \cdot, \vec{\xi})$ is periodically extended from Ω to $\mathbb{R}^3$. Then $\Psi^\delta g \longrightarrow g$ in $L^1(\Omega)$ and also pointwise a.e. in Ω as $\delta \longrightarrow 0+$ whenever $g \in L^1(\Omega)$. The physical interpretation of this gain operator Q^δ is that two particles may collide (or influence their trajectories) already at a positive distance $|\vec{x} - \vec{y}| < \delta$.

Let P be a particle which at the time t occupies the position $\vec{x}$ with the velocity $\vec{\xi}$, shortly $P(t, \vec{x}, \vec{\xi})$. The last time it went through a collision was $s \in (0, t)$ at which P changed its velocity from $\vec{\xi}_0$ to $\vec{\xi}$, shortly $P(t, \vec{x} - (t-s)\vec{\xi}, \vec{\xi} \longleftarrow \vec{\xi}_0)$. It collided with the particle P_1 whose velocity before the collision was $\vec{\xi}_1$ and after the collision $\vec{\eta}$, shortly $P_1(t, \vec{x}_1 - (t-s)\vec{\xi}, \vec{\eta} \longleftarrow \vec{\xi}_1)$ where $\vec{x}_1 - (t-s)\vec{\xi}$ was the position of P_1 at the time of the collision. Now consider the particles $P = P_0$ and P_1 at the time s (or sufficiently shortly before). The last time $P_\alpha (\alpha = 0, 1)$ went through a collision was $s_\alpha \in (0, s)$ at which P_α changed its velocity from $\vec{\xi}_{\alpha 0}$ to $\vec{\xi}_\alpha$, shortly

$$P_\alpha\left(s_\alpha, \vec{x}_\alpha - (t-s)\vec{\xi} - (s-s_\alpha)\vec{\xi}_\alpha, \vec{\xi}_\alpha \longleftarrow \vec{\xi}_{\alpha 0}\right)$$

where $\vec{x}_0 = \vec{x}$. It collided with the particle $P_{\alpha 1}$ whose velocity before the collision was $\vec{\xi}_{\alpha 1}$ and after the collision $\vec{\eta}_\alpha$, shortly

$$P_{\alpha 1}\left(s_\alpha, \vec{x}_{\alpha 1} - (t-s)\vec{\xi} - (s-s_\alpha)\vec{\xi}_\alpha, \vec{\eta}_\alpha \longleftarrow \vec{\xi}_{\alpha 1}\right)$$

where $\vec{x}_{\alpha 1} - (t-s)\vec{\xi} - (s-s_\alpha)\vec{\xi}_\alpha$ was the position of $P_{\alpha 1}$ at the time of the collision. Making use of this 0-1 multi-index notation consider the particle $P_{\alpha_1 \cdots \alpha_k}$ at the time $s_{\alpha_1 \cdots \alpha_{k-1}}$, where $\alpha_\ell \in \{0, 1\}$ for $1 \leq \ell \leq k$. The last time it went through a collision was $s_{\alpha_1 \cdots \alpha_k} \in (0, s_{\alpha_1 \cdots \alpha_{k-1}})$ (set $s_{\alpha_1 \cdots \alpha_k} = 0$ if it did not collide at all) at which $P_{\alpha_1 \cdots \alpha_k}$ changed its velocity from $\vec{\xi}_{\alpha_1 \cdots \alpha_k 0}$ to $\vec{\xi}_{\alpha_1 \cdots \alpha_k}$, shortly

$$P_{\alpha_1 \cdots \alpha_k}\big(s_{\alpha_1 \cdots \alpha_k}, \vec{x}_{\alpha_1 \cdots \alpha_k} - (t-s)\vec{\xi} - (s-s_\alpha)\vec{\xi}_\alpha - \cdots - $$
$$(s_{\alpha_1 \cdots \alpha_{k-1}} - s_{\alpha_1 \cdots \alpha_k})\vec{\xi}_{\alpha_1 \cdots \alpha_k}, \vec{\xi}_{\alpha_1 \cdots \alpha_k} \longleftarrow \vec{\xi}_{\alpha_1 \cdots \alpha_k 0}\big)$$

where $\vec{x}_{\alpha_1 \cdots \alpha_{k-1} 0} = \vec{x}_{\alpha_1 \cdots \alpha_{k-1}}$. It collided with the particle $P_{\alpha_1 \cdots \alpha_k 1}$ whose velocity before the collision was $\vec{\xi}_{\alpha_1 \cdots \alpha_k 1}$ and after the collision $\vec{\eta}_{\alpha_1 \cdots \alpha_k}$, shortly

$$P_{\alpha_1 \cdots \alpha_k 1}\big(s_{\alpha_1 \cdots \alpha_k}, \vec{x}_{\alpha_1 \cdots \alpha_k 1} - (t-s)\vec{\xi} - (s-s_\alpha)\vec{\xi}_\alpha - \cdots - $$
$$(s_{\alpha_1 \cdots \alpha_{k-1}} - s_{\alpha_1 \cdots \alpha_k})\vec{\xi}_{\alpha_1 \cdots \alpha_k}, \vec{\eta}_{\alpha_1 \cdots \alpha_k} \longleftarrow \vec{\xi}_{\alpha_1 \cdots \alpha_k 1}\big)$$

where $\vec{x}_{\alpha_1 \cdots \alpha_k 1} - (t-s)\vec{\xi} - \cdots - (s_{\alpha_1 \cdots \alpha_{k-1}} - s_{\alpha_1 \cdots \alpha_k})\vec{\xi}_{\alpha_1 \cdots \alpha_k}$ was the position of $P_{\alpha_1 \cdots \alpha_k 1}$ at the time of the collision. Finally set $P_{\alpha_1 \cdots \alpha_k 0} = P_{\alpha_1 \cdots \alpha_k}$ which, in particular, gives $P_{00 \cdots 0}\delta = P$.

To simplify our notation we set $A(k) \equiv \alpha_1 \cdots \alpha_k$ (a multi-index of order k) and $A(0) \equiv$ no index ($k = 0$), and denote by $\mathcal{A}(P)$ the set all multi-indices $A(k)$ obtained in the way described above for the particle P. We define an ordering "$<$" on $\mathcal{A}(P)$ as $A(k) < B(\ell)$ if and only if $0 \leq k \leq \ell$ and $A(k) = B(k)$, and call $\mathcal{A}(P)$ the *collision tree* of the particle P. A subset $A = \{A(0), A(1), \cdots, A(k)\} \subset \mathcal{A}(P)$ is called the *collision branch* of the particle P determined by $A(k)$ if $A(k)$ is a maximal element of the ordered set $\mathcal{A}(P)$. Hence, the collision tree $\mathcal{A}(P)$ is uniquely determined by its maximal elements, which form the subset $\mathcal{A}_{\max}(P)$, through the relation $\mathcal{A}(P) = \{A(\ell) : A(k) \in \mathcal{A}_{\max}(P) \text{ for some } k \geq \ell \geq 0\}$. Note that $A(k) \in \mathcal{A}_{\max}(P)$ if and only if $s_{A(k)} = 0$ and $s_{A(k-1)} > 0$, where $s_{A(-1)} = t$. Finally we set $\mathcal{A}'(P) = \mathcal{A}(P) \backslash \mathcal{A}_{\max}(P)$.

Next we replace Q, f_m and h_m in (20) and (21) by Q^δ, f_m^δ and h_m^δ, respectively, and calculate f_m^δ in terms of the multi-indices $A(k)$. As the first step we consider a particle $P_{A(k)}$, $A(k) \in \mathcal{A}(P)$, and denote by

$$(23) \qquad \vec{P}_{A(k)}^{\Omega}(\tau) = \vec{x}_{A(k)} - \sum_{\ell=0}^{k-1} (s_{A(\ell-1)} - s_{A(\ell)}) \vec{\xi}_{A(\ell)} - (s_{A(k-1)} - \tau) \vec{\xi}_{A(k)}$$

the position of this particle in Ω at the time $\tau \in [s_{A(k)}, s_{A(k-1)})$, where $k \geq 0$. Suppose that all $t > 0$, $s_{A(k)} \in [0, t]$, $\vec{x}_{A(k)} \in \Omega$, $\vec{\xi}_{A(k)} \in \mathbb{R}^3$ and $\vec{\eta}_{A(k)} \in \mathbb{R}^3$ are fixed for the moment, for each $A(k) \in \mathcal{A}(P)$. Then the density of the particles P which collided with other particles at fixed time moments and places with fixed velocities as described above is given by the expression

$$(24) \qquad d^\delta(P) = \Phi(P; \phi) \Lambda(P; B) \Psi(P; \delta) \exp(-\Sigma(P; R(g)))$$

where

$$(25) \qquad \Phi(P; \phi) = \prod_{A(k) \in \mathcal{A}_{\max}(P)} \phi(\vec{P}_{A(k)}^{\Omega}(0), \vec{\xi}_{A(k)}),$$

$$(26) \qquad \Lambda(P; B) = \prod_{A(k) \in \mathcal{A}'(P)} B(\theta_{A(k)}, |\vec{v}_{A(k)}|),$$

$$(27) \qquad \Psi(P; \delta) = \prod_{A(k) \in \mathcal{A}'(P)} c \delta^{-3} \psi(|\vec{x}_{A(k)} - \vec{x}_{A(k)1}|/\delta)$$

and

$$(28) \qquad \Sigma(P; R(g)) = \sum_{A(k) \in \mathcal{A}(P)} \int_{s_{A(k)}}^{s_{A(k-1)}} R(g)(\tau, \vec{P}_{A(k)}^{\Omega}(\tau), \vec{\xi}_{A(k)}) d\tau.$$

In these formulas we set $\vec{v}_{A(k)} = \vec{\eta}_{A(k)} - \vec{\xi}_{A(k)}$ and

$$(29) \qquad \begin{aligned} \vec{\xi}_{A(k)0} &= \vec{\xi}_{A(k)} + (\vec{v}_{A(k)} \cdot \vec{\omega}_{A(k)}) \vec{\omega}_{A(k)}, \\ \vec{\xi}_{A(k)1} &= \vec{\eta}_{A(k)} - (\vec{v}_{A(k)} \cdot \vec{\omega}_{A(k)}) \vec{\omega}_{A(k)} \end{aligned}$$

with the impact parameter $\vec{\omega}_{A(k)} = \vec{\omega}_{A(k)}(\theta_{A(k)}, \varsigma_{A(k)})$ whose polar and azimuthal angles are denoted by $\theta_{A(k)}$ and $\varsigma_{A(k)}$, respectively. In the second step we assume that only $t > 0$, $\vec{x} \in \Omega$, $\vec{\xi} \in \mathbb{R}^3$ and the collision tree $\mathcal{A}(P) = \mathcal{A}$ are fixed. Then the density of the particles $P(t, \vec{x}, \vec{\xi})$ whose collision tree is identical with $\mathcal{A}$ is given by the integral

$$(30) \qquad \hat{d}^\delta(t, \vec{x}, \vec{\xi}; \mathcal{A}) = \int_{\Delta(\mathcal{A})} \int_{\mathbb{R}^{3a}} \int_{\mathbb{R}^{3a}} \int_{D^a} d^\delta(P) \cdot d\theta^{\mathcal{A}} d\varsigma^{\mathcal{A}} d\vec{\eta}^{\mathcal{A}} d\vec{x}_1^{\mathcal{A}} ds^{\mathcal{A}}$$

where $a = |A'(P)| \equiv$ the number of elements of the set $A'(P)$,

$$\Delta(A) = \{s^A = (s_{A(k)})_{A(k)\in A'(P)} \in \mathbb{R}_+^a :$$
$$0 < s_{A(k)} < s_{A(k-1)} < t \text{ whenever } k \geq 1\},$$

$\vec{x}_1^A = (\vec{x}_{A(k)1})_{A(k)\in A'(P)} \in \mathbb{R}^{3a}, \vec{\eta}^A = (\vec{\eta}_{A(k)})_{A(k)\in A'(P)} \in \mathbb{R}^{3a}, \theta^A = (\theta_{A(k)})_{A(k)\in A'(P)} \in [0, \frac{\pi}{2}]^a$ and $\varsigma^A = (\varsigma_{A(k)})_{A(k)\in A'(P)} \in [0, 2\pi]^a$. Finally, in the third step we calculate the density of the particles $P(t, \vec{x}, \vec{\xi})$ which went through at most m collisions:

$$(31) \qquad f_m^\delta(t, \vec{x}, \vec{\xi}) = \sum_{\lambda(A) \leq m+1} \hat{d}^\delta(t, \vec{x}, \vec{\xi}; A)$$

where the summation is taken over all collision trees A whose *length* defined by $\lambda(A) = \max\{k+1 : A(k) \in A\}$ satisfies $\lambda(A) \leq m+1$. Formula (31) can be verified by (20) and induction on $m \geq 0$. Furthermore, (21) shows that the sum over all collision trees

$$(32) \qquad F^\delta(g)(t) = \sum_A \hat{d}^\delta(t; A)$$

exists pointwise a.e. in $\Omega \times \mathbb{R}^3$ for all $t \in \mathbb{R}_+$, and $F^\delta(g) \in M_+^*$ is the smallest solution of (11) with Q^δ in place of Q. Strictly speaking, formulas (24) and (30)-(32) hold only for $\delta > 0$. For $\delta = 0$ one has to take $\vec{x}_{A(k)} = \vec{x}$ for each $A(k) \in A$, $\Psi(P; 0) = 1$ in (27), and omit the integration with respect to $\vec{x}_1^A \in \mathbb{R}^{3a}$ in (30) to obtain the corresponding formulas for f_m and $F(g)$. Note that the only nonlinearity which appears in the formulas for F^δ and F is the exponential function in (24), and F^δ and F depend only on the function $R(g)$ rather than on g. We summarize these representation results as follows:

THEOREM 3.1. *Let $\phi \geq 0$, $B \geq 0$ and $g \in M_+^*$ satisfy (8)-(10), respectively, and let $\delta \in \mathbb{R}_+$. Then the smallest solution $f^\delta \in M_+^*$ of (11) with Q^δ in place of Q is given by the sum $F^\delta(g)$ defined by (32), with the corresponding amendment in (27) and (30) for $\delta = 0$, where $Q^0 = Q$ and $F^0 = F$. Furthermore, each F^δ is nonincreasing and convex as an operator from the set $M_+^R = \{g \in M_+^* : g \text{ satisfies } (10)\}$ into M_+^*, i.e. if $g_1 \in M_+^R$ and $g_2 \in M_+^*$ satisfy $0 \leq g_2 \leq g_1$ a.e., then $g_2 \in M_+^R$ and*

$$(33) \qquad F^\delta(g_2)(t) \geq F^\delta(g_1)(t) \qquad a.e. \text{ in } \Omega \times \mathbb{R}^3 \text{ for all } t \in \mathbb{R}_+,$$

and if $g_1, g_2 \in M_+^R$ and $\lambda \in [0, 1]$, then $\lambda g_1 + (1 - \lambda)g_2 \in M_+^R$ and

$$(34) \qquad F^\delta(\lambda g_1 + (1 - \lambda)g_2)(t) \leq \lambda F^\delta(g_1)(t) + (1 - \lambda)F^\delta(g_2)(t)$$
$$a.e. \text{ in } \Omega \times \mathbb{R}^3 \text{ for all } t \in \mathbb{R}_+.$$

4. PROOFS OF MAIN RESULTS

We first prove the following strengthened version of (10): By (9) we observe that $B(\theta, v) = B_1(\theta, v) + B_2(\theta, v)$ where

$$B_1(\theta, v) = B(\theta, v) \quad \text{if } 0 \le v \le 1, \text{ and } \quad B_2(\theta, v) = B(\theta, v) \quad \text{if } v > 1,$$

for a.e. $(\theta, v) \in [0, \frac{\pi}{2}] \times \mathbb{R}_+$. Then (9) entails

$$\text{(35)} \qquad b_1(v) = \int_0^{\pi/2} B_1(\theta, v) d\theta \le \gamma(v^{-\lambda_1} + 1) \quad \text{and}$$

$$\text{(36)} \qquad b_2(v) = \int_0^{\pi/2} B_2(\theta, v) d\theta \le \gamma(1 + v^{\lambda_2}) \quad \text{for a.e. } v > 0.$$

Furthermore, replacing the scattering kernel B in (4) by $B_i (i = 1, 2)$ we obtain a new operator R_i which is a convolution operator with respect to the $\vec{\xi}$-variable with the kernel $2\pi b_i(|\vec{\xi} - \vec{\eta}|)$. Since $b_1(v) = 0$ for $v > 1$, and $b_2(v) = 0$ for $0 \le v \le 1$, the following lemma is a straightforward consequence of (35) (with $0 < \lambda_1 < 3$) and (36) (with $0 < \lambda_2 < 2$):

LEMMA 4.1. *Let* M, E, H *and* $\Phi(M, E, H)$ *be as in Theorem 2.2, and let* $B \ge 0$ *satisfy* (9). *Define* b_1 *and* b_2 *by* (35) *and* (36). *Assume that* $g : \mathbb{R}_+ \longrightarrow L^1(\Omega \times \mathbb{R}^3)_+$ *is defined on all of* $\mathbb{R}_+$ *with* $g(t) \in \Phi(M, E, H)$ *for all* $t \in \mathbb{R}_+$ *and* $g \in \mathbb{M}_+^*$. *Then there exist constants* $C_1 = C_2(\gamma, \lambda_1) \ge 0$ *and* $C_2 = C_2(\gamma, \lambda_2) \ge 0$ *such that, for every* $\kappa \in [\lambda_2, 2]$, *we have*

$$\text{(37)} \qquad \| \int_0^h R_1^\#(g)(t + \tau) d\tau \|_\kappa \le C_1 \int_0^h \| g(t + \tau) \|_\kappa d\tau$$
$$\le C_1 h(M + E) \quad \text{and}$$

$$\text{(38)} \qquad (1 + |\vec{\xi}|^2)^{-\kappa/2} \int_\Omega \int_0^h R_2^\#(g)(t + \tau, \vec{x}, \vec{\xi}) d\tau d\vec{x}$$
$$\le C_2 \int_0^h \| g(t + \tau) \|_\kappa d\tau \le C_2 h(M + E)$$

for all $t, h \in \mathbb{R}_+$ *and* $\vec{\xi} \in \mathbb{R}^3$. ∎

For $\alpha > 0$, $\varrho > 0$ and $t, h \in \mathbb{R}_+$, $\vec{x} \in \Omega, \vec{\xi} \in \mathbb{R}^3$ we set $\chi_{\alpha,h}^\varrho(t, \vec{x}, \vec{\xi}) = 1$ if both $|\vec{\xi}| < \varrho$ and $\int_0^h R^\#(g)(t + \tau, \vec{x}, \vec{\xi}) d\tau > \alpha$, and $\chi_{\alpha,h}^\varrho(t, \vec{x}, \vec{\xi}) = 0$ otherwise. Combining (37) and (38) for $\kappa = 2$ with $R = R_1 + R_2$ we arrive at the following estimate:

$$\text{(39)} \qquad \| \chi_{\alpha,h}^\varrho(t) \|_0 \le C\alpha^{-1}(1 + \varrho^5)h \quad \text{for all } t, h \in \mathbb{R}_+$$

where the constant $C = C(\gamma, \lambda_1, \lambda_2, M + E) \ge 0$ is independent from α, ϱ and t, h. Now we are ready to prove the following lemma which entails Proposition 2.3:

LEMMA 4.2.   *Let all hypotheses in Lemma 4.1 be satisfied, and let $\chi_{\alpha,h}^{\varrho}$ be defined as above for $\alpha, \varrho > 0$ and $h \in \mathbb{R}_+$. Then there exists a constant $C' = C'(|\Omega|, \gamma, \lambda_1, \lambda_2, M, E, H) \geq 0$ independent from α, ϱ and h such that*

$$(40) \qquad \|g^{\#}(t)\chi_{\alpha,h}^{\varrho}(t)\|_0 \leq C'[(\log \sigma)^{-1} + \sigma\alpha^{-1}(1+\varrho^5)h]$$

for all $t \in \mathbb{R}_+$ and $\sigma > 1$.

Proof: For all $x \geq 0$ and $y > 0$ we have

$$(41) \qquad\qquad\qquad x \log x \geq -y + x \log y.$$

Thus, setting $x = g(t, \vec{x}, \vec{\xi})$ and $y = \exp(-|\vec{\xi}|^2)$ we obtain with regard to (14) and (16) that

$$(42) \qquad \int_\Omega \int_{\mathbb{R}^3} g(t) \log^- g(t) d\vec{x} d\vec{\xi} \leq |\Omega| \int_{\mathbb{R}^3} \exp(-|\vec{\xi}|^2) d\vec{\xi} + \mathcal{E}g(t)$$

$$\leq \pi^{3/2}|\Omega| + E \quad \text{for all } t \in \mathbb{R}_+.$$

Combining this estimate with (15) we arrive at

$$(43) \qquad \int_\Omega \int_{\mathbb{R}^3} g(t) \log^+ g(t) d\vec{x} d\vec{\xi} \leq C^+ = H + \pi^{3/2}|\Omega| + E \quad \text{for all } t \in \mathbb{R}_+.$$

Next we choose $\sigma > 1$ and set $\chi^\sigma(t, \vec{x}, \vec{\xi}) = 1$ if $g^{\#}(t, \vec{x}, \vec{\xi}) > \sigma$, and $\chi^\sigma(t, \vec{x}, \vec{\xi}) = 0$ if $g^{\#}(t, \vec{x}, \vec{\xi}) \leq \sigma$. Then (43) implies

$$(44) \qquad \|g^{\#}(t)\chi^\sigma(t)\|_0 \leq C^+(\log \sigma)^{-1} \quad \text{for all } t \in \mathbb{R}_+ \text{ and } \sigma > 1.$$

Finally we observe that $g^{\#}(t)\chi_{\alpha,h}^{\varrho}(t) \leq g^{\#}(t)\chi^\sigma(t) + \sigma\chi_{\alpha,h}^{\varrho}(t)$, and consequently (40) follows from (39) and (44) with $C' = \max\{C, C^+\}$.   ∎

Proof of Proposition 2.3: For $\alpha > 0$ and $h \in \mathbb{R}_+$ we set $\chi_{\alpha,h}(t, \vec{x}, \vec{\xi}) = 1$ if $\int_0^h R^{\#}(g)(t+\tau, \vec{x}, \vec{\xi})d\tau > \alpha$, and $\chi_{\alpha,h}(t, \vec{x}, \vec{\xi}) = 0$ otherwise. Combining (14) and (16) with (40) we arrive at

$$(45) \qquad \|g^{\#}(t)\chi_{\alpha,h}(t)\|_0 \leq \|g^{\#}(t)\chi_{\alpha,h}^{\varrho}(t)\|_0 + \int_\Omega \int_{|\vec{\xi}| \geq \varrho} g^{\#}(t)d\vec{x} d\vec{\xi} \leq$$

$$\|g^{\#}(t)\chi_{\alpha,h}^{\varrho}(t)\|_0 + \varrho^{-2}E \leq C''[(\log \sigma)^{-1} + \sigma\alpha^{-1}(1+\varrho^5)h + \varrho^{-2}]$$

for all $\alpha, \varrho > 0$, $\sigma > 1$ and $t, h \in \mathbb{R}_+$, where $C'' = \max\{C', E\}$. Now, given $0 < h < 1$, we choose α, ϱ and σ such that

$$(46) \qquad \alpha = \alpha(h) = (-\log h)^{-1} \text{ and } 1 + \varrho^5 = \sigma = h^{-1/3}.$$

Then (45) implies

$$(47) \qquad \|g^{\#}(t)\chi_{\alpha,h}(t)\|_0 \leq \tilde{C}(-\log h)^{-1}$$

for all $t \in \mathbb{R}_+$ and $0 < h < 1$, where $\tilde{C}$ depends only on C''. Consequently

$$(48) \qquad \|g^{\#}(t)[1 - \exp(-\int_0^h R^{\#}(g)(t+\tau)d\tau)]\|_0 \leq$$
$$\|g^{\#}(t)\|_0(1 - e^{-\alpha}) + \|g^{\#}(t)\chi_{\alpha,h}(t)\|_0 \leq (M + \tilde{C})(-\log h)^{-1}$$

by (12), (16), (46) and (47). Since by our hypothesis $0 \leq f(t) \leq g(t)$, we obtain

$$(49) \qquad \|f^{\#}(t)[1 - \exp(-\int_0^h R^{\#}(g)(t+\tau)d\tau)]\|_0 \leq \frac{1}{2}C_0(-\log h)^{-1}$$

for all $t \in \mathbb{R}_+$ and $0 < h < 1$, where $C_0 = 2(M + \tilde{C})$. Finally, since Mf is nonincreasing in time and (18) holds, we observe that (cf. (19))

$$\|f^{\#}(t+h) - f^{\#}(t)\|_0 \leq$$
$$\|f^{\#}(t+h) - f^{\#}(t)p^{\#}(t,t+h)\|_0 + \|f^{\#}(t)[1 - p^{\#}(t,t+h)]\|_0 =$$
$$\|f^{\#}(t+h)\|_0 - \|f^{\#}(t)p^{\#}(t,t+h)\|_0 + \|f^{\#}(t)[1 - p^{\#}(t,t+h)]\|_0 \leq$$
$$\|f^{\#}(t)\|_0 - \|f^{\#}(t)p^{\#}(t,t+h)\|_0 + \|f^{\#}(t)[1 - p^{\#}(t,t+h)]\|_0 =$$
$$2\|f^{\#}(t)[1 - p^{\#}(t,t+h)]\|_0 \leq C_0(-\log h)^{-1}$$

for all $t \in \mathbb{R}_+$ and $0 < h < 1$, by (49), which proves (17). Moreover, given $\kappa \in (0,2)$, we have

$$\|f^{\#}(t+h) - f^{\#}(t)\|_\kappa \leq (1 + \varrho^2)^{\kappa/2}\|f^{\#}(t+h) - f^{\#}(t)\|_0 +$$
$$(1 + \varrho^2)^{-(2-\kappa)/2}(\|f^{\#}(t+h)\|_2 + \|f^{\#}(t)\|_2) \leq$$
$$(1 + \varrho^2)^{\kappa/2}C_0(-\log h)^{-1} + 2(1 + \varrho^2)^{-(2-\kappa)/2}(M + E)$$

for all $\varrho \geq 0$ and $t \in \mathbb{R}_+$, $0 < h < 1$. Taking ϱ such that $1 + \varrho^2 = \max\{-\log h, 1\}$ we deduce from (50) that, for $0 < h \leq 1/e$,

$$(51) \qquad \|f^{\#}(t+h) - f^{\#}(t)\|_\kappa \leq \hat{C}(-\log h)^{-(2-\kappa)/2}$$

where $\hat{C} = C_0 + 2(M + E)$, which finishes our proof. ∎

Proof of Theorem 2.2: Consider a supersolution g of (7). It is not difficult to see that the function $f = F(g)$ defined by (21) is bounded above by any subsolution of (7) corresponding to g, since this is true of every function f_m defined by (20). By Theorem 3.1, the function f satisfies (11) a.e. in $\Omega \times \mathbb{R}^3$ for every $t \in \mathbb{R}_+$. Furthermore,

$f \in M_+^*$ and $0 \leq f(t) \leq g(t)$ a.e. for all $t \in \mathbb{R}_+$. Hence, f is a subsolution of (7) as soon as we show that $f^\# \in C_b(\mathbb{R}_+ \longrightarrow L^2(\Omega \times \mathbb{R}^3))$. But this claim follows from the fact that $Mf : \mathbb{R}_+ \longrightarrow \mathbb{R}_+$ is nonincreasing and Proposition 2.3. In order to show that Mf is nonincreasing, it suffices to show that every Mf_m is nonincreasing, by (21). It also suffices to show this claim for ϕ bounded with compact support in $\Omega \times \mathbb{R}^3$ and B bounded with compact support in $[0, \frac{\pi}{2}] \times \mathbb{R}_+$ in the gain operator Q while keeping the function $R(g)$ unchanged. Then by (20) the function $f_m : [0, T] \times \Omega \times \mathbb{R}^3 \longrightarrow \mathbb{R}_+$ is also bounded with compact support, for all $T \in \mathbb{R}_+$ and $m = -1, 0, 1, 2, \cdots$, and $f_m^\# : [0, T] \longrightarrow L^1(\Omega \times \mathbb{R}^3)$ is absolutely continuous with the weak derivative $\dot{f}_m^\#$ in time. Thus, (20) implies

$$(52) \qquad \dot{f}_{m+1} + f_{m+1}^\# \cdot R^\#(g) = Q^\#(f_m, f_m)$$

in $L^1(\Omega \times \mathbb{R}^3)$ and a.e. in $\mathbb{R}_+$. Integrating this equation with respect to $(\vec{x}, \vec{\xi}) \in \Omega \times \mathbb{R}^3$ we arrive at $\frac{d}{dt} Mf_{m+1} \leq 0$ a.e. in $\mathbb{R}_+$, since $f_m \leq f_{m+1} \leq g$ a.e. Hence, the function $Mf_{m+1} : \mathbb{R}_+ \longrightarrow \mathbb{R}_+$ is nonincreasing. For general ϕ and B, this result follows from the previous one by nondecreasing approximations of ϕ and B (in the gain operator Q) and the Lebesgue monotone convergence theorem. Consequently f is a subsolution of (7).

Next let g_n be a supersolution of (7) with B_n and ϕ_n in place of B and ϕ, respectively, and denote by $f_n = F(g_n; B_n, \phi_n)$ the corresponding smallest subsolution of (7). More precisely, f_n depends only on $R_n(g_n), B_n$ in the gain operator and ϕ_n, where the operator R_n is defined by (4) with B_n in place of B, i.e. $f_n = \tilde{F}(R_n(g_n); B_n, \phi_n) = F(g_n; B_n, \phi_n)$. Furthermore, our representation theory from Section 3 shows that the inequality

$$(53) \qquad f_n = \tilde{F}(R_n(g_n); B_n, \phi_n) \geq \tilde{F}(R(g_n); B_k, \phi_k)$$

holds for all $n \geq k$, since $0 \leq B_k \leq B_n \leq B$ and $0 \leq \phi_k \leq \phi_n \leq \phi$ by our hypotheses. Here we fix $k \geq 1$. In particular, we have

$$(54) \qquad g_n \geq \tilde{F}(R(g_n); B_k, \phi_k) \quad \text{for all } n \geq k.$$

Making use of the Dunford-Pettis weak compactness criterion in $L^1(\Omega \times \mathbb{R}^3)$ (cf. R. E. Edwards [8]) D. Morgenstern [13] and L. Arkeryd [1] showed that the set $\Phi(M, E, H)$ is sequentially weakly compact in $L^1(\Omega \times \mathbb{R}^3)$. Note that $\Phi(M, E, H)$ is convex and closed in $L^1(\Omega \times \mathbb{R}^3)$, and consequently also weakly closed. Since $\phi_n \leq \phi$, we conclude from (42) and (43) that $\phi_n \in \Phi^+ = \Phi(M, E, C^+)$ for all $n \geq 1$, and so $g_n(t) \in \Phi^+$ for all $t \in \mathbb{R}_+$ and $n \geq 1$. By our hypothesis the functions $g_n^\# : \mathbb{R}_+ \longrightarrow L^1(\Omega \times \mathbb{R}^3)$ are uniformly equicontinuous in time, and therefore by Ascoli's theorem (cf. R. E. Edwards [8]) there exists a subsequence of the sequence $\{g_n\}_{n=1}^\infty$, denoted again by $\{g_n\}_{n=1}^\infty$, such that $g_n^\#(t) \rightarrow g^\#(t)$ $(n \longrightarrow \infty)$ weakly in $L^1(\Omega \times \mathbb{R}^3)$ and uniformly for all $t \in \mathbb{R}_+$. Moreover, also $g(t) \in \Phi^+$ and $g^\# : \mathbb{R}_+ \longrightarrow L^1(\Omega \times \mathbb{R}^3)$ is uniformly continuous in time, by the weak lower semicontinuity of the L^1-norm. Since also $0 \le \phi_n \le \phi_{n+1} \le \phi \in \Phi(M, E, H)$ and $\phi_n \longrightarrow \phi$ a.e., it is easy to see that g satisfies all properties of a supersolution of (7) except for the existence of a subsolution f. We claim that $F(g) \le g$ a.e. which implies that g is a supersolution of (7). To prove this claim it suffices to show

$$(55) \qquad g \ge \tilde{F}(R(g); B_k, \phi_k) \quad \text{for all } k \ge 1,$$

and then apply the Lebesgue monotone convergence theorem with $k \longrightarrow \infty$.

We start with the observation that $g_n \rightharpoonup g$ $(n \longrightarrow \infty)$ weakly in $L^1([0, T] \times \Omega \times \mathbb{R}^3)$ for every $T \in \mathbb{R}_+$. Hence, there exists a sequence of finite convex combinations of the functions $g_n, g_{n+1}, \cdots$, for each $n = k, k+1, \cdots$, denoted again by $\{g_n\}_{n=k}^\infty$, such that $g_n \longrightarrow g$ $(n \longrightarrow \infty)$ strongly in $L^1([0, T] \times \Omega \times \mathbb{R}^3)$ for every $T \in \mathbb{R}_+$, and (54) still holds because the mapping $\tilde{F}(R(\cdot), B_k, \phi_k)$ is convex by our results from Section 3. Moreover, $g_n(t) \in \Phi^+$ for all $t \in \mathbb{R}_+$ and (14) imply $g_n \longrightarrow g$ strongly in $L_\kappa^1([0, T] \times \Omega \times \mathbb{R}^3) = L^1([0, T] \longrightarrow L_\kappa^1(\Omega \times \mathbb{R}^3))$ for all $T \in \mathbb{R}_+$ and $\kappa \in [0, 2)$. Thus, we may choose a subsequence of $\{g_n\}_{n=k}^\infty$, denoted again by $\{g_n\}_{n=k}^\infty$, such that $g_n \longrightarrow g$ a.e. in $\mathbb{R}_+ \times \Omega \times \mathbb{R}^3$ and $g_0 = \sup\{g_n : n \ge k\} \in L_\kappa^1([0, T] \times \Omega \times \mathbb{R}^3)$ for all $T \in \mathbb{R}_+$ and $\kappa \in [0, 2)$. Finally, we may assume that $g \le g_{n+1} \le g_n \le g_0$ for all $n \ge k$, since the sequence $\hat{g}_n = \sup\{g_n, g_{n+1}, \cdots\}$ also satisfies (54) because the mapping $\tilde{F}(R(\cdot), B_k, \phi_k)$ is nonincreasing. Note that also $R(g) \le R(g_{n+1}) \le R(g_n) \le R(g_0)$ for all $n \ge k$, where $R(g_0) \in L_{loc}^1(\mathbb{R}_+ \times \Omega \times \mathbb{R}^3)$ and $R(g_n) \longrightarrow R(g)$ a.e. in $\mathbb{R}_+ \times \Omega \times \mathbb{R}^3$ by (37) and (38). Hence, the same holds for the expressions in (28), i.e. $\Sigma(P; R(g_n))$ is nonincreasing with the limit $\Sigma(P; R(g))$ a.e. as $n \longrightarrow \infty$. Now we may apply the Lebesgue monotone convergence theorem to conclude that also $\tilde{F}(R(g_n); B_k, \phi_k) \longrightarrow \tilde{F}(R(g); B_k, \phi_k)$ $(n \longrightarrow \infty)$ strongly in $L^1([0, T] \times \Omega \times \mathbb{R}^3)$ for every $t \in \mathbb{R}_+$. In particular, (54) implies (55) which finishes our proof. ∎

5. SOME ADDITIONAL PROPERTIES OF SUPER- AND SUBSOLUTIONS

In the first part of the proof of Theorem 2.2 (cf. Section 4) we showed that $f = F(g)$ is the smallest subsolution of (7) whenever g is a supersolution of (7). Making use of similar techniques which involve (52) again, one can show that, for every $t \in \mathbb{R}_+$,

$$(56) \qquad M f(t) + \int_0^t \int_\Omega \int_{\mathbb{R}^3} R(f)(\tau) \cdot (g - f)(\tau) d\vec{\xi} d\vec{x} d\tau \le M\phi = Mg(t)$$

and

$$(57) \qquad \mathcal{E} f(t) + \int_0^t \int_\Omega \int_{\mathbb{R}^3} R(|\vec{\xi}|^2 f)(\tau) \cdot (g - f)(\tau) d\vec{\xi} d\vec{x} d\tau \le \mathcal{E}\phi.$$

Moreover, in the case when ϕ, f and g are space-homogeneous and B is bounded (Maxwellian molecules with a cut-off), an equality in (56) holds. Namely, since also

$R(f)$ is bounded, we may apply Gronwall's lemma to (56) to conclude that $f = g$, i.e. every supersolution g is a mild solution of the Boltzmann equation, and $g = F(g)$. We do not know whether this is the case in general. A special case of this problem is the uniqueness of the subsolution f which corresponds to a given supersolution g, i.e. whether $f = F(g)$.

Finally, using similar arguments as in the second part of the proof of Theorem 2.2, one can show that, given $\phi \geq 0$ and $B \geq 0$ satisfying (8) and (9), respectively, the set $\mathcal{G}$ of all supersolutions g of (7) is convex and sequentially weakly compact in $L^1_\kappa(\Omega \times \mathbb{R}^3)$ for each $\kappa \in [0, 2)$.

REFERENCES

1. L. Arkeryd, *On the Boltzmann equation. Part I: Existence.* Arch. Rat. Mech. Anal. **46**, 1-16(1972).

2. L. Arkeryd, *A time-wise approximated Boltzmann equation.* IMA J. of Appl. Math. **27**, 373-383(1981).

3. L. Arkeryd, *A non-standard approach to the Boltzmann equation.* Arch. Rat. Mech. Anal. **77**, 1-10(1981).

4. L. Arkeryd, *Loeb solutions of the Boltzmann equation.* Arch. Rat. Mech. Anal. **86**, 85-97(1984).

5. L. Arkeryd, R. Esposito and M. Pulvirenti, *The Boltzmann equation for weakly inhomogeneous data.* Commun. Math. Phys. **111**, 393-407(1987).

6. N. Bellomo and G. Toscani, *On the Cauchy problem for the nonlinear Boltzmann equation, global existence, uniqueness and asymptotic stability.* J. Math. Physics **26**, 334-338 (1985).

7. C. Cercignani, *Theory and application of the Boltzmann equation.* Elsevier, New York, 1975.

8. R. E. Edwards, *Functional Analysis.* Holt-Rinehart-Winston, New York, 1965.

9. W. Greenberg, J. Voigt and P. F. Zweifel, *Discretized Boltzmann equation: lattice limit and non-Maxwelian gases.* J. Stat. Phys. **21**(6), 649-657(1979).

10. K. Hamdache, *Existence in the large and asymptotic behaviour for the Boltzmann equation.* Japan J. of Appl. Math. **2**, 1984.

11. R. Illner and M. Shinbrot, *The Boltzmann equation: global existence for a rare gas in an infinite vacuum.* Comm. Math. Physics **95**, 217-226 (1984).

12. S. Kaniel and M. Shinbrot, *The Boltzmann equation, I: Uniqueness and local existence.* Comm. Math. Physics **58**, 65-84 (1979).

13. D. Morgenstern, *Analytical studies related to the Maxwell-Boltzmann equation.* J. Rat. Mech. Anal. **4**, 533-555 (1955).

14. M. Shinbrot, *The Boltzmann equation: Flow of a cloud of gas past a body.* Transport Theory and Statistical Physics **15**(3), 317-332(1986).

15. Y. Shizuta and R. Asano, *Global solutions of the Boltzmann equation in a bounded convex domain.* Proc. Japan. Acad. Ser. A Math. Sci. **53**, 3-5(1977).

16. G. Toscani, *On the non-linear Boltzmann equation in unbounded domains.* Arch. Rat. Mech. Anal. **95**, 37-49 (1986).

17. C. Truesdell and R. G. Muncaster, *Fundamentals of Maxwell's kinetic theory of a simple monatomic gas.* Academic Press, New York, 1980.

18. S. Ukai, *On the existence of global solutions of mixed problems for the non-linear Boltzmann equation.* Proc. Japan. Acad. Ser. A Math. Sci. **50**, 179-184(1974).

19. E. Wild, *On Boltzmann's equation in the kinetic theory of gases.* Proc. Cambr. Phil. Soc. **47**, 602-609(1951).

Comparison Theorems and Their Use in the Analysis of Some Nonlinear Diffusion Problems*

HANS G. KAPER and MAN KAM KWONG** Mathematics and Computer Science Division, Argonne National Laboratory, Argonne, Illinois

This article is concerned with concavity and monotonicity properties of radial solutions of nonlinear diffusion equations of the type $u_t = \Delta u - \lambda u^q$ in $\mathbf{R}^N$; λ is a positive constant and $0 < q < 1$. It is shown that the Sturm-Picone theorem of classical linear oscillation theory can be used effectively in the study of nonlinear differential equations.

1 INTRODUCTION

Let Ω be a bounded domain in $\mathbf{R}^N$ ($N \geq 1$) with smooth boundary $\partial\Omega$. The following analysis is motivated by the study of boundary value problems of the type

$$u_t = \Delta u + f(u), \quad x \in \Omega, t > 0; \tag{1.1a}$$

$$u(x, t) = 0, \quad x \in \partial\Omega, t > 0; \tag{1.1b}$$

*This work was supported by the Applied Mathematical Sciences subprogram of the Office of Energy Research, U. S. Department of Energy, under contract nr. W-31-109-Eng-38.
**Permanent address: Department of Mathematical Sciences, Northern Illinois University, DeKalb, IL 60115-2888.

$$u(x, 0) = \phi(x), \quad x \in \Omega. \tag{1.1c}$$

Here, f is a nonlinear function, which we assume to be twice continuously differentiable; ϕ is a continuous nonnegative function on $\overline{\Omega}$, which vanishes on $\partial\Omega$.

Equation (1.1a) models heat conduction in media with constant thermal conductivity and a temperature-dependent volumetric heat source or sink. The coefficient of thermal conductivity has been normalized to one. In general, the function f can have positive as well as negative values; it models a heat source whenever $f(u)$ is positive, a heat sink whenever $f(u)$ is negative. A fundamental question in the study of boundary value problems of the type (1.1) concerns the existence of a *ground state,* i.e., a nontrivial solution that does not change sign.

The boundary value problem (1.1) is a rich source of mathematical problems. It incorporates many interesting phenomena, including *blow-up, quenching,* and *extinction* at finite times.

Blow-up occurs, for example, when f represents a superlinear heat source. If $f(u) = \lambda u^q$, where $q > 1$ and $\lambda > 0$, then there exists a finite value T such that

$$\lim_{t \to T, \, t<T} \|u(t)\| = \infty. \tag{1.2}$$

Here, $\| \cdot \|$ denotes the sup-norm over Ω, $\|u(t)\| = \sup\{u(x, t) : x \in \Omega\}$. The phenomenon of blow-up at finite time has received considerable attention in the literature on nonlinear diffusion equations. We mention only the early article of Fujita [1966] and the recent publications of Weissler [1984], Friedman and McLeod [1985], and Giga and Kohn [1985].

Quenching is a more subtle phenomenon, where the solution u remains finite, but one of its derivatives blows up in finite time. It occurs, for example, when f stands for a heat source whose strength increases indefinitely as the temperature approaches a critical value. A typical case is $f(u) = \lambda(1 - u)^{-1}$, where $\lambda > 0$. In this case, there exists a finite value T such that $\|u(t)\|$ remains finite, but

$$\lim_{t \to T, \, t<T} \|u_t(t)\| = \infty. \tag{1.3}$$

The phenomenon of quenching at finite time was first analyzed by Kawarada [1975]. Some relevant publications are Walter [1976], Acker and Walter [1976, 1978], Levine and Montgomery [1980], Chan and Kaper [1987], and Chan and Kwong [1987].

Extinction is a phenomenon in the opposite direction. It occurs, for example, when f represents a sublinear heat sink. If $f(u) = -\lambda u^q$, where $0 < q < 1$ and $\lambda > 0$, then the solution of (1.1) vanishes identically after a finite time T,

$$\lim_{t \to T, \, t<T} \|u(t)\| = 0. \tag{1.4}$$

The infimum of the set of such T is called the *extinction time.* The phenomenon of extinction at finite time was first analyzed by Kalashnikov [1974] More recently, it has been investigated by Friedman, Friedman, and McLeod [1986], Friedman and Herrero [1987], and Kaper and Kwong [1987].

In this article, we review some of the arguments of Kaper and Kwong [1987]. These arguments are of intrinsic interest, because they use classical Sturm comparison theorems from the theory of *linear* ordinary differential equations to derive monotonicity properties of solutions of some *nonlinear* differential equations.

In Section 2, we summarize and prove the classical Sturm (or rather, Sturm-Picone) comparison theorem and state some of its relevant consequences. In Sections 3, we address the question of extinction in more detail. In Sections 4 and 5 we show how comparison theorems are used in some typical cases. In Section 6 we summarize our conclusions for the extinction problem and indicate further generalizations.

2 COMPARISON THEOREMS

Sturm's comparison theorem concerns solutions of second-order linear differential equations of the form

$$(p(x)y')' + q(x)y = 0. \tag{2.1}$$

Its history goes back more than 150 years, to the original work of Sturm [1836]. A more recent discussion is found, for example, in the monographs of Coddington and Levinson [1955, Section 8.1] and Hartman [1973, Section XI.3]. The theorem is a statement about the relative oscillation of two functions that are solutions of (2.1) for different, but comparable, coefficients q. More precisely, suppose that y_1 is a solution on (a, b) of

$$(p(x)y')' + q_1(x)y = 0 \tag{2.2}$$

and y_2 a solution on (a, b) of

$$(p(x)y')' + q_2(x)y = 0, \tag{2.3}$$

where $p(x) > 0$ and $q_1(x) \leq q_2(x)$ for all $x \in (a, b)$. Sturm's comparison theorem states that, if x_1 and x_2 are two successive zeros of y_1 on (a, b), then y_2 must have at least one zero between x_1 and x_2. Loosely speaking, y_1 oscillates less than y_2 on (a, b).

Sturm's theorem was generalized by Picone [1908], who considered the possibility of different, but comparable coefficients p as well. The resulting theorem is commonly referred to as the Sturm-Picone comparison theorem. The following theorem is a stronger version of it.

THEOREM 1. *Suppose that y_1 is a solution on (a, b) of*

$$(p_1(x)y')' + q_1(x)y = 0 \tag{2.4}$$

and y_2 a solution on (a, b) of

$$(p_2(x)y')' + q_2(x)y = 0, \tag{2.5}$$

where $p_1(x) \geq p_2(x) > 0$ and $q_1(x) \leq q_2(x)$ for all $x \in (a, b)$. Suppose that y_2 does not vanish on (a, b). If

$$\frac{p_1(a)y_1{}'(a)}{y_1(a)} \geq \frac{p_2(a)y_2{}'(a)}{y_2(a)}, \tag{2.6}$$

then

$$\frac{p_1(x)y_1'(x)}{y_1(x)} \geq \frac{p_2(x)y_2'(x)}{y_2(x)}, \quad x \in (a, b). \tag{2.7}$$

Proof. Let the functions r_1 and r_2 be defined by the expressions $r_1(x) = -p_1(x)y_1'(x)/y_1(x)$ and $r_2(x) = -p_2(x)y_2'(x)/y_2(x)$, respectively. Then

$$r_1' = q_1 + (1/p_1)r_1^2, \quad r_2' = q_2 + (1/p_2)r_2^2, \tag{2.8}$$

pointwise everywhere on (a, b). The second equality can be converted to a differential inequality with the same coefficients as the first equality,

$$r_1' = q_1 + (1/p_1)r_1^2, \quad r_2' \geq q_1 + (1/p_1)r_2^2. \tag{2.9}$$

It is a well-known result from the theory of differential inequalities that, if r_1 and r_2 satisfy (2.9) on (a, b) and $r_2(a) \geq r_1(a)$, then $r_2(x) \geq r_1(x)$ for all $x \in (a, b)$; see, for example, Walter [1970, Section II.8]. The assertion of the theorem follows. ∎

In the above form, the Sturm-Picone theorem is a statement about the *relative degree of oscillation* of two functions that satisfy second-order linear differential equations with comparable coefficients. When the conclusion of the theorem holds, we say that y_1 *oscillates less than* y_2 or, equivalently, that y_2 *oscillates more than* y_1 on (a, b).

A particular consequence of the theorem is that, if y_1 and y_2 are two distinct solutions on (a, b) of the *same* differential equation (2.1) and $y_1'(a)/y_1(a) \geq y_2'(a)/y_2(a)$, then y_1 oscillates less than y_2 on (a, b).

The differential equations (2.4) and (2.5) are in selfadjoint form. This is, of course, not necessary, as an arbitrary second-order differential equation can always be transformed to selfadjoint form by means of an integrating factor. Thus, we obtain the following form of the Sturm-Picone comparison theorem.

THEOREM 2. *Suppose that y_1 is a solution on (a, b) of*

$$y'' + p(x)y' + q_1(x)y = 0 \tag{2.10}$$

and y_2 a solution on (a, b) of

$$y'' + p(x)y' + q_2(x)y = 0, \tag{2.11}$$

where $q_1(x) \leq q_2(x)$ for all $x \in (a, b)$. Suppose that y_2 does not vanish on (a, b). If $y_1'(a)/y_1(a) \geq y_2'(a)/y_2(a)$, then y_1 oscillates less than y_2 on (a, b).

The following theorem shows that there is an interplay between the inequality for the coefficients q and the inequality for the solutions at the starting point a.

THEOREM 3. *Suppose that y_1 is a solution on (a, b) of (2.10) and y_2 a solution on (a, b) of (2.11). Suppose that y_2 does not vanish on (a, b). If there exists a constant $\lambda \in (0, 1]$, such that*

$$q_1(x) \le \lambda q_2(x), \quad \frac{y_1{'}(a)}{y_1(a)} \ge \lambda \frac{y_2{'}(a)}{y_2(a)}, \quad x \in (a, b), \tag{2.12}$$

then y_1 *oscillates less than* y_2 *on* (a, b).

Proof. The comparison is made in two steps. First, we compare y_1 with the solution y_3 on (a, b) of

$$y'' + p(x)y' + \lambda q_2(x)y = 0, \tag{2.13}$$

for which $y_3{'}(a)/y_3(a) = y_1{'}(a)/y_1(a)$. According to Theorem 2, y_1 oscillates less than y_3 on (a, b). Then we compare y_2 and y_3 on (a, b). Using the integrating factor $\exp\!\int p(x)\mathrm{d}x$ and dividing the resulting equation for y_3 by λ, we obtain a pair of equations of the form (2.1) with comparable coefficients p and identical coefficients q. Theorem 1 applies and, because $\lambda \le 1$, we find that y_3 oscillates less than y_2 on (a, b). Combining the results of the two steps, we conclude that y_1 oscillates less than y_2 on (a, b). ∎

Theorems 2 and 3 concern cases where the coefficients p are identical and the coefficients q are comparable. The following theorem deals with the reverse case. As we will see, an additional sign condition needs to be satisfied.

THEOREM 4. *Suppose that* y_1 *is a solution on* (a, b) *of*

$$y'' + p_1(x)y' + q(x)y = 0 \tag{2.14}$$

and y_2 *a solution on* (a, b) *of*

$$y'' + p_2(x)y' + q(x)y = 0, \tag{2.15}$$

where $p_1(x) \ge p_2(x)$ *for all* $x \in (a, b)$. *Suppose that* y_2 *does not vanish on* (a, b) *and* $y_2{'}(x)/y_2(x) \le 0$ *for all* $x \in (a, b)$. *If* $y_1{'}(a)/y_1(a) \ge y_2{'}(a)/y_2(a)$, *then* y_1 *oscillates less than* y_2 *on* (a, b).

Proof. The proof is similar to the proof of Theorem 1. Let r_1 and r_2 be defined by the expressions $r_1(x) = -y_1{'}(x)/y_1(x)$ and $r_2(x) = -y_2{'}(x)/y_2(x)$, respectively. Then

$$r_1{'} = q - p_1 r_1 + r_1^2, \quad r_2{'} = q - p_2 r_2 + r_2^2. \tag{2.16}$$

The second equality can be converted to an inequality with the same coefficients as the first equality,

$$r_1{'} = q - p_1 r_1 + r_1^2, \quad r_2{'} \ge q - p_1 r_2 + r_2^2. \tag{2.17}$$

(Notice that we need the positivity of r_2.) Again it follows from the theory of differential inequalities that $r_2(x) \ge r_1(x)$ for all $x \in (a, b)$ whenever $r_2(a) \ge r_1(a)$. ∎

With the above comparison theorems in hand, we now turn to the extinction problem introduced in Section 1.

3 EXTINCTION

Let Ω be an open ball B_R of radius R centered at the origin in $\mathbf{R}^N$, and assume that ϕ is a function of $|x|$ only. Then all solutions of (1.1) are radially symmetric if f is sufficiently regular; cf. Gidas, Ni, and Nirenberg [1979]. Hence, the boundary value problem (1.1) reduces to

$$u_t = u_{rr} + \frac{N-1}{r} u_r + f(u), \quad 0 < r < R, t > 0; \tag{3.1a}$$

$$u_r(0, t) = 0, \quad u(R, t) = 0, \quad t > 0; \tag{3.1b}$$

$$u(r, 0) = \phi(r), \quad 0 \le r \le R. \tag{3.1c}$$

Consider the particular case $f(u) = -\lambda u^q$, where $\lambda > 0$ and $0 < q < 1$. This is the case of a sublinear heat sink, for which the phenomenon of extinction in finite time was observed by Kalashnikov [1974]. Clearly, at the extinction time, the solution of (3.1) vanishes identically in B_R, but until that time the solution is still positive in a subdomain of positive measure. Intuitively, we expect that the support of the solution separates from the boundary as t approaches the extinction time T, and that the solution extinguishes at the origin. This is indeed the case, at least if ϕ is concave on $(0, R)$ and decreasing from a positive value at the origin to the value 0 at R. We are thus led to investigate the behavior of $u(r, t)$ near $(0, T)$. This behavior can be analyzed by the techniques of Giga and Kohn [1985], who studied the opposite problem of blow-up at finite time in the case of a superlinear heat source. Without going into the details, we mention that

$$u(r, t) \sim (\lambda(T - t))^{1/(1-q)} y(r(T - t)^{-1/2}), \tag{3.2}$$

as $(r, t) \to (0, T)$, provided $r = O((T - t)^{1/2})$. Here, y is a solution of the initial value problem

$$y'' + \left(\frac{N-1}{x} - \frac{1}{2}x\right)y' + \frac{1}{1-q}y - y^q = 0, \quad x > 0; \quad y'(0) = 0. \tag{3.3}$$

One solution of (3.3) is immediately obvious, namely the constant solution,

$$y_0(x) \equiv \eta = (1 - q)^{1/(1-q)}. \tag{3.4}$$

The question is, whether (3.3) admits any other solutions besides (3.4). (We consider only classical solutions that are bounded, nonnegative, and nontrivial.) In the remainder of this article we shall prove that the answer is negative. Because the arguments for $N = 1$ are somewhat more straightforward than for $N > 1$, we discuss the two cases separately.

4 ONE-DIMENSIONAL CASE

We begin with the one-dimensional case, $N = 1$. We take $q = \frac{1}{2}$ for the sake of simplicity, although we emphasize that the analysis carries over to any $q \in (0, 1)$ without essential modifications.

If $N = 1$ and $q = \frac{1}{2}$, the differential equation in (3.3) reduces to

$$y'' - \frac{1}{2}xy' + 2y - y^{1/2} = 0. \tag{4.1}$$

The constant solution (3.4) is $y_0(x) \equiv \eta = 1/4$.

THEOREM 5. *Any solution y_1 of (4.1) for which $y_1(0) > \eta$ is concave as long as it is positive.*

Proof. Let $z = -y''$. The differential equation for z follows from (4.1),

$$z'' - \frac{1}{2}xz' + (1 - \frac{1}{2}y^{-1/2})z - \frac{1}{4}y^{-3/2}(y')^2 = 0. \tag{4.2}$$

Because we want to apply the comparison theorems of Section 2, we rewrite (4.1) and (4.2) in "linear" form. That is, instead of (4.1) we write

$$y'' - \frac{1}{2}xy' + q_1(x)y = 0, \tag{4.3}$$

where

$$q_1(x) \equiv q_1(y(x)) = 2 - (y(x))^{-1/2}, \tag{4.4}$$

and instead of (4.2) we write

$$z'' - \frac{1}{2}xz' + q_2(x)z = 0, \tag{4.5}$$

where

$$q_2(x) \equiv q_2(y(x)) = 1 - \frac{1}{2}(y(x))^{-1/2} + \frac{1}{4}(y(x))^{-3/2}(y'(x))^2(y''(x))^{-1}. \tag{4.6}$$

In (4.6), y is defined in terms of z, $y(x) = y(0) - \int_0^x (x - t)z(t) \, dt$.

Let y_1 be a solution of (4.3) for which $y_1(0) > \eta$, and let $z_1 = -y_1''$. Then z_1 is a solution of (4.5). Furthermore, $z_1(0) = 2y_1(0) - (y_1(0))^{1/2}$, which is positive, and $z_1'(0) = 0$.

Since $z_1(0)$ is positive, z_1 is positive near 0 by continuity, so y_1 is concave near 0. Suppose that there is a first point $b > 0$, where y_1 is still positive, but y_1 ceases to be concave. As long as y_1 is concave, z_1 is positive, so $z_1(x) > 0$ for all $x \in [0, b)$. When y_1 ceases to be concave, z_1 vanishes, so $z_1(b) = 0$. On the other hand, $y_1(x) > 0$ for all $x \in [0, b]$.

Because z_1 is positive on $(0, b)$, y_1'' is negative there, so

$$q_2(y_1(x)) \leq 1 - \frac{1}{2}(y_1(x))^{-1/2} = \frac{1}{2}q_1(y_1(x)), \quad x \in (0, b). \tag{4.7}$$

Hence, according to Theorem 3, z_1 oscillates less than y_1 on $[0, b]$.

But here we have a contradiction. Because $y_1(b) > 0$, it would follow that $z_1(b) > 0$, whereas we have assumed that $z_1(b) = 0$. Hence, as long as y_1 is positive, it is concave, as claimed. ∎

THEOREM 6. *Let y_1 be a solution of (4.1) for which $0 < y_1(0) < \eta$. There is a point $a \in (0, \infty)$, such that y_1 is monotonically increasing on $(0, a)$ to a maximum at a, where $y_1(a) > \eta$. Beyond a, y_1 is concave as long as it is positive.*

Proof. Since $0 < y_1(0) < \eta$, we have $y_1''(0) = -(2y_1(0) - (y_1(0))^{1/2}) > 0$. By continuity, y_1'' is positive near 0, so y_1' is increasing there. Because $y_1'(0) = 0$, it follows that $y_1' > 0$ near 0. We claim that there is a finite point a, where y_1' vanishes.

We prove the claim by contradiction. Suppose that $y_1' > 0$ on $(0, \infty)$. Because y_1 is bounded, it must be the case that $y_1(x)$ approaches a finite limit as x tends to infinity, and $\lim_{x \to \infty} y_1'(x) = 0$. Consider the functional

$$E_1(x) = \frac{1}{2}(y_1'(x))^2 + \int_0^{y_1(x)} (2s - s^{1/2})\, ds. \tag{4.8}$$

Since y_1 satisfies (4.1), we have $E_1'(x) = \frac{1}{2}x(y_1'(x))^2 \geq 0$, so E_1 is nondecreasing along the solution path. Then $\lim_{x \to \infty} E_1(x) \geq E_1(0)$. This inequality is satisfied only if $\lim_{x \to 0} y_1(x) > \eta$, from which we deduce that $y_1(x)$ exceeds η for all sufficiently large x. Therefore, $2y_1(x) - (y_1(x))^{1/2} > 0$ for all sufficiently large x. Since y_1 satisfies (4.1), we have the identity

$$y_1'(x) = \frac{2}{x}y_1''(x) + \frac{2}{x}(2y_1(x) - (y_1(x))^{1/2}). \tag{4.9}$$

Upon integration, (4.9) yields the identity

$$y_1(x) = y_1(1) - 2y_1'(1) + \frac{2}{x}y_1'(x) + 2\int_1^x y_1'(t)\, \frac{dt}{t^2} + 2\int_1^x (2y_1(t) - (y_1(t))^{1/2})\, \frac{dt}{t}. \tag{4.10}$$

The first two terms in the right member of (4.10) are constant, while the third and fourth term converge to 0 and some positive constant, respectively, as $x \to \infty$. There exists therefore a constant C, such that

$$y_1(x) \geq C + 2\int_1^x (2y_1(t) - (y_1(t))^{1/2})\, \frac{dt}{t}, \tag{4.11}$$

for all sufficiently large x. But the integral in the lower bound diverges as x tends to infinity, so (4.11) implies that $y_1(x)$ grows beyond bounds. This conclusion contradicts the assumption that y_1 is bounded, so the supposition that y_1' is positive everywhere is false. There is therefore a finite point a, where $y_1'(a) = 0$, as claimed.

As we have seen in the preceding paragraph, E_1 is a nondecreasing function along the solution path, so $E_1(a) \geq E_1(0)$. This inequality necessarily implies that $y_1(a) > \eta$.

Since $y_1(a) > \eta$ and $y_1'(a) = 0$, we can repeat the arguments given in the proof of Theorem 5, considering y_1 beyond a instead of 0, to show that, as long as y_1 is positive, it is concave. ∎

THEOREM 7. *The only nontrivial solution of (4.1) that satisfies the initial condition $y'(0) = 0$ and is nonnegative everywhere is the constant solution y_0.*

Proof. Suppose that y_1 is a solution that is different from y_0. Whatever the (positive) initial value $y_1(0)$, y_1 is concave beyond some finite point a, where it has a maximum that exceeds η. The point a may be at the origin., Because of the concavity, $y_1(x)$ must reach the value 0 at some finite point $b > a$, where $\lim_{x \to b,\, x < b} y_1(x) < 0$. Beyond b, y_1 vanishes identically, so y_1 fails to be twice continuously differentiable at b. This contradicts the definition of a classical solution, so we must conclude that (4.1) admits no other solution than y_0. ∎

5 MULTIDIMENSIONAL CASE

The arguments in the multidimensional case are somewhat more delicate, because of the singularity in the coefficient of y'. In fact, in the multidimensional case the solutions of (3.3) are not necessarily concave. But, as we will see, they have certain monotonicity properties that enable us to prove that (3.3) has no other solution besides the constant solution.

As in the one-dimensional case, we take $q = \frac{1}{2}$. The differential equation in (3.3) reduces to

$$y'' + \left(\frac{N-1}{x} - \frac{1}{2}x\right)y' + 2y - y^{1/2} = 0. \tag{5.1}$$

The constant solution (3.4) is $y_0(x) \equiv \eta = 1/4$, as in Section 4.

THEOREM 8. *Any solution y_1 of (5.1) for which $y_1(0) > \eta$ is monotonically decreasing as long as it is positive.*

Proof. Let $z = -y'/x$. The differential equation for z follows from (5.1),

$$z'' + \left(\frac{N+1}{x} - \frac{1}{2}x\right)z' + \left(1 - \frac{1}{2}y^{-1/2}\right)z = 0. \tag{5.2}$$

We rewrite (5.1) and (5.2) in "linear" form:

$$y'' + \left(\frac{N-1}{x} - \frac{1}{2}x\right)y' + q_1(x)y = 0, \tag{5.3}$$

where

$$q_1(x) \equiv q_1(y(x)) = 2 - (y(x))^{-1/2}, \tag{5.4}$$

and

$$z'' + \left(\frac{N+1}{x} - \frac{1}{2}x\right)z' + q_2(x)z = 0, \tag{5.5}$$

where

$$q_2(x) \equiv q_2(y(x)) = 1 - \frac{1}{2}(y(x))^{-1/2}. \tag{5.6}$$

In (5.6), y is defined in terms of z, $y(x) = y(0) - \int_0^x tz(t)\,dt$.

Let y_1 be a solution of (5.3) for which $y_1(0) > \eta$, and let $z_1 = -y_1'/x$. Then z_1 is a solution of (5.5). To obtain information about the limiting values of z_1 and z_1' at the origin, we proceed as follows.

Applying l'Hospital's rule, we find

$$\lim_{x \to 0} [y_1''(x) + (\frac{N-1}{x} - \frac{1}{2}x)y_1'(x)]$$

$$= [1 + \{\lim_{x \to 0} ((\frac{N-1}{x} - \frac{1}{2}x)^{-1})')\}^{-1}] \lim_{x \to 0} y_1''(x) = N \lim_{x \to 0} y_1''(x). \tag{5.7}$$

Because y_1 is continuous at the origin, it follows from (5.1) and (5.7) that

$$y_1''(0) = \lim_{x \to 0} y_1''(x) = -\frac{1}{N}(2y_1(0) - (y_1(0))^{1/2}). \tag{5.8}$$

The quantity in the right member of (5.8) is negative, so the same is true of $y_1''(0)$. Using l'Hospital's rule, one verifies that $\lim_{x \to 0} z_1(x) = -\lim_{x \to 0} y_1''(x)$. Hence, $z_1(0) = \lim_{x \to 0} z_1(x)$ is well-defined and $z_1(0) > 0$.

Similarly, by considering the once differentiated equation (5.1), we show that

$$\lim_{x \to 0} y_1'''(x) = \frac{1}{N} \lim_{x \to 0} (\frac{N+1}{x^2} + \frac{1}{2})y_1'(x). \tag{5.9}$$

As we have seen in the previous paragraph, y_1'' is negative near 0; hence, y_1' is decreasing there. Given that $y_1'(0) = 0$, we conclude that y_1' is, in fact, negative near 0. The expression under the limit sign in the right member of (5.9) is therefore negative near 0. Taking the limit, we find that $\lim_{x \to 0} y_1'''(x) \leq 0$. (The limit may be $-\infty$.) Applying l'Hospital's rule to the identity $z_1'(x) = -y_1''(x)/x + y_1'(x)/x^2$, we find $\lim_{x \to 0} z_1'(x) = -\frac{1}{2}\lim_{x \to 0} y_1'''(x)$. Hence, $\lim_{x \to 0} z_1'(x) \geq 0$.

Since z_1 is positive near 0, y_1' is negative there, so y_1 must be decreasing. Suppose that there is a first point $b > 0$, where y_1 is still positive, but y_1 ceases to be decreasing. As long as y_1 is decreasing, z_1 is positive, so $z_1(x) > 0$ for all $x \in [0, b)$. When y_1 ceases to be decreasing, z_1 vanishes, so $z_1(b) = 0$. On the other hand, $y_1(x) > 0$ for all $x \in [0, b]$.

Because z_1 is positive and nondecreasing near 0, there exists an $a \in [0, b)$, where z_1 has a maximum. The point a may be at the origin. We compare y_1 with z_1 on the interval $[a, b]$. The comparison will be made in two steps.

First, we compare y_1 with the solution w_1 of the equation

$$w'' + (\frac{N+1}{x} - \frac{1}{2}x)w' + q_3(x)w = 0, \tag{5.10}$$

where q_3 is defined in terms of y_1,

$$q_3(x) \equiv q_3(y_1(x)) = 2 - (y_1(x))^{-1/2}, \tag{5.11}$$

and w_1 satisfies the initial conditions $w_1(a) = y_1(a)$ and $w_1'(a) = 0$. Because $y_1'(a) < 0$, it follows from Theorem 4 that w_1 oscillates less than y_1 on $[a, b]$. It must therefore be the case that $w_1(b) > 0$, as $y_1(b) > 0$.

Next, we compare w_1 with z_1. We have

$$q_2(y_1(x)) = \frac{1}{2}(2 - (y_1(x))^{-1/2}) = \frac{1}{2}q_3(y_1(x)). \tag{5.12}$$

Hence, according to Theorem 3, z_1 oscillates less than w_1 on $[a, b]$.

But here we have a contradiction. After the foregoing comparison we found that $w_1(b) > 0$, so here it would follow that $z_1(b) > 0$. But we have assumed that $z_1(b) = 0$. Hence, as long as y_1 is positive, it is monotonically decreasing, as claimed. ∎

THEOREM 9. *Let y_1 be a solution of (5.1) for which $0 < y_1(0) < \eta$. There is a point $a \in (0, \infty)$, such that y_1 is monotonically increasing on $(0, a)$ to a maximum at a, where $y_1(a) > \eta$. Beyond a, y_1 is monotonically decreasing as long as it is positive.*

Proof. The proof is similar to the proof of Theorem 6. We begin by observing that $y_1''(0)$ is positive, so $y_1' > 0$ near 0. We claim that there is a finite point a, where y_1' vanishes.

We prove the claim by contradiction. Suppose that $y_1' > 0$ on $(0, \infty)$. Then $y_1(x)$ approaches a finite limit as $x \to \infty$, and $\lim_{x \to \infty} y_1'(x) = 0$. Also, y_1 is monotone, so the inverse y_1^{-1} exists. Consider the functional

$$E_1(x) = \frac{1}{2}(x^{N-1}y_1(x))^2 + \int_0^{y_1(x)} (y_1^{-1}(s))^{2(N-1)}(2s - s^{1/2}) \, ds. \tag{5.13}$$

Since y_1 satisfies (5.1), we have $E_1'(x) = \frac{1}{2}x^{2N-1}(y_1(x))^2 \geq 0$, so E_1 is nondecreasing along the solution path. Then $\lim_{x \to \infty} E_1(x) \geq E_1(0)$, from which we conclude that $2y_1(x) - (y_1(x))^{1/2} > 0$ for all sufficiently large x. From (5.1) we obtain, upon integration,

$$y_1(x) = y_1(1) - 2y_1'(1) + \frac{2}{x}y_1'(x) + 2N\int_1^x y_1'(t)\frac{dt}{t^2} + 2\int_1^x (2y_1(t) - (y_1(t))^{1/2})\frac{dt}{t}. \tag{5.14}$$

Given this expression, we show as in the proof of Theorem 6 that $y_1(x)$ grows beyond bounds as x tends to infinity, and we have arrived at a contradiction.

From the inequality $E_1(a) \geq E_1(0)$ we conclude that $y_1(a) > \eta$. Since $y_1'(a) = 0$, we can now repeat the arguments given in the proof of Theorem 8, considering y_1 beyond a instead of 0, to show that, as long as y_1 is positive, it is monotonically decreasing. ∎

The analog of Theorem 7 is much more difficult to prove, because solutions of (5.1) do not necessarily have the concavity property. Nevertheless, given the monotonicity property, we can again prove that the constant solution is the only solution that is nonnegative everywhere. Our proof involves the comparison theorems at several points.

THEOREM 10. *The only nontrivial solution of (5.1) that satisfies the initial condition $y'(0) = 0$ and is nonnegative everywhere is the constant solution y_0.*

Proof. Suppose that y_1 is a solution that is different from y_0. According to Theorems 8 and 9, whatever the (positive) initial value $y_1(0)$, y_1 is monotonically decreasing beyond some finite point a, where it has a maximum that exceeds η. The point a may be at the origin.

Suppose that y_1 is positive everywhere on $[0, \infty)$. Then $\lim_{x \to \infty} y_1(x) = 0$.

Let z_1 be defined by $z_1(x) = -y_1'(x)/x$. Then $z_1(x) > 0$ for all $x \in (a, \infty)$. As in the proof of Theorem 8, we show that $z_1(a) = \lim_{x \to a, \, x > a} z_1(x) > 0$ and $\lim_{x \to a, \, x > a} z_1'(x) \geq 0$.

Let $b \geq a$ be the first point where z_1' vanishes. Then $y_1(b) > 0$ and $y_1'(b) \leq 0$. We compare y_1 with the solution w_1 of the equation

$$w'' + \left(\frac{N+1}{x} - \frac{1}{2}x\right)w' + q_3(x)w = 0, \qquad (5.13)$$

where

$$q_3(x) \equiv q_3(y_1(x)) = 2 - (y_1(x))^{1/2}, \qquad (5.14)$$

which satisfies the initial conditions $w_1(b) = y_1(b)$ and $w_1'(b) = 0$. According to Theorem 4, w_1 oscillates less than y_1 beyond b.

Choose an arbitrary point $c > b$. Then $w_1'(c)/w_1(c) \geq y_1'(c)/y_1(c)$. Let y_2 be a second solution of (5.1), for which $y_2(c) > 0$ and $y_2'(c)/y_2(c) = w_1'(c)/w_1(c)$. The last condition implies that $y_2'(c)/y_2(c) \geq y_1'(c)/y_1(c)$. Since y_1 and y_2 are two linearly independent solutions of the same equation (5.1) and $y_1(x)$ tends to 0 as x tends to infinity, it must be the case that y_2 is unbounded at infinity. In fact, since $y_2(c)$ is positive and y_2 oscillates less than y_1 beyond c, it is easy to see that $y_2(x)$ tends to $+\infty$ as $x \to \infty$.

Next, we observe that w_1 oscillates less than y_2 beyond c. Also, z_1 oscillates less than w_1, so z_1 oscillates less than y_2 beyond c. Since $y_2(x)$ tends to ∞ as $x \to \infty$, it must be the case that $z_1(x)$ tends to ∞. But then $y_1'(x)$ tends to $-\infty$, which implies that y_1 becomes unbounded. This conclusion contradicts the definition of a solution. We must therefore conclude that (5.1) admits no other solution than y_0. ∎

6 CONCLUSIONS AND GENERALIZATIONS

The conclusions of Theorems 7 and 10 can be generalized to cover the case of equation (3.3). We formulate the resulting conclusion in the following theorem.

THEOREM 11. *The only nonnegative nontrivial solution of (3.3) is the constant solution y_0, given in (3.4).*

The theorem has the following consequence for the extinction problem described in Section 3.

COROLLARY. *There exists a finite constant T, such that, for any positive constant C, the solution u of (3.1) satisfies the limiting relation*

$$\lim_{t \to T} (T - t)^{-1/(1-q)} u(r, t) = (\lambda(1 - q))^{1/(1-q)}, \tag{6.1}$$

uniformly for $0 \le r \le C(T - t)^{1/2}$.

The result formulated in Theorem 11 can also be found in the article by Peletier and Troy [1987], while Friedman and Herrero [1987] were the first authors to give the result formulated in the Corollary. However, these authors used rather different methods, which applied to the equation (3.3), but not necessarily to other similar equations. The advantage of the methods presented here is that they do apply to a broad class of equations of the type

$$y'' + p(x)y' + f(y) = 0. \tag{6.2}$$

The coefficient p may be singular near the origin, as in Section 5; for example, the case $p(x) = \alpha x^{-q} - x/2$, where $\alpha > 0$ and $0 < q \le 1$, can be handled without significant changes. The nonlinear function f must have the general property that $f(y) < 0$ for $y > \eta$ and $f(\eta) = 0$ for some positive η, so (6.2) admits at least the constant solution $y_0(x) \equiv \eta$, but otherwise the conditions on f are rather weak. For example, the case $f(y) = y/(1 - q) - |y|^q \text{sign}(y)$, where $q > 1$, can be handled by the same methods, and the case $f(y) = 1 - e^y$ requires only a minor modification of the arguments. We refer the interested reader to our recent manuscript, Kaper and Kwong [1987] for details.

REFERENCES

A. Acker and W. Walter [1976], "The Quenching Problem for Nonlinear Parabolic Differential Equations," Lecture Notes in Mathematics **564**, 1-12. Springer-Verlag, New York.

A. Acker and W. Walter [1978], "On the Global Existence of Solutions of Parabolic Differential Equations with a Singular Nonlinear Term," Nonlinear Analysis **2**, 499-505.

C. Y. Chan and Man Kam Kwong [1987], "On a Result on the Problem of Quenching," Nonlinear Analysis (to appear).

C. Y. Chan and H. G. Kaper [1987], "Quenching for Semilinear Singular Parabolic Problems," preprint.

E. A. Coddington and N. Levinson [1955], *Theory of Ordinary Differential Equations*, McGraw-Hill, New York.

A. Friedman, J. Friedman, and B. McLeod [1986], "Concavity of Solutions of Nonlinear Ordinary Differential Equations," Technical Report No. 47, Center for Applied Mathematics, Purdue University.

A. Friedman and M. A. Herrero [1987], "Extinction Properties of Semilinear Heat Equations with Strong Absorption," J. Math. Anal. Applic. **124**, 530-546.

A. Friedman and B. McLeod [1985], "Blow-Up of Positive Solutions of Semilinear Heat Equations," Indiana Univ. Math. J. **34**, 425-447.

J. Fujita [1966], "On the Blowing Up of Solutions of the Cauchy Problem $u_t = \Delta u + u^{1+\alpha}$," J. Fac. Sci. Univ. Tokyo, Sect. J **13**, 109-124.

B. Gidas, Wei-Ming Ni, and L. Nirenberg [1979], "Symmetry and Related Properties via the Maximum Principle," Comm. Math. Phys. **68**, 209-243.

Y. Giga and R. Kohn [1985], "Asymptotically Self-Similar Blow-Up of Semilinear Heat Equations," Comm. Pure Appl. Math. **38**, 297-319.

P. Hartman [1973], *Ordinary Differential Equations*, Baltimore.

A. S. Kalashnikov [1974], "The Propagation of Disturbances in Problems of Non-Linear Heat Conduction with Absorption," USSR Comp. Math. Math. Phys. **14** (4), 70-85.

H. G. Kaper and Man Kam Kwong [1987], "Concavity and Monotonicity Properties of Solutions of Emden-Fowler Equations," Differential and Integral Equations, to appear.

H. Kawarada [1975], "On Solutions of Initial-Boundary Problem for $u_t = u_{xx} + (1 - u)^{-1}$," Publ. RIMS, Kyoto Univ. **10**, 729-736.

H. A. Levine and J. T. Montgomery [1980], "The Quenching of Solutions of Some Nonlinear Parabolic Equations," SIAM J. Math. Anal. **11**, 842-847.

L. A. Peletier and W. C. Troy [1987], "On Nonexistence of Similarity Solutions," preprint.

M. Picone [1908], "Su un problema al contorno nelle equazioni differenziali lineari ordinarie del secondo ordine," Ann. R. Scuola Norm. Sup. Pisa **10** (4).

C. Sturm [1836], "Sur les équations différentielles linéaires du second ordre," J. de Math. **1**, 106-186.

W. Walter [1970], *Differential and Integral Inequalities*, Springer Verlag, New York.

W. Walter [1976], "Parabolic Differential Equations with a Singular Nonlinear Term," Funkcial. Ekvac. **19**, 271-277.

F. B. Weissler [1984], "Single Point Blow-Up for a Semilinear Initial Value Problem," J. Diff. Equations **55**, 204-224.

Bubbles and Bends and Bio-Transport

B. R. WIENKE   Applied Theoretical Physics Division, Los Alamos National Laboratory, Los Alamos, New Mexico

B. A. HILLS   Hyperbaric Physiology, Marine Biomedical Institute, Galveston, Texas

1. INTRODUCTION

The biophysical complexities underlying the systematics[1-14] of decompression are so great that some feel attempts to formulate gas transport theories only succeed by accident. Details of exact size, shape, and location of bubbles, in most cases of decompression sickness, remain largely speculative. Research probing such matters is global in scope. Nonetheless, computational approaches to decompression based on gas supersaturation[1-11] and phase equilibration[10-13] have been successfully employed, even though underlying microphysics, interpretations of experimental data, and theoretical correlations differ. While, historically, researchers regarded critical pressures as limit points, separating bubble formation from non-bubble formation, the data is more properly a statistical predictor of bends thresholds for a given set of predisposing conditions. Decompression tables and procedures, as extracted from experiment and parameterized with any transfer model, serve to separate bends provoking exposures from non-bends provoking exposures within limits, so that bubble formation can be viewed as symptomatic or sub-symptomatic, instead of present or non-present. Transport and response functions then gain efficacy by their ability to track data independent of interpretation. In this sense, the bottom line for computational models is operational reliabilty and reproducibilty. In underwater practice, continuous model application can be handled by digital wrist computers (decomputers), which monitor activity at regular time intervals (3 to 5 sec). Precise feedback loops between depth sensors and elapsed time optimize calculations. Our focus is the bases of bio-transport models, with discussion keyed to nitrogen. Minor parameter modifications accommodate other inert gases. Pressures are conveniently measured in feet-of-sea water (*fsw*).

The following section details three equations for gas transport. After that, the supersaturation (multi-tissue) and phase equilibration (thermodynamic) models are contrasted, and a short development of bubble mechanics is included. Symptomology and tissue gas kinetics are also linked. The fifth short section compares limiting forms of elimination gradients in the two extremes of supersaturation and separation, and introduces a phenomenological parameter (fraction) scaling contributions from each component. In the next to last section, comparisons of model predictions for a decompression dive are detailed. Validation of the computational models is also discussed.

2. TRANSPORT EQUATIONS

Bio-media is usually separated into intravascular (circulatory) and extravascular (cellular) regions for modeling purposes. Blood containing dissolved inert and metabolic gases passes through the intravascular zone, providing both initial and boundary conditions for subsequent gas transport through the extravascular zone. Three equations are applied to model gas transport, namely, a diffusion equation, a rate equation, and a combined diffusion-rate equation. All three attempt to quantify extravascular tensions in time, given a set of intravascular initial and boundary conditions, as follows.

Defining the instantaneous gas tension, p, given any constant tension, p_0, with the relative difference, $\Pi = p - p_0$, the diffusion equation can be conveniently written,

$$\nabla \cdot (D \ \nabla) \Pi = \frac{\partial \Pi}{\partial t} \ , \tag{1}$$

for ∇ the gradient operator, t the time, and D the diffusion coefficient. The arbitrary tension, p_0, can always be chosen to yield homogeneous initial or boundary conditions, that is, $\Pi = 0$ initially or on the boundary. Solutions to Eq. (1) obviously depend upon boundary conditions, initial conditions, and geometry.

The bulk rate equation for gas transfer depends only on time and the initial condition and takes the form,

$$\frac{\partial \Pi}{\partial t} = \lambda \ \Pi \ , \tag{2}$$

with λ some characteristic time constant for buildup, or decay. This rather simple equation has been widely employed in decompression applications. It was originally suggested by Haldane[1] and coworkers for caisson studies at the turn of the century. Spatially-dependent models can be averaged to eliminate explicit positional dependence, thus providing bulk-equivalent representations. Resulting expressions can then be compared with the simpler solutions to Eq. (2). In the bulk approach, extravascular diffusion is assumed to take place rapidly compared to time scales on the order of λ^{-1}.

If diffusion, perfusion, and metabolic assimilation are included in the balance, the transport equation generalizes to the Fick-Fourier expression,

$$\nabla \cdot (D \, \nabla) \, \Pi = \frac{\partial \Pi}{\partial t} + \kappa \, \Pi + Z \;\; , \tag{3}$$

with Z the metabolic consumption rate and κ a perfusion time constant. Clearly, the Fick-Fourier expression contains the diffusion and bulk rate equations as limiting subsets, but, for historical and technical reasons, both predecessors enjoy greater utility. In perfusion-dominated situations, the rate equation is useful, while in diffusion-controlled applications, the diffusion equation has greater utility. Perfusion-versus-diffusion controversies have emerged following attempts to correlate theory and experiment, and in such analyses proponents invoke one or the other transfer equations. If the the media is homogeneous in composition, the diffusion coefficient, D, can be pulled outside of the divergence operator. Accordingly, in one-dimensional planar geometries, we have,

$$\nabla \cdot (D \, \nabla) \, \Pi = D \frac{\partial^2 \Pi}{\partial x^2} \;\; , \tag{4}$$

while in one-dimensional cylindrical geometries,

$$\nabla \cdot (D \, \nabla) \, \Pi = D \frac{\partial^2 \Pi}{\partial^2 r} + \frac{D}{r} \frac{\partial \Pi}{\partial r} \;\; , \tag{5}$$

with x and r the appropriate coordinates. Planar and cylindrical coordinates are natural geometries for modeling gas transport in the body.

The solutions to Eqs. (1), (2), and (3), can always be put into a general form,

$$(p - p_0) = (p_i - p_0) \, F \;\; , \tag{6}$$

with p_0 and p_i approriate boundary and initial conditions, and the *response* function, F, bounded in space and time,

$$0 \leq |F| \leq 1 \;\; . \tag{7}$$

More specifically,

$$\lim_{t \to \infty} F = 0 \; ,$$

$$\lim_{t \to 0} F = 1 \; . \tag{8}$$

In a series of step excursions, p_0 and p_i represent extremes for the excursion, usually the alveolar tension and the initial tension of the gas at the step. Steps are treated sequentially, with finishing partial pressures at one step representing initial partial pressures for the following step, and so on. In neglecting transit times between steps, such queuing is also termed superposition. If transit times between stages are not neglected, the same transfer functions can be averaged over the distance and absolute pressure. Or, an exact equation with velocity-dependent terms appended can be employed to account for exposures between steps.

3. SUPERSATURATION MULTI-TISSUE MODEL

The Haldane approach to staged decompression is based on fundamental assumptions of supersaturation, uniform gas transfer, exponential tissue response functions, and limit points to nitrogen tensions as follows.

Transport of matter by random molecular motion across regions of varying concentration or, equivalently by Henry's law, pressure is driven by the local gradient.[8] Most early approaches to decompression rest on this assumption. For a uniform body system, the time rate of change of gas tension is proportional to the difference between the inspired (initial) partial pressure of the inert gas, p_i, and the instantaneous tension, p, assuming equilibration of ambient and dissolved arterial gases,

$$\frac{\partial p}{\partial t} = -\lambda(p - p_i) \; , \tag{9}$$

with λ the particular tissue decay parameter. Integrating Eq. (9), taking, $p = p_0$ at $t = 0$, yields,

$$(p - p_i) = (p_0 - p_i) \exp(-\lambda t) \; . \tag{10}$$

The above equations often receive the misnomer, *diffusion*, from investigators, probably because of semblance to heat flow expressions. Actually, Eqs. (9) and (10) are basic rate equations, neither collisional diffusion nor streaming transport equations, and *transfer* is the more apppropriate adjective.

The time for $p - p_i$ to half, or double, its initial value is the tissue half-life, τ. Nine tissues with 2.5, 10, 20, 40, 80, 120, 180, and 240 minute half-lives are typically employed in applications, and half-lives are assumed to be independent of p_i. A one-to-one correspondence between the nine computational tissues and real anatomical tissues is neither established, nor implied. Specification of the tissue half-life, τ, permits immediate evaluation of the tissue decay constant, λ, from Eq. (10),

$$\lambda = \frac{\ln 2}{\tau} \ . \tag{11}$$

An underlying assumption built into Eqs. (9) through (11) requires that arterial blood, saturated with inert gas at ambient pressure, and tissues relax to equilibration with successive rounds of circulation. Although one appreciates that the distribution of blood flow throughout the body is not uniform, so that better perfused tissues saturate faster than poorly perfused tissues, the continuous spectrum of perfusion-controlled response times is thus approximated by nine compartments with time constants differing in roughly constant ratio. In this way, perfusion limiting is implicit in the multi-tissue model. Intercellular spatial diffusion of inert gas is also assumed to occur rapidly in the supersaturation scheme, so as to not significantly limit, nor effect, the pressure gradients of Eq. (9). Such is obviously not the case in all bio-mass.

Denoting the ambient (absolute) pressure by P, Haldanian theory[1,6,7] postulates that any tissue compartment is limited in its ability to hold saturated nitrogen. Permissible supersaturation at ambient pressure, P, is limited by the ratio, R,

$$\frac{p}{P} \le R \ , \tag{12}$$

with $1.10 \le R \le 3.20$, determined by many trial and error experiments. Alternatively, according to fixed gradient theory,[2,14] tissue saturation is limited by a critical difference, L, such that,

$$p - .79 P \le L \ . \tag{13}$$

In the fixed gradient approach, $L \approx 24 \ fsw$, for all pressures. Obviously R and L depend on many factors, not always discernible. The different values of p for which the equalities hold in Eq. (12) or Eq. (13) are the critical pressures, M, (maximum-values). Determinations of M and L, or the corresponding critical ratios, R, have been a phenomenological thrust of theory and bases for construction of staged tables.[9-12,15] Early analyses employed fixed values for R, L, and M, but today critical parameters generally vary across compartments, and with depth. Supersaturation theories regard the critical parameters as trigger points for bubble formation, while phase equilibration theories might view the critical parameters as signal points for clinical manifestation of symptoms, following bubble growth to damaging size. As will be seen, the focus of thermodynamic theory is the volume of gas separated from solution, rather than R, L, or M.

The critical pressures collected by Workman[6] and Buhlmann[7] for the tissue compartments at various depths, as well as the later compilation of Schreiner[5] and Spencer,[10] close the loop for the Haldane approach with experiment. Critical ratios, R, versus depth, x, are plotted in Figure 1 for six representative compartments at sea level, $(P_0 = 33 \ fsw)$. Corresponding critical nitrogen pressures, M, are also exhibited in Figure 2. Surfacing ratios, R_0, critical surface nitrogen pressures, $M_0 = R_0 P_0$, and depth ratios, R_∞, $(x \to \infty)$, are shown in Table 1. As a function of altitude, z, measured in feet, atmospheric pressure, P_0, is conveniently represented[15,16] by the barometer-like expression,

$$P_0 = 33 \exp(-\gamma z) \ , \tag{14}$$

with,

$$\gamma = 0.000038119 \ ft^{-1} \ . \tag{15}$$

The extension of the critical surface pressures and ratios in Table 1 to altitude has been an object of continuing study in itself. Linear[17] and exponential[16] extrapolations of critical pressures back to zero absolute pressure, have been proposed within the framework of the classical model, with the linear extrapolation yielding infinite critical surfacing ratios and the exponential extrapolation producing constant critical surfacing ratios.

Table 1. Surfacing Ratios And Critical Nitrogen Pressures $(P_0 = 33 \ fsw)$

half-life (min)	surface ratio	nitrogen pressure (*fsw*)	depth ratio
5	3.15	104	2.27
10	2.67	88	2.01
20	2.18	72	1.67
40	1.76	58	1.34
80	1.58	52	1.26
120	1.55	51	1.19

The critical ratios are larger for faster tissues and lesser absolute pressures, yet range of variation is not large, especially within compartments. Depending on the method used to extend critical pressures to reduced pressure (altitude), however, the surface values can change dramatically while the values at depth remain unchanged. Indeed, near and surfacing values are the principal concerns in many staging regimens. Blood rich, well-perfused, aqueous tissues are usually assumed to be *fast* (small values of τ), while blood poorer, scarcely-perfused, lipid tissues are assumed to be *slow* (large values of τ).,

Today, as many as sixteen compartments have been been proposed in some applications, with variable ratios at 10 *ft* increments producing 736 degrees of freedom (fit parameters) to characterize the data. The proliferation of computational tissues, not physiologically linked to bio-mass, underscores the lack of true transport and decompression microphysics reflected by the supersaturation model. Additionally, uptake and elimination

FIGURE 1. Critical Ratios (*R* -values)

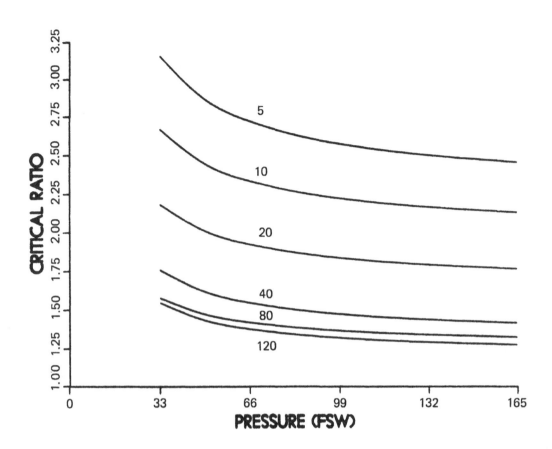

The critical ratios, *R* , are hyperbolic functions of pressure. For each of the six tissue compartments (5, 10, 20, 40, 80, 120 minutes), the above ratios permit greater degrees of saturation for the faster compartments. Similarly, the ratios increase with decreasing pressure. The curves typify saturation profiles employed in classical Haldane approaches to decompression. An equivalent, and less cumbersome, representation of classical limit points for decompression can be gleaned directly from the critical tensions (Figure 2). Denoting ambient pressure as P and dissolved inert gas tension as p , the Haldane criteria requires,

$$\frac{p}{P} \leq R \ .$$

Today, we tend to interpret the critical ratios as trigger points for clinical manifestations of symptoms, rather than limit points for bubble formation.

FIGURE 2. Critical Tensions (*M*-values)

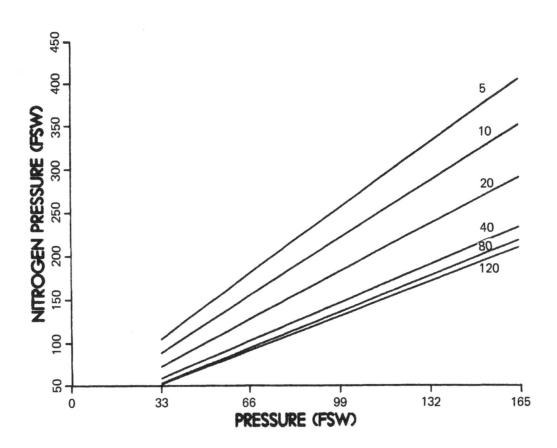

The critical tensions are linear functions of pressure, obviously increasing with ambient pressure. As before, the faster tissues permit greater degrees of dissolved nitrogen. At all times during a dive, maximum tensions in compartments must stay below the depicted curves for given absolute pressures in the Haldane approach. Dividing the M-values by the ambient pressure yields the critical ratios shown in Figure 1. Extensions of the curves to altitude (ambient pressure less than 33 fsw), can be effected linearily or exponentially. In the linear case, the zero pressure intercepts are positive, while in the exponential case, the intercepts are zero. Thus, in the former case, the critical ratios at zero ambient pressure become unbounded, while in the later case, they remain finite.

symmetry of the tissue response function, Eq. (11), remains valid only if the biological system maintains a true state of supersaturation, and there is no separation of gas from solution. If inert gas separates (as it inevitably does according to nucleation theory), the driving force for elimination can differ significantly from the uptake gradient, destroying the symmetry. The classical supersaturation assumptions are tantamount to a *bubble –free* hypothesis, but the presence or absence of bubbles does not affect the computational model, since expressions ultimately link λ to experimental R, L, or M. However, for the record, the preponderance of measurement and theory soundly support mechanisms of nucleation, gas separation, and bubble growth, even below observable bends thresholds.

4. BUBBLE MECHANICS AND SYMPTOMS

Establishment of a free gas phase and subsequent bubble trouble involves a number of distinct, but overlapping, steps:

(1) *nucleation* and *cavitation*, or free gas inception and siting;
(2) *transport* and *saturation*, or dissolved gas buildup;
(3) *growth*, or transfer of gas from solution to the free phase;
(4) *coalescence*, or bubble aggregation by fracture or diffusion;
(5) *deformation* and *occulsion*, of local tissues, nerves, and blood vessels causing possible damage and pain.

Many mechanisms possess the potential to excite gas pockets in the body. Quantification of these seemingly random processes, however, continues to elude accurate description and mathematical representation. Attempts at siting usually invoke relative probability and statistical arguments, in the absence of deterministic input. The statistical mechanics of nucleation treat the excitation mechanisms stochastically, often relying on Monte Carlo computational techniques. A number of candidate mechanisms are obvious.

Cosmic radiation, charged particles, and trace radioactive elements (U_{235}, U_{238}) cause micro-explosions in the body and are possible sources of gas nuclei. The fluids we drink, containing dissolved gases (and certainly micronuclei), are another source. Lymph, draining tissues into veins, is a strong candidate for micronuclei. Statistical analyses of the random movements of dissolved gases point to non-vanishing, mutual coalescence probabilities when two or more gas molecules collide. Exercise, generating huge cavitation forces along muscles and tendons, most certainly is another source of gas nuclei. Turbulence and vortex formation in the blood, possibly associated with exercise or other vigorous endeavor, could also produce gas pockets. Another source of gas micronuclei is the lungs, where there exists an enormous air-blood interface separated by a thin endothelium. Nucleation, in all cases, appears to be a random process, with neither benchmarks nor pathways linking formation to decompression criteria such as R, L, or M. Quantitative representations of nucleation processes are lacking in the thermodynamic formulation, and that shortcoming constrains theoretical development in part.

Once a stable, free, gas phase has been established at a nucleation site, that micropocket will grow if surrounded by dissolved gas at very high tension. Unlike nucleation, bubble growth[18] is governed by well-understood, physical laws with precise mathematical representation. To this end, first denoting the partial pressures of free gas (oxygen, nitrogen, carbon dioxide, and water vapor), P_j, and tensions of dissolved gas, p_j, with $j = O_2, N_2, CO_2, H_2O$, the total free gas pressure, P_t, and total dissolved gas tension, p_t, are obviously given by,

$$P_t = P_{O_2} + P_{N_2} + P_{CO_2} + P_{H_2O} \quad , \tag{16}$$

and,

$$p_t = p_{O_2} + p_{N_2} + p_{CO_2} + p_{H_2O} \quad . \tag{17}$$

Trace gases are neglected. The interplay between ambient pressure, P_t, p_t, and bubble growth, or collapse, becomes a matter of pressure balances against surface tension and tissue elasticity in the following way.

Defining ambient pressure, P, as before, the pressure (free gas) within any bubble must exceed the ambient pressure by an amount just equal to the sum of the surface tension, $2\gamma/r$, plus tissue compliance, δ, at equilibrium, that is,

$$P_t = P + \frac{2\gamma}{r} + \delta = P + B = p_t \quad , \tag{18}$$

with r the bubble radius, and γ, δ experimental constants. Tissue compliance is a phenomenological factor in determining thermodynamic bends thresholds, while at the same time accounting for any elastic help exerted by the site in holding the bubble together. The compliance can be estimated using Boyle's law and the bulk modulus for tissue expansion, K,

$$\delta = K\frac{v}{V} = K\chi \quad , \tag{19}$$

with v the volume of separated gas in tissue volume V, and $\chi = v/V$ the separation fraction. The hydrostatic balance of pressures is depicted in Figure 3.

FIGURE 3. Bubble Pressure Balance.

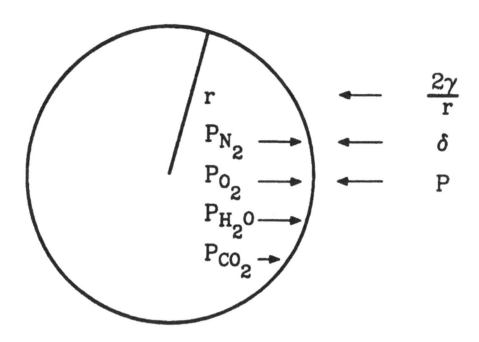

So long as the free gas pressure within a bubble equals the sum of the ambient pressure, P, plus surface tension, $2\gamma/r$, plus compliance, δ, the bubble enjoys instantaneous equilibrium,

$$P_{O_2} + P_{N_2} + P_{H_2O} + P_{CO_2} = P + \frac{2\gamma}{r} + \delta \; .$$

At increasingly small radii, the surface tension becomes very large, while for large radii the surface tension vanishes. The tissue compliance also serves as a decompression pain indicator in the thermodynamic approach, replacing R, L, and M as trigger points. Gradients, between separated and dissolved gases, will drive the bubble (by diffusion) to new configurations, re-establishing hydrostatic equilibrium.

Although bubble formation does not necessarily lead to the bends in all cases, there is little doubt that decompression sickness can be avoided if the amount of gas separating out of solution is absolutely minimized. The classical viewpoint postulates zero gas separation if critical tensions are not exceeded, so that the critical values represent decompression criteria in themselves. The phase viewpoint, however, maintains that gas separation is inevitable following compression, so that vital questions center on how much gas separates, what are permissible levels, what are the trigger mechanisms for pain, and how do we define a decompression criteria consistent with the amount of separated gas. The starting point is the tissue compliance, δ.

Inman and Saunders[19] argued that they could induce bends-like pain in connective tissue using a local injection of salt solution. Pain was not correlated with the volume of injected fluid, but rather with the injection pressure. Thresholds varied between 0.4323 *fsw* and 1.1286 *fsw*, as injection tissue pressure differentials. A criteria for pain might be provided by the pressure differential between separated gas and surrounding tissue. If the differential pressure, δ, exceeds the critical threshold, δ', pain occurs,

$$\delta > \delta' \ , \tag{20}$$

or, from Eq. (19),

$$K\chi \geq \delta' \ , \tag{21}$$

with,

$$0.43 \leq \delta' \leq 1.13 \ fsw \ .$$

Accepting this threshold as a criteria for bending nerve endings, one bubble, or coalesced composite, would suffice to cause pain. Such thresholds would correspond, from Eq. (19), to separated gas fractions,

$$0.0039 \leq \chi \leq 0.0093 \ ,$$

which are small, yet not insignificant.

As is well known, arterial oxygen and carbon dioxide tensions differ from venous values. Separated metabolic gas (O_2, CO_2) and water vapor pressures are, however, always near venous values, rapidly reverting to venous levels following any change in external pressure. Given normal resting conditions for decompression, therefore, we quote the Harris[20] result,

$$P_{O_2} + P_{H_2O} + P_{CO_2} = 5.47 \ fsw \ , \tag{22}$$

While inert gas tissue tensions vary dramatically with depth, oxygen, carbon dioxide, and water vapor tensions are fairly constant under normal conditions. Table 2, for comparison sake, lists nominal surface, arterial and venous, gas tensions.[11] Arterial, venous, and tissue partial tensions, of course, reflect the inherent unsaturation by summing up to less than ambient pressure in all three cases.

Table 2. Dissolved Gas Partial Pressures In Arterial And Venous Blood ($P_0 = 33$ *fsw*)

gas	arterial tension (*fsw*)	venous tension (*fsw*)
oxygen	4.10	1.74
carbon dioxide	1.71	1.99
water vapor	2.04	2.04
nitrogen	24.75	24.75
total	32.60	30.62

Similarly, a representative elastic-mechanical factor, B, can be approximated by,

$$B = \frac{2\gamma}{r} + \delta' = 8.68 \; fsw \tag{23}$$

from the analysis of Hills.[12] Combining Eqs. (16) through (18), (22), and (23), gives the separated nitrogen pressure, P_{N_2}, as a function of ambient pressure,

$$P_{N_2} = P + 3.21 \; fsw \quad . \tag{24}$$

Separated nitrogen essentially takes up the difference between total hydrostatic pressure and the sum of metabolic gas and water vapor pressures. Arterial nitrogen equilibrates with alveolar nitrogen in less than a minute. At equilibrium, the nitrogen tissue tension, p_{N_2}, venous, and arterial tensions are all equal to the alveolar partial pressure, which can be written, allowing for dilution by water vapor,

$$p_{N_2} = f \, [\, P - P_{H_2O} \,] = f \, [\, P - 2.04 \,] \; fsw \quad , \tag{25}$$

with f the nitrogen mole fraction (.79). The pressure difference, Δ, between separated nitrogen and the sum of dissolved nitrogen plus mechanical pressure, B, is termed the *inherent unsaturation*,[13] or the *oxygen window*[21],

$$\Delta = P_{N_2} - p_{N_2} - B = P - P_{O_2} - P_{CO_2} - P_{H_2O} - p_{N_2} \quad , \tag{26}$$

and can be simplified using Eqs. (22) through (25),

$$\Delta = (1 - f)P + f \, 2.04 - 5.47 \; fsw \quad . \tag{27}$$

The inherent unsaturation quantifies the amount of tissue undersaturation (relative to ambient) due to separated gas, and is an important factor in thermodynamic staging. Staging is designed to take up the inherent saturation while maintaining the driving force for elimination as a maximum. The tissues are thus maintained on the brink of further phase change, until any additional gas separation caused by direct surfacing can be safely tolerated.

Bubble growth, or collapse, ultimately depends on the dissolved gas tension, p_t. and the hydrostatic pressure of the bubble, as seen in Figure 4. A gas phase obeying Eq. (18) enjoys instantaneous equilibrium, but will grow, or collapse, by gas diffusion from, or to, surroundings according to the relationship,

$$\frac{rC}{DS}\frac{\partial r}{\partial t} \approx p_t - P - \frac{2\gamma}{r} - \delta = p_t - P - B \quad . \tag{28}$$

for C the concentration, D the diffusion coefficient, and S the gas solubility. Defining the so-called critical radius, r_c,

$$r_c = \frac{2\gamma}{p_t - P - \delta} \quad , \tag{29}$$

growth occurs when the radial derivative increases in time,

$$\frac{\partial r}{\partial t} > 0 \quad \text{for } r > r_c \quad , \tag{30}$$

and collapse when the derivative decreases in time,

$$\frac{\partial r}{\partial t} < 0 \quad \text{for } r < r_c \quad , \tag{31}$$

until a new mechanical equilibration point satisfying Eq. (18) is reached. Characteristic gas phases with radii less than r_c contract by diffusion, while phases with radii greater than r_c expand. Clearly, the tissue tension, p_t, establishes diffusion gradients for gas transfer across the bubble surface (dissolved-to-free, and vice-versa), while the mechanical term, B, is a crucial control factor for expansion, or contraction, of gas micronuclei. For non-vanishing bubble radii, Eq. (29) predicts asymptotic growth as $p_t \rightarrow \infty$ and collapse as $P \rightarrow \infty$, consistent with intuition and observation. Similarly, as radii increase, the mechanical effects of surface tension have less impact on bubble dynamics, while, as radii decrease, surface tension effects on bubble dynamics become pronounced.

Representative values and ranges of biophysical parameters, deduced from experiment and theory,[12,13] offer contrast for relative strengths of competing terms. Some ranges and typical values are:

$$0 \le B \le 10 \; fsw \quad ;$$

FIGURE 4. Bubble Gas Diffusion

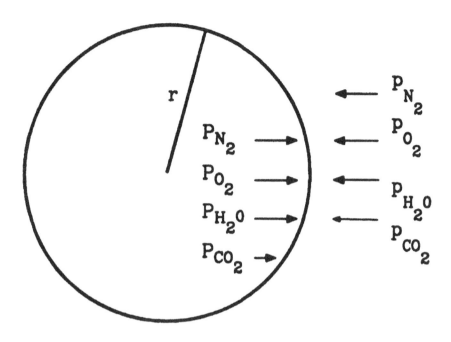

A bubble in hydrostatic equilibrium will grow or contract, depending on its size, and any relative gradients between the free gas in the bubble and dissolved gas in tissues. Gradients are inward if the dissolved tensions exceed the hydrostatic bubble pressure, and outward if the hydrostatic pressure exceeds the surrounding tissue tensions. The critical radius, r_c, separates growing from contracting bubbles for a given set of pressures. The critical radius depends on the gas gradient and tissue compliance,

$$r_c = \frac{2\gamma}{p_t - P - \delta} \ ,$$

and growth requires $r > r_c$, while contraction requires $r < r_c$. The presence of gas micronuclei in the body and pressure gradients between dissolved gases in tissues and free gases in the micronuclei, under conditions of local hydrostatic equilibrium, are central tenets of thermodynamic decompression theory.

$$0.40 \leq \delta \leq 1.20 \; fsw \quad ;$$

$$0.049 \leq r \leq 3.5 \; \mu m \quad ;$$

$$\gamma \approx 50 \; dyne/cm \quad ;$$

$$D \approx .022 \; \mu m^2/sec \quad ;$$

$$K \approx 3.7 \times 10^6 \; dyne/cm^2 \quad ; \tag{32}$$

with $1 \; \mu m = 10^{-6} \; m$. The radii are reasonable, when compared with capillary diameters near $8 \; \mu m$ and inter-capillary separations of 20-40 μm. The surface tension of water, $72 \; dyne/cm$, is only slightly higher than the tissue value. The value of D is close to some transient measurements of the nitrogen cellular diffusion coefficient, while the bulk modulus, K, lies within experimental limits.

5. PHASE EQUILIBRATION MODEL

The thermodynamic approach of Hills[12,13] is more complex than the supersaturation model of Haldane. However, it does address the same issues. The thermodynamic model is based on phase equilibration of dissolved and separated gases, radial perfusion-diffusion of capillary gases across cell membranes, Bessel-exponential tissue response functions, and limit points for separated gas volumes. While these segments parallel developments in the supersaturation model, fundamentals differ considerably.

Temporal uptake and elimination of inert gas into the tissue bulk is limited by perfusion and diffusion. The transport model invoked is one of a fully stirred extended vascular zone from which venous blood leaves in equilibrium with respect to all gases, while arterial blood either diminishes or replenishes gases in the zone. From the extended vascular zone, gases then diffuse into cellular zone. Cylindrical, one-dimensional, symmetry, as seen in Figure 5, is also assumed for simplicity in the extended zone. The radial equation for gas diffusion into cellular material is written,

$$\frac{\partial^2 p}{\partial r^2} + \frac{1}{r} \frac{\partial p}{\partial r} = \frac{1}{D} \frac{\partial p}{\partial t} \quad , \tag{33}$$

with p the instantaneous nitrogen tension, as before. Separating variables, for ε^2 the separation constant, applying initial conditions,

$$p = p_0 \quad \text{for } t = 0 \tag{34}$$

with boundary conditions,

FIGURE 5. Cylindrical Capillary Symmetry For Radial Diffusion Model

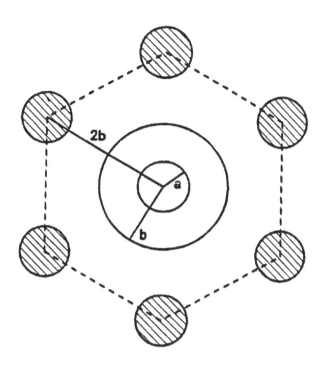

Radial diffusion, across capillaries of radii a and mean separation 2b, is the main transport mechanism for intercellular gas penetration. Irregular boundaries are approximated by the cylindrically symmetric configuration shown. The zone of influence for each capillary is an annulus of inner radius a and outer radius b. The diffusion equation with $D = .22 \ \mu m^2/sec$, for initial conditions given by,

$$p = p_0 \quad \text{for } t = 0 \ ,$$

and boundary conditions,

$$p = p_\infty \quad \text{for } r = a \ ,$$

$$\frac{\partial p}{\partial r} = 0 \quad \text{for } r = b \ ,$$

yields a Bessel-exponential tissue response function similar to the Haldane expression, but varying with position. Capillary parameters, a and b, have approximate range, $4.73 \le b/a \le 5.29 \ \mu m$, with $a \approx 4 \ \mu m$.

$$p = p_i \quad at \quad r = a \quad ,$$

$$\frac{\partial p}{\partial r} = 0 \quad at \quad r = b \quad , \tag{35}$$

and solving[22] yields, in the purely diffusive case,

$$(p - p_i) = (p_0 - p_i)E(r,t) \quad , \tag{36}$$

with,

$$E(r,t) = \pi \sum_{n=1}^{\infty} \frac{J_1^2(b\varepsilon_n)[J_0(r\varepsilon_n)N_0(a\varepsilon_n) - N_0(r\varepsilon_n)J_0(a\varepsilon_n)]\exp(-\varepsilon_n^2 Dt)}{[J_0^2(a\varepsilon_n) - J_1^2(b\varepsilon_n)]} \tag{37}$$

The eigenvalues, ε_n, are the roots (real and simple) of the expression,

$$J_0(a\varepsilon_n)N_1(b\varepsilon_n) = N_0(a\varepsilon_n)J_1(b\varepsilon_n) \quad , \tag{38}$$

for J and N Bessel and Neumann functions. The form of Eq. (36) is analogous to Eq. (10) in the Haldane model. The tissue response function, E, serves the same purpose as the multi-tissue exponential in Eq. (10) and the pressure coefficients are the same in both cases. In terms of Eq. (25), in fsw,

$$p_0 = f[P_0 - 2.04] \quad , \tag{39}$$

$$p_i = f[P_i - 2.04] \quad . \tag{40}$$

The total gas tension, p_t, at any point in the tissue can be written, using Eqs. (17), (22), (36), (39), and (40), again in fsw,

$$p_t = f[P_i - 2.04] + f[P_0 - P_i]E(r,t) + 5.47 \quad . \tag{41}$$

Alternatively, perfusion limiting can be applied to the radial diffusion equation by first invoking a mass balance across both the vascular and cellular regions at $r = a$,

$$S_b \frac{\partial q}{\partial t} = -\kappa S_b[q - p_a] - \frac{3}{a}S_c D\left[\frac{\partial p}{\partial r}\right]_{r=a} \quad , \tag{42}$$

for S_c, S_b the nitrogen cellular, blood solubilities, p_a the arterial tension, and then employing the perfusion limited tension, q, as the inner boundary condition, replacing p_i in Eqs. (34) through (36). In this case, we can estimate the volume fraction, χ, of separated gas under worse case conditions, and program any decompression away from bends provocation as required by Eq. (20). With cellular nitrogen solubility, S_c, as before, the separated gas fraction, under the worse possible case of zero gas elimination, is obtained from the balance equation,

$$\chi P_{N_2} = S_c [p - P_{N_2}] \ ,$$ (43)

which states that the amount of separated gas is the difference between the original amount of nitrogen in solution before decompression and the amount left in solution. Accordingly, using Eqs. (24) and (36), there finally results,

$$\chi [P + 3.21] = S_c [q + 2(p_0 - q)(b^2 - a^2)^{-1} \int_a^b E(r,t) r dr - P - 3.21] \ ,$$ (44)

which permits estimation of the separated gas fraction, χ, for arbitrary exposures and staging. Estimates within sub-regions of the extended vascular zone can also be affected by volume averaging $E(r,t)$ over the appropriate limits, that is, by replacing a and b with the region boundaries.

6. ELIMINATION GRADIENTS

When there is no nitrogen separation in a particular tissue, with the tension p still given by Eq. (36), the driving force for gas elimination, Δp_{N_2}, via the blood, is provided by the tissue-arterial gradient,

$$\Delta p_{N_2} = p - f [P - 2.04] \ ,$$ (45)

incorporating Eq. (25) for the arterial tension. The rate of elimination is thus *greatest* for the *lowest* ambient pressure, as historically assumed in Haldane procedures to decompression. However, if gas has separated from solution, tissue tension does not drive gas elimination, and, as seen, all elimination and uptake functions based on tissue tension will be asymmetric. Because of the inherent unsaturation, separated gas partial pressures exceed tissue tensions and, at phase equilibration, the driving force for elimination, ΔP_{N_2}, now depends on the gradient between separated gas and arterial tension,

$$\Delta P_{N_2} = P_{N_2} - f [P - 2.04] \ ,$$ (46)

which, using Eqs. (24) and (27), gives,

$$\Delta P_{N_2} = (1 - f)P + f 2.04 + 3.21 = \Delta + 8.68 \ .$$ (47)

Gas elimination rates are now *greatest* at the *highest* ambient pressure, driven by the inherent unsaturation. As such, Eq. (47) represents a radical change for traditional staging procedures, reversing the usual surface haul.

When the full inherent unsaturation gradient is not employed to eliminate inert gas, a representative dissolved gas term can be included in an effective elimination function, Γ. Separated and dissolved gradients are then split by a simple weighting factor, α,

$$\Gamma = \alpha(P_{N_2} - p) + (p - f [P - 2.04]) ,$$

(48)

with $0 \le \alpha \le 1$. Rewriting Eq. (48), using Eqs. (37) and (46), there results,

$$\Gamma = \alpha P_{N_2} + (1 - \alpha)p - f [P - 2.04] = \alpha \Delta P_{N_2} + (1 - \alpha)\Delta p_{N_2} ,$$

(49)

serving to link model extremes by a phenomenological fraction. Use of a split elimination gradient implies that only a portion of the maximum allowable tissue gas has separated, leaving the remainder of the gas dissolved.

7. STAGING AND VALIDATION

A comparison of the optimal ($\alpha = 1$) thermodynamic and supersaturation approaches to staged decompression staging appear in Figure 6. Both the supersaturation US Navy and Royal Navy dive profiles are plotted against the phase equilibration profile for a dive to 150 *ft* for 40 *minutes*. Clearly, a greater amount of staging time is spent at greater depth according to the thermodynamic format, followed by an earlier direct ascent to the surface. Total thermodynamic staging time is roughly 2/3 of the US Navy and 1/2 the Royal Navy staging time, in this case, 39, 59 and 75 *minutes* according to the thermodynamic, US Navy, and Royal Navy methods, respectively. The US Navy schedule requires 5 *minutes* at 30 *ft*, 19 *minutes* at 20 *ft*, and 33 *minutes* at 10 *ft*. The Royal Navy schedule specifies 5 *minutes* at 40 *ft*, 10 *minutes* at 30 *ft*, 20 *minutes* at 20 *ft*, and 40 *minutes* at 10 *ft*. In the thermodynamic scheme, the inherent unsaturation is rapidly taken up by direct ascent to 115 *ft*, followed by continuous decompression, in roughly 1 *minute* intervals, to drop-out at 30 *ft*. Figure 6 typifies differences between thermodynamic and supersaturation decompression formats. If the full gradient of the inherent unsaturation is not employed, ($\alpha < 1$), so that the thermodynamic approach is not driven optimally, staging procedures approaching the supersaturation formats are recoverable as a subset. Further reductions in the elimination gradients yield more conservative schedules, such as the popular "no-bubble" tables advocated for sport diving.[10]

The supersaturation model, of course, is correlated strongly with many tests over the past seventy years, serving as bases for decompression tables and staging formats around the world. Most existent tables employ the classical hypothesis in some form or another. It is doubtful that these conventional tables ever prevented bubble formation, or ever really will, though efforts aimed at measuring venous gas emboli with Doppler techniques sometimes still make a *bubble free* claim. The M-value approach is, however, easy to use, statistically safe, and generally fit for mass consumption. Little more need be said about the classical model from a valida-

FIGURE 6. Thermodynamic And Supersaturation Decompression Profiles

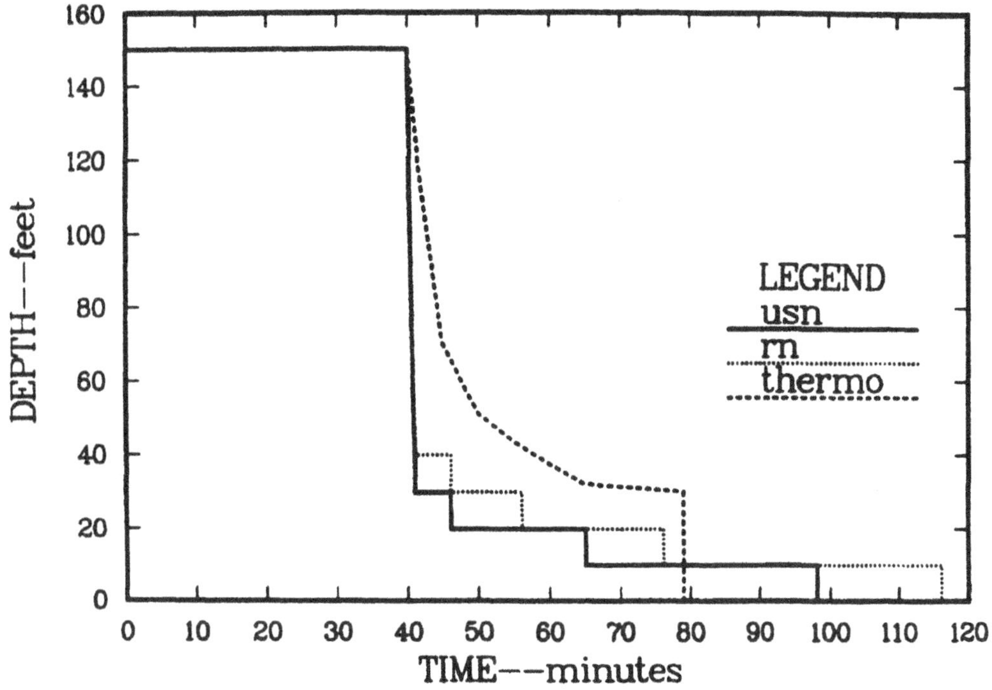

Viewed from the thermodynamic standpoint, the conventional tables are safe, but far from optimal. Decompression schedules for a dive to 150 ft for 40 minutes are depicted according to the supersaturation and thermodynamic formats. Both the US Navy and Royal Navy schedules (supersaturation) are compared with the Hills staging (thermodynamic). The major differences involve:

> *1--more time spent deeper to avoid phase inception and hence providing a greater driving force to remove excess gas, followed by;*
>
> *2--drop out at 20-30 fsw to dump the remainder of sub-symptomatic gas.*

Tests and diving experiments have provided validation for the thermodynamic model over the past 10 years, and research continues. Models extending the microphysics will provide greater insight and more optimal staging techniques.

tion viewpoint.

On the thermodynamic side, the conventional tables are safe but not optimal. They might constitute therapeutic treatment for a large amount of free gas, separated on the first long decompressive haul upwards, which must be eliminated from the system. Although the foregoing suggests that supersaturation schedules are not optimal, the vast body of experimental diving data must be computationally correlated with the thermodynamic model. Suffice to say that Hills[11] has provided massive computational correlations to back the precepts of the thermodynamic scheme. Additionally, others have tested the model under working conditions, and in laboratory experiments. Starck[12] employed the model for short, deep, air dives off the Bahamas for depths up to 300 *ft* without bends incidence in near 200 exposures. Hempleman[23] introduced deeper stops cautiously to the Royal Navy regimen by adding the 10 *ft* stop time to the 20 *ft* level, followed by direct surfacing. Bennett and Vann[24] used a version of the scheme to produce a profile for a dive to 500 *ft* for 30 *minutes*. Krasberg[12], in the ocean, in over 800 dives down to 600 *ft* for up to 1 *hour*, recorded only 4 bends cases employing the thermodynamic scheme.

8. SUMMARY

Inert gas transport in the body of a diver breathing compressed air or exotic mixture is governed by many factors, such as gas diffusion, blood perfusion, phase separation, nucleation and cavitation, membrane permeation, fluid shifts, and combinations thereof. Owing to the complexity of biological systems, multiplicity of tissues and media, and diversity of boundary conditions, it is difficult to solve the nitrogen transport problem directly *in vivo*. Early decompression studies adopted the supersaturation viewpoint. A closer look at the microphysics of bubble formation and phase separation in the mid-70s, provided considerable insight into gas transfer mechanisms, culminating in the thermodynamic theory. Integration of the two approaches is proceeding from the computational side because calculational techniques can be made operationally equivalent in staging. The phase model is more general than the supersaturation model, incorporating its predictive capability as a subset.

REFERENCES

(1)   A.E. Boycott, G.C.C. Damant, and J.S. Haldane, J. Hyg. (London) 8, 342 (1908).

(2)   A.R. Behnke, USN Med. Bull. 35, 219 (1937).

(3)   H.V. Hempleman, "Investigation Into The Diving Tables", Medical Research Council Report UPS 131, London (1952).

(4)   M. Des Granges, "Standard Air Decompression Tables", USN Experimental Diving Unit Research Report 5-57, Washington, D.C. (1957).

(5) H.R. Schreiner and P.L. Kelley, "A Pragmatic View Of Decompression", Proc. Fourth Symp. Underwater Physiol., Academic Press, New York (1971).

(6) R.D. Workman, "Calculation Of Decompression Schedules For Nitrogen-Oxygen And Helium-Oxygen Dives", USN Experimental Diving Unit Research Report 6-65, Washington, D.C. (1965).

(7) A.A. Buhlmann, P. Frey, and H. Keller, J. Appl. Physiol. 23, 458 (1967).

(8) A. Fick, Ann. Phys. (Leipzig) 170, 59 (1855).

(9) J.S. Haldane and J.G. Priestly, "Respiration", Yale University Press, New Haven (1922).

(10) M.P. Spencer, J. Appl. Physiol. 40, 231 (1976).

(11) P.B. Bennett and D.H. Elliott, "The Physiology And Medicine Of Diving And Compressed Air Work", Williams And Wilkins Company, Baltimore (1982).

(12) B.A. Hills, "Decompression Sickness", John Wiley And Sons, New York (1977).

(13) B.A. Hills, "A Thermodynamic And Kinetic Approach To Decompression Sickness", Library Board Of South Australia, Adelaide (1966).

(14) H.V. Hempleman, "Investigation Into The Decompression Tables: Further Basic Facts On Decompression Sickness", Medical Research Council Report UPS 168, London (1957).

(15) B.R. Wienke, Il Nuovo Cimento 8D, 417 (1986).

(16) B.R. Wienke, Comp. Phys. Comm. 40, 327 (1986).

(17) R.L. Bell and R.E. Borgwardt, Undersea Biomed. Res. 3, 1 (1976).

(18) P.S. Epstein and M.S. Plesset, J. Chm. Phys. 18, 1505 (1950).

(19) V.T. Inman and J.B. Saunders, J. Nerv. Ment. Dis. 99, 660 (1944).

(20) M. Harris, W.E. Berg, D.M. Whitaker, V.C. Twitty, and L.R. Blinks, J. Gen. Physiol. 28, 225 (1945).

(21) A.R. Behnke, "The Isobaric (Oxygen Window) Principle Of Decompression", Trans. Third Annual Conf. Marine Tech. Soc. 3, 213 (1967).

(22) B.R. Wienke, Int. J. Bio-Med. Comp. 21, 205 (1987).

(23) H.V. Hempleman, "Present State Of The Art", Proc. Workshop On Decompression Procedures For Depths In Excess Of 400 Feet, Undersea Biomedical Society, Washington, D.C. (1975).

(24) P.B. Bennett and R.D. Vann, "Theory And Development Of Sub-Saturation Decompression Procedures For Depths In Excess Of 400 Feet," Proc. Workshop On Decompression Procedures For Depths In Excess Of 400 Feet, Undersea Biomedical Society, Washington, D.C. (1975).

A Semigroup for Lagrangian 1D Isentropic Flow

G. H. PIMBLEY   Los Alamos National Laboratory, Los Alamos, New Mexico

1. INTRODUCTION

The purpose of this contribution is to solve the problem of isentropic hydro-dynamic flow in one space dimension, using methods of functional analysis and semigroups. These are the methods chosen by J. Lehner and G. M. Wing in 1953 to solve the linear problem of Neutron Transport. Here, however, we employ a much later development of these analytic methods to treat this non-linear problem, that was also important at Los Alamos in 1953. Heretofore, most hydrodynamics problems, including this one, have been handled mainly at the numerical level. The existence of the semigroup solution operator, developed here, has previously been guessed but not known.

We give only sketchy proofs in this paper, so as to demonstrate the methodology. More detail will be provided in later publications.

In the space $B_1 = L^1(R) \times L^1(R)$, $R : -\infty < x < \infty$, we study the system:

$$v_t - u_x = 0$$
$$u_t - \Phi(v)_x = 0 \,, \tag{1}$$
$$\text{or} \qquad \frac{dw}{dt} - Aw = 0 \,, \qquad w = \begin{pmatrix} v \\ u \end{pmatrix} \,,$$

with initial conditions $v(x,0) = v_0(x), u(x,0) = u_0(x)$, where $v_0, u_0 \in L^1(R)$. Here, v, u stand for fluid specific volume and velocity. We assume that $\Phi(v)$ has a couple of derivatives, and that $0 < m \leq \Phi'(v) \leq M$, where m, M are given constants. The function Φ is derived from an equation of state.

Equations (1) are discussed by Smoller, [1, p. 306], in the context of the Riemann problem, and called the p-System, and in a differentiated form by Courant and Friedrichs, [2, p. 38]. DiPerna, [3, p. 55], discusses Eqs. (1) as the equations of elasticity, and in [4, Vol. I, pp. 267–287] uses a compensated compactness result and vanishing viscosity to obtain a global weak solution.

The would-be generator of a semigroup of $L^1(R)$-continuous solution operators in B_1, that would solve the initial value problem for Eqs. (1) in the mild sense, [5, pp. 311-315], is seen to be

$$A \begin{pmatrix} v \\ u \end{pmatrix} = \begin{pmatrix} u_x \\ \Phi(v)_x \end{pmatrix}, \qquad v, u \in W^{1,1}(R) \subset L^1(R), \tag{2}$$

with $D(A)$ consisting of vectors in B_1 with components having derivatives in $L^1(R)$. Thus, $\overline{D(A)} \equiv B_1$. Operator (2) belongs to a class of prospective generators of the form $A_L \Phi$ defined in B_1, where A_L is linear, closed, and densely defined, and Φ is Frechet differentiable on B_1 into B_1. Operators of this class have the property that the resolvent $J_\lambda = (I - \lambda A)^{-1}, \lambda > 0$, is Frechet differentiable everywhere in B_1, [6, pp. 68, 69].

2. AN APPROXIMATE DE

We recount here some yet to be published work, [7], on the Yosida Approximation, [8, p. 447]:

$$A_\lambda = \frac{1}{\lambda}[(I - \lambda A)^{-1} - I], \qquad \lambda > 0, \tag{3}$$

of the operator A in (2), defined on $\overline{D(A)} \equiv B_1$. The resolvent J_λ is obtained by solving the coupled system:

$$v - \lambda u_x = f$$
$$u - \lambda \Phi(v)_x = g, \qquad f, g \in L^1(R), \qquad \lambda > 0. \tag{4}$$

Solution of (4) in B_1 is enabled by the m-accretiveness of the operator $-\Phi(\cdot)_{xx}$, (see [9, p. 256, Lemma 6.1], which is valid in R). A Lipschitz constant for J_λ can then be computed. Then for operator (3) we have

$$\left\| A_\lambda \begin{pmatrix} f_1 \\ g_1 \end{pmatrix} - A_\lambda \begin{pmatrix} f_2 \\ g_2 \end{pmatrix} \right\| \leq \frac{1}{\lambda} \left(2 + \frac{1}{\sqrt{M}} + \sqrt{M} + \frac{M}{m} \right) \left\| \begin{pmatrix} f_1 \\ g_1 \end{pmatrix} - \begin{pmatrix} f_2 \\ g_2 \end{pmatrix} \right\| \tag{5}$$

for arbitrary vectors $\begin{pmatrix} f_1 \\ g_1 \end{pmatrix}, \begin{pmatrix} f_2 \\ g_2 \end{pmatrix} \in B_1$. We had $0 < m \leq \Phi'(v) \leq M$. Since -A is not accretive in B_1, the constant in (5) is not the usual $\frac{1}{\lambda}$; nor is it necessarily the least constant. We can show nevertheless that, with B_1 normed convergence, [7, Thm. 7],

$$\lim_{\lambda \to 0} A_\lambda \begin{pmatrix} f \\ g \end{pmatrix} = A \begin{pmatrix} f \\ g \end{pmatrix} \quad \text{provided} \quad \begin{pmatrix} f \\ g \end{pmatrix} \in D(A) \,. \tag{6}$$

It is useful now to consider the Yosida approximate DE:

$$\frac{dw}{dt} - A_\lambda w = 0 \,, \qquad w(0) = w_0 = \begin{pmatrix} v_0 \\ u_0 \end{pmatrix} \in B_1 \,. \tag{7}$$

Theorem 1: Eq. (7) has a semigroup solution $w_\lambda = S_\lambda(t)w_0$, which is uniformly Lipschitz; i.e, we have

$$\|S_\lambda(t)w_1 - S_\lambda(t)w_2\| \le M_1 \|w_1 - w_2\| \,, \qquad 0 \le t < \infty \,, \qquad w_1, w_2 \in B_1 \,, \tag{8}$$

where M_1 is independent of $\lambda > 0$ and $t > 0$.

Sketch of the Proof: The generator A_λ is Lipschitz continuous, (see (5)). With reference to a paper of Brezis and Pazy, [10, p. 63], $A_\lambda \in \mathcal{A}(L_\lambda)$, where L_λ is the constant in (5). Indeed, with $0 < \varsigma < L_\lambda^{-1}$, we have

$$\|(w_1 - \varsigma A_\lambda w_1) - (w_2 - \varsigma A_\lambda w_2)\|$$
$$\ge \|w_1 - w_2\| - \varsigma\|A_\lambda w_1 - A_\lambda w_2\| \qquad w_1, w_2 \in B_1$$
$$\ge \|w_1 - w_2\| - \varsigma L_\lambda\|w_1 - w_2\| = (1 - \varsigma L_\lambda)\|w_1 - w_2\| \,.$$

Moreover, $R(I - \varsigma A_\lambda) \supseteq \overline{D(A_\lambda)} \equiv B_1, 0 < \varsigma < L_\lambda^{-1}$, since we can solve the equation $w - \varsigma A_\lambda w = f, f \in B_1$, by showing that the operator $\psi_f(\cdot) \equiv f + \varsigma A_\lambda(\cdot)$ is everywhere contracting in B_1. By a known result, [10, p. 64, Thm. 1], $S_\lambda(t)$ exists, and $\|S_\lambda(t)w_1 - S_\lambda(t)w_2\| \le e^{L_\lambda t}\|w_1 - w_2\|$, $w_1, w_2 \in B_1$. It turns out, however, that a Lipschitz constant exists for $S_\lambda(t)$, independent of λ, t.

Since the generator $A_\lambda = \frac{1}{\lambda}(J_\lambda - I)$ is Frechet differentiable in B_1, it can be shown that $S_\lambda(t)$, for fixed $t > 0$, is also Frechet differentiable. Then by Taylor's theorem, with $0 \le \theta \le 1$, we have

$$S_\lambda(t)w_1 - S_\lambda(t)w_2 = \int_0^1 S'_{\lambda,(1-\theta)w_1 + \theta w_2}(t)(w_1 - w_2)d\theta \,, \qquad w_1, w_2 \in B_1 \,, \tag{9}$$

where $S'_{\lambda,w}(t)$ is the Frechet derivative of $S_\lambda(t)$ at $w \in B_1$. The task is now to show that $S'_{\lambda,(1-\theta)w_1 + \theta w_2}(t)$, as a linear operator for fixed $t > 0$, is bounded independently of λ, θ, t. This latter operator happens to be the solution operator for the following family of linear initial value problems, parameterized by $0 \le \theta \le 1$:

$$\frac{dh}{dt} - A'_{\lambda, S_\lambda(t)[(1-\theta)w_1 + \theta w_2]}h = 0 \,, \qquad h(0) = h_0 \,, \tag{10}$$

where $A'_{\lambda,w}$ is the Frechet derivative of A_λ at $w \in B_1$.

Now, if we can show that the following linear initial value problem, (Note: $0 \le m \le \Phi'(\xi) \le M$):

$$\frac{dk}{d\tau} - A'_z k = 0 \,, \qquad k(0) = k_0 \,; \qquad A'_z \binom{k_1}{k_2} = \binom{k_{2x}}{[\Phi'(\xi)k_1]_x} \,, \qquad (11)$$

(where $z = J_\lambda S_\lambda(t)[(1-\theta)w_1 + \theta w_2]$, $z = (\xi, \eta)$, $0 \le \theta \le 1$, fixed $t > 0$, $\tau =$ time), has a linear solution operator bounded uniformly in λ, θ, t, say by M_1, then likewise the solution operator of the following linear initial value problem has the same bound M_1:

$$\frac{dk_\lambda}{d\tau} - A'_{\lambda, S_\lambda(t)[(1-\theta)w_1 + \theta w_2]} k_\lambda = 0 \,, \qquad k(0) = k_0 \,, t > 0 \quad \text{fixed} \,. \qquad (12)$$

This is because, since $A'_{\lambda, w} = (A'_z)_\lambda$, (12) is the Yosida approximate problem for (11). Then, (see [11, p. 243; 12, p. 85] for notation):

$$A'_{\lambda, S_\lambda(t)[(1-\theta)w_1 + \theta w_2]} \in G(B_1, M_1, 0) \,, \qquad 0 < t < \infty \,, \text{ fixed } \theta$$

forms a trivially stable operator family, for which Kato's existence theorem, [11, p. 246, Thm. 4.1; 12, pp. 130-139] can be trivially applied to show existence of a unique system $U_\lambda(t, s, \theta), 0 \le \theta \le 1$, of evolution operators that solve evolution equations (10) with common bound M_1. Thus in (9), we have

$$\|S'_{\lambda, (1-\theta)w_1 + \theta w_2}(t)h_0\| \le M_1 \|h_0\| \,, \qquad M_1 \text{ uniform in } \lambda, \theta, t \,, \qquad (13)$$

which is what we intended to show. The result follows from (9).

That problem (11), which is a conventional linear wave problem, has its solution operator with such a uniform bound is best seen by applying Fourier transform analysis in the locally convex Schwartz space $\mathscr{S} \times \mathscr{S}$ of vectors with rapidly decreasing components, with its semi-norm topology, [8, p. 146]. The resulting solution operator in the transform space has unit spectral radius. It is then shown that the inverse transformed solution operator that results for Eq. (11), continuous in the semi-norm topology of $\mathscr{S} \times \mathscr{S} \subset B_1$, is also continuous in the normed topology of B_1, uniformly in λ, θ, t. ∎

Corollary 2: There exists a constant M_2 such that for $w_1, w_2 \in B_1$,

$$\|J_\nu^k w_1 - J_\nu^k w_2\| \le M_2 \|w_1 - w_2\| \,, \qquad k = 1, 2, 3, \cdots, \text{etc.} \qquad (14)$$

where $J_\nu^k = (I - \nu A)^{-k}$, and $\nu > 0$.

Sketch of the Proof: By (8), since A_λ is the unique generator of $S_\lambda(t)$,

$$\lim_{k \to \infty} \left\| \left(I - \frac{t}{k}A_\lambda\right)^{-k} w_1 - \left(I - \frac{t}{k}A_\lambda\right)^{-k} w_2 \right\| \le M_1 \|w_1 - w_2\| \,, 0 \le t < \infty, \qquad (15)$$

where the limit is uniform on compact t-sets, using the exponential formula for

$S_\lambda(t)$, [10, p. 64, (1.6)]. Then there exists an integer k_1, such that

$$\left\|\left(I - \frac{t}{k}A_\lambda\right)^{-k}w_1 - \left(I - \frac{t}{k}A_\lambda\right)^{-k}w_2\right\|$$
$$\leq \widehat{M_1}\|w_1 - w_2\|, \qquad k = k_1, k_1 + 1, \cdots, \text{etc.}, \qquad (16)$$

$0 \leq t < \infty$, for a constant $\widehat{M_1} > M_1$. Now, for given $\nu > 0$, we can insert a sequence $\{t_k\} : t_k = \nu k, k = k_1, k_1 + 1, \cdots \text{etc.}$, for $t > 0$ in (16). This gives

$$\|(I - \nu A_\lambda)^{-k}w_1 - (I - \nu A_\lambda)^{-k}w_2\| \leq \widehat{M_1}\|w_1 - w_2\|, \ k = k_1, \ k_1 + 1, \cdots, \text{etc.} \ (17)$$

Using (6), we now let $\lambda \to 0$ in (17). Then, recalling (5) and (3), whatever be the remaining constants $M_1^{(k)}$ such that for $k = 1, 2, \cdots, k_1 - 1$,
$\|(I - \nu A)^{-k}w_1 - (I - \nu A)^{-k}w_2\| \leq M_1^{(k)}\|w_1 - w_2\|$, there are only finitely many; we may set $M_2 = \max\{\widehat{M_1}, M_1^{(k)}, k = 1, \cdots, k_1 - 1\}$. The result follows. ∎

3. BANACH SPACE RE-METRIZATION

The Lipschitz condition on $J_\nu^k = (I - \nu A)^{-k}, k = 1, 2, \cdots$, etc., $0 \leq \nu < \infty$, given in Corollary 2, namely (14), is of little use in current nonlinear semigroup generation methods, unless we can make $M_2 = 1$. This requires "re-metrization," as opposed to simple re-norming often used in linear problems. To develop a re-metrizing method, we show how Banach space B_1 can be considered a manifold.

The elements of this proposed B_1-manifold comprise B_1 itself. The tangent bundle, [6, pp. 150-161] is $B_1 \times B_1$, and the fiber space over $x \in B_1$ is just $\{x\} \times B_1$. Each fiber space $T_x B_1$ is a translation of B_1 to have origin at x. This B_1-manifold inherits the linear structure of B_1. The Finsler norm $\|\cdot\|_x$ of the fiber $T_x B_1 = \{x\} \times B_1$ over x is given by the distance in the fiber, from its origin x, measured by the original norm $\|\cdot\|$ of B_1.

In sect. 1, we mentioned that $A = A_L\Phi$ is an abstract form of operator (2), which we are considering a prospective generator of a nonlinear semigroup. This type of operator has a resolvent $J_\nu = (I - \nu A)^{-1}$ which is Frechet differentiable. The mapping $x \to J_\nu x$ takes the B_1-manifold into itself, and has a continuous Frechet derivative $D_x J_\nu$ at any point $x \in B_1$, that also maps this flat B_1-manifold into itself.

Formally, a manifold-type metric, corresponding to the Finsler structure in $B_1 \times B_1$, does exist in B_1. For $x, y \in B_1$, we write

$$\rho(x, y) = \inf_{\{C(s)\}} \int_0^1 \left\|\frac{d}{ds}C(s)\right\|_{C(s)} ds \qquad (18)$$

where $C(s), 0 \leq s \leq 1$, is a C^1 curve joining x and y. In (18), $\|\cdot\|_{C(s)}, 0 \leq s \leq 1$, are the Finsler norms encountered along path $C(s)$. Since the shortest path is known: $C_0(s) = (1-s)x + sy$, (18) reduces to $\rho(x,y) = \|x-y\|$.

In the terminology of (18), the result of Corollary 2, namely (14), is stated as follows:

$$\rho\left(J_\nu^n x, J_\nu^n y\right) \leq M_2 \rho(x,y) , \qquad n = 1, 2, \cdots, \text{etc.} \tag{19}$$

Statement (19) is tantamount to having

$$\|TJ_\nu^n(x) \cdot v_x\|_{J_\nu^n x} \leq M_2 \|v_x\|_x , \qquad n = 1, 2, \cdots, \text{etc.} , \tag{20}$$

where $TJ_\nu^n(x) = T\left(I - \nu A_L \Phi\right)^{-n}(x)$, evaluated at $x \in B_1$, is called a "tangent bundle map" taking the fiber over x linearly into the fiber over $J_\nu^n x$. It consists of two parts: (i) $x \to J_\nu^n x$, and (ii) $v_x \to DJ_\nu^n(x) \cdot v_x$, where v_x is the generic element in the fiber over x, [6, pp. 150-161].

The following is easy to prove:

Lemma 3: (20) implies, and is implied by (19).

We are now in a position to state our re-metrization result:

Theorem 4: Let $A = A_L \Phi : D(A) \subset B_1 \to B_1$ be a nonlinear operator, with Frechet-differentiable resolvent $J_\nu : B_1 \to D(A) \subset B_1$, $0 < \nu < \infty$. Conditions on $A_L \Phi$ were given in Sect. 1. Suppose, for any $x, y \in B_1$,

$$\|J_\nu^n x - J_\nu^n y\| \leq M_2 \|x - y\| , \qquad n = 1, 2, \cdots, \text{etc.} , \tag{21}$$

where $M_2 \geq 1$ is independent of ν. Then there exists a metric d in B_1, based on a Finsler structure equivalent to $\|\cdot\|_x$ on which ρ in (18) was based, such that for $x, y \in B_1$,

$$d\left(J_\nu^n x, J_\nu^n y\right) \leq d(x,y) , \qquad n = 1, 2, \cdots, \text{etc.} \tag{22}$$

Thus, in terms of metric d, $J_\nu^n = (I - \nu A)^{-n}$ is contracting for every n.

Sketch of the Proof: We follow the linear case proved by Eilenberg and Feller, as related by A. Pazy, [12, pp. 17-18]. We first define an intermediate Finsler structure. Let $\mu > 0$, and put

$$\|v_x\|_{\mu,x} = \sup_{p \geq 0} \|TJ_\mu^p(x) \cdot v_x\|_{J_\mu^p x} , \text{ integer } p . \tag{23}$$

Then by first setting $p = 0$, and then using (20), we get

$$\|v_x\|_x \leq \|v_x\|_{\mu,x} \leq M_2 \|v_x\|_x , \tag{24}$$

so that our intermediate Finsler structure is equivalent to the old. Using now the Composite Mapping Theorem, [6, pp. 76, 154], refering also to (23),

$$\|TJ_\mu^n(x) \cdot v_x\|_{\mu,J_\mu^n x} = \sup_{p \geq 0} \|TJ_\mu^{p+n}(x) \cdot v_x\|_{J_\mu^{p+n} x} \leq \|v_x\|_{\mu,x} . \tag{25}$$

We claim now that

$$\|TJ_\nu^n(x)\cdot v_x\|_{\mu,J_\nu^n x} \le \|v_x\|_{\mu,x}\,, \qquad 0 < \mu \le \nu\,. \tag{26}$$

For this, the resolvent "identity:" $J_\nu x = J_\mu\left[\frac{\mu}{\nu}x + \frac{\nu-\mu}{\nu}J_\nu x\right]$, [13, p. 268], is used, together with (25), to produce the following inequality when $n = 1$:

$$\|TJ_\nu(x)\cdot v_x\|_{\mu,J_\nu x} \le \frac{\mu}{\nu}\|v_x\|_{\mu,x} + \frac{\nu-\mu}{\nu}\|TJ_\nu(x)\cdot v_x\|_{\mu,J_\nu x}\,, \tag{27}$$

whence we get

$$\|TJ_\nu(x)\cdot v_x\|_{\mu,J_\nu x} \le \|v_x\|_{\mu,x}\,. \tag{28}$$

Iteration on (28), n times, gives (26) as claimed.

Then from (24) and (26), it follows that for $0 < \mu \le \nu$,

$$\|TJ_\nu^n(x)\cdot v_x\|_{J_\nu^n x} \le \|TJ_\nu^n(x)\cdot v_x\|_{\mu,J_\nu^n x} \le \|v_x\|_{\mu,x}\,. \tag{29}$$

Taking the sup in (29) over all $n > 0$ implies by (23) that

$$\|v_x\|_{\nu,x} \le \|v_x\|_{\mu,x}\,, \qquad 0 < \mu \le \nu\,. \tag{30}$$

This monotonicity enables us to define the new Finsler norm:

$$\llbracket v_x \rrbracket_x = \lim_{\mu\to 0}\|v_x\|_{\mu,x}\,. \tag{31}$$

Then we see the equivalence of this new Finsler structure, (i.e. (31)), with that used in (18), by taking the μ-limit in (24), $\big($viz (30)$\big)$:

$$\|v_x\|_x \le \llbracket v_x \rrbracket_x \le M_2\|v_x\|_x\,. \tag{32}$$

Using now the right inequality of (29), and taking $\mu \to 0$, there results

$$\llbracket TJ_\nu^n(x)\cdot v_x \rrbracket_{J_\nu^n x} \le \llbracket v_x \rrbracket_x\,, \qquad n = 1, 2\cdots, \text{etc.} \tag{33}$$

Now we can prove (22) if we base metric d on the Finsler structure with norms $\llbracket\cdot\rrbracket_x$. Let $\big\{C^{(q)}(s), 0 \le s \le 1\big\}$ be a sequence of C^1 curves joining $x, y \in B_1$, such that the corresponding sequence of lengths:

$$\ell_q = \int_0^1 \left\|\frac{d}{ds}C^{(q)}(s)\right\|_{C^{(q)}(s)} ds\,, \qquad q = 1, 2, \cdots, \text{etc.}$$

tends to an infimum as $q \to \infty$. Then, using (33), we get:

$$d\left(J_\nu^n x, J_\nu^n y\right) \le \int_0^1 \left\|\frac{d}{ds} J_\nu^n C^{(q)}(s)\right\|_{J_\nu^n C^{(q)}(s)} ds$$

$$= \int_0^1 \left\| T J_\nu^n [C^{(q)}(s)] \cdot \frac{d}{ds} C^{(q)}(s)\right\|_{J_\nu^n C^{(q)}(s)} ds$$

$$\le \int_0^1 \left\|\frac{d}{ds} C^{(q)}(s)\right\|_{C^{(q)}(s)} ds, \qquad q = 1, 2, \cdots, \text{etc.} \qquad (34)$$

Taking the limit in (34) as $q \to \infty$, noting that the inequality is preserved, we obtain (22). ■

4. SEMIGROUP GENERATION

The standard semigroup generation theorem for use with accretive operators in real, not necessarily reflexive Banach space is that proved by Crandall and Liggett, [13, pp. 266-273]. The idea of accretiveness has not yet been extended to semigroup generation in Banach manifolds, not even to our B_1-manifold. For general Banach manifolds, we mention the semigroup existence theorem of J. Marsden, [14, pp. 53-60], which has hypotheses, however, that are somewhat stronger than we should like here.

Inequality (22), when $n = 1$, states an accretiveness for $-A = -A_L \Phi$, expressed in terms of the non-homogeneous metric d of our special manifold. To use this accretiveness notion, namely that J_ν be contractive, we must generalize the Crandall-Liggett result.

Theorem 5: Let $-A$, with domain $D(A) \subset B_1$, be "accretive" in terms of the metric d, so that with $f, g \in D(J_\nu)$,

$$d(J_\nu f, J_\nu g) \le d(f, g), \qquad J_\nu = (I - \nu A)^{-1}, \qquad \nu > 0, \qquad (35)$$

where d is defined as in (18), but with Finsler norm $\|\cdot\|_x$ satisfying (32). Suppose also a range condition: $R(I - \nu A) \supset \overline{D(A)}, 0 \le \nu < \infty$. Then for $f \in D(A)$ and $t > 0$, the following limit exists:

$$S(t)f = \lim_{n \to \infty} J_{t/n}^n f \in B_1, \qquad (36)$$

in terms of the metric d. $S(t)$ is a semigroup operator, (see [13, p. 265, (4)]). Moreover, $S(t)$ is contracting:

$$d\left(S(t)f, S(t)g\right) \le d(f, g), \qquad f, g \in D(A). \qquad (37)$$

Definition of $S(t)$ is readily extended to $\overline{D(A)}$.

Before sketching the proof of Theorem 5, we re-state the four points investigated by Crandall-Liggett prior to their proof, [13, p. 268, Lemma 1.2] in terms of the metric d, with indication of proof.

Lemma 6: Setting $J_\nu f = (I - \nu A)^{-1} f$, we assert that

(i) J_ν is a function, and for $f, g \in D(J_\nu), d(J_\nu f, J_\nu g) \le d(f, g)$,

(ii) $d(J_\nu f, f) \le \nu \inf_{g \in Af} d(0, g)$ for $f \in D(A) \cap D(J_\nu)$,

(iii) $d(J_\nu^n f, f) \le n d(J_\nu f, f)$, $f \in D(J_\nu^n)$, $n = 1, 2, \cdots$, etc.

(iv) $J_\nu f \in J_\mu \left(\frac{\mu}{\nu} f + \frac{\nu - \mu}{\nu} J_\nu f \right)$, $f \in D(J_\nu)$.

Sketch of the Proof: (i) is being assumed as our definition of accretiveness for A. The Crandall-Liggett proof of (iv) will suffice here for the B_1-manifold. We can use elements of their proof of (ii) also: taking $[f_1, g_1] \in -A$ and $[f, g] \in -A$ such that $f_1 + \nu g_1 = f$, we obtain $J_\nu f = f_1$, as they do. Then

$$d(f_1, f) = d\Big(J_\nu(f_1 + \nu g_1), J_\nu(f + \nu g) \Big) \le d(f_1 + \nu g_1, f + \nu g) = d(f, f + \nu g) . \tag{38}$$

Now let $\left\{ C_0^{(q)}(s), q = 1, 2, \cdots, \text{etc.} \right\}$ be a family of C^1 paths in the B_1-manifold, that connect f with itself, with lengths tending to zero as $q \to \infty$. Let $\{ C_1^{(q)}(s), q = 1, 2, \cdots, \text{etc.} \}$ be a family of C^1 paths connecting the origin of B_1 with $g \in B_1$, with lengths that approach the distance between 0 and g, as $q \to \infty$ in the B_1 manifold, i.e. $d(0, g)$. Then $C_0^{(q)}(s) + \nu C_1^{(q)}(s)$ is a family of C^1 paths connecting f with $f + \nu g, q = 1, 2, \cdots$, etc., with lengths approaching an infimum. Therefore,

$$d(f, f + \nu g) \le \int_0^1 \left\| \frac{d}{ds} \left[C_0^{(q)}(s) + \nu C_1^{(q)}(s) \right] \right\|_{C_0^{(q)}(s) + \nu C_1^{(q)}(s)} ds$$

$$\le \int_0^1 \left\| \frac{d}{ds} C_0^{(q)}(s) \right\|_{C_0^{(q)}(s)} ds + \nu \int_0^1 \left\| \frac{d}{ds} C_1^{(q)}(s) \right\|_{C_1^{(q)}(s)} ds \tag{39}$$

where the second inequality is justified using the theory of interpolated Banach spaces, [15, p. 57, (2.2); pp. 64-74]. Going to the limit in (39) as $q \to \infty$, the inequality is preserved, and we obtain (ii). For (iii) we have

$$d(J_\nu^n f, f) \le \sum_{i=0}^{n-1} d(J_\nu^{i+1} f, J_\nu^i f) \le \sum_{i=0}^{n-1} d(J_\nu f, f) = n d(J_\nu f, f) , \tag{40}$$

since $f \in D(J_\nu)$, and we use the triangle inequality for d. The second inequality in (40) results from: $d(J_\nu^{i+1} f, J_\nu^i f) \le d(J_\nu^i f, J_\nu^{i-1} f) \le \cdots \le d(J_\nu f, f)$. ■

Sketch of the Proof of Theorem 5: Let $f \in D(J_\nu^n) \cap D(J_\mu^n)$. Now define $\alpha_{k,j} = d(J_\mu^j f, J_\nu^k f)$. Then using Lemma 6 (iv), (noting that by (35) J_ν is single valued), we have

$$\alpha_{k,j} = d\left(J_\mu^j f, J_\nu\left(J_\nu^{k-1}f\right)\right) = d\left(J_\mu^j f, J_\mu\left(\frac{\mu}{\nu}J_\nu^{k-1}f + \frac{\nu-\mu}{\nu}J_\nu^k f\right)\right)$$

$$\leq d\left(J_\mu^{j-1}f, \left(\frac{\mu}{\nu}J_\nu^{k-1}f + \frac{\nu-\mu}{\nu}J_\nu^k f\right)\right) \qquad (41)$$

by Lemma 6, (i). We wish now to show that the quantity on the right in (41) is less than the following quantity:

$$\frac{\mu}{\nu}\alpha_{k-1,j-1} + \frac{\nu-\mu}{\nu}\alpha_{k,j-1} = \frac{\mu}{\nu}d\left(J_\mu^{j-1}f, J_\nu^{k-1}f\right) + \frac{\nu-\mu}{\nu}d\left(J_\mu^{j-1}f, J_\nu^k f\right) . \qquad (42)$$

Let $\left\{C_1^{(q)}(s)\right\}$, respectively $\left\{C_2^{(q)}(s)\right\}$, $q = 1, 2, \cdots$, etc., $0 \leq s \leq 1$, be two families of C^1 paths, which for fixed q join $J_\mu^{j-1}f$ and $J_\nu^{k-1}f$, respectively $J_\mu^{j-1}f$ and $J_\nu^k f$, on our B_1-manifold. Then $\frac{\mu}{\nu}C_1^{(q)}(s) + \frac{\nu-\mu}{\nu}C_2^{(q)}(s)$, for given integer q, is a C^1 path joining $J_\mu^{j-1}f$ and $\frac{\mu}{\nu}J_\nu^{k-1}f + \frac{\nu-\mu}{\nu}J_\nu^k f$. We are supposing that the lengths of these paths approach the infimum as $q \to \infty$. Then, with reference to (41):

$$d\left(J_\mu^{j-1}f, \left(\frac{\mu}{\nu}J_\nu^{k-1}f + \frac{\nu-\mu}{\nu}J_\nu^k f\right)\right)$$

$$\leq \int_0^1 \left\|\frac{d}{ds}\left[\frac{\mu}{\nu}C_1^{(q)}(s) + \frac{\nu-\mu}{\nu}C_2^{(q)}(s)\right]\right\|_{\frac{\mu}{\nu}C_1^{(q)}(s)+\frac{\nu-\mu}{\nu}C_2^{(q)}(s)} ds$$

$$\leq \frac{\mu}{\nu}\int_0^1 \left\|\frac{d}{ds}C_1^{(q)}(s)\right\|_{C_1^{(q)}(s)} ds + \frac{\nu-\mu}{\nu}\int_0^1 \left\|\frac{d}{ds}C_2^{(q)}(s)\right\|_{C_2^{(q)}(s)} ds \qquad (43)$$

where the last inequality is justified again in this B_1-manifold on the basis of the theory of interpolated Banach spaces and interpolation norms, [15, p. 57, (2.2); pp. 64-73]. Going to the limit in (43), as $q \to \infty$, preserves the inequality, and gives us (42) on the right in (41), as desired.

The inequality that results for $\alpha_{k,j} = d\left(J_\mu^j f, J_\nu^k f\right)$ from (41)-(43):

$$\alpha_{k,j} \leq \frac{\mu}{\nu}\alpha_{k-1,j-1} + \frac{\nu-\mu}{\nu}\alpha_{k,j-1} \qquad (44)$$

is of the very same form as the one that results in the proof of Crandall-Liggett, [13, p. 270]. It can be used in the same way, in Liggett's triple induction process to estimate $\alpha_{m,n}$ in terms of $\alpha_{i,0}$ and $\alpha_{0,j}$, $i, j = 1, \cdots, n$, and thus to arrive at a B_1-manifold version of Crandall-Liggett inequality (1.9), [13, p. 271].

In a forthcoming book, [16], J. Goldstein of Tulane plans to discuss Liggett's triple induction process for (44). Such a discussion is omitted in [13, p. 270] unfortunately, though there is something similar in print, [8, pp. 454-457].

Moreover, Goldstein, [16], may also mention an alternative way of getting from inequality (44) to an inequality similar to, but simpler than, Crandall-Liggett (1.9), [13, p. 271], by probability theory and random walk; namely

$$d\left(J_\mu^n f, J_\nu^m f\right) \le \left\{(m\nu - n\mu)^2 + n\mu(\nu - \mu)\right\}^{1/2} \cdot \inf_{g \in Af} d(0, g), \qquad (45)$$

where $n \ge m$. As mentioned, this is more simple, but similar, serving the same purpose. Letting $\nu = t/k$ and $\mu = t/j$ in (45), we see that the sequence $d\left(J_{t/j}^j f, J_{t/k}^k f\right)$ is Cauchy as $j, k \to \infty$. Using (32), one sees that $\left\{J_{t/j}^j\right\}$ converges to a limit in B_1 as $j \to \infty$, and (36) results. Then conventional methods of Crandall-Liggett, [13, p. 272], used with (45), suffice for the rest of the proof. ∎

Corollary 7: The problem of 1D Isentropic Lagrangian Flow, namely (1), has a global unique mild solution in $L^1(R) \times L^1(R)$ given by a semigroup operator: $w = S(t)w_0$. Moreover, we have $d\left(S(t)w_1, S(t)w_2\right) \le d(w_1, w_2)$, which through (32) and (18) becomes $\|S(t)w_1 - S(t)w_2\| \le M_2\|w_1 - w_2\|, w_1, w_2 \in B_1$.

Proof: This rests on Theorem 5, where an effort has been made to include the generality of Crandall-Liggett. We need only note that with problem (1), $\overline{D(A)} \equiv B_1$ and $D(J_\nu) \equiv B_1$, so that $R(I - \nu A) \equiv B_1, 0 < \nu < \infty$, is the range condition.

5. SUMMARY

The existence of $S(t)$ directly implies a large data global existence, uniqueness, and stability theorem in terms of the $L^1(R)$ normed topology for the mild solution of Eqs. (1). Most previous results in hydrodynamics have required either small initial data (in some sense), or a time restriction indicating local validity only.

The problem treated above, namely that of one dimensional isentropic flow in Lagrange coordinates, is only the most simple, nontrivial, physically meaningful problem in hydrodynamics. One direction for generalization is that of adding more space dimensions, though this is not simply a matter of inserting the divergence and gradient for the space derivatives in Eqs. (1), (see [2, Chapter I, Section 1]). One could also consider a finite geometry with boundary conditions. Another direction of generalization is to treat non-isentropic, flow using semigroup methods. Since energy is then conserved, an energy balance equation must be added to the coupled system of DE's. Moreover, the equation of state must then be included explicitly. For these generalizations, it may be desirable to use Eulerian coordinates instead of Lagrangian coordinates. The equivalent of Eqs. (1) in Eulerian coordinates, [2, p. 37, (20.01)], has not yet been studied using these direct functional analysis methods.

ACKNOWLEDGMENT

The author is grateful to Professors J. Goldstein, J. Marsden, and A. Pazy of respectively Tulane, Berkeley, and Jerusalem, each of whom have contributed ideas along the line. Also, he is thankful to Professors J. Lehner and G. M. Wing for having encouraged the author in semigroup thinking, over 30 years ago, in Neutron Transport.

REFERENCES

1. J. Smoller, **Shock Waves and Reaction Diffusion Equations**, Springer-Verlag, Grundlagen No. 258, New York, Heidelberg, Berlin, 1982.

2. R. Courant and K. Friedrichs, **Supersonic Flow and Shock Waves**, Springer-Verlag, Appl. Math. Ser. No. 21, New York, Heidelberg, Berlin, 1976.

3. R. DiPerna, "Convergence of Approximate Solutions to Conservation Laws," Arch. Rat. Mech. and Anal., Vol. 82 (1983), pp. 27-70.

4. R. DiPerna, "Nonlinear P.D.E. and the Weak Topology," Lectures in Applied Math., AMS, Vol. No. 23 (1986), pp. 267-281.

5. M. Crandall, "Nonlinear Semigroups and Evolution Governed by Accretive Operators," Proc. of Symp. on Pure Math.," Vol. 45 (1986), I, pp. 305-337.

6. R. Abraham, J. Marsden, and T. Ratiu, **Manifolds, Tensor Analysis, and Applications**, Addison-Wesley Publishing Co., Reading, Mass, 1983.

7. G. Pimbley, "The Yosida Approximation of an Operator Found in Hydrodynamics and Elasticity Theory," unpublished.

8. K. Yosida, **Functional Analysis, 6th ed.**, Springer-Verlag, Grundlagen No. 123, New York, Heidelberg, Berlin, 1980.

9. A. Pazy, "The Lyapunov Method for Semigroups of Nonlinear Contractions in Banach Spaces," J. D'Analyse Math., Vol. 40 (1981), pp. 239- 262.

10. H. Brezis and A. Pazy, "Convergence and Approximation of Semigroups of Nonlinear Operators in Banach Spaces," J. Funct. Anal., Vol. 9 (1972), pp. 63-74.

11. T. Kato, "Linear Evolution Equations of 'Hyperbolic' Type," J. Fac. Sci., U. of Tokyo, Sect. I, Vol. 17 (1970), pp. 241-258.

12. A. Pazy, **Semigroups of Linear Operators and Applications to Partial Differential Equations**, Springer-Verlag, Appl. Math. Sci. No. 44, 1983.

13. M. Crandall and T. Liggett, "Generation of Semi-Groups of Nonlinear Transformations on General Banach Spaces," Amer. J. Math., Vol. 93 (1971), pp. 265-298.

14. J. Marsden, "On Product Formulas for Nonlinear Semigroups," J. Funct. Anal., Vol. 13 (1973), pp. 51-72.

15. N. Aronszahn and E. Gagliardo, "Interpolation Spaces and Interpolation Methods," Annali Mat., pura. and appl., Ser. 4, Vol. 68 (1965), pp. 51-118.
16. J. Goldstein, **Semigroups of Nonlinear Operators**, to be published, Oxford Mathematical Monographs, Oxford University Press.

Mass Transfer, Heat Transfer, and Brownian Particle Deposition onto Two Spheres in Contact

H. W. CHIANG   Whiteshell Nuclear Research Establishment, Atomic Energy of Canada Limited, Pinawa, Manitoba, Canada

ABSTRACT

This paper describes the deposition of Brownian particles from a fluid stream onto two unequal-size spheres in contact. The deposition process is assumed to be controlled by convective diffusion as well as the surface interaction potentials between the spheres and Brownian particles. In the absence of a repulsive surface potential, the problem becomes identical to that of mass or heat transfer between spheres and surrounding fluid. The convective diffusion equation is solved using a similarity transformation. The results of the present study show that, for a fixed Peclet number, the mass transfer (or heat transfer) rate is the smallest for two equal-size spheres and increases monotonically as the size of one of the spheres is decreased. Surface interaction potentials are shown to have a significant effect on the Brownian deposition rate.

1   INTRODUCTION

Fibrous and packed-bed filters have been used for removing aerosols and hydrosols from fluid streams. Practical applications for these filters can be found in many industries: wastewater treatment [1, 2], respirator masks, product and process streams in coal conversion processes, emergency filtration systems for radioactive particle leaks, etc. [3]. The process of Brownian particle deposition onto various media is important in understanding and modelling of these filters. A fibrous or packed-bed filter can be considered as a number of unit bed elements connected in series. For a homogeneous filter, the unit bed elements are uniform in thickness and are statistically similar. Each unit bed element contains a number of unit cells with a given size distribution [4]. The shape of the unit cells can be modelled using a well-defined geometry of a single sphere, a sphere-in-cell, a cylinder, a constricted tube, bi-spheres, etc. Thus the filtration of Brownian particles by a filter of complex geometry can be studied in terms of accumulation of particles by a collector of well-defined geometry. Unlike other methods, this approach provides a more fundamental insight into the physical phenomena of filtration process.

Brownian particle deposition onto collectors of various geometries, rotating disks [5], spheres [6, 7], spheres-in-cell [8, 9], cylinders [6], and constricted tubes [10], in the absence or in the presence of repulsive surface interaction potential has been studied previously. In the present work, we are interested in Brownian deposition onto a bi-sphere collector.

In the absence of any repulsive interaction potential barrier between particles and collector, the Brownian deposition problem is identical to that of mass transfer of convective-diffusion species to a medium. Mass or heat transfer involving particles is important in many fields, e.g., combustion of dispersed fuels, spray drying, meteorological studies, ion exchange, and gas chromatography [11]. As a first step towards understanding of the combustion of fuel droplets in the presence of forced convection, Tal and his co-workers [12] studied heat and momentum transfer around a pair of solid spheres separated by a certain distance in viscous flow at intermediate Reynolds numbers. Chen and Pfeffer [13] and Aminzadeh et al. [11] studied mass transport around two separated spheres at low Reynolds numbers. All these studies showed that the overall Sherwood number (or Nusselt number in heat transfer) for either sphere was less than that of a single isolated sphere. As the distance between the two spheres was decreased, the increased particle-to-particle interaction resulted in a further reduction of Sherwood number. Because of the limitations of bipolar coordinate system used by the above investigators used, it was not possible to study the case of the two touching spheres.

This is a case we have considered in our study, we have also investigated the effect of sphere size on Sherwood number. As mentioned earlier, in the absence of repulsive interaction potential barrier between particles and collector, the Brownian deposition problem is identical to that of mass transfer of convective-diffusion species to a medium. Thus, the results from Brownian deposition can be directly applied to this study.

As pointed out by Bird et al. [14], a direct analogy can be drawn between mass transfer and heat transfer if the following six conditions are met: constant physical properties, relatively small rate of mass transfer, no chemical reaction in the fluid, no viscous dissipation, no emission or absorption of radiant energy, and no pressure diffusion or thermal diffusion. In the present paper these six conditions are assumed satisfied, and for convenience, only mass transfer will be presented in the following sections. However, it should be kept in mind that results can be applied to heat transfer as well.

2 THEORETICAL ANALYSIS

2.1 Bi-sphere Geometry

As shown in Fig. 1a. we consider a Newtonian, incompressible fluid with a particle concentration of c_o and approach velocity of u_o past two touching spheres of sizes a_A and a_B. The main flow is parallel to the line joining the centres of the spheres. The cylindrical coordinate system (p, z) has the origin at the centre of sphere B. In the boundary layer coordinate system (x, y), x is measured from the stagnation point and along the tangential direction to the surface of the spheres, and y is in the direction normal to the surface of the spheres. These two coordinate systems are related to each other by

$$p = \begin{cases} (y + a_A) \sin(x/a_A) & \text{for } x \leq \pi a_A \\ (y + a_B) \sin[(x - \pi a_A)/a_B] & \text{for } x > \pi a_A \end{cases} \quad (1a)$$

$$z = \begin{cases} (y + a_A) \cos(x/a_A) & \text{for } x \leq \pi a_A \\ (y + a_B) \cos[(x - \pi a_A)/a_B] & \text{for } x > \pi a_A \end{cases} \quad (1b)$$

Another coordinate system convenient to describe two spheres in contact is the tangent-sphere coordinates (η, ξ) shown in Fig. 1b. This coordinate system is related to the cylindrical coordinates by [15]:

$$\eta = \frac{2 p}{(z + a_B)^2 + p^2} \quad (2a)$$

$$\xi = \frac{?(z + a_B)}{(z + a_B)^2 + p^2} \quad (2b)$$

The contour of sphere B is given by $\xi = 1/a_B$ and that of sphere A by $\xi = -1/a_A$. The contact point between the two spheres corresponds to $\eta = \infty$. The convenience of using tangent-sphere coordinates is that the flow is bounded within the region $-1/a_A < \xi < 1/a_B$ and $0 < \eta < \infty$; therefore, the flow boundary conditions can be easily specified at contours of spheres in terms of the tangent-sphere coordinates.

2.2 Governing Equations and Solutions

At high Peclet numbers, a concentration boundary layer is formed near the surface of spheres. Thus it is more convenient to use boundary-layer coordinates to express the governing equations. The following dimensionless

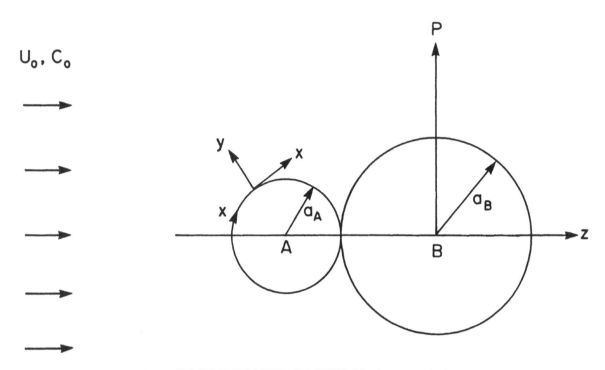

FIG. 1a COORDINATE SYSTEMS (p, z) AND (x, y)

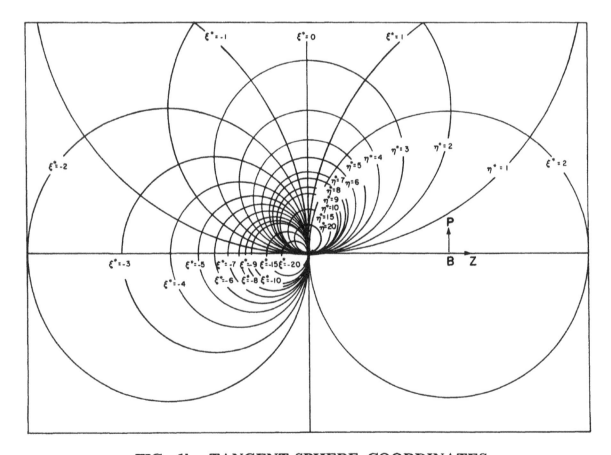

FIG. 1b TANGENT-SPHERE COORDINATES

continuity and transport equations can be used to describe mass transfer or Brownian deposition onto an axisymmetric body (e.g., see [16]):

$$\partial(p^* u_x^*)/\partial x^* + \partial(p^* u_y^*)/\partial y^* = 0 \tag{3}$$

$$u_x^*(\partial c^*/\partial x^*) + u_y^*(\partial c^*/\partial y^*) = (\partial^2 c^*/\partial y^{*2})/N_{Pe} \tag{4}$$

Boundary conditions can be specified as follows:

(i)   at $x^* = 0$, $c^* = 1$ $\tag{5}$

(ii)   as $y^* \to \infty$, $c = 1$ $\tag{6}$

(iii) at $y^* = 0$, $\begin{cases} \partial c^*/\partial y^* = k^* c^* & \text{for Browninan deposition in the} \\ & \text{presence of a repulsive barrier} \tag{7a} \\ \text{or } c^* = 0 & \text{for mass transfer} \tag{7b} \end{cases}$

where $x^* = x/2a_B$, $y^* = y/2a_B$, $p^* = p/2a_B$, $c^* = c/c_o$

$$u_x^* = u_x/u_o, \qquad u_y^* = u_y/u_o$$

$$N_{Pe} = (2a_B u_o)/D, \quad k^* = k/u_o \tag{8}$$

and u_x and u_y are velocity components along x and y direction, c is the concentration of diffusion species or Brownian particles, and D is the diffusion coefficient. Equation (7a) is based on the derivation of Spielman and Friedlander [8] in which the effect of surface interaction potential can be accounted for in the boundary condition instead of transport equation. Such an effect can be characterized by a pseudo first-order rate constant, k, which is related to surface interaction potential, Φ, by

$$k = \frac{D}{\displaystyle\int_o^\infty (e^{\Phi/k_b T} - 1)\, dy} \tag{9}$$

For particles of finite size, the boundary conditions, eqs. (7a, b), strictly speaking, should be specified at $y = a_p$ instead at $y = 0$. However, since we are interested in deposition of Brownian particles the size of which can be assumed negligible, boundary conditions can be specified at $y = 0$.

Solution for the Case of Mass Transfer

The solution of eqs. (3) and (4) with the boundary conditions, eqs. (5), (6) and (7b), can be obtained with the use of Lighthill's technique [17] (a similarity transformation for mass transfer at high Peclet numbers). The overall Sherwood number can be found as

$$N_{Sh} = 2a_B m / D$$

$$= \frac{1}{4(a_A^{*2} + a_B^{*2})} \frac{3^{1/3}}{\Gamma(4/3)} N_{Pe}^{1/3} \left[\int_0^{x_f^*} p_w^* \{ p_w^* (\partial u_x^* / \partial y^* |_w) \}^{1/2} dx^* \right]^{2/3} \tag{10}$$

where

$$p_w^* = \begin{cases} a_A^* \sin(x/a_A) & \text{for } x \leq \pi a_A \\ a_B^* \sin[(x - \pi a_A)/a_B] & \text{for } x > \pi a_A \end{cases} \tag{11}$$

and m is the average mass transfer coefficient, x_f^* the arc length, which equals to $\pi(a_A + a_B)$.

Solution for Brownian Deposition in the Presence of Repulsive Potential Barrier

Similarly, the solution of eqs. (3) and (4) with the boundary conditions, eqs. (5), (6) and (7a), valid for high Peclet numbers, can be obtained by using Lighthill's technique [17] and Duhamel's theorem [18]. The concentration at the surface of spheres is found to be

$$c_w^*(\sigma) = \frac{D}{2a_B k} \frac{N_{Pe}^{1/3}}{9^{1/3} \Gamma(4/3) \sigma^{1/3}} \{ p_w^* (\partial u_x^* / \partial y^* |_w) \}^{1/2}$$

$$\cdot \left[1 - c_w^*(0) - \sigma^{1/3} \int_0^\sigma \frac{(dc_w^*/d\lambda)}{(\sigma - \lambda)^{1/3}} d\lambda \right] \tag{12}$$

with

$$\sigma = \int_0^{x^*} p_w^{*2/3} (\partial u_x^* / \partial y^* |_w)^{1/2} dx^* \tag{13}$$

where

$$c_w^*(0) = B/(B + 1) \tag{14}$$

$$B = \frac{D}{2a_B k} \frac{N_{Pe}^{1/3}}{9^{1/3} \Gamma(4/3)} \lim_{x \to o} \{ p_w^{*1/2} (\partial u_x^* / \partial y^* |_w)^{1/2} / \sigma^{1/3} \} \tag{15}$$

The integral on the right-hand side of eq. (12) can be approximated by appropriate quadrature, and then through rearrangement of terms, eq. (12) can be expressed as

$$
c_w^*(\sigma_i) = \cfrac{M_i}{1 + \cfrac{1.5\sigma^{1/3}M_i}{(\sigma_i - \sigma_{i-1})^{1/3}}} \left[\frac{1}{1+B} + 1.5\sigma_i^{1/3} \left\{ \frac{c_w^*(\sigma_{i-1})}{(\sigma_i - \sigma_{i-1})^{1/3}} \right. \right.
$$

$$
\left. \left. + \sum_{j=1}^{i-1} \frac{c_w^*(\sigma_j) - c_w^*(\sigma_{j-1})}{(\sigma_j - \sigma_{j-1})} \left((\sigma_i-\sigma_j)^{2/3} - (\sigma_i-\sigma_{j-1})^{2/3} \right) \right\} \right]
$$

$$(16)$$

where

$$
M_i = \frac{1}{k^*} \frac{N_{Pe}^{-2/3}}{9^{1/3}\Gamma(4/3)\sigma^{1/3}} \, p_w^{*1/2}(\partial u_x^*/\partial y^*|_w)^{1/2}
$$

With $c_w^*(0)$ given by eqs. (14) and (15), c_w^* can be evaluated incrementally using eq. (16) with specified values of N_{Pe} and k^*.

The collection efficiency, β, is defined as the ratio of particles collected by spheres per unit time to total amount of particles approaching the collectors. The expression for β can be given as

$$
\beta = \frac{\displaystyle\int_0^{x_f} 2\pi p_w k c_w \, dx}{\pi \, a_B^2 \, u_o \, c_o}
$$

$$
= 8k^* \int_0^{x_f^*} p_w^* c_w^* \, dx^*
$$

$$(17)$$

For the special case of infinitely large k, namely, when there is no repulsive potential barrier, the surface concentration vanishes (i.e., $c_w = 0$). The physical problem corresponding to this limiting situation becomes identical to that of mass transfer. The solution can be reduced to that shown in section (2.2.a). The Sherwood number is given by eq. (10) and collection efficiency can be expressed as

$$
\beta_{k\to\infty} = \frac{4 \cdot 3^{1/3}}{\Gamma(4/3)} \, N_{Pe}^{-2/3} \left[\int_0^{x_f^*} p_w^* \{p_w^*(\partial u_x^*/\partial y^*|_w)\}^{1/2} dx^* \right]^{2/3}
$$

$$(18)$$

The other limiting case, when k is small, corresponds to the situation in which the deposition process is controlled by the surface interaction. The collection efficiency then becomes

$$\beta_{k \to o} = 16 \ (a_A^{*2} + a_B^{*2})k^*$$ (19)

3 NUMERICAL COMPUTATIONS

The stream function for two contact spheres moving in a Newtonian fluid has been obtained by Cooley and O'Neill [19]. Chiang and Tien [20] have further derived velocity components in p and z directions. The results were expressed in tangent-sphere coordinates and in terms of infinite integrals. In trajectory calculations for particle deposition where inertial and interception mechanisms dominated Brownian diffusion, evaluation of these integrals was found to be tedious and time-consuming, and therefore, new procedures have been developed to overcome this difficulty [20]. The infinite integrals could be rewritten into two integrals. These two integrals had different integrands and different upper integration bounds. One of these two integrals was integrated from zero to infinity, but, analytical expression for this infinite integral could be obtained through a lengthy derivation. The other integral was integrated from zero to a finite value and could be evaluated numerically. The detail of these procedures was described previously by Chiang and Tien [20]. These procedures are also used in the present work to evaluate flow velocities.

The Sherwood number and collection efficiency are evaluated in accordance with eqs. (10) and (17). The range of integration argument, x^* from zero to x_f^*, is divided into several segments. Then p_w and $(\partial u_x^*/\partial y^*)|_w$ at each node are obtained through appropriate coordinate transformation and with the available flow field solution. The wall concentration, c_w, as a function of σ (or x^*) is calculated according to eq. (16).

The pseudo first-order rate constant k (or $k^* = k/u_o$) can be evaluated from eq. (9). The surface interaction potential for the Brownian particle case include those arising from the London-van der Waals and double-layer forces, given as

$$\Phi = \Phi_{Lo} + \Phi_{DL}$$ (20)

where

$$\Phi_{DL} = \frac{\epsilon a_p(\psi_1^2 + \psi_2^2)}{4}\left[\frac{2\psi_1\psi_2}{\psi_1^2 + \psi_2^2}\ln\frac{1 + \exp(-\kappa y_s)}{1 - \exp(-\kappa y_s)} + \ln\{1 - \exp(-2\kappa y_s)\}\right]$$ (21)

and

$$\Phi_{Lo} = H\left[\frac{1}{6}\ln\left(\frac{y_s + 2a_p}{y_s}\right) - \frac{a_p}{3y_s}\frac{y_s + a_p}{y_s + 2a_p}\right]$$ (22)

Evaluation of the improper integral of eq. (9) presents some difficulty, which is circumvented by using the expressions suggested by Ruckenstein and Prieve [7].

4 RESULTS AND DISCUSSION
4.1 Mass Transfer

The results for mass transfer to an axisymmetric body or in packed beds at low Reynolds numbers have usually been expressed in the form of

$$N_{Sh} = C \ N_{Pe}^{1/3} \tag{23}$$

where C is a coefficient dependant on the geometry of the body and the flow field around the body. Fig. 2 shows that coefficient C (or $N_{Sh}/N_{Pe}^{1/3}$) exhibits the smallest value, 0.67, when spheres are of equal size, and increases monotonically as the size of sphere A (or a_A/a_B) is decreased. As the size of sphere A approaches zero, the value of $N_{Sh}/N_{Pe}^{1/3}$ approaches that of single isolated sphere given by Friedlander [6] or Levich [21]. The reduction of $N_{Sh}/N_{Pe}^{1/3}$ with the size increase of sphere A may be attributed to the change of flow field around two spheres. This is consistent with the observation of Chen and Pfeffer [13] and Aminzadeh et al. [11]. In their results, at a fixed Peclet number, the Sherwood number for two spheres decreased as the spacing between them decreased. Chen and Pfeffer [13] and Aminzadeh et al. [11] explained that this phenomenon is caused by increased sphere-sphere hydrodynamic interaction as the distance between two spheres decreases.

In the study of Chen and Pfeffer [13], the highest Peclet number used was 1000 and the closest spacing between the centres of two spheres was $2.162a_B$. The value of $N_{Sh}/N_{Pe}^{1/3}$ corresponding to this condition was 0.78. In the study of Aminzadeh et al. [11], the Peclet numbers were in the range of 0 to 50, and the closest spacing between the centres of two spheres was $2.0402a_B$. The value of Sherwood number obtained from their studies, at maximum Peclet number (i.e. 50) and at closest spacing, was 3.8 for sphere A and 2.7 for sphere B, respectively. The average value of Sherwood number for two spheres was 3.25, and the average value of $N_{Sh}/N_{Pe}^{1/3}$ became 0.88. Furthermore, their calculated results indicated that $N_{Sh}/N_{Pe}^{1/3}$ decreased with the increase of Peclet numbers or with the decrease of the spacing between the centres of two spheres. Comparison of the values of $N_{Sh}/N_{Pe}^{1/3}$ from the above two studies, i.e., 0.78 and 0.88, with that obtained from the present work, 0.67, may suggest that the present work is the limiting case of studies conducted by Chen and Pfeffer [13] and Aminzadeh et al. [11], i.e., at high Peclet number and zero separation distance between two sphere surfaces.

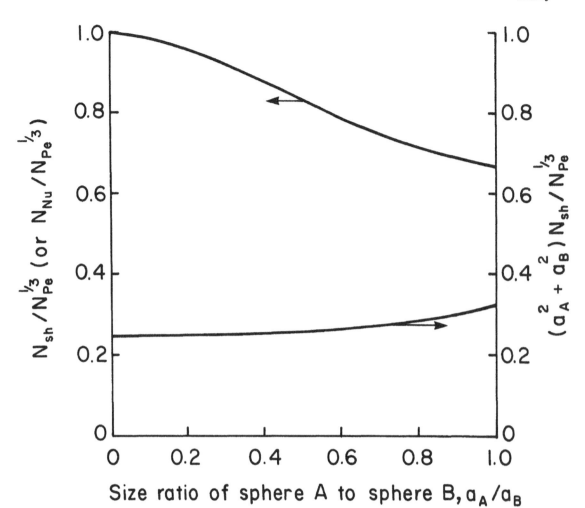

FIG. 2 $N_{Sh}/N_{Pe}^{1/3}$ AND $(a_A^{*2}+a_B^{*2})N_{Sh}/N_{Pe}^{1/3}$ VS.
SPHERE SIZE RATIO a_A/a_B

4.2 Brownian Deposition in the Presence of Repulsive Potential Barrier

When the surface interaction is repulsive, the results can be best expressed using the collection efficiency, β. Fig. 3 shows that β increases with k^* and, for sufficiently large k^*, approaches an asymptotic value. The asymptotic value of β at high value of k^* corresponds to the case of pure mass transfer. The value of β increases with size ratio a_A/a_B, which seems to contracdict the trend of $N_{Sh}/N_{Pe}^{1/3}$ with a_A/a_B at large value of k^*. This difference is due to the definition of β where the deposition rate is divided by the projected area of sphere B, whereas in the definition of N_{Sh}, the deposition rate is divided by the entire surface area of bi-sphere. It can be shown that

$$\beta_{k\to\infty} N_{Pe}^{2/3} = 16(a_A^{*2} + a_B^{*2})N_{Sh}/N_{Pe}^{1/3} \tag{24}$$

The values of $(a_A^{*2} + a_B^{*2})N_{Sh}/N_{Pe}^{1/3}$ shown in Fig. 2 increase with the a_A/a_B ratio.

The effect of the surface interaction forces on Brownian deposition is expressed through the pseudo first-order rate constant k^*, which in turn can be characterized by a number of relevant dimensionless groups [10]:

$$N_{Lo} = H/(k_b T)$$

$$N_{DL} = \kappa a_p$$

$$N_{E1} = 2\psi_1\psi_2/(\psi_1^2 + \psi_2^2)$$

$$N_{E2} = \epsilon a_p \, (\psi_1^2 + \psi_2^2)/(4k_b T)$$

$$N_{Ret} = 2\pi a_p/\lambda_e \tag{25}$$

A case study was undertaken to demonstrate the effect of these dimensionless groups (plus N_{Pe} and N_R) on β. A set of base values for the pertinent system and operating variables used by Chiang and Tien [10] was selected first, and then corresponding dimensionless groups were obtained (see Table 1). The effect of each dimensionless group was determined by varying the value of that particular group while keeping all other groups at their base values. In general, β was found to be very sensitive to the change in surface interaction groups as shown in Figs. (4)-(7). Thus, a careful control of the surface conditions of filter media and Brownian particles is very important in filtration of Brownian particles.

For Brownian deposition in the presence of repulsive potential barrier, the analysis presented in this study makes it possible to evaluate the wall concentration and, in turn, the collection efficiency. These evaluations, however, need successive numerical computations. This inconvenience can be

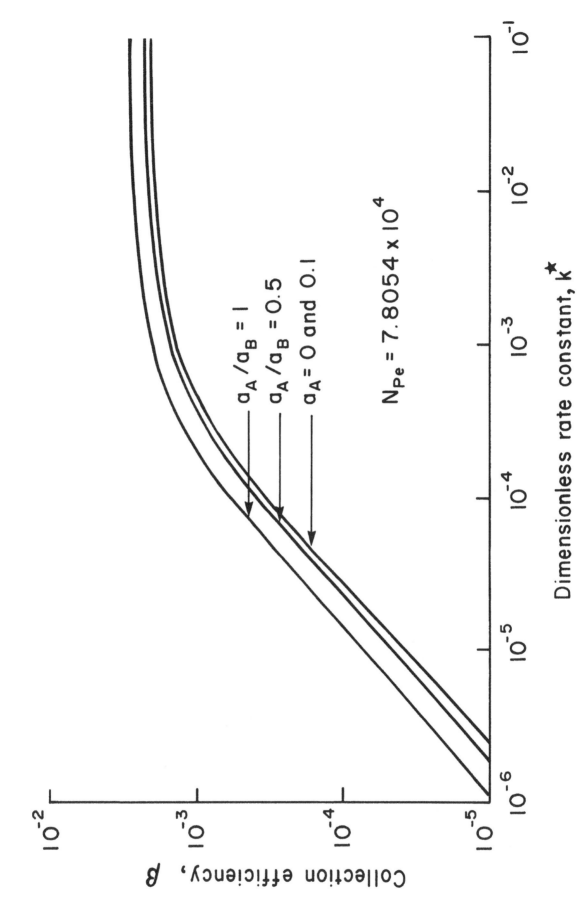

FIG. 3   COLLECTION EFFICIENCY VS. DIMENSIONLESS RATE CONSTANT

Table 1   Parameter Values Used in Our Case Study

a_B 0.0275 cm

a_p 3.05×10^{-6} cm

D 8×10^{-8} cm^2/sec

u_o 0.1135 cm/sec

T   298 K

μ 0.0008937 Pa•s

ϵ 81

κ 1.02×10^7 cm^{-1}

         (Electrolyte conc. = 0.1 M)

λ_e 10^{-5} cm

H 10^{-20} J

ψ_1 - 26.2 mV

ψ_2 - 26.2 mV

ρ 1 g/cm^3

N_{DL} 31.134

N_{E1} 1

N_{E2} 22.903

N_{Lo} 2.431

N_{Pe} 78054

N_{Ret} 1.916

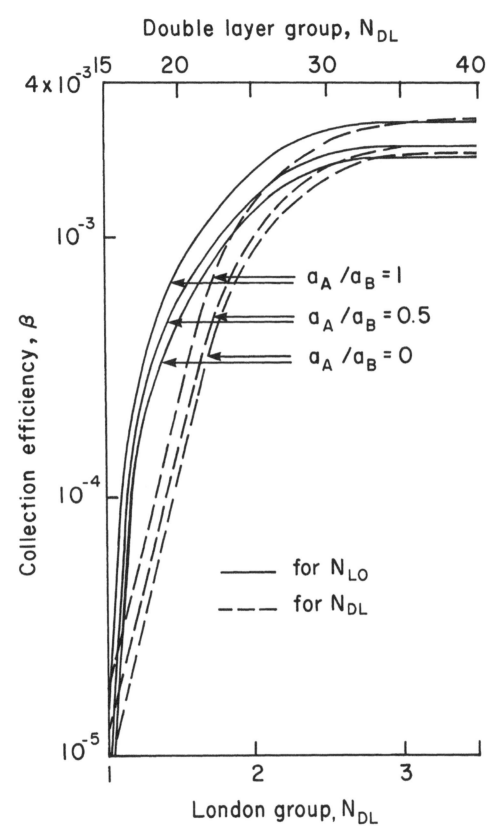

FIG. 4 EFFECT OF LONDON GROUP N_{LO} AND
DOUBLE LAYER GROUP N_{DL} ON β

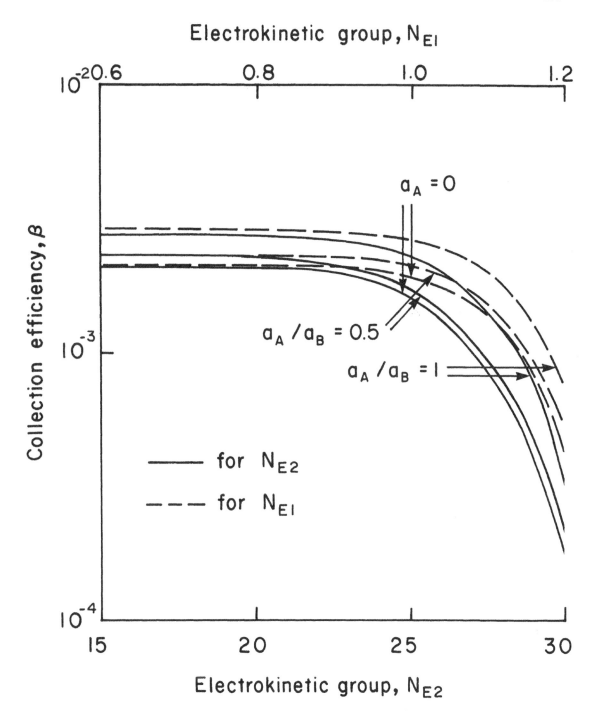

FIG. 5 EFFECT OF ELECTROKINETIC GROUPS, N_{E1} AND N_{E2}, ON β

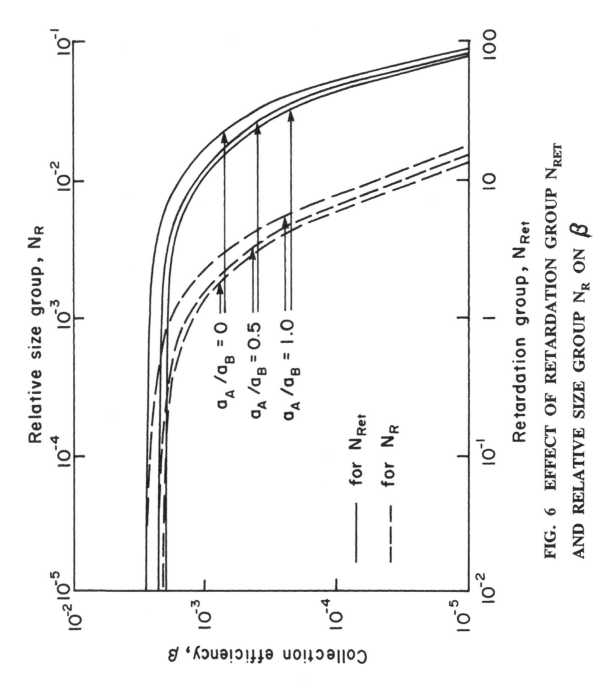

FIG. 6 EFFECT OF RETARDATION GROUP N_{RET}
AND RELATIVE SIZE GROUP N_R ON β

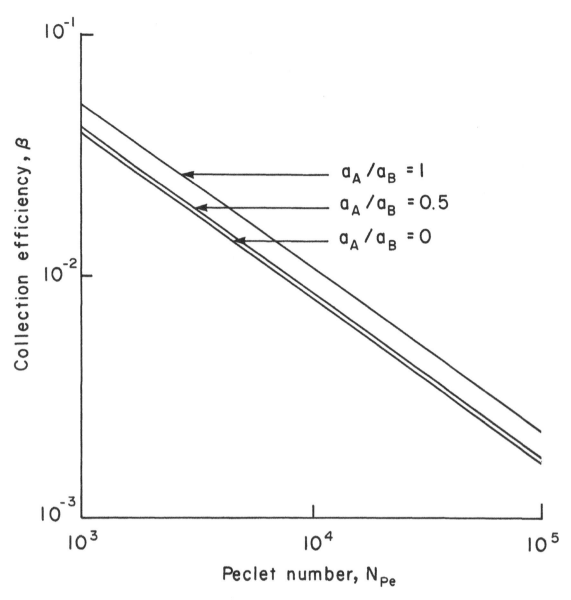

FIG. 7 EFFECT OF PECLET NUMBER N_{Pe} ON β

circumvented by the use of the following empirical expression developed from numerical results of the exact calculation:

$$\beta = \frac{16 \ (a_A^{*2} + a_B^{*2})k^*}{1 + 16 \ (a_A^{*2} + a_B^{*2})k^*/\beta_{k\to\infty}} \qquad (26)$$

The results using eq. (26) are within 7% of the exact calculations.

5 SUMMARY

A theoretical analysis of mass transfer, heat transfer and Brownian particle deposition onto two spheres in contact has been presented. For mass (or heat) transfer, the present analysis complements two earlier studies [11, 13] in which two spheres were separated. For Brownian deposition, the study presented in this paper shows that collection efficiency (or filter efficiency) is very sensitive to the surface conditions of filter media and Brownian particles.

NOMENCLATURE

a_A radius of sphere A

a_B radius of sphere B

a_p radius of Brownian particles

c particle concentration

c_o particle concentration far from spheres

c_w particle concentration on surface of spheres

D diffusion coefficient

H Hamaker's constant

k_* pseudo first-order rate constant

k^* k/u_o

k_b Boltzmann's constant

m mass transfer coefficient

N_{DL} κa_p, double layer parameter (dimensionless)

N_{E1} $2\psi_1\psi_2/(\psi_1^2+\psi_2^2)$, the first dimensionless electrokinetic group

N_{E2} $\epsilon a_p(\psi_1^2+\psi_2^2)/4k_bT$, the second dimensionless electrokinetic group

N_{Lo} H/k_bT, the London parameter, (dimensionless)

N_{Pe} $2a_Bu_o/D$, Peclet number

N_R a_p/a_B, relative size group

N_{Ret} $2\pi a_p/\lambda_e$, retardation parameter (dimensionless)

N_{Sh} $2a_bm/D$, Sherwood number

p radial coordinate in cylindrical coordinate system

p_w radial distance of sphere surface from centre-to-centre line

T absolute temperature

u_x velocity component along x
u_y velocity component along y
u_o approaching velocity
x coordinate measured along tangential direction of sphere surface
y coordinate measured along normal direction of sphere surface
y_s normal separatin distance between particel and collector surface
z axial coordinate in cylindrical coordinate system

Greek symbols

β collection efficiency
ϵ dielectric constant
η tangent sphere coordinate
ξ tangent sphere coordinate
κ Debye's reciprocal length parameter
ν kinematic viscosity
λ_e wavelength of the electron oscillation
σ defined by eq. (13)
Φ net surface interaction potential
Φ_{DL} interaction potential due to double layer force
Φ_{Lo} interaction potential due to London-van der Waals force
ψ_1 surface potential of suspended particle
ψ_2 surface potential of collector

Superscript

* dimensionless variable of its counterpart

REFERENCES

[1] G.M. Allison, Summary of results of graphite filter tests in high temperature water systems to January 1976- A filter task force study, unrestricted, unpublished Whiteshell Nuclear Research Establishment Report, WNRE-241-4, 1976.
[2] C. Tien and A.C. Payatakes, Advances in deep bed filtration, AIChE J., 25(1979), pp. 385-395.
[3] A.C. Payatakes and C. Tien, Particle deposition in fibrous media with dendrite-like pattern: A preliminary model, J. Aerosol Sci., 7(1976), pp. 85-100.
[4] H. Pendse, H.W. Chiang, and C. Tien, Analysis of transport processes with granular media using the constricted tube model, Chem. Eng. Sci., 38(1983), pp. 1137-1150.
[5] W.J. Wnek, D. Gadispow and D.T. Wasan, The deposition of colloid particels onto the surface of a rotating disk, J. Colloid Interface Sci., 59(1977), pp. 1-6.

[6] S.K. Friedlander, Mass and heat transfer to single sphere and cylinder at low Reynolds number, AIChE J., 3(1957), pp. 43-48.

[7] E. Ruckenstein and D.C. Prieve, Rate of deposition of Brownian particles under the action of London and double layer forces, J. Chem. Soc. (London), Faraday Trans., 69(1973), pp. 1522-1536.

[8] L.A. Spielman and S.K. Friedlander, Role of the electrical double layer in particle deposition by convective diffusion, J. Colloid Interface Sci., 46(1974), pp. 22-31.

[9] R. Rajagopalan and T.E. Karis, Effect of electrokinetics on removal of colloidal particles from liquids, AIChE Symposium Ser. No., 190, Vol. 75(1977), pp. 73-79.

[10] H.W. Chiang and C. Tien, Deposition of Brownian particels in packed beds, Chem. Eng. Sci., 37(1982), pp. 1159-1171.

[11] K. Aminzadeh, T.R. Al Taha, A.R.H. Cornish, M.S. Kolansky and R. Pfeffer, Mass transport around two spheres at low Reynolds numbers, Int. J. Heat Mass Transfer, 17(1974), pp. 1425-1436.

[12] R. Tal (Thau), D.N. Lee, and W. A. Sirignano, Heat and momentum transfer around a pair of spheres in viscous flow, Int. J. Heat Mass Transfer, 27(1984), pp. 1953-1962.

[13] W.C. Chen and R. Pfeffer, Mass transfer rates with first-order homegeneous chemical reaction around two spheres, AIChE Symposium Ser. No. 105, Volume 66(1970), pp. 109-122.

[14] R.B. Bird, W.E. Stewart, and E.N. Lightfoot, Transport phenomena, Wiley, New York, 1960.

[15] J. Happel and H. Brenner, Low Reynolds number hydrodynamics, Nordfoff Internatinal Publishing, Leydeon, Neitherlands, 1973.

[16] H. Schlichting, Boundary-layer theory, McGraw-Hill Book Co., New York, 1968.

[17] M.J. Lighthill, Contributions to the theory of heat transfer through a laminar boundary layer, Proc. Roy. Soc. London, A202(1950), pp. 359-377.

[18] H.C. Carslaw and J.C. Jaeger, Conduction of heat in solids, Oxford, Clarendon, 1959.

[19] M.D.A. Cooley and M.E. O'Neill, On the slow motion of two spheres in contact along their line of centers through a viscous fluid, Proc. Phil. Soc., 66(1969), pp. 407-412.

[20] H.W. Chiang and C. Tien, Analysis of particle deposition on bi-sphere collectors, Chem. Eng. Communication, 20(1983), pp. 291-309.

[21] V.G. Levich, Physical hydrodynamics, Prentice-Hall, Englewood Cliffs, N.J., 1962.

Eigenmodes of a Random Walk on a 2D Lattice Torus with One Trap

DAVID C. TORNEY   Theoretical Biology and Biophysics, Los Alamos National Laboratory, Los Alamos, New Mexico

The complete set of eigenvalues and eigenmodes is elaborated for a nearest-neighbor random walk on a two-dimensional $N \times N$ square lattice torus with one trap site. The eigenvalue closest to unity is found to have an asymptotic expansion in N; the leading behavior is

$$\lambda \sim 1 - \pi \, / \, (2N^2 \, log \, N) + O(N \, log \, N)^{-2}$$

In general, the eigenvalues and eigenmodes for this problem are constructed from those for a random walk on the same lattice containing no trap. The degeneracy of the latter eigenmodes is a prominent feature of this construction, and a formula is derived for this degeneracy.

1. INTRODUCTION

The problem of determining the eigenvalues and eigenmodes for a random walk on an $N \times N$ square lattice torus with one trap site is motivated by the analogous continuum problem – the determination of the complete set of eigenvalues and eigenfunctions for a boundary value problem:

$$(\nabla^2 + k^2)\psi(\underline{x}) = 0 \quad .$$

The analogous boundary conditions are that $\psi(\underline{x})$ vanish on a circle (trap) centered in a square on which the normal derivative of $\psi(\underline{x})$ vanishes (periodicity). Although one can make progress in determining the smallest eigenvalues, k^2, and corresponding eigenfunctions for this problem, so long as the ratio of the radius of the circle to the side of the square is small [1], it is difficult to determine the complete set of eigenfunctions. For this reason, one formulates the intrinsically interesting lattice problem — involving analysis with points of similarity and departure from that pertaining to the analogous continuum problem — as follows.

Consider an $N \times N$ square lattice torus. This torus can be thought of as a planar square lattice having N sites on a side and unit spacing between these sites. However, the periodicity of the torus means that one of the nearest-neighboring sites to a non-corner site on the boundary is the equivalent point on the opposite boundary, and, similarly, the sites at the corners of the boundary have the adjacent corner sites for two of their nearest-neighbors. The random walk on this lattice is a symmetric

Work supported by the U. S. Department of Energy.

nearest-neighbor random walk. That is, one step of the walk has the transition probability $p(\underline{m} - \underline{m}')$ of going from site $\underline{m}' = (m_1', m_2')$ to site $\underline{m} = (m_1, m_2)$;

$$p(\underline{m} - \underline{m}') = \frac{1}{4}\{\delta_{m_1 - m_1'}[\delta_{m_2 - m_2' - 1} + \delta_{m_2 - m_2' + 1}] + \delta_{m_2 - m_2'}[\delta_{m_1 - m_1' - 1} + \delta_{m_1 - m_1' + 1}]\} \quad .$$
(1.1)

The transition probability is modified at the boundary points as discussed, to enforce the periodicity of the lattice torus. The δ_x that appears in Eq. (1.1) and in subsequent equations is unity if x is zero and is zero otherwise. By analogy with the continuum equation appearing above, one solves for the complete set of *eigenvalues*, λ, and *eigenmodes*, $\psi(\underline{m}; \lambda)$, satisfying the equation

$$\sum_{\underline{m}' \in L'} p(\underline{m} - \underline{m}')\psi(\underline{m}'; \lambda) - \lambda\psi(\underline{m}; \lambda) = 0 \quad ,$$
(1.2)

where L' is a specified subset of the N^2 sites. In this paper, L' equals either L, all N^2 sites, or L^1, all N^2 sites but one. The latter site is a *trap site*, one from which the random walk does not exit. Clearly, $p(\underline{m} - \underline{m}')$ is closely related to a discrete Laplacian operator. Although there are points of contact with previous analyses of this trapping problem, particularly in the use of Green's functions [2] and in the asymptotic treatment of the largest eigenvalue [3], the following exposition of the eigenmodes is new.

2. EIGENVALUES AND EIGENMODES

First, let L' be L; *i.e.*, there is no trap site. It is easily established [4] that the complete set of eigenmodes satisfying Eq. (1.2), with $p(\underline{m} - \underline{m}')$ given in Eq. (1.1), can be taken to be the set of N^2 harmonic functions

$$exp\{\frac{2\pi i}{N}(k_1 m_1 + k_2 m_2)\} \quad ,$$

$$k_1, k_2 \in \{0, 1, \dots N - 1\} \quad ,$$
(2.1)

having the eigenvalues

$$\frac{1}{2}\left(cos\frac{2\pi k_1}{N} + cos\frac{2\pi k_2}{N}\right) \quad .$$
(2.2)

Unlike the analogous problem in the continuum, the complete set of eigenmodes is finite in number, and the eigenvalues lie in a finite interval, namely the closed interval $[-1, 1]$.

Now let L' be L^1; *i.e.*, there is one trap site that is taken to be at the origin. One can obtain solutions of Eq. (1.2) using the Green's function $\Gamma_N(\underline{n}; \lambda)$ satisfying

$$\sum_{\underline{m}' \in L} p(\underline{m} - \underline{m}')\Gamma_N(\underline{m}' - \underline{m}''; \lambda) - \lambda\Gamma_N(\underline{m} - \underline{m}''; \lambda) = \delta_{m_1 - m_1''}\delta_{m_2 - m_2''} \quad .$$
(2.3)

By rewriting Eq. (1.2) with L' equal L^1 as follows,

$$\sum_{\underline{m}' \in L} p(\underline{m} - \underline{m}')\psi(\underline{m}'; \lambda) - \lambda\psi(\underline{m}; \lambda) = p(\underline{m})\psi(\underline{0}; \lambda) \quad ,$$

one can use the Green's function to write [2]

$$\psi(\underline{m}; \lambda) = \psi(\underline{0}; \lambda) \sum_{\underline{m}' \in L} \Gamma_N(\underline{m} - \underline{m}'; \lambda) p(\underline{m}') \quad . \tag{2.4}$$

Using a convenient representation of the Green's function,

$$\Gamma_N(\underline{m} - \underline{m}'; \lambda) = \frac{1}{N^2} \sum_{\underline{k} \in L} \frac{exp\left\{\frac{2\pi i}{N}\left[k_1(m_1 - m_1') + k_2(m_2 - m_2')\right]\right\}}{\frac{1}{2}\left(cos\frac{2\pi k_1}{N} + cos\frac{2\pi k_2}{N}\right) - \lambda} \quad , \tag{2.5}$$

and rearranging, Eq. (2.4) gives

$$\psi(\underline{m}; \lambda) = \psi(\underline{0}; \lambda) \left\{ \delta_{m_1}\delta_{m_2} + \frac{\lambda}{N^2} \sum_{\underline{k} \in L} \frac{exp\left\{\frac{2\pi i}{N}(k_1 m_1 + k_2 m_2)\right\}}{\frac{1}{2}\left(cos\frac{2\pi k_1}{N} + cos\frac{2\pi k_2}{N}\right) - \lambda} \right\} \quad . \tag{2.6}$$

The factor $\psi(\underline{0}; \lambda)$ reflects the normalization of $\psi(\underline{m}; \lambda)$ and is of no import so long as it does not vanish. The set of eigenmodes that vanish at the origin is discussed subsequently.

There are non-trivial solutions of Eq. (2.6) only for exceptional values of λ; the *eigenvalues*. By evaluating this equation at the origin, one finds the following equation determining the eigenvalues:

$$\sum_{\underline{k} \in L} \left[\frac{1}{2}\left(cos\frac{2\pi k_1}{N} + cos\frac{2\pi k_2}{N}\right) - \lambda \right]^{-1} = 0 \quad . \tag{2.7}$$

Equation (2.7) is readily converted into a polynomial in λ of order $D - 1$ where D is the number of distinct values of $cos\frac{2\pi k_1}{N} + cos\frac{2\pi k_2}{N}$ for k_1 and k_2 in the set $\{0, 1, \dots N - 1\}$. Taking the derivative of Eq. (2.7) reveals that between every two adjacent values of $cos\frac{2\pi k_1}{N} + cos\frac{2\pi k_2}{N}$ there is one real value of λ satisfying Eq. (2.7). Thus, all of the $D - 1$ eigenvalues are real. Therefore, Newton's method with Eq. (2.7) readily finds these $D - 1$ eigenvalues. Figure 1 gives the result for $N = 20$. Each triangle below the x axis points to one of the values of $\frac{1}{2}\left(cos\frac{2\pi k_1}{20} + cos\frac{2\pi k_2}{20}\right)$. Each triangle above the x axis points to a root of Eq. (2.7), that is, an eigenvalue. The right half of the interval $[-1, 1]$ is shown because the distribution is symmetric about zero for even N. The eigenvalues near zero and unity are very much closer to one of the adjacent values of $\frac{1}{2}(cos\frac{2\pi k_1}{N} + cos\frac{2\pi k_2}{N})$. As will be seen, this proximity can be used to make an initial guess in a perturbation analysis of these eigenvalues. In general, however, the eigenvalues are not close to a value of $\frac{1}{2}(cos\frac{2\pi k_1}{N} + cos\frac{2\pi k_2}{N})$, clear evidence for a non-standard perturbation problem. Equivalently, most of the eigenmodes of the problem with one trap do not closely resemble any one of the eigenmodes of the problem with no trap.

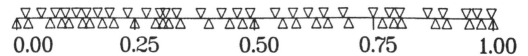

0.00   0.25   0.50   0.75   1.00

Figure 1. Eigenvalues for $N = 20$.

A *complete set* for the problem on L^1 consists of $N^2 - 1$ *orthogonal* eigenmodes, one for each non-trap site. The definition of orthogonality is

$$\sum_{\underline{m} \in L^1} \psi(\underline{m}; \lambda) \psi^*(\underline{m}; \lambda') = 0 \quad , \tag{2.8}$$

for $\lambda \neq \lambda'$. Furthermore, the *degenerate* eigenmodes, different eigenmodes with the same eigenvalue, can be made orthogonal by the Gram-Schmidt process. In addition to the $D - 1$ eigenmodes that one obtains from Eqs. (2.6) and (2.7), orthogonal on L^1, it is readily appreciated that any linear combination of the degenerate eigenmodes for the problem on L given in Eq. (2.1), vanishing at the origin (trap), will be an eigenmode of Eq. (1.2), with unchanged eigenvalue, given in Eq. (2.2).

The cardinality of the degeneracy of the eigenmodes of Eq. (2.1), $d(k_1, k_2; N)$, abbreviated as d, is the number of pairs of numbers j_1, j_2 that one can choose from the set $\{0, 1, \ldots, N - 1\}$ such that $cos \frac{2\pi j_1}{N} + cos \frac{2\pi j_2}{N}$ equals $cos \frac{2\pi k_1}{N} + cos \frac{2\pi k_2}{N}$. The quantity d could be obtained in a novel lattice-point problem where one places N^2 points on a square lattice with integer x, y coordinates between zero and $N - 1$ and enumerates the points on the curve

$$cos\left(\frac{2\pi x}{N}\right) + cos\left(\frac{2\pi y}{N}\right) = cos\left(\frac{2\pi k_1}{N}\right) + cos\left(\frac{2\pi k_2}{N}\right) \quad .$$

By methods given in the Appendix, it is determined that

$$d(k_1, k_2; N) = 8 \sum_{j=0}^{[x]} \frac{\delta_{q^N - (-)^j}}{(1 + \delta_j)(1 + \delta_{j-x})} \tag{2.9}$$

with

$$q = \frac{cos\,\theta + i\sqrt{cos^2 \frac{j\pi}{N} - cos^2 \theta}}{cos \frac{j\pi}{N}}$$

$$cos\,\theta = \frac{1}{2}\left(cos \frac{2\pi k_1}{N} + cos \frac{2\pi k_2}{N}\right)$$

$$x = \left(N\,cos^{-1}\sqrt{|cos\,\theta|}\right) / \pi \quad ,$$

and $[x]$ denotes the integer part of x. The numerator of the summand is unity if q^N equals $(-)^j$ and is zero otherwise. Equation (2.9) is not intended to apply if $cos\,\theta$ equals $+1, 0$, or -1; the degeneracy for these cases is $1, 2N - 2$, and 1, respectively. Evaluation of Eq. (2.9) for N up to 1000 and equal 3600 and 8640 leads to the conjecture that for odd N, d is either 1, 4, or 8, and for even N, d is either 1, 4, 8, 12, 16, 20, 24, or $2N - 2$. The latter degeneracy corresponds to eigenmodes with eigenvalue zero.

In any case, the Vandermonde matrix of order d with elements v_{jk} can be used to create an orthogonal set of $d - 1$ eigenmodes vanishing at the origin. * The elements

$$v_{jk} = exp\{2\pi i (j - 1)(k - 1)/d\} \quad . \tag{2.10}$$

---

* For d a multiple of 4, many appropriate matrices with real elements exist [5].

Denote the members of sets of degenerate eigenmodes of Eq. (2.1) with $\phi_k(\underline{m}; \lambda)$; $1 \leq k \leq d$. Define a new set of degenerate eigenmodes by

$$\psi_j(\underline{m}; \lambda) = \sum_{k=1}^{d} v_{jk}\phi_k(\underline{m}; \lambda) \ . \tag{2.11}$$

It follows from the orthogonality properties of the $\phi_k(\underline{m}; \lambda)$ and the v_{jk}, namely

$$\sum_{\underline{m}} \phi_k(\underline{m}; \lambda)\phi_{k'}^*(\underline{m}; \lambda') = \delta_{\lambda-\lambda'}\delta_{k-k'} \ ,$$

$$\sum_{k=1}^{d} v_{jk} \, v_{j'k}^* = d\delta_{j-j'} \ ,$$

that the $\psi_j(\underline{m}; \lambda)$ are mutually orthogonal on L. Furthermore, all but $\psi_1(\underline{m}; \lambda)$ vanish at the origin, and therefore the remaining $d-1$ are orthogonal eigenmodes for the problem on L^1 with unchanged eigenvalue. One can see from Eq. (2.6) that, essentially, only the $\psi_1(\underline{m}; \lambda)$ are used to construct eigenmodes corresponding to "new eigenvalues," those determined by Eq. (2.7). Since the remaining $d-1$ eigenmodes are orthogonal to $\psi_1(\underline{m}; \lambda)$, they are also orthogonal on the eigenmodes resulting from Eq. (2.6) on L^1.

The results for the problem on L^1 are now summarized. $D-1$ nondegenerate eigenmodes are determined using Eqs. (2.6) and (2.7). Furthermore, $N^2 - D$ eigenmodes are determined using (2.10) and (2.11); these have the respective eigenvalues and degeneracies $d-1$ for the problem on L. The $N^2 - 1$ eigenmodes are mutually orthogonal and therefore constitute a complete set of eigenmodes for the problem in L^1. The number of distinct eigenvalues is readily seen to be $2D - 3$; all the triangles in Fig. 1, except for the triangle at unity point to an eigenvalue of the trapping problem.

This section focused on the evaluation of d, and the interesting problem of determining the dependence of D upon N was not addressed. Clearly

$$D(N) = \sum_{\underline{k}\epsilon L} 1/d(k_1 k_2; N) \ .$$

Numerical evaluation of $D(N)$ for $1 \leq N \leq 800$ shows

$$D(N) \sim \frac{N^2}{8} + O(N) \ ,$$

due to the preponderance of eigenmodes with degeneracy eight.

3. ASYMPTOTIC BEHAVIOR OF THE EIGENVALUES AS $N \to \infty$

The "new eigenvalue" closest to 1, and its eigenmode, have substantial physical significance. Therefore, it is useful to carry out the following perturbation analysis along the lines given for the continuum problem [6].

Denote the largest eigenvalue $1 - \epsilon$. From Eq. (2.7) one gets

$$\frac{1}{\epsilon} = \sum_{\underline{k}\in L^1} \left[1 - \frac{1}{2}\left(cos\frac{2\pi k_1}{N} + cos\frac{2\pi k_2}{N} \right) \dot{-}\ \epsilon \right]^{-1} \ . \tag{3.1}$$

As before, the summation over L^1 is over all lattice points $k_1, k_2 \in \{0, 1, \ldots, N-1\}$ except the point (0,0). Expanding the right-hand side in a geometric series in powers of ϵ enables the successive approximation of ϵ in terms of the values of the series

$$S_j = \sum_{\underline{k} \in L^1} \left[1 - \frac{1}{2} \left(cos \frac{2\pi k_1}{N} + cos \frac{2\pi k_2}{N} \right) \right]^{-j} \quad . \tag{3.2}$$

To determine the first three terms in the asymptotic expansion of ϵ for large N, one needs the following behavior of S_1 and S_2:

$$S_1 \sim AN^2 \log N + BN^2 + O(1) \quad ,$$

$$S_2 \sim CN^4 + o(N^4),$$

with

$$A = \frac{2}{\pi} \quad ,$$

$$B = \frac{1}{\pi} \left\{ \frac{\pi}{3} + 2\gamma - 2\log \pi + \log 2 + 4 \sum_{k=1}^{\infty} \frac{e^{-2\pi k}}{k(1 - e^{-2\pi k})} \right\} \doteq 0.1950625\ldots \quad ,$$

$$C = \frac{4}{\pi^4} \left\{ \sum_{i,j=1}^{\infty} (i^2 + j^2)^{-2} + \frac{\pi^4}{90} \right\} \doteq 0.061871145\ldots \quad ,$$

with

$$\gamma = 0.577216\ldots \quad .$$

The techniques used to do the asymptotic analysis of the S_j can be found in the appendices of Ref. 3. The series appearing in these expressions arose, not surprisingly, in the expansion of the smallest eigenvalue of the analogous continuum problem [1]. In terms of these constants,

$$\epsilon \sim \frac{1}{AN^2 \log N} - \frac{B}{A^2 N^2 \log^2 N} + \frac{B^2 - C}{A^3 N^2 \log^3 N} + O\left(\frac{1}{N^2 \log^4 N} \right) \quad . \tag{3.3}$$

It is expected that a similar analysis applies for the other "new eigenvalues" near 1, 0 (when N is even), and -1.

A quantitative·comparison between the expansion of the continuum problem's smallest eigenvalue [1], k^2, and its lattice analogue, 4ϵ, reveals the following. If one takes L, the center-to-center spacing between circular traps, equal to N, the leading-order terms in the expansions are equal. One can then adjust the free parameter, r, the radius of the circular traps, to make the second-order terms in the expansions equal. This is achieved with

$$r = 0.1985\ldots \quad .$$

This is a smaller value than found when one enforces asymptotic equality of the second-order terms for the random walk and continuum diffusion problems with one trap in an infinite domain [7], where $r = 0.2807\ldots$. However, in both cases, having picked an r to make the second-order terms equal results in equality of the third-order terms being equal as well. One suspects that all of the pairs of terms $O(N^{-2}(log\, N)^{-J})$ will be equal, having made the indicated choice of r. This would be so if, roughly speaking, these logarithmic terms all reflect the size of the trap but not its shape.

Finally, a line of analysis is sketched that, among other things, may lead to simplification of the determination of the "new eigenvalues" in the limit that N goes to infinity. It is noted that Eq. (2.7), determining the "new eigenvalues," is equivalent to the vanishing of $\Gamma_N(\underline{0}; \lambda)$. Now, the Green's function can be given an alternative representation in terms of a series of Green's functions $\Gamma(\underline{m}; \lambda)$ for an infinite domain,

$$\Gamma_N(\underline{m}; \lambda) = \sum_{k_1, k_2 = -\infty}^{\infty} \Gamma(m_1 + k_1 N, m_2 + k_2 N; \lambda) \ , \tag{3.4}$$

with

$$\Gamma(\underline{m}; \lambda) = \frac{1}{\pi^2} \iint_0^\pi d\theta_1 d\theta_2 \ \frac{\cos\, m_1 \theta_1 \ \cos\, m_2 \theta_2}{\frac{1}{2}(\cos \theta_1 + \cos \theta_2) - \lambda} \ . \tag{3.5}$$

Equation (3.5) results from Eq. (2.5) in the limit that N goes to infinity [4]. The integral is a Cauchy principal value double integral, which has an obvious value that one obtains if, for example, one approaches the curve $\cos \theta_1 + \cos \theta_2 = 2\lambda$ at a normal distance δ, and takes the limit $\delta \to 0$.

$\Gamma(\underline{0}; \lambda)$ obeys the same second-order differential equation as the complete elliptic integral $K(\lambda)$,

$$K(\lambda) = \int_0^{\frac{\pi}{2}} d\theta (1 - \lambda^2 sin^2 \theta)^{-\frac{1}{2}} \ ,$$

with

$$1 > |\lambda| > 0 \ .$$

This is demonstrated, for example, when one uses the identity

$$(1-x)^{-3} = \sum_{l=2}^{L} \binom{l}{2} x^{l-2} + x^{L-1}(1-x)^{-3} \left\{ \binom{L+1}{2} - (L^2 - 1)x + \binom{L}{2} x^2 \right\} \ ,$$

with $x = (\cos \theta_2 + \cos \theta_2)/2\lambda$, and lets L go to infinity. It follows from the limiting behavior of $\Gamma(\underline{0}; \lambda)$ as $|\lambda|$ goes to unity that

$$\Gamma(\underline{0}; \lambda) = -\frac{2}{\pi} K(\lambda) sgn(\lambda) \ . \tag{3.6}$$

The leading order behavior of $\Gamma(m_1 N, m_2 N; \lambda)$ arises from the neighborhood of the principal value curve. The principle of stationary phase [8] can be used along the direction of the curve to obtain

$$\Gamma(m_1 N, m_2 N; \lambda) \sim 2\pi^{\frac{3}{2}} M^{-\frac{1}{2}} \left[sin(M\xi) - cos(M\xi) \right]$$

$$\times \left[\left\{ cos(\phi)(1 - (\lambda + (1-\lambda)\varsigma)^2)^{-\frac{1}{2}} + sin(\phi)(1 - (\lambda - (1-\lambda)\varsigma)^2)^{-\frac{1}{2}} \right\} \right. \tag{3.7}$$

$$\left. \lambda(1 - \lambda^2 + (1-\lambda)^2 \varsigma^2) \right]^{-\frac{1}{2}} \;,$$

with

$$\lambda > 0 \;,$$

$$\varsigma = \left\{ \lambda - (\lambda^2 + cos^2(2\phi)(1 - \lambda^2))^{\frac{1}{2}} \right\} / \left[cos(2\phi)(1 - \lambda) \right] \;,$$

$$\xi = cos(\phi) \, cos^{-1}(\lambda + (1-\lambda)\varsigma) + sin(\phi) \, cos^{-1}(\lambda - (1-\lambda)\varsigma) \;,$$

$$M = N\sqrt{m_1^2 + m_2^2} \;,$$

and

$$\phi = tan^{-1} \left| m_2/m_1 \right| \;.$$

For $\lambda < 0$, one uses $\Gamma(\underline{m}; -\lambda) = -\Gamma(\underline{m}; \lambda)$. Eq. (3.7) does not apply if either m_1 or m_2 equals zero, in which case the leading order behavior is of lower order in N. With an exact form for $\Gamma(\underline{0}; \lambda)$ and the leading order behavior in N, given above for $\Gamma(m_1 N, m_2 N; \lambda)$, one should be able to obtain approximations for the "new eigenvalues."

4. APPENDIX: DERIVATION OF THE FORMULA FOR $d(k_1, k_2; N)$

Define

$$d(k_1, k_2; N; a) =$$

$$a \sum_{j_1, j_2 = 1}^{N} \int_0^\infty dx \; exp - x \left\{ a + i \left(cos \frac{2\pi j_1}{N} + cos \frac{2\pi j_2}{N} - cos \frac{2\pi k_1}{N} - cos \frac{2\pi k_2}{N} \right) \right\} \;,$$

$$\tag{A.1}$$

with a real; then the quantity defined in Eq. (2.9),

$$d(k_1, k_2; N) = lim \, a \downarrow 0 \; d(k_1, k_2; N; a) \;. \tag{A.2}$$

Several formulas are used to transform Eq. (A.1). First, the exponentials involving j_1 and j_2 are expanded in a series of Bessel functions [9]

$$exp - \{ix \; cos\,\theta\} = J_0(x) + 2 \sum_{k=1}^{\infty} exp - \left\{ \frac{i\pi k}{2} \right\} J_k(x) \, cos\,k\theta \;.$$

Then, one performs the j_1 and j_2 series. Next, one does the integral over x, using the result [10]

$$\int_0^\infty dx \; exp-(cx) J_\mu(x) J_\nu(x)$$

$$= \frac{2}{\pi} \int_0^{\frac{\pi}{2}} d\phi \; \frac{cos(\mu - \nu)\phi}{(c^2 + 4 cos^2 \phi)^{\frac{1}{2}}} \left[\frac{(c^2 + 4 cos^2 \phi)^{\frac{1}{2}} - c}{2 cos \phi} \right]^{\mu + \nu} .$$

The branch of the square root is chosen so that the modulus of the quantity in the square brackets does not exceed unity.

This gives:

$$d(k_1, k_2; N; a) =$$

$$\frac{8aN^2}{\pi} \sum_{j_1, j_2 = 0}^\infty \frac{exp-\{i(j_1 + j_2)\frac{\pi N}{2}\}}{(1 + \delta_{j_1})(1 + \delta_{j_2})} \qquad (A.3)$$

$$\int_0^{\frac{\pi}{2}} d\phi \frac{cos(j_1 - j_2)N\phi}{(A^2 + 4 cos^2 \phi)^{\frac{1}{2}}} \left\{ \frac{(A^2 + 4 cos^2 \phi)^{\frac{1}{2}} - A}{2 cos \phi} \right\}^{(j_1 + j_2)N} ,$$

with

$$A = a + i\left(cos \frac{2\pi k_1}{N} + cos \frac{2\pi k_2}{N} \right) .$$

One observes that the summations over j_1 and j_2 are geometric series. Performing these summations gives

$$d(k_1, k_2; N; a) = \frac{2aN^2}{\pi} \int_0^{\frac{\pi}{2}} \frac{d\phi}{R} \left[\frac{1 + 2p^N \; cos \, N\phi + p^{2N}}{1 - 2p^N \; cos \, N\phi + p^{2N}} \right] , \qquad (A.4)$$

with

$$R = (A^2 + 4 cos^2 \phi)^{\frac{1}{2}} ,$$

and

$$p = \frac{i(R - a) + cos \frac{2\pi k_1}{N} + cos \frac{2\pi k_2}{N}}{2 cos \phi} .$$

The final step is to let $a \downarrow 0$. Examination of the zeros of the denominator of the integrand with $a = 0$ at points ϕ_0, and expansion of the integrand in power series in ϕ and a about these points leads to the result given in Eq. (2.9), when one uses the value of the integral

$$\int_{-\infty}^\infty dx[(\alpha + i\beta x)^2 + x^2]^{-1} = \pi H(1 - |\beta|)/\alpha ,$$

with

$$\alpha = 1 \bigg/ \left(4\cos^2\phi_0 - \left(\cos\frac{2\pi k_1}{N} + \cos\frac{2\pi k_2}{N}\right)^2 \right)^{\frac{1}{2}} \ ,$$

$$\beta = \alpha \left(\cos\frac{2\pi k_1}{N} + \cos\frac{2\pi k_2}{N}\right)\tan\phi_0 \ ,$$

and

$$
\begin{aligned}
H(x) &= 0 && x < 0 \\[2mm]
&= \frac{1}{2} && x = 0 \\[2mm]
&= 1 && x > 0
\end{aligned}
$$

In this derivation there is a justifiable interchange of the order of integration and infinite summation.

5. ACKNOWLEDGMENTS

Dr. Yves Elskens provided valuable criticisms of a draft of this manuscript. Ms. Patricia Reitemeier produced the camera-ready manuscript.

6. REFERENCES

[1] DAVID C. TORNEY AND BYRON GOLDSTEIN, *Rates of diffusion-limited reaction in periodic systems*, J. Stat. Phys, 49 (1987), pp. 725-750.

[2] ELLIOTT W. MONTROLL, *Random walks on lattices containing traps*, J. Phys. Soc. Japan, Suppl., 26 (1969), pp. 6-10.

[3] ELLIOTT W. MONTROLL, *Random walks on lattices, III*, J. Math. Phys., 10 (1969), pp. 753-765.

[4] ELLIOTT W. MONTROLL, *Random walks on lattices*, Proc. Symp. Appl. Math. Am. Math. Soc., 16 (1964), pp. 193-220.

[5] M. J. HADAMARD, *Resolution d'une question relative aux déterminants*, Bull. Sci. Math., 17 (1893), pp. 240-246.

[6] ENZA ORLANDI, *An integral equation method for solving boundary value problems in regions with small holes*, Dipartimento di Matematica, Universita di Roma, La Sapienza, Italy (1984) (unpublished manuscript).

[7] D. C. TORNEY AND H. M. McCONNELL, *Diffusion-limited reaction rate theory for two-dimensional systems*, Proc. Roy. Soc. Lond., A387 (1983), pp. 147-170.

[8] A. ERDÉLYI, *Asymptotic Expansions*, Dover Publications, New York, 1956, p. 51.

[9] MILTON ABRAMOWITZ AND IRENE A. STEGUN, *Handbook of Mathematical Functions*, U.S. Government Printing Office, Washington, D.C., 1964, p. 361.

[10] G. N. WATSON, *An infinite integral involving Bessel functions*, London Math. Soc., 9 (1934), pp. 16-22.

A Stability Analysis of the Chandrasekhar Reflection Equation in Radiative Transfer

T. W. MULLIKIN   Department of Mathematics, Purdue University, West Lafayette, Indiana

Dedicated to Milt Wing in his sixty-fifth year and
S. Chandrasekhar in his seventy-seventh year.

1 INTRODUCTION

Several years ago I published a paper containing a one-parameter family of solutions to Chandrasekhar's reflection equation for radiative transfer in a homogeneous, isotropically scattering slab. The study was motivated by an interest in stability of Chandrasekhar's solution and implications for numerical calculations being done at the RAND Corporation.

The solutions I constructed were of a specific form obtained by forcing Chandrasekhar's representation of the reflection function in terms of X- and Y-functions. These solutions had interesting behavior with increasing slab thickness and could be related to the reflection function for anisotropically scattering slabs.

In response to the invitation to talk at this conference in honor of Milt, I could not resist the urge to return to the reflection equation, which is typical of those derived by invariant embedding methods. As a Riccati operator equation, with slab thickness as parameter, the equation can be analyzed in the context of a dynamical equation. I will investigate equilibrium solutions and their stability in this more general context. An infinity of new equilibrium solutions will be defined and related to radiative transfer problems.

2 THE REFLECTION EQUATION AND EQUILIBRIUM SOLUTIONS

I will first give the reflection equation and some well-known solutions, and later relate them to physical problems in radiative transfer. The Chandraskhar equation [6] is

$$\frac{\partial R}{\partial \tau} + (\frac{1}{\mu} + \frac{1}{\nu})R = [1 + \frac{c}{2}\int_0^1 R(\tau, \mu, \sigma)\frac{d\sigma}{\sigma}][1 + \frac{c}{2}\int_0^1 R(\tau, \sigma, \nu)\frac{d\sigma}{\sigma}] \qquad (2.1)$$

for $0 \le \mu, \nu \le 1$ and $\tau > 0$. The parameter c measures the local albedo and will be restricted to the interval $(0, 1]$. The initial condition

$$R(0, \mu, \nu) \equiv 0 \qquad (2.2)$$

for the standard physical problem will be generalized.

In the terminology of dynamical equations, the steady-state solutions to (2.1) satisfy

$$(\frac{1}{\mu} + \frac{1}{\nu})S(\mu, \nu) = [1 + \frac{c}{2}\int_0^1 S(\mu, \sigma)\frac{d\sigma}{\sigma}][1 + \frac{c}{2}\int_0^1 S(\sigma, \nu)\frac{d\sigma}{\sigma}]. \qquad (2.3)$$

This implies that

$$S(\mu, \nu) = \frac{\mu\nu}{\mu + \nu}L(\mu)M(\nu), \qquad (2.4)$$

and leads to the equivalent problem of solving the system

$$\begin{aligned}
L(\mu) &= 1 + L(\mu)\frac{\mu c}{2}\int_0^1 \frac{M(\sigma)d\sigma}{\mu + \sigma}, \\
M(\mu) &= 1 + M(\mu)\frac{\mu c}{2}\int_0^1 \frac{L(\sigma)d\sigma}{\mu + \sigma}.
\end{aligned} \qquad (2.5)$$

For most physical applications, S is analytic in μ and ν for $\mathrm{Re}(\mu) > 0$ and $\mathrm{Re}(\nu) > 0$ and satisfies a reciprocity principle

$$S(\mu, \nu) = S(\nu, \mu), \qquad (2.6)$$

which reduces (2.5) to a single equation with $L = M$. In Section 4 we will see that there is an infinity of distribution solutions to (2.3), some of which do not satisfy (2.6).

Solutions to (2.5) which are continuous on $[0, 1]$ have been known for a long time and are expressed in terms of Chandrasekhar's H-function [3,6,7,13]

$$L_1(\mu) = M_1(\mu) = H(\mu) = \frac{1 + \mu}{1 + k\mu}\exp[\int_0^1 \frac{\mu\Theta(\sigma)d\sigma}{\sigma(\mu + \sigma)}], \qquad (2.7)$$

and

$$L_2(\mu) = M_2(\mu) = \frac{1 + k\mu}{1 - k\mu}H(\mu). \qquad (2.8)$$

The parameter k is the positive solution to

$$2k = c \ln \frac{1+k}{1-k} \qquad (0 \le k \le 1 \text{ for } 0 \le c \le 1), \qquad (2.9)$$

and $\Theta(\sigma)$, ranging over the interval $[0,1]$, is given by

$$\theta(\sigma) = \frac{1}{\pi} \cos^{-1} \left[\frac{\lambda(\sigma)}{\lambda^2(\sigma) + (\frac{\pi \sigma c)}{2})^2} \right] \qquad (2.10)$$

with

$$\lambda(\sigma) = 1 - \frac{c\sigma}{2} \ln \frac{1+\sigma}{1-\sigma} \qquad (0 \le \sigma \le 1). \qquad (2.11)$$

When $c = 1$, the solutions in (2.7) and (2.8) are equal, since then $k = 0$.

The function $H(\mu)$ is analytic and non-vanishing for complex μ with $\text{Re}(\mu) > 0$. It also satisfies the nonlinear Chandrasekhar equation [6] (equation (2.5) with $L = M$)

$$H(\mu) = 1 + H(\mu) \frac{\mu c}{2} \int_0^1 \frac{H(\sigma) d\sigma}{\mu + \sigma}, \qquad (2.12)$$

and the constraint

$$1 = \frac{c}{2} \int_0^1 \frac{H(\sigma) d\sigma}{1 - k\sigma}. \qquad (2.13)$$

For simplicity of notation, we will not denote dependence of the H-function on the parameter c.

It is well known also that the H-function provides a Wiener-Hopf factorization [9]

$$H(\mu) H(-\mu) \left[1 - \frac{c\mu}{2} \int_{-1}^1 \frac{d\sigma}{\mu - \sigma} \right] = 1 \qquad (\mu \notin [-1, 1]). \qquad (2.14)$$

This relates to the Fourier transform analysis of a Wiener-Hopf integral equation for the source function $J(x, \nu)$ which is easily derived from the following problem for radiative transfer in an isotropically scattering half-space.

The function

$$S_1(\mu, \nu) = \frac{\mu \nu H(\mu) H(\nu)}{\mu + \nu}, \qquad (2.15)$$

as a solution of (2.3), is related to the intensity of radiation satisfying the transfer equation

$$\mu \frac{\partial}{\partial x} I(x, \mu, \nu) + I(x, \mu, \nu) = J(x, \nu) \qquad (2.16)$$

with

$$J(x, \nu) = \frac{c}{2} \int_{-1}^1 I(x, \sigma, \nu) d\sigma \qquad (2.17)$$

for $0 \leq x < \infty$, $-1 \leq \mu \leq 1$, $0 \leq \nu \leq 1$. Boundary conditions are

$$I(0, \mu, \nu) = \frac{2}{c}\delta(\mu - \nu) \qquad \text{for } 0 \leq \mu, \nu \leq 1$$

$$\lim_{x \to \infty} I(x, \mu, \nu) = 0 \qquad \text{for } -1 \leq \mu \leq 0 \leq \nu \leq 1. \tag{2.18}$$

The reflection function S is related to the intensity by

$$S(\mu, \nu) = \mu I(0, -\mu, \nu) \qquad \text{for } 0 \leq \mu, \nu \leq 1 \tag{2.19}$$

and to the source function J by

$$S(\mu, \nu) = \int_0^\infty J(x, \nu)e^{-x/\mu}dx. \tag{2.20}$$

The integrodifferential equation (2.1) is derived as a necessary condition on the solution $I(x, \mu, \nu)$ to (2.15) and (2.16) for a slab of thickness τ ($0 \leq x \leq \tau$) with boundary conditions on the faces of the slab

$$I(0, \mu, \nu; \tau) = \frac{2}{c}\delta(\mu - \nu) \qquad \text{for } 0 \leq \mu, \nu \leq 1$$

$$I(\tau, \mu, \nu; \tau) = 0 \qquad \text{for } -1 \leq \mu \leq 0 \leq \nu \leq 1. \tag{2.21}$$

The reflection function $R(\tau, \mu, \nu)$ is related to this finite slab problem by

$$R(\tau, \mu, \nu) = \mu I(0, -\mu, \nu; \tau) \qquad \text{for } 0 \leq \mu, \nu \leq 1. \tag{2.22}$$

Equation (2.1) relates differential changes in R to differential changes in slab thickness [1,6].

In Section 4 the second solution to (2.5), given in (2.8), will be related to a radiative transfer problem. That (2.5) has only these two Hölder continuous solutions can be seen as follows. Operate on the first equation to obtain

$$\frac{c}{2}\mu \int_0^1 \frac{L(\nu)d\nu}{\mu + \nu} = \frac{c}{2}\mu \ln\frac{1 + \mu}{\mu} + (\frac{c}{2})^2 \int_0^1 \int_0^1 \frac{\mu\nu L(\nu)M(\sigma)d\sigma\, d\nu}{(\mu + \nu)(\sigma + \nu)}. \tag{2.23}$$

Use of a partial fraction decomposition of the integrand in the double integral, together with equations (2.5), gives the linear singular integral equation as a necessary condition

$$\lambda(\mu)M(\mu) = 1 + \frac{\mu c}{2} \int_0^1 \frac{M(\sigma)d\sigma}{\sigma - \mu}. \tag{2.24}$$

The integral is computed as a Cauchy principal value. It is well known [3,12,13,16] that (2.24) has a one parameter family of Hölder continuous solutions

$$M(\mu) = \frac{1 + a\mu}{1 - k\mu}H(\mu) \qquad \text{for arbitrary } a. \tag{2.25}$$

A necessary condition on L similar to (2.24) is satisfied by

$$L(\mu) = \frac{1 + b\mu}{1 - k\mu} H(\mu) \qquad \text{for arbitrary } b. \tag{2.26}$$

The nonlinear equations (2.5) restrict the constants a and b to the values $a = b = \pm k$, and determine the solutions in (2.7) and (2.8).

For use in the following sections, (2.24) is written in distribution notation as

$$\int_0^1 \psi_\mu(\sigma) M(\sigma) d\sigma = 1, \tag{2.27}$$

where the distribution, parameterized by μ, is

$$\psi_\mu(\sigma) = \lambda(\sigma)\delta(\sigma - \mu) - \frac{c}{2} PV \frac{\mu}{\sigma - \mu} \qquad \text{for } 0 \leq \mu, \sigma \leq 1. \tag{2.28}$$

Equation (2.27) is satisfied for both

$$M(\sigma) = H(\sigma) \tag{2.29}$$

and for

$$M(\sigma) = \frac{1 + k\sigma}{1 - k\sigma} H(\sigma). \tag{2.30}$$

Both of these functions also satisfy

$$M(\mu) = 1 + M(\mu) \frac{\mu c}{2} \int_0^1 \frac{M(\sigma) d\sigma}{\mu + \sigma}, \tag{2.31}$$

which is (2.5) with $L = M$.

Multiple solutions of (2.3) can also be obtained by a standard bifurcation analysis, common in dynamical studies. We set $c = 1 - \epsilon$ and $S = S_1 + D$ and use (2.12) to transform (2.3) to

$$(\frac{1}{\mu} + \frac{1}{\nu})D(\mu, \nu) - \frac{H(\mu)}{2} \int_0^1 \frac{D(\sigma, \nu) d\sigma}{\sigma} - \frac{H(\nu)}{2} \int_0^1 \frac{D(\mu, \sigma) d\sigma}{\sigma}$$
$$= N(\epsilon, D) \tag{2.32}$$

where the terms linear in ϵ and quadratic in ϵ and D are simply denoted by $N(\epsilon, D)$. Bifurcation occurs since the linear operator, given by neglecting N, has 0 in its spectrum with eigenfunction

$$D(\mu, \nu) = \mu\nu H(\mu)H(\nu). \tag{2.33}$$

This follows by dropping $N(\epsilon, D)$ from (2.32) and using (2.13) for $c = 1$ and $k = 0$.

In the next three sections spectral theory will be used extensively. Distribution solutions to (2.5) will result from this analysis and be related to the Case singular eigenfunctions [5]

$$I(x, \mu, \nu) = \Phi_\nu(\mu) \exp[-x/\nu] \tag{2.34}$$

and

$$I(x, \mu, \pm k) = \frac{c}{2(1 \pm k\mu)} \exp[\pm kx] \tag{2.35}$$

as generalized solutions to

$$\mu \frac{\partial I}{\partial x} + I = \frac{c}{2} \int_{-1}^{1} I(x, \sigma, \nu) d\sigma. \tag{2.36}$$

The distributions $\Phi_\nu(\mu)$ are defined by

$$\Phi_\nu(\mu) = \lambda(\mu)\delta(\mu - \nu) - \frac{c}{2} PV \frac{\nu}{\mu - \nu} \qquad \text{for } -1 \le \mu, \nu \le 1, \tag{2.37}$$

and are related to (2.28) by

$$\Phi_\nu(\mu) = \Phi_{-\nu}(-\mu) = \Psi_\nu(\mu) \quad \text{for } 0 \le \mu, \nu \le 1. \tag{2.38}$$

3 STABILITY ANALYSIS

We have observed bifurcation of the steady-state equation (2.3) at $c = 1$. For $c < 1$, we will investigate stability of the steady-state solutions

$$S(\mu, \nu) = \frac{\mu\nu}{\mu + \nu} M(\mu)M(\nu) \tag{3.1}$$

for the two choices of M given in (2.29) and (2.30).

We substitute in (2.1) the representation

$$R(\tau, \mu, \nu) = S(\mu, \nu) - \mu\nu M(\mu)M(\nu)T(\tau, \mu, \nu) \tag{3.2}$$

and use (2.3) to obtain

$$\frac{\partial T}{\partial \tau} + (\frac{1}{\mu} + \frac{1}{\nu})T = \frac{c}{2\mu} \int_0^1 \frac{T(\sigma, \nu)dm(\sigma)}{\sigma} +$$
$$\frac{c}{2\nu} \int_0^1 \frac{T(\mu, \sigma)dm(\sigma)}{\sigma} - (\frac{c}{2})^2 \int_0^1 \int_0^1 \frac{T(\mu, \sigma)T(\rho, \nu)dm(\sigma)dm(\rho)}{\sigma\rho}. \tag{3.3}$$

The measure in (3.3) is given by

$$dm(\sigma) = \sigma M(\sigma)d\sigma \tag{3.4}$$

and is non-negative since M is positive on $[0, 1]$.

We interpret T as the kernel of a linear integral operator (possibly unbounded) with domain in the real Hilbert space

$$\mathcal{H} = \{f \mid \int_0^1 |f(\sigma)|^2 dm(\sigma) < \infty\}. \tag{3.5}$$

For M in (2.29) or in (2.30), $\mathcal{H}$ is norm equivalent to the Hilbert space $L_2[0, 1; \mu \, d\mu]$. Equation (3.3) is expressed as a functional Riccati equation

$$\frac{\partial \mathcal{T}}{\partial \tau} = -\mathcal{A}\mathcal{T} - \mathcal{T}\mathcal{A} - \mathcal{T}\mathcal{B}\mathcal{T}. \tag{3.6}$$

The unbounded operator $\mathcal{A}$ has the representation

$$\mathcal{A} = \mathcal{M} - \mathcal{M}\mathcal{C}\mathcal{M} \tag{3.7}$$

where $\mathcal{M}$ is the unbounded, self-adjoint, multiplication operator

$$\mathcal{M}[f](\mu) = f(\mu)/\mu \tag{3.8}$$

and C is the bounded, self-adjoint operator

$$C[f] = \frac{c}{2} \int_0^1 f(\sigma) dm(\sigma). \tag{3.9}$$

The unbounded, self-adjoint operator $\mathcal{B}$ is represented by

$$\mathcal{B} = \frac{c}{2}\mathcal{M}\mathcal{C}\mathcal{M}. \tag{3.10}$$

The real Hilbert space

$$\mathcal{H}_0 = \{f \mid \int_0^1 |f(\sigma)|^2 dm_0(\sigma) < \infty\} \tag{3.11}$$

with

$$dm_0(\sigma) = M(\sigma) d\sigma$$

is norm equivalent to the ordinary Lebesgue space $L_2[0, 1]$ for either choice of M, and is dense in the space $\mathcal{H}$. The operators $\mathcal{A}$, $\mathcal{B}$ and $\mathcal{M}$ have a common domain dense in $\mathcal{H}_0$ with ranges in $\mathcal{H}$. The following result extends these operators to $\mathcal{H}_0$ but with ranges in the dual space $\mathcal{H}_0'$, containing $\mathcal{H}$, as extensions of the embedding of $\mathcal{H}$ into $\mathcal{H}_0'$ defined for g in $\mathcal{H}$ and f in $\mathcal{H}_0$ by $\langle f, g \rangle_{\mathcal{H}}$,

$$|\langle f, g \rangle_{\mathcal{H}}| = |\langle f, \mu g \rangle_{\mathcal{H}_0}| \leq \|f\|_{\mathcal{H}_0} \|g\|_{\mathcal{H}} \tag{3.13}$$

as determining a continuous linear functional on $\mathcal{H}_0$.

LEMMA 1 *The operators $\mathcal{A}$, $\mathcal{B}$ and $\mathcal{M}$ define continuous linear mappings of the Hilbert space $\mathcal{H}_0$ into the dual space $\mathcal{H}_0'$. The operators $\mathcal{A}$ and $\mathcal{M}$ have bounded inverses when $c < 1$.*

Proof: We have the estimate for f in $\mathcal{H}_0$ and h in the domain of $\mathcal{A}$

$$|\langle f, \mathcal{A}h \rangle_{\mathcal{H}}| = |\int_0^1 f[h - \frac{c}{2}\int_0^1 h\, dm_0]\, dm_0|$$

$$\leq \|f\|_{\mathcal{H}_0}\|h\|_{\mathcal{H}_0}(1 + \frac{c}{2}\int_0^1 dm_0), \tag{3.14a}$$

and define the extension of $\langle f, \mathcal{A}h \rangle_{\mathcal{H}}$ to h in $\mathcal{H}_0$ by the integral in (3.14a). The mapping from $\mathcal{H}_0$ to $\mathcal{H}_0'$ is bounded. Similar results hold for $\mathcal{B}$ and $\mathcal{M}$. For g in $\mathcal{A}(\mathcal{H}_0)$ an easy calculation gives

$$\mathcal{A}^{-1}[g](\mu) = \mu g(\mu) + \frac{c}{2}\langle 1, \mu g \rangle_{\mathcal{H}_0} / (1 - \frac{c}{2}\int_0^1 dm_0) \tag{3.14b}$$

with nonzero denominator for $c < 1$ as follows from (2.13) for M given either by (2.29) or by (2.30). The inverse of the operator $\mathcal{M}$ is merely multiplication by μ. ☐

The choice of $M = H$ is particularly important.

LEMMA 2 *When $M = H$ and $c < 1$, the extended operator $\mathcal{A}$ is positive definite.*

Proof: We use the Schwarz inequality and equation (2.13) to obtain the inequality for f in $\mathcal{H}_0$

$$\langle f, \mathcal{A}f \rangle_{\mathcal{H}} = \int_0^1 f^2\, dm_0 - \frac{c}{2}[\int_0^1 f\, dm_0]^2$$

$$\geq [1 - \frac{c}{2}\int_0^1 dm_0]\|f\|_{\mathcal{H}_0}^2 = \sqrt{1-c}\|f\|_{\mathcal{H}_0}^2. \tag{3.15a}$$

We have used the integral of (2.12) to determine that

$$1 - \frac{c}{2}\int_0^1 dm_0 = \sqrt{1-c}. \quad ☐ \tag{3.15b}$$

For $M = H$ and $c < 1$, the spaces $\mathcal{H}_0$ and $\mathcal{H}$ and the operator $\mathcal{A}$ satisfy some of the hypotheses in the paper by Temam [17] on Riccati equations. One of his conditions is satisfied by (3.6) since, from (3.15a) equivalent norms on $\mathcal{H}_0$ are defined by $\langle f, f \rangle_{\mathcal{H}_0}$ and $\langle f, \mathcal{A}f \rangle_{\mathcal{H}}$, namely

$$\|f\|_{\mathcal{H}_0}^2 \geq \langle f, \mathcal{A}f \rangle_{\mathcal{H}} \geq \sqrt{1-c}\|f\|_{\mathcal{H}_0}^2. \tag{3.16}$$

The natural injection of $\mathcal{H}_0$ into $\mathcal{H}$, has a continuous inverse which is not compact. Temam's analysis does not apply because of this lack of compactness and the unboundedness of $\mathcal{B}$ in (3.6).

The spectral analysis in $\mathcal{H}$ of the operator $\mathcal{A}$ is well known [5,8,10,11], and is most conveniently expressed in terms of the spectrum of the bounded, self-adjoint operator $\mathcal{A}^{-1}$. The continuous spectrum of $\mathcal{A}^{-1}$ determines eigendistributions as solutions to

$$\mathcal{A}\psi_\alpha = \frac{1}{\alpha}\psi_\alpha, \tag{3.17}$$

that is

$$(\alpha - \mu)\psi_\alpha(\mu) - \frac{c\alpha}{2}\int_0^1 \psi_\alpha(\sigma)dm_0(\sigma) = 0. \tag{3.18}$$

We summarize well known results that follow easily from (3.18), (2.27) and (2.13).

LEMMA 3 *For the continuous spectrum,* $0 \le \alpha \le 1$, *eigendistributions of* $\mathcal{A}^{-1}$ *are given by*

$$\psi_\alpha(\mu) = \lambda(\mu)\delta(\mu - \alpha) - \frac{c}{2}P.V.\frac{\alpha}{\mu - \alpha}. \tag{3.19}$$

(a) When $M(\sigma) = H(\sigma)$, *the point spectrum of* $\mathcal{A}^{-1}$ *consists of one eigenvalue* $\alpha = 1/k$ *with eigenfunction*

$$\psi_k(\mu) = \frac{c}{2(1 - k\mu)} \tag{3.20}$$

(b) When $M(\sigma) = (1 + k\sigma)H(\sigma)/(1 - k\sigma)$, *the point spectrum of* $\mathcal{A}^{-1}$ *consists of the eigenvalue* $\alpha = -1/k$ *with eigenfunction*

$$\psi_{-k}(\mu) = \frac{c}{2(1 + k\mu)}. \tag{3.21}$$

These generalized eigenfunctions have been normalized so that by (2.13) and (2.27)

$$\int_0^1 \psi_\alpha(\sigma)dm_0(\sigma) = 1. \tag{3.22}$$

Two easy consequences of the spectral results are the following.

THEOREM 1 *The steady-state solution of (2.1) given in (3.1) is unstable in* $\mathcal{H}$ *when M is given by (2.30).*

Proof: We use the representation

$$T(\tau, \mu, \nu) = -a(\tau)\psi_{-k}(\mu)\psi_{-k}(\nu) \tag{3.23}$$

in (3.3) and (3.22) to obtain the scalar Riccati equation

$$\frac{da}{d\tau} = 2ka + \left(\frac{ca}{2}\right)^2. \tag{3.24}$$

For every initial value $a(0) > 0$ the solution $a(\tau)$ goes to infinity in a finite τ-interval. □

THEOREM 2 *For $M(\sigma) = H(\sigma)$, the semigroup in $\mathcal{H}$ generated by $\mathcal{A}$ expresses the solution to (3.6) and bounded initial conditions*

$$T(0) = T_0 \tag{3.25}$$

by the bounded operator

$$T(\tau) = e^{-\mathcal{A}\tau} T_0 [I + \int_0^\tau e^{-\mathcal{A}s} \mathcal{B} e^{-\mathcal{A}s} ds T_0]^{-1} e^{-\mathcal{A}\tau}. \tag{3.26}$$

For every bounded T_0 this solution exists for a τ-interval, and the solution exists for $0 \leq \tau < \infty$ provided $0 < c < 1$ and

$$\|T_0\| \leq \frac{4\sqrt{1-c}}{c} \tag{3.27}$$

For such T_0 the solution $T(\tau)$ has

$$\lim_{\tau \to \infty} T(\tau) = 0, \tag{3.28}$$

and the solution of (2.1) given in (2.15) is asymptotically stable.

Proof: For f in $\mathcal{H}$, the norm equivalence in (3.16) gives

$$
\begin{aligned}
\langle f, e^{-\mathcal{A}s} \mathcal{B} e^{-\mathcal{A}s} f \rangle_{\mathcal{H}} &= [\frac{c}{2} \int_0^1 e^{-\mathcal{A}s} f dm_0]^2 \\
&\leq (\frac{c}{2})^2 \int_0^1 dm_0 \|e^{-\mathcal{A}s} f\|^2_{\mathcal{H}_0} \\
&\leq \frac{c}{2\sqrt{1-c}} \langle f, \mathcal{A} e^{-2\mathcal{A}s} f \rangle_{\mathcal{H}}.
\end{aligned}
\tag{3.29}
$$

The integral of the inequality is

$$\langle f, \int_0^\tau e^{-\mathcal{A}s} \mathcal{B} e^{-\mathcal{A}s} ds f \rangle_{\mathcal{H}} \leq \frac{c}{4\sqrt{1-c}} [\|f\|^2_{\mathcal{H}} - \|e^{-\mathcal{A}\tau} f\|^2_{\mathcal{H}}], \tag{3.30}$$

and self-adjointness gives

$$\| \int_0^\tau e^{-\mathcal{A}s} \mathcal{B} e^{-\mathcal{A}s} ds \| \leq \frac{c}{4\sqrt{1-c}}. \tag{3.31}$$

(This estimate will be improved in Section 4 so that the right-hand side is bounded as c goes to 1 and k goes to 0). If (3.27) is satisfied, the inverse in (3.26) exists as a Neumann series for all τ, and it follows from the spectrum of $\mathcal{A}$ that

$$\|T(\tau)\| \leq \text{ constant } e^{-2k\tau}. \tag{3.32}$$

The standard representation in (3.26) is easily checked to be the solution to (3.6) and (3.25) (c.f. [15]). □

These two results relate to large τ results in Theorem 2 of my paper [14] in which special solutions to (2.1) were constructed with the representation

$$R(\tau, \mu, \nu) = \frac{\mu\nu}{\mu + \nu}[X(\tau, \mu)X(\tau, \nu) - Y(\tau, \mu)Y(\tau, \nu)] \tag{3.33}$$

in terms of the general solutions to Chandrasekhar's nonlinear X- and Y- integral equations without constraints [12,13]. Initial data was of the form

$$R(0, \mu, \nu) = 2kE(\frac{1}{Ea - 1} + k^2\mu\nu)\frac{\mu\nu}{[1 - (k\mu)^2][1 - (k\nu)^2]} \tag{3.34}$$

with E arbitrary and $a = 1 - \frac{c}{k}\ln(1 + k)$. I showed that for a special value E_0

$$\lim_{\tau \to \infty} R(\tau, \mu, \nu) = \frac{\mu\nu M(\mu)M(\nu)}{\mu + \nu} \tag{3.35}$$

with M given by (2.30), and that for $E \neq E_0$ in an interval containing E_0, the solution $R(\tau, \mu, \nu)$ goes to infinity for finite τ. This is related to the results in Theorem 1. For $Ea < 1$ it was shown that the solution exists for all τ, $0 \leq \tau < \infty$, and the limit in (3.35) is true with $M = H$ for all such initial data. The relationship with Theorem 2 can be made more direct by obtaining another domain of validity for the representation (3.26).

The above stability results are for initial data of $R(\tau, \mu, \nu)$ in a neighborhood of $S(\mu, \nu)$, and hence of initial data of $T(\tau, \mu, \nu)$ in a neighborhood of 0. The physical problem posed by (2.1) and boundary conditions (2.2), as well as initial data in (3.34), impose for T_0 in (3.25) an integral operator with kernel near the kernel of the operator T_1 given by

$$T_1(\mu, \nu) = \frac{1}{\mu + \nu} \tag{3.36}$$

with

$$||T_1||^2 \leq \int_0^1 \int_0^1 \frac{\mu\nu H(\mu)H(\nu)d\mu d\nu}{(\mu + \nu)^2} \leq [\frac{1}{2}\int_0^1 dm_0]^2 \leq (\frac{1}{c})^2. \tag{3.37}$$

For c near 1, one can show that $||T_1|| \geq 0.4$ in violation of (3.27). We can show validity of (3.26) for $T_0 = T_1$ by observing that T_1 is a compact, self-adjoint, positive semidefinite operator on $\mathcal{H}$ to $\mathcal{H}$ as follows from (3.37) and

$$\langle f, T_1 f \rangle_{\mathcal{H}} = \int_0^\infty [\int_0^1 e^{-x/\mu}f(\mu)dm_0(\mu)]^2 dx \geq 0. \tag{3.38}$$

THEOREM 3 *Equation (3.26) gives the solution of (3.6) for $0 \leq \tau < \infty$ and $0 < c < 1$ if initial data T_0 is bounded, self-adjoint, and positive semi-definite. In this case the solution is bounded, self-adjoint with*

$$\|T(\tau)\| \leq \|T(0)\| \exp(-2k\tau) \tag{3.39}$$

and so approaches 0 as τ goes to infinity.

Proof: Operator inversion in (3.26) is expressed by solving the equation for f in $\mathcal{H}$

$$f + \int_0^\tau e^{-As} B e^{-As} ds\, T_0 f = g. \tag{3.40}$$

For the representation (3.26), the term $T_0 f$ can be expressed for T_0 (and $\sqrt{T_0}$) bounded, self-adjoint, and positive semi-definite as

$$T_0 f = \sqrt{T_0} h, \tag{3.41}$$

where h solves

$$h + \sqrt{T_0} \int_0^\tau e^{-As} B e^{-As} ds \sqrt{T_0} h = \sqrt{T_0} g. \tag{3.42}$$

This last equation has an unique solution since the operator is bounded by (3.31), and positive semi-definite by the first line in (3.29), i.e.

$$\langle h, \sqrt{T_0} \int_0^\tau e^{-As} B e^{-As} ds \sqrt{T_0} h \rangle_{\mathcal{H}} = \int_0^\tau [\frac{c}{2} \int_0^1 e^{-As} \sqrt{T_0} h\, dm_0]^2 ds \geq 0. \tag{3.43}$$

It follows that

$$h = \mathcal{L} \sqrt{T_0} y \tag{3.44}$$

for $\mathcal{L}$ bounded and self-adjoint with

$$\|\mathcal{L}\| \leq 1. \tag{3.45}$$

From (3.41) we have

$$T(\tau) = e^{-A\tau} \sqrt{T_0} \mathcal{L} \sqrt{T_0} e^{-A\tau}, \tag{3.46}$$

and, by spectral properties of A, that (3.39) holds. $\square$

This last result shows that in a neighborhood of T_1, with kernel in (3.36), all bounded, self-adjoint, positive semi-definite initial data determine solutions $T(\tau)$ asymptotic to 0. This relates to initial data in (3.34) for $Ea < 1$, which can have large norm. In particular, the solution to (2.1) and (2.2) defines an integral operator in $\mathcal{H}$ to $\mathcal{H}$

$$\mathcal{R}_1(\tau) = T_1 - e^{-A\tau} T_1 [I + \int_0^\tau e^{-As} B e^{-As} ds\, T_1]^{-1} e^{-A\tau}. \tag{3.47}$$

The operator inversion required in this representation is related, through the spectral analysis of A^{-1}, to Fredholm equations derived in [13] by applying the theory of singular integral equations and used to compute extensive tables of Chandrasekhar's X- and Y functions [4]. This connection will be made more obvious in the last section.

A perturbation of the initial data T_1 can be made to give existence for all τ with some initial data which is compact but not necessarily self-adjoint.

THEOREM 4 *Equation (3.6) is solved by (3.26) for $0 \leq \tau < \infty$ if initial data T_0 is given by*

$$T_0 = T_1 + \mathcal{P}T_1 \tag{3.48}$$

for $\mathcal{P}$ bounded and satisfying

$$\|\mathcal{P}\| \leq 4\sqrt{1-c}. \tag{3.49}$$

Proof: By use of (3.41), (3.44) and (3.48) it follows that

$$
\begin{aligned}
&T_0[I + \int_0^\tau e^{-\mathcal{A}s}\mathcal{B}e^{-\mathcal{A}s}ds T_0]^{-1} = \\
&(I + \mathcal{P})[I + \sqrt{T_1}\mathcal{L}\sqrt{T_1}\int_0^\tau e^{-\mathcal{A}s}\mathcal{B}e^{-\mathcal{A}s}ds\mathcal{P}]^{-1}\sqrt{T_1}\mathcal{L}\sqrt{T_1}.
\end{aligned}
\tag{3.50}
$$

The estimates (3.31), (3.37), and (3.45) show that for all τ, $0 \leq \tau < \infty$, the inverse exists as a Neumann series if (3.49) is satisfied. □

Existence of $T(\tau)$ for $0 \leq \tau < \infty$ and for arbitrary initial data T_0 near T_1 has not been established. Another approach is suggested in the next section.

The motivation for my paper [14] was stability of the physical solution as being relevant to errors inherent in numerical calculations [2]. I showed that nonphysical solutions could be near the physical solution for large τ and then go to infinity for even larger τ. This is, of course, shown above by the result that the physical solution for large τ is near the stable steady-state solution which in turn is near the unstable steady-state solution for c near 1 and k near 0.

4 DISTRIBUTION SOLUTIONS

The last two sections have determined bounded operators as solutions to (3.6). It is an interesting fact that there are operators with distribution kernels that satisfy the steady-state equation (2.3). We will determine a set of these, relate them to radiative transfer problems, and interpret a subset as operators, not on $\mathcal{H}$, but on a space of Hölder continuous functions.

In the remainder of this paper we restrict to the Hilbert space $\mathcal{H}$ defined in (3.5) and (3.4) for the choice $M = H$. Some results hold also for the choice of M in (2.30), e.g. Theorem 5 below is true with $H(\mu)$ replaced by $(1 + k\mu)/(1 - k\mu)H(\mu)$.

We define distributions kernels in terms of the eigendistributions of $\mathcal{A}$ given in Lemma 3

$$\psi_{\alpha\beta}(\mu, \nu) = \psi_\alpha(\mu)\psi_\beta(\nu) \qquad \text{for } 0 \leq \mu, \nu \leq 1. \tag{4.1}$$

Equations (3.17) and (3.22) readily show that the $\psi_{\alpha\beta}$ are generalized eigenfunctions of the linear part of (3.6) associated with spectral values $(1/\alpha + 1/\beta)$. They are also generalized eigenfunctions of the nonlinear part of (3.6), i.e.

$$\int_0^1 \int_0^1 \psi_{\alpha\beta}(\mu, \sigma)\psi_{\alpha\beta}(\rho, \nu)dm_0(\sigma)dm_0(\rho) = \psi_{\alpha\beta}(\mu, \nu) \tag{4.2}$$

as follows from (3.22).

THEOREM 5 *The steady-state equation (2.3) has distribution solutions*

$$S_{\alpha\beta}(\mu,\nu) = \mu\nu H(\mu)H(\nu)[\frac{1}{\mu+\nu} + (\frac{2}{c})^2(\frac{1}{\alpha}+\frac{1}{\beta})\psi_{\alpha\beta}(\mu,\nu)] \qquad (4.3)$$

for all α, β *in the spectrum of* $\mathcal{A}^{-1}$. *For* $\alpha \neq \beta$ *the reciprocity relation (2.6) is violated. An equivalent representation is*

$$S_{\alpha\beta}(\mu,\nu) = \frac{\mu\nu}{\mu+\nu}L_{\alpha\beta}(\mu)M_{\alpha\beta}(\nu), \qquad (4.4)$$

where

$$L_{\alpha\beta}(\mu) = H(\mu)[\frac{2}{c}(\frac{\beta}{\alpha}+1)\psi_{\alpha}(\mu) - \frac{\beta}{\alpha}]$$
$$M_{\alpha\beta}(\nu) = H(\nu)[\frac{2}{c}(\frac{\alpha}{\beta}+1)\psi_{\beta}(\nu) - \frac{\alpha}{\beta}]. \qquad (4.5)$$

The choices $\alpha = \beta = 1/k$ *and* ψ_k *of (3.20) give the solution in (2.8) multiplied by* -1.

Proof: That $S_{\alpha\beta}$ in (4.3) is a solution of (2.3) follows from the easily verified fact that $\mathcal{T}$ with kernel

$$T(\mu,\nu) = -(\frac{2}{c})^2(\frac{1}{\alpha}+\frac{1}{\beta})\psi_{\alpha\beta}(\mu,\nu) \qquad (4.6)$$

satisfies (3.6) because of (4.2). The factorization (4.5) follows from (4.3) by repeated use of (3.18) in the form

$$\mu\psi_{\alpha}(\mu) = \alpha[\psi_{\alpha}(\mu) - \frac{c}{2}]. \quad \square \qquad (4.7)$$

In Theorem 1 it was observed that for $\alpha = \beta = 1/k$ the solution of (2.3) given above is unstable. We extend this result.

THEOREM 6 *The steady-state solutions* $S_{\alpha\beta}$ *are unstable.*

Proof: We set $T(\tau) = a(\tau)\psi_{\alpha\beta}$ in (3.3) to obtain the scalar Riccati equation

$$\frac{da}{d\tau} = -(\frac{1}{\alpha}+\frac{1}{\beta})a - (\frac{ca}{2})^2. \qquad (4.8)$$

The equilibrium solution $a = -(\frac{2}{c})^2(\frac{1}{\alpha}+\frac{1}{\beta})$, which gives (4.6), is unstable. $\square$

The steady-state solution in (3.1) with $M = H$ is determined by the radiative transfer problem stated in (2.16)-(2.18). The set of solutions $S_{\alpha\beta}(\mu,\nu)$ are related to radiative transfer problems in which the second boundary condition in (2.18) is replaced by an exponential growth condition as x tends to infinity. All of these equilibrium solutions can also be obtained as reflection functions for finite slabs with the boundary condition (2.21) at $x = \tau$ replaced by correctly chosen input radiation. We can see this most easily by using the Case representation of solutions to (2.36).

It is most convenient to express the Case representation in terms of the spectral resolution determined by the operator $\mathcal{A}^{-1}$. We use the notation for the spectrum of $\mathcal{A}^{-1}$ with domain in $\mathcal{H}$ with $M(\mu) = H(\mu)$

$$sp = \{[0,1], \frac{1}{k}\}. \tag{4.9}$$

For Hölder continuous f it is known [5,8,10,11] that

$$f(\mu) = \int_{sp} \mathcal{F}[f](\alpha)\psi_\alpha(\mu)dn(\alpha) \tag{4.10}$$

with the measure defined on the spectrum of $\mathcal{A}^{-1}$ by

$$dn(\alpha) \; = \; \delta(\alpha - \frac{1}{k})\frac{d\alpha}{\Delta(k)} + \chi_{[0,1]}(\alpha)\frac{d\alpha}{\Delta(\alpha)}, \tag{4.11}$$

and

$$\begin{aligned}\Delta(k) \; &= \; (\frac{c}{2})^2 \int_0^1 \frac{dm(\sigma)}{(1-k\sigma)^2}, \\ \Delta(\alpha) \; &= \; \alpha H(\alpha)[\lambda^2(\alpha) + (\frac{\alpha c\pi}{2})^2].\end{aligned} \tag{4.12}$$

The transform of Hölder continuous functions

$$\mathcal{F}[f](\alpha) = \int_0^1 f(\mu)\psi_\alpha(\mu)dm(\mu) \tag{4.13}$$

can be extended to a unitary map from the Hilbert space $\mathcal{H}$ to the Hilbert space

$$\Sigma = \{F| \int_{sp} |F(\alpha)|^2 dn(\alpha) < \infty\}. \tag{4.14}$$

The generalized eigenfunctions are given in (3.19) and (3.20).

A dense subspace of Σ is defined by $\Sigma_0 = \mathcal{F}(\mathcal{H}_0)$. Multiplication by $1/\alpha$ in $\mathcal{F}(D(\mathcal{A}))$, dense in Σ_0, extends to a mapping of Σ_0 into its dual Σ_0' by Lemma 1 and the definition

$$\langle \mathcal{F}f, \mathcal{F}f/\alpha \rangle_\Sigma \; = \; \langle g, \mathcal{A}f \rangle_\mathcal{H} \tag{4.15}$$

for f and g in $\mathcal{H}_0$. The unitary map $\mathcal{F}^*$ from Σ to $\mathcal{H}$ defines a unitary map $\mathcal{F}^*$ on Σ_0' to $\mathcal{H}_0'$ by

$$\langle g, \mathcal{F}^*F \rangle_\mathcal{H} \; = \; \langle g, \mathcal{A}\mathcal{F}^*(\alpha F) \rangle_\mathcal{H} \tag{4.16}$$

for F in $\frac{1}{\alpha}\Sigma_0$ and g in $\mathcal{H}$.

LEMMA 4 *Transform pairs are given by*

(a) $\mathcal{F}[1](\alpha) = \alpha\sqrt{1-c}$

(b) $\mathcal{F}^*[1](\mu) = \mathcal{M}[1](\mu) = 1/\mu$ (4.17)

(c) $\mathcal{F}[\frac{H(\mu)}{\mu+\cdot}](\alpha) = \frac{\alpha}{\mu+\alpha} = \frac{2}{c}\Phi_\alpha(-\mu)$ for $0 \le \mu \le 1$.

It follows that the measure in (4.11) satisfies

$$\sqrt{1-c}\int_{sp}\alpha\,dn(\alpha) = \frac{2}{1+\sqrt{1-c}}. \qquad (4.18)$$

Proof: The result in (4.17a) follows by the use of (4.7), (3.22) and (3.15b) in

$$\int_0^1 \psi_\alpha(\mu)\mu H(\mu)d\mu = \int_0^1 \alpha[\psi_\alpha(\mu) - \frac{c}{2}]H(\mu)d\mu. \qquad (4.19)$$

The result in (4.17b) follows from (4.16) and (4.17a) since

$$\langle g, \mathcal{F}^*(1)\rangle_{\mathcal{H}} = \langle g, \mathcal{A}\mathcal{F}^*(\alpha)\rangle_{\mathcal{H}}$$
$$= \langle g, \mathcal{A}(1)\rangle_{\mathcal{H}}/(1 - \frac{c}{2}\int_0^1 dm_0) = \langle g, \mathcal{M}(1)\rangle_{\mathcal{H}}. \qquad (4.20)$$

We also use (4.7) to compute

$$\int_0^1 \frac{\mu}{\mu+\nu}\psi_\alpha(\nu)\nu H(\nu)d\nu = \alpha[\int_0^1 (1 - \frac{\nu}{\mu+\nu})\psi_\alpha(\nu)H(\nu)d\nu$$
$$- \frac{c}{2}\mu\int_0^1 \frac{H(\nu)d\nu}{\mu+\nu}]. \qquad (4.21)$$

By (2.12) and (3.22) it follows from (4.21) that (4.17c) is a restatement of

$$\int_0^1 \frac{\mu+\alpha}{\mu+\nu}\psi_\alpha(\nu)dm(\nu) = \alpha[1 - \frac{c\mu}{2}\int_0^1 \frac{H(\nu)d\nu}{\mu+\nu}] = \frac{\alpha}{H(\mu)}. \qquad (4.22)$$

The result in (4.18) follows from (3.15b) and the extended Parseval relation

$$\langle 1, \mathcal{A}(1)\rangle_{\mathcal{H}} = \langle \mathcal{F}(1), \mathcal{F}(1)/\alpha\rangle_\Sigma \qquad (4.23)$$

expressed in terms of the measures previously defined as

$$(1 - \frac{c}{2}\int_0^1 dm_0)\int_0^1 dm_0 = \int_{sp}\alpha\,dn(\alpha)(1 - \frac{c}{2}\int_0^1 dm_0)^2. \qquad (4.24)$$

This last result can also be established [13] by applying the residue theorem from complex analysis to the Wiener-Hopf factorization (2.14). □

The Case representation as a superposition of generalized solutions given in (2.34) and (2.35) can be expressed by

$$I(x, \mu; \tau) = \langle \Phi_\alpha(\mu) e^{-x/\alpha}, \mathcal{F}(f) \rangle_\Sigma - \langle \Phi_\alpha(-\mu) e^{-(\tau - x)/\alpha}, \mathcal{F}(g) \rangle_\Sigma. \tag{4.25}$$

Using (2.38), (4.10) and (4.13), we can also express this for $0 \le \mu \le 1$ as

$$I(x, \mu; \tau) = e^{-Ax}[f](\mu) - \langle \Phi_\alpha(-\mu) e^{-(\tau - x)/\alpha}, \mathcal{F}(g) \rangle_\Sigma, \tag{4.26}$$

and for $-1 \le -\mu \le 0$ as

$$I(x, -\mu; \tau) = \langle \Phi_\alpha(-\mu) e^{-x/\alpha}, \mathcal{F}(f) \rangle_\Sigma - e^{-A(\tau - x)}[g](\mu), \tag{4.27}$$

where

$$\Phi_\alpha(-\mu) = \frac{c\alpha}{2(\mu + \alpha)} \qquad \text{for } 0 \le \mu \le 1 \text{ and } \alpha \text{ in } sp. \tag{4.28}$$

The functions f and g will be determined by boundary conditions, and the reflection function expressed in terms of (4.27) for $x = 0$

$$I(0, -\mu; \tau) = \langle \mathcal{F}^*[\frac{c\alpha}{2(\mu + \alpha)}], f \rangle_{\mathcal{H}} - e^{-A\tau}[g](\mu), \tag{4.29}$$

which by (4.17c) can be written as

$$I(0, -\mu; \tau) = \frac{cH(\mu)}{2} \langle \frac{1}{\mu + \cdot}, f(\cdot) \rangle_{\mathcal{H}} - e^{-A\tau}[g](\mu). \tag{4.30}$$

By a limiting process it follows that (4.30) is meaningful for f determined by the boundary condition (2.21) equated to (4.26) for $x = 0$

$$\begin{aligned} I(0, \mu; \tau) &= \frac{2}{c} \delta(\mu - \nu) \\ &= f(\mu) - \frac{c}{2} \langle \frac{\alpha e^{-\tau/\alpha}}{\mu + \alpha}, \mathcal{F}(g) \rangle_\Sigma. \end{aligned} \tag{4.31}$$

THEOREM 7 *The determination of f in terms of g from (4.31) gives the reflection function in terms of g*

$$R(\tau, \mu, \nu) = \frac{\mu \nu H(\mu) H(\nu)}{\mu + \nu} - \mu H(\mu) \mathcal{F}^*[\frac{e^{-\tau/\alpha}}{H(\alpha)} \mathcal{F}(g)](\mu). \tag{4.32}$$

Proof: We substitute f from (4.31) into (4.30) to get

$$\begin{aligned} I(0, -\mu, \nu; \tau) &= \frac{\nu H(\nu) H(\mu)}{\mu + \nu} - e^{-A\tau}[g](\mu) \\ &+ (\frac{c}{2})^2 H(\mu) \langle \langle \frac{1}{\mu + \cdot}, \frac{\alpha}{\alpha + \cdot} \rangle_{\mathcal{H}}, e^{-\tau/\alpha} \mathcal{F}(g) \rangle_\Sigma. \end{aligned} \tag{4.33}$$

This last expression is simplified by use of a partial fraction decomposition and (2.12) to compute

$$(\frac{c}{2})^2 \langle \frac{1}{\mu + \cdot}, \frac{\alpha}{\alpha + \cdot} \rangle_{\mathcal{H}} = \frac{\alpha}{H(\alpha)} \frac{c}{2} \frac{H(\mu) - H(\alpha)}{\mu - \alpha}$$
$$= \frac{1}{H(\alpha)} [H(\alpha)\psi_\alpha(\mu) - H(\mu)\psi_\alpha(\mu)]. \tag{4.34}$$

We combine the last two equations to obtain (4.32) from (2.22) and

$$I(0, -\mu, \nu; \tau) = \frac{\nu H(\nu) H(\mu)}{\mu + \nu} - e^{-A\tau}[g](\mu) +$$
$$\mathcal{F}^*[e^{-\tau/\alpha}\mathcal{F}(g)](\mu) - H(\mu)\mathcal{F}^*[\frac{e^{-\tau/\alpha}}{H(\alpha)}\mathcal{F}(g)](\mu), \tag{4.35}$$

since the second and third terms of the right hand side cancel. □

Input radiation at $x = \tau$ is expressed from (4.27) as

$$I(\tau, -\mu; \tau) = \frac{c}{2} \langle \frac{\alpha e^{-\tau/\alpha}}{\mu + \alpha}, \mathcal{F}(f) \rangle_\Sigma - g(\mu). \tag{4.36}$$

The choice of $g \equiv 0$ and $f = \frac{2}{c}\delta(\mu - \nu)$ in this expression and in (4.32) gives the following.

COROLLARY 1 *The finite slab radiative transfer problem with boundary condition (4.31) at $x = 0$ and boundary condition*

$$I(\tau, -\mu, \nu; \tau) = \nu H(\nu) \mathcal{F}^*[\frac{\alpha e^{-\tau/\alpha}}{\mu + \alpha}](\nu) \tag{4.37}$$

at $x = \tau$ has reflection function independent of τ for $0 \leq \tau < \infty$

$$R(\tau, \mu, \nu) = \frac{\mu\nu H(\mu) H(\nu)}{\mu + \nu}. \tag{4.38}$$

The input radiation in (4.37) is expressed in terms of $h(\tau, \mu, \nu)$, where the function h is defined by

$$h(x, \mu, \nu) = \mathcal{F}^*[\frac{\alpha e^{-x/\alpha}}{\mu + \alpha}](\nu). \tag{4.39}$$

This input radiation, giving the reflection function in (4.38), vanishes as τ goes to infinity, in agreement with the transfer problem (2.15)-(2.19) for a half-space.

Other equilibrium solutions $S_{\alpha\beta}$ of (2.1) require choices of g in (4.31) and (4.36) that are exponentially increasing with τ. We equate $S_{\alpha\beta}$ in (4.3) to $R(\tau, \mu, \nu)$ in (4.32) to obtain an equation for g

$$\nu H(\nu)(\frac{2}{c})^2(\frac{1}{\alpha} + \frac{1}{\beta})\psi_\alpha(\mu)\psi_\beta(\nu) = -\mathcal{F}^*[\frac{e^{-\tau/\alpha}}{H(\alpha)}\mathcal{F}(g)](\mu) \tag{4.40}$$

with solution

$$g_{\alpha\beta}(\tau,\mu,\nu) = -\nu H(\nu)H(\alpha)(\frac{2}{c})^2(\frac{1}{\alpha}+\frac{1}{\beta})\psi_\alpha(\mu)\psi_\beta(\nu)e^{\tau/\alpha} \qquad (4.41)$$

which, as a function of μ, is in $\mathcal{H}$ when $\alpha = 1/k$ but is a distribution otherwise.

For these, and other, choices of g, input radiation at $x = \tau$ is given by (4.31), (4.36) and (4.39) as

$$\begin{aligned}
I(\tau,-\mu,\nu;\tau) = h(\tau,\mu,\nu) - g(\tau,\mu,\nu)+ \\
\langle\langle h(\tau,\mu,\cdot),h(\tau,\cdot,\rho)\rangle_\mathcal{H}, g(\tau,\rho,\nu)\rangle_\mathcal{H}.
\end{aligned} \qquad (4.42)$$

COROLLARY 2 *The finite slab radiative transfer problem with boundary conditions (4.31) at $x = 0$ and (4.42) with $g = g_{\alpha\beta}$ at $x = \tau$ has the reflection function $S_{\alpha\beta}(\mu,\nu)$ independent of τ, $0 \le \tau < \infty$. Input radiation at $x = \tau$ increases exponentially as $\exp(\alpha\tau)$. When $\alpha = 1/k$, these unstable solutions $S_{1/k,\beta}$ of equation (2.1) are physically reasonable.*

We support the last claim in this corollary by the following observations. With $g = g_{\alpha\beta}$ for $\alpha = 1/k$ and any β in sp, both boundary conditions (4.31) and (4.42) are distributions in the parameter ν. The finite slab radiative transfer problem with boundary conditions specified in terms of a Hölder continuous function $F(\nu)$ from (4.31) by

$$I(0,\mu;\tau) = \int_0^1 F(\nu)\delta(\mu-\nu)d\nu \qquad \text{for } 0 \le \mu \le 1, \qquad (4.43)$$

and from (4.42) for $g_{1/k,\beta}$ by

$$I(\tau,-\mu;\tau) = \int_0^1 F(\nu)I(\tau,-\mu,\nu;\tau)d\nu \qquad \text{for } 0 \le \mu \le 1 \qquad (4.44)$$

is, in principle, physically realizable since input radiation at the boundaries then depends continuously on input angle, and the reflection

$$\rho_{1/k,\beta}(\mu) = \mu H(\mu)\int_0^1 [\frac{1}{\mu+\nu}+\frac{2}{c}(k+\frac{1}{\beta})\frac{1}{1-k\mu}\psi_\beta(\nu)]F(\nu)dm(\nu) \qquad (4.45)$$

depends continuously on output angle.

Any other solution $R(\tau,\mu,\nu)$ to (2.1) can be used to determine $g(\tau,\mu,\nu)$ from (4.32) and then input radiation at $x = \tau$ from (4.42). The solutions suggested in (3.33) and computed in detail in [14] can, therefore, be generated by specifying input radiation at $x = \tau$ as a function of τ. In [14] I showed that these solutions could also be determined by models of anisotropic scattering which change with τ and have zero input radiation at $x = \tau$. These interpretations provide two models for the same observations, and have implications for the inverse problem of determining model parameters from observations of only the reflection function.

5 AN EQUIVALENT REPRESENTATION

We briefly discuss another representation of $T(\tau)$ in (3.26) as a solution of (3.6). This results from the unitary map $\mathcal{F}$ from $\mathcal{H}$ to Σ which diagonalizes the semigroup $\exp(-\mathcal{A}\tau)$.

For a give initial value T_0 an equivalent operator acting on the Hilbert space Σ is defined by

$$\hat{T}_0 = \mathcal{F}T_0\mathcal{F}^*. \tag{5.1}$$

We define a multiplication operator on Σ by unitary equivalence with the semigroup

$$\mathcal{E}(\tau)[F](\alpha) = \mathcal{F}e^{-\mathcal{A}\tau}\mathcal{F}^*[F](\alpha) = e^{-\tau/\alpha}F(\alpha), \tag{5.2}$$

and a self-adjoint integral operator $Q(\tau)$ on Σ with kernel

$$q(\tau, \alpha, \beta) = (\frac{c}{2})^2 \frac{\alpha\beta}{\alpha + \beta}[1 - e^{-\tau(\frac{1}{\alpha} + \frac{1}{\beta})}]. \tag{5.3}$$

THEOREM 8 *For τ sufficiently small, the solution to (3.6) and initial conditions T_0 is given by*

$$T(\tau) = \mathcal{F}^*\mathcal{E}(\tau)\hat{T}_0[I + Q(\tau)\hat{T}_0]^{-1}\mathcal{E}(\tau)\mathcal{F}. \tag{5.4}$$

Proof: The representation will follow from (3.26) written as

$$T(\tau) = \mathcal{F}^*\mathcal{E}(\tau)\hat{T}_0[I + \mathcal{F}\int_0^\tau e^{-\mathcal{A}s}Be^{-\mathcal{A}s}ds\mathcal{F}^*\hat{T}_0]^{-1}\mathcal{E}(\tau)\mathcal{F}. \tag{5.5}$$

A degenerate rank-one integral operator on $\mathcal{H}$ is given by (3.10) and (4.17b) as

$$\begin{aligned}
e^{-\mathcal{A}s}Be^{-\mathcal{A}s}f &= (\frac{c}{2})^2 e^{-\mathcal{A}s}\mathcal{F}^*[1]\langle\mathcal{F}^*[1], e^{-\mathcal{A}s}f\rangle_\mathcal{H} \\
&= (\frac{c}{2})^2 \mathcal{F}^*[e^{-s/\alpha}]\langle\mathcal{F}^*[e^{-s/\alpha}], f\rangle_\mathcal{H}.
\end{aligned} \tag{5.6}$$

It follows by (4.17b) that

$$\begin{aligned}
\mathcal{F}^*[e^{-s/\alpha}](\mu) &= \mathcal{F}^*[e^{-s/\alpha} - e^{-s/\mu}](\mu) + e^{-s/\mu}\mathcal{F}^*[1](\mu) \\
&= \frac{c}{2}\int_{sp}\frac{e^{-s/\alpha} - e^{-s/\mu}}{\alpha - \mu}\alpha dn(\alpha) + \frac{1}{\mu}e^{-s/\mu}
\end{aligned} \tag{5.7}$$

is in $\mathcal{H}$ (even $\mathcal{H}_0$) as a continuous positive function on $[0, 1]$ for $s > 0$. It follows that on the space Σ

$$\mathcal{F}e^{-\mathcal{A}s}Be^{-\mathcal{A}s}\mathcal{F}^*[F](\alpha) = (\frac{c}{2})^2\int_{sp}e^{-s(\frac{1}{\alpha} + \frac{1}{\beta})}F(\beta)dn(\beta), \tag{5.8}$$

and hence that

$$\int_0^\tau \mathcal{F}e^{-As}Be^{-As}\mathcal{F}^*ds = Q(\tau) \tag{5.9}$$

with kernel in (5.3). □

That $Q(\tau)$ is self-adjoint, positive semi-definite, follows from unitary equivalence with the operator shown to be self-adjoint and positive semidefinite in the proofs of Theorems 2 and 3. Compactness was not proved before, but it now follows from square integrability of the kernel q given in (5.3). We can also improve the norm estimate in (3.31) by use of (2.13), (4.11) and (4.24) to obtain

$$\begin{aligned}
\|\int_0^\tau e^{-As}Be^{-As}ds\| &= \|Q(\tau)\| = (\frac{c}{2})^2 \sup_F \int_0^\tau \langle e^{-s/\alpha}, F\rangle_\Sigma^2 ds/\langle F, F\rangle_\Sigma \\
&\le \frac{c^2}{8}\frac{1-e^{-2k\tau}}{k\Delta(k)} + \frac{c^2}{8}\int_0^1 \alpha dn(\alpha) \\
&= \frac{c^2}{8}\frac{1-e^{-2k\tau}}{k\Delta(k)} + \frac{\int \frac{\mu dm}{1-(k\mu)^2} - \frac{c}{2}\int\frac{dm}{1-k\mu}\int\frac{dm}{(1-k\mu)^2}}{2\int\frac{dm}{1-k\mu}\int\frac{dm}{(1-k\mu)^2}}.
\end{aligned} \tag{5.10}$$

It follows from (2.9), (2.12), (2.13) and (4.12) that the limit $c = 1$ and $k = 0$ in (5.10) gives

$$\begin{aligned}
\lim_{c\uparrow 1}\|\int_0^\tau e^{-As}Be^{-As}ds\| &= \lim_{c\uparrow 1}\|Q(\tau)\| \\
&\le \frac{\sqrt{3}\tau}{2} + \frac{\int\mu dm - \frac{1}{2}(\int dm)^2}{2(\int dm)^2}.
\end{aligned} \tag{5.11}$$

The Rayleigh quotient gives the lower bound

$$\begin{aligned}
\|\int_0^\tau e^{-As}Be^{-As}ds\| &\ge \int_0^\tau \langle e^{-As}Be^{-As}\psi_k, \psi_k\rangle_\mathcal{H}ds/\langle\psi_k, \psi_k\rangle_\mathcal{H} \\
&= \frac{c^2}{8}\frac{1-e^{-2k\tau}}{k\Delta(k)}
\end{aligned} \tag{5.12}$$

and the limit

$$\lim_{c\uparrow 1}\|\int_0^\tau e^{-As}Be^{-As}ds\| = \lim_{c\uparrow 1}\|Q(\tau)\| \ge \frac{\sqrt{3}\tau}{2}. \tag{5.13}$$

From Theorem 3 and unitary equivalence it follows that the operator inversion in (5.4) exists for all bounded, self-adjoint, positive semi-definite $\hat{T}_0$. In particular, the kernel of interest for the determination of the solution (3.47) to (2.1) and (2.2) is given by (4.17c) and (3.36) to be

$$\begin{aligned}
\hat{T}_1(\alpha, \beta) &= \int_0^1 \psi_\alpha(\mu)\frac{\beta dm(\mu)}{(\mu+\beta)H(\mu)} \\
&= \frac{\alpha\beta}{\alpha+\beta}[1 - \frac{c\alpha}{2}\ln\frac{1+\alpha}{\alpha} - \frac{c\beta}{2}\ln\frac{1+\beta}{\beta}].
\end{aligned} \tag{5.14}$$

This kernel is nonnegative on the spectrum of $\mathcal{A}^{-1}$. This fact can likely be exploited to determine a different set of conditions on $\hat{T}_0$ sufficient for the inverse to exist in (5.4) for all τ. We have not done such an analysis.

ACKNOWLEDGMENT: I wish to thank Professor C. T. Kelly for helpful conversations concerning this work and Professor E. Burniston for making them possible during a visit to the Mathematics Department at North Carolina State University.

REFERENCES

[1] R. BELLMAN AND G. M. WING, *An Introduction to Invariant Imbedding*, Wiley, New York, 1975.

[2] R. BELLMAN, R. KALABA, AND M. PRESTRUD, *Invariant Imbedding and Radiative Transfer in Slabs of Finite Thickness*, American, Elsevier, New York, 1963.

[3] I. W. BUSBRIDGE, *On solutions of Chandrasekhar's integral equation*, Trans. Amer. Math. Soc., 105 (1962), pp. 112-117.

[4] J. L. CARLSTEDT AND T. W. MULLIKIN, *Chandrasekhar's X and Y Functions*, Astrophys. J. Suppl XII, No 113 (1966), pp. 449-586.

[5] K. M. CASE AND P. F. ZWEIFEL, *Linear Transport Problems*, Addison Wesley, Reading, Mass., 1967.

[6] S. CHANDRASEKHAR, *Radiative Transfer*, Oxford Univ. Press, London, 1950.

[7] C. FOX, *A solution of Chandrasekhar's integral equation*, Trans. Amer. Math. Soc., 99 (1961), pp. 285-291.

[8] R. J. HANGELBROEK, *A functional analytic approach to the linear transport equation*, Transport Theory and Statis. Phys., 5 (1976), pp. 1-85.

[9] E. HOPF, *Mathematical Problems of Radiative Equilibrium*, Cambridge Tracts No. 31, Cambridge University Press, 1934.

[10] H. G. KAPER, C. G. LEKKERKERKER, AND J. HEJTMANEK, *Spectral Methods in Linear Transport Theory*, Birkhauser Verlag, Basel, 1982.

[11] E. W. LARSEN AND G. J. HABETLER, *A functional analytic derivation of Case's full and half-range formulas*, Comm. Pure and Appl. Math., 26 (1973), pp. 525-537.

[12] T. W. MULLIKIN, *A complete solution of the X and Y equations of Chandrasekhar*, Astrophys. J., 136 (1962), pp. 627-635.

[13] T. W. MULLIKIN, *Chandrasekhar's X and Y equations*, Trans. Amer. Math. Soc., 113 (1964), pp. 316-332.

[14] T. W. MULLIKIN, *A nonlinear integrodifferential equation in radiative transfer*, SIAM J. Appl. Math, 13 (1965), pp. 388-409.

[15] W. T. REID, *Riccati Differential Equations*, Academic Press, New York, 1972.

[16] V. V. SOBOLEV, *A Treatise on Radiative Transfer*, Van Nostrand, Princeton, 1963.

[17] R. TEMAM, *Sur l'equation de Riccati associee a des operateurs non bornes en dimension infinie*, J. Fcnl. Anal., 7 (1971), pp. 85-115.

[18] G. M. WING, *An Introduction to Transport Theory*, Wiley, New York, 1962.

A Discrete Model of Transport and Invariant Imbedding in Two or More Dimensions

JAMES CORONES Applied Mathematical Sciences, Ames Laboratory—USDOE, and Department of Mathematics, Iowa State University, Ames, Iowa

1 INTRODUCTION

This paper presents an attempt to generalize the ideas of invariant imbedding to problems involving two or more space dimensions. The one dimensional theory has been extensively developed, however there has been little effort in the literature to extend these ideas to problems that do not contain a single distinguished parameter: a parameter that allows the ordering of the regions in which the physical processes of interest take place. The absence of an ordering does not affect the basic approach of invariant imbedding, which can be thought of as a sort of perturbation theory that perturbs the structure of the system under study [1], it merely makes the analysis more complicated.

In this work a simple discrete model is investigated. Most of the conceptual problems associated with higher dimensional invariant imbedding are present in the model but the technical problems, since the model is finite dimensional, are vastly reduced from those that need to be solved in a full, continuous, model. The key results are the extension, or perhaps better application, of the Redheffer * - product [2, 3] to this model and the derivation of the simplest generalization of the Ambarzumian equations [4, 5]. The reader is referred to the original literature or to the very useful monograph [1] for a discussion of these now standard results. There are several sets of results that are similar in approach to the work present here [6, 7, 8] are some representative examples.

2 THE MODEL

Consider one type of particle swarming about in a source and sink free medium that can scatter these particles. Assume the system of particles and medium to be in a stationary state. A differential characterization of the medium is given by describing how the particles are scattered at a point. A macroscopic characterization is given by (mentally) delineating a region of the medium, A, (assumed to be connected and finite) and describing how the fluxes of particles traversing the boundary of A into the region are transformed into the fluxes of particles traversing the boundary of A out of the region A. This characterization depends on A, the medium with A, the orientation of the boundary of A and the ability to determine the fluxes of particles with and against this orientation.

The oriented boundary of A is thus used to establish which are the natural dependent and which are the natural independent variables of the problem, e.g. with respect to A. The object of primary physical interest is the operator that carries the incoming fluxes to the outgoing fluxes of A. This operator is called the <u>scattering operator</u> for the region.

If no approximations are made this operator acts on an infinite dimensional space, since even at one point on the boundary of A particles may enter from a continuum of directions. To provide a tractable model that presents all the conceptual problems of the full system assume that the boundary of A consists of a finite number of patches, called ports, $a_1,\ldots,a_n$. Let the net flux of particles entering A through the jth port be $i(a_j)$ and the net flux of particles leaving A at the jth port be $o(a_j)$. The totality of particles fluxing into A is given by $I(A) = (i(a_1),\ldots i(a_n))^T$, where T denotes transpose. This is called the <u>input vector</u> of A. The totality of particles fluxing out of A is given by $O(A) = (o(a_1),\ldots o(a_n))^T$. This is called the output vector of A. If the system is linear then the scattering operator for A is an n × n matrix $F(A)$. The scattering is explicitly given by

$$O(A) = F(A)I(A). \qquad (2.1)$$

If $F(A)$ is local in the sense that it depends only on properties of A and not of properties of the complement of A it is natural to suppose that if A and B are contiguous regions that there is a scattering operator that can be built from the constituents of $F(A)$ and $F(B)$ that completely determines the scattering operator of the join of A and B. This in fact can be done, as will now be shown.

3   COMBINATORICS

Consider two regions A and B such that A has n ports and B has m ports. Assume that A and B are contiguous and that they share p ports. Particles leaving A through one of these ports directly enter B and particles leaving B through one of these ports directly enter A.

If the elements of the n × n and m × m matrices $F(A)$ and $F(B)$ are

$$[F(A)]_{ik} = A_{ik} \qquad (3.1a)$$

$$[F(B)]_{ik} = B_{ik} \qquad (3.1b)$$

respectively then the scattering associated with A is

$$o(a_k) = \sum_{j=1}^{n} A_{kj} i(a_j) \qquad (3.2a)$$

and that of B

$$o(b_k) = \sum_{j=1}^{m} B_{kj} i(b_j). \qquad (3.2b)$$

Moreover since A and B are contiguous there are relations between some of the components of I(A), I(B), O(A) and O(B).

To state these relations number the ports of A and B such that particles leaving a_n enter b_1, those leaving a_{n-1} enter b_2 and so on up to those leaving a_{n-p} enter b_p. The reverse process is also assumed to hold so that

$$o(b_1) = i(a_n)$$

$$o(b_2) = i(a_{n-1})$$

$$\cdot$$
$$\cdot \qquad (3.3a)$$
$$\cdot$$

$$o(b_p) = i(a_{n-p})$$

and

$$o(a_n) = i(b_1)$$

$$o(a_{n-1}) = i(b_2)$$

$$\cdot$$
$$\cdot \qquad (3.3b)$$
$$\cdot$$

$$o(a_{n-p}) = i(b_p)$$

Now the scattering operator for the join of A and B (denote this by C) can be obtained from the full description (3.2) subject to the restrictions (3.3) by observing that there are n+m equations in (3.2) and 2p additional relations among the particle fluxes. These 2p components can be systematically eliminated from (3.2) by the following argument.

First write (3.2) as

$$
\begin{bmatrix} o(A) \\ o(B) \end{bmatrix} = \begin{bmatrix} F(A) & 0 \\ 0 & F(B) \end{bmatrix} \begin{bmatrix} I(A) \\ I(B) \end{bmatrix}. \tag{3.4}
$$

The region C has $m+m-2p$ ports, with input and output vectors $I(C)$ and $o(C)$. Let S denote the vector of shared components in some particular order. Now let P_0 be a permutation matrix such that

$$
P_0 \begin{bmatrix} I(A) \\ I(B) \end{bmatrix} = \begin{bmatrix} I(C) \\ S \end{bmatrix}. \tag{3.5}
$$

That is P_0 simply rearranges the components of $[I(A), I(B)]^T$ into first the input vector of C and then the remaining (shared) fluxes, e.g.,

$$
S = \begin{bmatrix} i(a_n) \\ i(b_1) \end{bmatrix} = \begin{bmatrix} o(b_1) \\ o(a_n) \end{bmatrix}.
$$

Further, let P_1 be a permutation matrix such that

$$
P_1 \begin{bmatrix} o(A) \\ o(B) \end{bmatrix} = \begin{bmatrix} o(C) \\ S \end{bmatrix}. \tag{3.6}
$$

From these definitions it readily follows that

$$
\begin{bmatrix} o(C) \\ S \end{bmatrix} = P_1 \begin{bmatrix} F(A) & 0 \\ 0 & F(B) \end{bmatrix} P_0^{-1} \begin{bmatrix} I(C) \\ S \end{bmatrix}. \tag{3.7}
$$

A bit of cumbersome notation (that has a transient role) is now needed. Let block components of $\hat{F}(C)$ be defined as

$$
\hat{F}(C) = P_1 \begin{bmatrix} F(A) & 0 \\ 0 & F(B) \end{bmatrix} P_0^{-1}
$$

$$
\tag{3.8}
$$

$$
= \begin{bmatrix} F(C,C) & F(C,S) \\ F(S,C) & F(S,S) \end{bmatrix}.
$$

In this notation (3.7) is

$$o(\underline{C}) = F(C,C)I(C) + F(C,S)S$$

$$S = F(S,C)I(C) + F(S,S)S.$$

(3.9)

The scattering matrix for C is that matrix which connects I(C) to O(C). This matrix can easily be found from (3.9) by eliminating S from the system. The result is

$$F(C) = F(C,C) + F(C,S)[1-F(S,S)]^{-1} F(S,C).$$

(3.10)

This is the generalization of the Redheffer * - product to the case when an n-port and an m-port have p shared ports. That is

$$F(A) * F(B) = F(C).$$

(3.11)

Consider the case when exactly one port is common, say the nth port of the n-port A and the 1st port of the m-port B. In this case the general formula (3.10) (or as easily in this example direct manipulation of the components of the linear system) yield the relations

$$o(a_k) = \sum_{q=1}^{n-1} \{A_q + A_{kn}B_{11}R_{n1}A_{nq}\}i(a_q)$$

$$+ \sum_{q=2}^{m} \{A_{kn}R_{1n}B_{1q}\}i(b_q)$$

(3.12a)

for k = 1,...n-1 and

$$o(b_k) = \sum_{q=1}^{n-1} \{B_{k1}R_{n1}A_{nq}\}i(a_q)$$

$$+ \sum_{q=2}^{m} \{B_{kq} + B_{k1}A_{nn}R_{1n}B_{1q}\}i(b_q)$$

(3.12b)

for k = 2,3,...m. The notation

$$R_{n1} = [1 - A_{nn}B_{11}]^{-1} \qquad (3.13a)$$

$$R_{1n} = [1 - B_{11}A_{nn}]^{-1} \qquad (3.13b)$$

has also been used.

This case is of interest for a derivation of the simplest generalization of the Ambarzumian equations, see [1]. This derivation will now be given.

4 DIFFERENTIAL EQUATIONS

In this section the simplest differential equation for F(A) with respect to one of its ports is derived. There are more complicated differential equations that can, and must, be found for F(A) if the full freedom of two or more dimensions is to be exploited. However the present case is important in itself and serves as a useful guide to the more intricate derivation.

The sense of the derivation is as follows. Consider a scattering region A and its scattering matrix F(A). Adjoin to this region another region B with scattering matrix F(B). Let $F_V(B)$ be the scattering matrix of B if all the scattering medium is removed i.e. $F_V(B)$ is the scattering matrix for the vacuum. Further, let β be a measure of the volume of B. The difference between F(A) * F(B) and F(A) * $F_V(B)$ measures the effect of the medium on the scattering. Differential changes are found if $\beta \rightarrow 0$ and it is the ratio of the influence of the medium to its size that provides the differential equation needed. Thus

$$\frac{dF(A)}{d\beta} \equiv \lim_{\beta \rightarrow 0} \left[\frac{F(A) * F(B) - F(A) * F_v(B)}{\beta} \right]. \qquad (4.1)$$

Numerous remarks are in order. It is enough for the moment to point out two major points of flexibility in this definition. First the definition of $F_V(B)$ need not be that of the true vacuum. The definition (4.1) is in some sense a covariant derivative with the standard of propagation provided by $F_V(B)$. Second the derivative should also be indexed by B and not merely β. In the discrete model treated here B may be a 2,3...m-port region independent of unique empty n-port scattering matrix. More precisely stating that B is an empty n-port region is not giving sufficient information to determine the scattering matrix of B. In the 2-port case there is a natural definition.

If B has two ports, 1 and 2, and is empty it is natural to say that F(B) takes the total incoming flux at 1 to 2 and the total incoming flux at 2 to 1. For 2-ports this is equivalent to saying that there are no reflections. It also says something about the geometry. For n-ports this is not the case. The incoming flux at one port may, depending on say the geometry of B, be distributed in many ways into outgoing fluxes. A full discussion of (4.1) can only be given when a complete elaboration of these points is in hand. For the moment only the unambiguous case will be treated. Thus consider the case when B is a 2-port and write

$$F(B) = \begin{bmatrix} b_{-+}\beta & 1+b_{--}\beta \\ 1+b_{++}\beta & b_{+-}\beta \end{bmatrix} \tag{4.2}$$

$$F_v(B) = \begin{bmatrix} 0 & 1 \\ 1 & 0 \end{bmatrix}. \tag{4.3}$$

In (4.2) it has been assumed that the measure of B is small so that (4.2) is correct to first order in β.

It is a tedious but straightforward computation to use (4.2) and (4.3) in (4.1). With the aid of (3.12) restricted to the case when m=2 it follows that

$$\frac{dA_{nn}}{d\beta} = b_{-+} + b_{--}A_{nn} + A_{nn}b_{++} + A_{nn}b_{+-}A_{nn} \tag{4.4a}$$

$$\frac{dA_{nk}}{d\beta} = (b_{--} + A_{nn}b_{+-})A_{nk} \tag{4.4b}$$

$$\frac{dA_{kn}}{d\beta} = A_{kn}(b_{++} + b_{+-}A_{nn}) \tag{4.4c}$$

$$\frac{dA_{jk}}{d\beta} = A_{qn}b_{+-}A_{nk}. \tag{4.4d}$$

These are the differential equations for the components of F(A) with respect to an infinitesimal 2-port variation of the nth-port of A. They are the generalization of Ambarzumian's equations.

If such a deformation is done at each port the total 2-port variation of F(B) is found. The general case of m-port variations will be treated in a subsequent paper.

ACKNOWLEDGMENT

I wish to thank Prof. Paul Nelson for bringing several references to my attention. This work was supported by the United States Department of Energy, Contract No. W-7405-Eng-82.

REFERENCES

1. R. Bellman and G. M. Wing, <u>An Introduction to Invariant Imbedding</u>, Wiley - Interscience, New York, 1975.
2. R. Redheffer, J. Math. Phys. <u>28</u>, 1950, 237-248.
3. R. Redheffer, J. Rat. Mech. Anal. <u>3</u>, 1954, 271-279.
4. V. A. Ambarzumian, C. R. Acad. Sci. SSR <u>38</u>, 1943, 229-235.
5. V. A. Ambarzumian, <u>Theoretical Astrophysics</u>, Pergamon, New York, 1958.
6. M. Ribariĉ, "Functional-Analytic Concepts and Structures of Neutron Transport Theory," Vols 1 and 2, Slovene Academy of Sciences and Arts, Ljubljana (1973).
7. Z. Weiss, "Some Basic Properties of the Response Matrix Equations," <u>Nucl. Sci. Engng.</u>, <u>63</u>, 457-492 (1977).
8. N. Nakata and W. R. Martin, "The Finite Element Response Matrix Method," <u>Nucl. Sci. Engng.</u>, <u>85</u>, 289-305 (1983).

Invariant Imbedding in Two Dimensions

V. FABER, DANIEL L. SETH, and G. MILTON WING   Computer Research and Applications, Los Alamos National Laboratory, Los Alamos, New Mexico

INTRODUCTION

J. Corones [1] has noted that the doubling and addition formulas of invariant imbedding can be extended conceptually to very general situations. All that is needed is a black box "process" with n "ports." The ith port has vector input I_i and vector output J_i. (In Fig. 1, such a process is represented as an n-gon, but this is only a visual device.)

Addition formulas result when two or more of these processes are joined together to form a new process in some regular way. For example, four congruent squares can be juxtaposed to form a larger square. (This program is carried out in some detail in Sec. 2.) At each join, the output of one process becomes the input of the other and vice versa. (We always suppose the join to occur at one or more ports.) Addition formulas result from the combination of these shared quantities. Corones has thus pointed out that invariant imbedding is not, as is sometimes asserted, an inherently one-dimensional (1-D) method, but works conceptually in any number of dimensions; some previous work that is conceptually along these lines, with references to other such works, can be found in Refs. 2-4. The details can, of course, become very complicated. We shall show that the method is computationally feasible for certain two-dimensional (2-D) problems. To conform to the thrust of these proceedings, we shall usually phrase our discussions in terms of transport theory rather than speaking of more abstract processes.

This work was performed under the auspices of the U.S. Department of Energy.

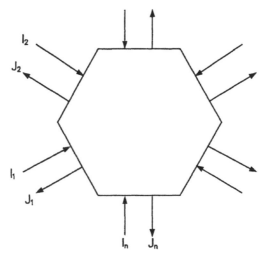

Figure 1. N - port process.

1 INVARIANT IMBEDDING IN 1-D

1.1 The Addition Formulas

Consider an infinite slab of material of thickness z. Linear transport of particles occurs in the slab in a manner independent of the x and y coordinates. We suppose no internal particle sources. The left face of the slab may be considered a "port," as may the right side. The input of particles at the left face is described by a vector I_1, which may have as components energies, directions, etc. These are assumed to be discrete quantities. The vector I_1 will not be allowed to depend on the x and y coordinates. (We are considering the entire left face to be a port). For instance, I_1 might be the vector whose components give the number of particles per unit area entering the left face at six different angles. Similarly, I_2, J_1, and J_2 are input and output vectors as shown in Fig. 2. (Note that the indexing does not agree with Fig. 1.)

The components of the input and output vectors are sometimes referred to as <u>states</u> (see [5]). Although energies and directions are often the physically interesting states, such descriptors as color, size, and generation number are quite admissible.* Observe that the symbol I is used in Fig. 2 for "input," with J for "output." Also, we do not mention anything about the particles inside the slab, save to emphasize that all processes are linear.

We associate with this process a <u>transition matrix</u>

$$A = A(z) \ , \tag{1.1}$$

which has the property that its entries (α,β) give the probability that an input particle in state α will lead to an output particle in state β. Thus

---

*The number of states can become very large easily. Suppose input particles can move in any of eight directions at any of 12 energies. There are then 96 possible states, and the input (and output) vectors will have 96 components.

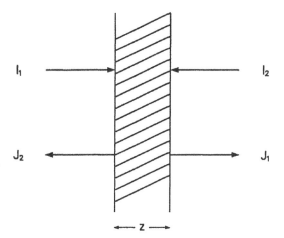

Figure 2. Slab of width z.

$$\begin{bmatrix} J_1 \\ J_2 \end{bmatrix} = A \begin{bmatrix} I_1 \\ I_2 \end{bmatrix} . \tag{1.2}$$

Because I and J are vectors, A is not a simple two-by-two matrix. It is customarily written in block form:

$$A = \begin{bmatrix} T_+ & R_+ \\ R_- & T_- \end{bmatrix} . \tag{1.3}$$

To understand the meaning of the T and R matrices, observe that (1.2) gives

$$J_1 = T_+ I_1 + R_+ I_2 . \tag{1.4}$$

If $I_2 = 0$ this yields

$$J_1 = T_+ I_1 . \tag{1.5a}$$

Thus T_+ provides the information about particles <u>transmitted</u> through the slab <u>to the right</u> when the only input is at the left face.

Again, suppose $I_1 = 0$. Then

$$J_1 = R_+ I_2 . \tag{1.5b}$$

Hence R_+ provides information about particles <u>reflected</u> from the slab <u>to the right</u> when the only input is at the right face; T_- and R_- may similarly be understood.

In many cases of interest the material is isotropic – it does not know its right from its left.* Then

$$R = R_+ = R_- , \quad T = T_+ = T_- . \tag{1.6}$$

---

*Note that the word <u>isotropic</u> here applies to the material's "sense of direction." It has nothing to do with the details of scattering processes.

In some of what follows we assume (1.6) holds. Suppose we now have two slabs of thicknesses z_1 and z_2. Assume that we know the transition matrices $A(z_1)$ and $A(z_2)$. Let us juxtapose the two slabs forming one of thickness $z_1 + z_2$. Can we find the resulting transition matrices, which we denote by $A(z_1 + z_2)$?

This problem was first addressed by R. Redheffer (see [6]) and the bibliography therein) but in a somewhat different context. We shall sketch the fundamental derivation.

Refer to Fig. 3. Obviously we wish, for example, to consider the output vector J_1 from the slab of thickness z_1 as an input vector K_1 for the slab of thickness z_2. This requires a certain compatability amongst the input and output vectors. The details are best left to the reader. Also, we must require that when we form the slabs, J_1 does indeed become K_1, etc. No particles are created or absorbed. This is a fundamental continuity assumption physically appropriate to transport phenomena (and many other processes).

For notational convenience, write

$$A(z_i) = \begin{bmatrix} T_+(i) \, R_+(i) \\ R_-(i) \, T_-(i) \end{bmatrix} , \quad i = 1,2 . \tag{1.7}$$

We seek the matrix

$$A(z_1 + z_2) = C = \begin{bmatrix} C_{11} \, C_{12} \\ C_{21} \, C_{22} \end{bmatrix} \tag{1.8}$$

such that

$$\begin{bmatrix} J_2 \\ L_1 \end{bmatrix} = C \begin{bmatrix} I_1 \\ K_2 \end{bmatrix} . \tag{1.9}$$

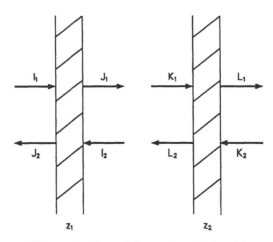

Figure 3. Two slabs to be "joined."

From Fig. 3 we have (see (1.2) and (1.3))

$$J_1 = T_+(1)I_1 + R_+(1)I_2$$
$$J_2 = R_-(1)I_1 + T_-(1)I_2$$
$$L_1 = T_+(2)K_1 + R_+(2)K_2$$
$$L_2 = R_-(2)K_1 + T_-(2)K_2 \ .$$

(1.10)

The continuity requirement implies

$$J_1 = K_1 \ , \quad I_2 = L_2 \ .$$

(1.11)

Eliminating J_1, I_2, K_1, L_2 from (1.10) and (1.11) finally gives (observe that the unsubscripted I denotes the identity)

$$J_2 = T_-(1) \left[R_-(2)(I - R_+(1)R_-(2))^{-1}R_+(1)T_-(2) \right] K_2$$
$$+ \left[T_-(1)R_-(2)(I - R_+(1)R_-(2))^{-1}T_+(1) + R_-(1) \right] I_1$$
$$L_1 = \left[T_+(2)(I - R_+(1)R_-(2))^{-1}R_+(1)T_-(2) \right.$$
$$\left. + R_+(2) \right] K_2 + \left[T_+(2)(I - R_+(1)R_-(2))^{-1}T_+(1) \right] I_1 \ .$$

(1.12)

We now easily identify the elements of C (Eq. (1.8)), which we write as $R_\pm(z_1 + z_2)$ and $T_\pm(z_1 + z_2)$. Reverting to the R and T notation yields

$$C_{11} = T_+(z_1 + z_2) = T_+(z_2)(I - R_-(z_1)R_+(z_2))^{-1}T_+(z_1)$$
$$C_{12} = R_+(z_1 + z_2) = T_+(z_2)(I - R_-(z_1)R_+(z_2))^{-1}R_-(z_1)T_-(z_2) + R_-(z_2)$$
$$C_{21} = R_-(z_1 + z_2) = T_-(z_1)R_+(z_2)(I - R_-(z_1)R_+(z_2))^{-1}T_+(z_1) + R_+(z_1)$$
$$C_{22} = T_-(z_1 + z_2) = T_-(z_1)[R_+(z_2)(I - R_-(z_1)R_+(z_2))^{-1}R_-(z_1)T_-(z_2) + T_-(z_2)] \ .$$

(1.13)

Equation (1.13) is often written in the form

$$A(z_1 + z_2) = A(z_1) * A(z_2) \ ,$$

(1.14)

where the "*product" is obtained by forming the elements of $A(z_1 + z_2)$ by use of (1.13). Clearly, the inverse of the particular matrix $(I - R_-(z_1)R_+(z_2))$ must exist. In the transport theory of neutrons, the existence or non-existence of this inverse is closely connected with the problem of physical criticality of the system.

1.2 The Doubling Formulas

Finally, if the medium is isotropic (see Eq. (1.6)) and if $z_1 = z_2 = z$, we find

$$T(2z) = T(z)(I - R^2(z))^{-1}T(z)$$
$$R(2z) = T(z)(I - R^2(z))^{-1}R(z)T(z) + R(z) \ .$$

(1.15)

These so-called "doubling" formulas obviously make it possible to calculate the R and T functions for "thick" slabs starting only with knowledge of a single "thin" slab. Equations (1.15) were apparently first noted by the astrophysicist van der Hulst [7].

These formulas were very effectively used by G. Hughes et al. [5] who made extensive calculations involving direction dependent neutron transport. The continuous angle dependence was first discretized by careful "binning," a process conceptually easy but nontrivial in practice. As many as 256 bins were considered. The initial slab was taken sufficiently thin to make it possible to employ the single scattering transport approximation. Excellent results were obtained.

1.3 The *-Product

The properties of the *-product invite investigation, and a considerable amount of work has been done. The operation is associative and has the identity

$$Iden = \begin{bmatrix} I & 0 \\ 0 & I \end{bmatrix} . \tag{1.16}$$

Under certain conditions there is an inverse. To pursue the properties of the *-product in detail would take us far afield.

2 INVARIANT IMBEDDING IN TWO DIMENSIONS

2.1 Discussion of the Problem

We have implied that invariant imbedding in more than 1-D is conceptually possible, but gives promise of being quite complicated. In this section we shall face some of those complications. Transport in 2-D appears to be a somewhat unrealistic process. To see that it is actually meaningful we simply consider an infinitely long cylinder and confine our investigations to a cross section of the cylinder. (This is the analogue of what was done with the infinite slab in Sec. 1). To keep matters as simple as possible, we consider that cross section to be a square, or, at worst, a rectangle. Later, we shall note that more complicated regions, which can be approximated by unions of squares, can be handled.

Continuing our emphasis on particle transport, we intuitively see at once that particle direction is going to be the most difficult matter to handle. Energy, particle size or kind, or other types of states usually offer much less complication. We shall therefore consider particle directions as the only states.

Consider the case of a square. A particle impinging on one side of the square can result in outputs on the adjacent sides as well as on the opposite side. Clearly, more than the simple R and T functions are needed. As another complication, we note that location of the point of entry on the square side is probably going to play a role, a difficulty avoided in the slab geometry by our assumption that there was no (x,y) dependence.

From the 1-D analysis, we see that it is probably reasonable to start with a very small fundamental square in the hope of getting, in some way, results for this "element." The addition and doubling formulas (Eqs. (1.13), (1.15)) suggest that we may then be able to "add" (juxtapose) these elements to obtain "thin" rectangles and then "add" these rectangles to eventually "double" the original element. This is the essence of the program we shall describe.

2.2 Notation and Operators

We shall not at this point describe how to obtain results for the fundamental square element Σ. That matter will be taken up in Sec. 3. We suppose all that information is available. For convenience we assume the square Σ is isotropic. It does not distinguish its top from its sides or bottom. More accurately, it is insensitive to 90° rotations.

Figure 4 schematically shows Σ with input vectors denoted by u and outputs by v. When four Σ elements are joined–hence a doubling–the picture looks like Fig. 5. Here the heavy arrows labeled V denote overall outputs resulting from the inputs u to each of the four fundamental squares Σ_i. Figure 6 indicates the situation after five doublings. In none of these diagrams is any quantitative relationship among the u's and V's implied.

Let us now introduce the necessary reflection and transmission operators. Figure 7 indicates an input vector u on the left side of the square Σ. This results in <u>four</u> output vectors,

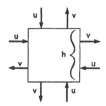

Figure 4. Fundamental square of width h.

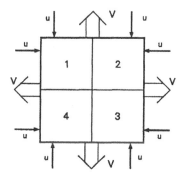

Figure 5. Four fundamental squares "joined"
together resulting in a 2hx2h square
with known flux values.

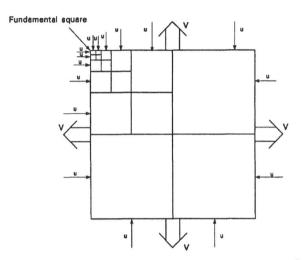

Figure 6. Result after five doublings (32h×32h, 2^5h×2^5h).

one from the left face, one from the right, and one from each of the two sides. The last are new. There are no analogues in 1-D. We denote the required operators by R, T, S, $\tilde{S}$. If we call the output vectors V_E, V_W, V_S, V_N, as shown in Fig. 8 (note that the indexing corresponds to the customary way of labeling wind direction), and similarly write $u = u_W$, we have (assuming momentarily that $U_S = U_E = U_N = 0$)

$$V_S = SU_W$$
$$V_W = TU_W$$
$$W_N = \tilde{S}U_W \qquad\qquad (2.1)$$
$$V_E = RU_W \ .$$

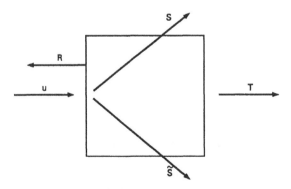

Figure 7. Possible exit vector operators R, S, $\tilde{S}$, and T associated with an incident vector u.

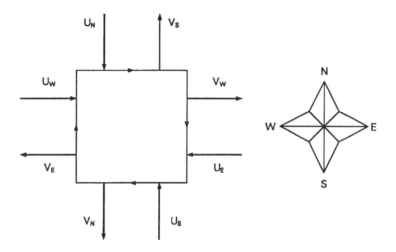

Figure 8. Incident and exit vectors U_N, U_E, U_S, U_W and
V_N, V_E, V_S, V_W, respectively.

Now recall the fact that Σ is isotropic. We see at once that

$$\tilde{S} = S^* \ , \tag{2.2}$$

where * indicates adjoint (transpose in this real case). Also we see that if the input is only from the south, the same operators (permuted) apply. Thus, finally,

$$\begin{bmatrix} V_N \\ V_E \\ V_S \\ V_W \end{bmatrix} = \begin{bmatrix} T & S & R & S^* \\ S^* & T & S & R \\ R & S^* & T & S \\ S & R & S^* & T \end{bmatrix} \begin{bmatrix} U_N \\ U_E \\ U_S \\ U_W \end{bmatrix} \ . \tag{2.3}$$

The four-by-four matrix operator (recall the elements themselves are matrices) is the precise analogue of the transition matrix A of Sec. 1.1.

2.3 Addition

We now wish to adjoin two of the fundamental squares Σ. We do so as shown in Fig. 9. Using the continuity assumption introduced in Sec. 1.1 we see

$$\begin{aligned} V_{W1} &= U_{W2} \\ U_{E1} &= V_{E2} \ . \end{aligned} \tag{2.4}$$

The R, T, S, S^* are the same for the individual squares Σ_1 and Σ_2. We now need to find the transition matrix $A_{\mathcal{R}}$ relating the input vector

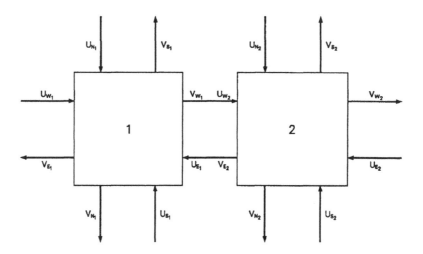

Figure 9. "Joining" two squares.

$$U = \begin{bmatrix} U_{N1} \\ U_{N2} \\ U_{E2} \\ U_{S2} \\ U_{S1} \\ U_{W1} \end{bmatrix} \tag{2.5}$$

to the output vector

$$V = \begin{bmatrix} V_{N1} \\ V_{N2} \\ V_{E2} \\ V_{S2} \\ V_{S1} \\ V_{W1} \end{bmatrix} . \tag{2.6}$$

Obviously, Eq. (2.3) yields four equations for each square, a total of eight. Two of these may be eliminated by use of (2.4). After a very considerable amount of algebra we find

$$V = A_{\mathcal{R}} \mathcal{U} , \tag{2.7}$$

where

$$
A_R = \left(
\begin{array}{cc:cc:cc}
T+S(I-R^2)^{-1}RS & S(I-R^2)^{-1}S^* & S(I-R^2)^{-1}T & S(I-R^2)^{-1}S & R+S(I-R^2)^{-1}RS^* & S^*+S(I-R^2)^{-1}RT \\
S^*(I-R^2)^{-1}S & T+S^*(I-R^2)^{-1}RS^* & S+S^*(I-R^2)^{-1}RT & R+S^*(I-R^2)^{-1}RS & S^*(I-R^2)^{-1}S^* & S^*(I-R^2)^{-1}T \\[2pt]
\hdashline\\[-8pt]
S^*+T(I-R^2)^{-1}RS & T(I-R^2)^{-1}S^* & T(I-R^2)^{-1}T & T(I-R^2)^{-1}S & S+T(I-R^2)^{-1}RS^* & R+T(I-R^2)^{-1}RT \\
S(I-R^2)^{-1}S & R+S(I-R^2)^{-1}RS^* & S^*+S(I-R^2)^{-1}RT & T+S(I-R^2)^{-1}RS & S(I-R^2)^{-1}S^* & S(I-R^2)^{-1}T \\[2pt]
\hdashline\\[-8pt]
R+S^*(I-R^2)^{-1}RS & S^*(I-R^2)^{-1}S^* & S^*(I-R^2)^{-1}T & S^*(I-R^2)^{-1}S & T+S^*(I-R^2)^{-1}RS^* & S+S^*(I-R^2)^{-1}RT \\
T(I-R^2)^{-1}S & S+T(I-R^2)^{-1}RS^* & R+T(I-R^2)^{-1}RT & S^*+T(I-R^2)^{-1}RS & T(I-R^2)^{-1}S^* & T(I-R^2)^{-1}T
\end{array}
\right)
\qquad (2.8)
$$

Clearly, $A_\mathcal{R}$ is an analogue of $A(z_1 + z_2)$. Rewrite (2.8)

$$A_\mathcal{R} = \begin{bmatrix} A_{11} & A_{12} & A_{13} & A_{14} \\ A_{21} & A_{22} & A_{23} & A_{24} \\ A_{31} & A_{32} & A_{33} & A_{34} \\ A_{41} & A_{42} & A_{43} & A_{44} \end{bmatrix} , \tag{2.9}$$

where the A_{ij} are the block matrices indicated by the dotted lines in (2.8). Also set

$$\tilde{U}_N = \begin{bmatrix} U_{N1} \\ U_{N2} \end{bmatrix} , \quad \tilde{U}_E = (U_{E2}) , \quad \tilde{U}_S = \begin{bmatrix} U_{S2} \\ U_{S1} \end{bmatrix} , \quad \tilde{U}_W = (U_{W1})$$

$$\tilde{V}_N = \begin{bmatrix} V_{N1} \\ V_{N2} \end{bmatrix} , \quad \tilde{V}_E = (V_{E1}) , \quad \tilde{V}_S = \begin{bmatrix} V_{S2} \\ V_{S1} \end{bmatrix} , \quad \tilde{V}_W = (V_{W2}) . \tag{2.10}$$

Equation (2.7) can be rewritten

$$A_\mathcal{R}\tilde{U} = \tilde{V} . \tag{2.11}$$

For example,

$$\tilde{V}_N = A_{11}\tilde{U}_N + A_{12}\tilde{U}_E + A_{13}\tilde{U}_S + A_{14}\tilde{U}_W . \tag{2.12}$$

From this equation we see that A_{11} is a T-like matrix, A_{12} is S-like, A_{13} is R-like, and A_{14} is again S-like. Making this identification for the other A_{ij} and noting that some certain blocks in $A_\mathcal{R}$ occur more than once, we get

$$A_\mathcal{R} = \begin{bmatrix} T_1 & S_1 & R_1 & S_2 \\ \hat{S}_1 & T_2 & \hat{S}_2 & R_2 \\ R_1 & S_2 & T_1 & S_1 \\ \hat{S}_2 & R_2 & \hat{S}_1 & T_2 \end{bmatrix} . \tag{2.13}$$

We now have results for a rectangle $\mathcal{R}$. The operator can be repeated to obtain $A_{2\mathcal{R}}$, the transition matrix for a rectangle twice as long, but of the same width. Instead we wish to "build" a square 2Σ, the double of Σ.

2.4 Doubling

We double the original square element by juxtaposing two rectangles, as indicated in Fig. 10. The reasoning is straightforward but the algebra is messy. We simply write the final result.

$$A_{2\Sigma} = \begin{pmatrix}
T_1(I-R_1^2)^{-1}T_1 & T_1(I-R_1^2)^{-1}S_1 & T_1(I-R_1^2)^{-1}S_2 \\[4pt]
\bar{S}_1+\bar{S}_2(I-R_1^2)^{-1}R_1T_1 & T_2+\bar{S}_2(I-R_1^2)^{-1}R_1S_1 & R_2+\bar{S}_2(I-R_1^2)^{-1}R_1S_2 \\[4pt]
R_1+T_1(I-R_1^2)^{-1}R_1T_1 & S_2+T_1(I-R_1^2)^{-1}R_1S_1 & S_1+T_1(I-R_1^2)^{-1}R_1S_2 \\[4pt]
\bar{S}_2(I-R_1^2)^{-1}T_1 & \bar{S}_2(I-R_1^2)^{-1}S_1 & \bar{S}_2(I-R_1^2)^{-1}S_2 \\[4pt]
\bar{S}_2+\bar{S}_1(I-R_1^2)^{-1}R_1T_1 & R_2+\bar{S}_1(I-R_1^2)^{-1}R_1S_1 & T_2+\bar{S}_1(I-R_1^2)^{-1}R_1S_2 \\[4pt]
\bar{S}_1(I-R_1^2)^{-1}T_1 & \bar{S}_1(I-R_1^2)^{-1}S_1 & \bar{S}_1(I-R_1^2)^{-1}S_2 \\
\end{pmatrix} \qquad (2.14)$$

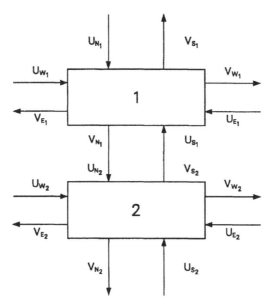

Figure 10. Joining two rectangles.

In Eq. (2.14)

$$A_{2\Sigma}\begin{bmatrix}U_{N1}\\U_{E1}\\U_{E2}\\U_{S2}\\U_{W1}\\U_{W1}\end{bmatrix}=\begin{bmatrix}V_{W2}\\V_{E1}\\V_{E2}\\V_{S1}\\V_{W2}\\V_{W1}\end{bmatrix}\,,\tag{2.15}$$

or again in block notation

$$A_{2\Sigma}\begin{bmatrix}\tilde{U}_N\\\tilde{U}_E\\\tilde{U}_S\\\tilde{U}_W\end{bmatrix}=\begin{bmatrix}\tilde{V}_W\\\tilde{V}_E\\\tilde{V}_S\\\tilde{V}_W\end{bmatrix}\,,\tag{2.16}$$

where the U's and V's with tildes take their meanings from (2.15) and are not the same as vectors defined by (2.10). The operator $A_{2\Sigma}$ may be written

$$A_{2\Sigma}=\begin{bmatrix}T_1&S_1&R_1&S_2\\\hat{S}_1&T_2&\hat{S}_2&R_2\\R_1&S_2&T_1&S_1\\\hat{S}_2&R_2&\hat{S}_1&T_2\end{bmatrix}\,.\tag{2.17}$$

The elements of $A_{2\Sigma}$ take their meanings from (2.14) and are not to be confused with those in (2.13).

We may now treat 2Σ as the fundamental square, and, by iterating, find $A_{2^n\Sigma}$. Clearly, the matrices (and the input and output vectors) increase in order at each step. We shall say more about this and its consequences shortly.

3 MORE ABOUT THE FUNDAMENTAL ELEMENT S

3.1 General Comments

In Sec. 2, we have observed that to start the doubling process, it is necessary to begin with a fundamental square Σ for which the transition matrix is completely known. It was implied that Σ would probably be "small." This is not really necessary if, by one device or another, we have all of the needed R, T, and S information. However, in Sec. 1, we observed that for the calculations made in [5], a thin slab initiated the process, and for it the single scattering approximation sufficed. This suggests the probable usefulness of a small Σ, but the approximation is much less obvious on simple physical grounds. In Sec. 3.2, we shall outline one way of getting started. It is by no means the only way.

3.2 Details for a "Small" Σ Calculation

We begin with a specific transport problem in mind. This will provide enough information so that generalizations are fairly obvious, albeit perhaps unpleasant. The fundamental square Σ, of side h, is oriented with respect to x and y axes as shown in Fig. 11. Scattering is supposed isotropic so that the classical transport equation is, assuming macroscopic cross section unity,

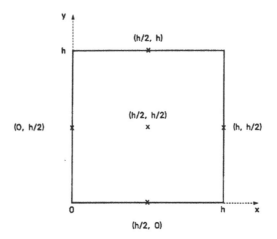

Figure 11. Stencil for difference scheme applied.

$$\mu \, \frac{\partial \psi}{\partial x} \, (x,y\mu,\eta) + \eta \, \frac{\partial \psi}{\partial y} + \psi(x,y,\mu,\eta)$$

$$= \frac{c}{2} \int_{-1}^{1} \psi(x,y,\mu',\eta)d\mu' \, , \quad c = constant \, . \tag{3.1}$$

Here

$$\mu = \cos \theta \, , \quad \eta = \sin \theta \, , \tag{3.2}$$

where θ is the angle the particle direction makes with respect to the x-axis. In this example we discretize θ so that

$$\theta_j = (2(j-1)+1)\frac{\pi}{4} \, , \quad j = 1,2,3,4 \, .$$

(In the terminology of nuclear engineering, this is an S_2 approximation. See [8].)

Now we consider Σ as the fundamental cell in a finite difference scheme using a five-point stencil as indicated in Fig. 11. The discretized version of (3.1) becomes

$$\mu_j(\psi^j(h,\frac{h}{2}) - \psi^j(0,\frac{h}{2}) + \eta_j\psi^j(\frac{h}{2},h)$$

$$- \psi^j(\frac{h}{2},0)) + \psi^j(\frac{h}{2},\frac{h}{2}) \tag{3.3}$$

$$= \frac{c}{2} \sum_{k=1}^{4} \psi^k(\frac{h}{2},\frac{h}{2}) \, , \quad j = 1,2,3,4 \, ;$$

the simplest quadrature formula approximates the integral in (3.1).

To determine the known quantities and the unknowns in (3.3), we examine Fig. 12. Clearly some of the ψ's are inputs (known) and some are outputs (unknowns). The function values $\psi^j(\frac{h}{2},\frac{h}{2})$ appear to be uncalculable. To overcome this we use a kind of central difference scheme often referred to by nuclear experts as the "diamond difference." Numerical analysts refer to it as the Crank-Nicholson method. We set

$$\psi^j(\frac{h}{2},\frac{h}{2}) = \frac{1}{2}(\psi^j(0,\frac{h}{2}) + \psi^j(h,\frac{h}{2})) \tag{3.4a}$$

and

$$\psi^j(\frac{h}{2},\frac{h}{2}) = \frac{1}{2}(\psi^j(\frac{h}{2},0) + \psi^j(\frac{h}{2},h)) \, . \tag{3.4b}$$

In the solution of the transport equation, this approximation is known to be of accuracy $O(h^2)$.

We may now write (3.3) in a somewhat different form. For specificity we select $\theta = \theta_1$. After some algebra we find, using the equations (3.4),

$$\frac{\sqrt{2}}{h} \left(\psi^1(h, \frac{h}{2}) - \psi^1(0, \frac{h}{2}) \right)$$

$$+ \frac{1}{2} \left(\psi^1(\frac{h}{2}, h) + \psi^1(0, \frac{h}{2}) \right) \tag{3.5a}$$

$$= \frac{c}{4} \sum_{k=1}^{4} \left(\psi^K(0, \frac{h}{2}) + \psi^k(\frac{h}{2}, \frac{h}{2}) \right)$$

and also

$$\frac{\sqrt{2}}{h} \left(\psi^1(h, \frac{h}{2}) - \psi^1(\frac{h}{2}, 0) \right)$$

$$+ \frac{1}{2} \left(\psi^1(h, \frac{h}{2}) + \psi^1(0, \frac{h}{2}) \right) \tag{3.5b}$$

$$= \frac{c}{4} \sum_{k=1}^{4} \left(\psi^k(0, \frac{h}{2}) + \psi^k(h, \frac{h}{2}) \right) .$$

Clearly, six further equations can be obtained by choosing $\theta = \theta_2, \theta_3, \theta_4$.

We have eight equations and sixteen ψ's. However, a glance at Figs. 12a and b reveal that eight of the ψ's are known inputs. It is convenient to write (3.5) and its analogues in the form

$$MS = NS , \tag{3.6}$$

where M and N are eight-by-eight matrices, and the u and v are input and output vectors, respectively. Specifically, we have

$$V = \begin{bmatrix} \psi^2(h, \frac{h}{2}) \\ \psi^1(\frac{h}{2}, h) \\ \psi^1(h, \frac{h}{2}) \\ \psi^4(h, \frac{h}{2}) \\ \psi^4(\frac{h}{2}, 0) \\ \psi^3(\frac{h}{2}, 0) \\ \psi^3(0, \frac{h}{2}) \\ \psi^2(0, \frac{h}{2}) \end{bmatrix} \tag{3.7a}$$

$$
U = \begin{bmatrix} \psi^2(\dfrac{h}{2},0) \\[6pt] \psi^1(\dfrac{h}{2},0) \\[6pt] \psi^1(0,\dfrac{h}{2}) \\[6pt] \psi^4(0,\dfrac{h}{2}) \\[6pt] \psi^4(\dfrac{h}{2},h) \\[6pt] \psi^3(\dfrac{h}{2},h) \\[6pt] \psi^3(h,\dfrac{h}{2}) \\[6pt] \psi^2(h,\dfrac{h}{2}) \end{bmatrix} . \qquad\qquad (3.7b)
$$

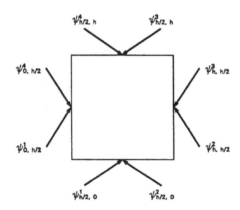

Figure 12a. Incident flux.

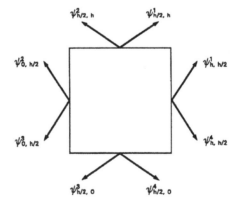

Figure 12b. Exit flux.

Also,

$$
M = \frac{1}{h}
\begin{bmatrix}
\sqrt{2}\,\hat{I} & 0 & 0 & 0 \\
0 & \sqrt{2}\,\hat{I} & 0 & 0 \\
0 & 0 & \sqrt{2}\,\hat{I} & 0 \\
0 & 0 & 0 & \sqrt{2}\,\hat{I}
\end{bmatrix}
$$

$$
+
\begin{bmatrix}
\beta & 0 & \gamma & 0 \\
0 & \beta & 0 & \gamma \\
\gamma & 0 & \beta & 0 \\
0 & \gamma & 0 & \beta
\end{bmatrix} ,
\tag{3.8}
$$

where the entries are two-by-two block matrices:

$$
\hat{I} =
\begin{bmatrix}
1 & 0 \\
0 & 1
\end{bmatrix}
\tag{3.9a}
$$

$$
\beta = \frac{1}{2}\hat{I} + \gamma
\tag{3.9b}
$$

$$
\gamma = -\frac{c}{4}
\begin{bmatrix}
1 & 1 \\
1 & 1
\end{bmatrix} .
\tag{3.9c}
$$

(The somewhat strange choice of the components of V has been made so that M is diagonally dominant for small h). Also

$$
N =
\begin{bmatrix}
-\beta & \dfrac{\delta}{h} & -\gamma & \dfrac{\delta*}{h} \\[2mm]
\dfrac{\delta*}{h} & -\beta & \delta & -\gamma \\[2mm]
-\gamma & \dfrac{\delta*}{h} & -\beta & \dfrac{\delta}{h} \\[2mm]
\dfrac{\delta}{h} & -\gamma & \dfrac{\delta*}{h} & -\beta
\end{bmatrix} ,
\tag{3.10}
$$

where

$$
\delta =
\begin{bmatrix}
0 & 0 \\
\sqrt{2} & 0
\end{bmatrix} .
\tag{3.11}
$$

To find V we must determine $M^{-1}N$. Write using (3.8)

$$
hM = \sqrt{2}\,I + hP .
\tag{3.12}
$$

Thus

$$
\frac{M^{-1}}{h} = \frac{1}{\sqrt{2}}\,(I + \frac{h}{\sqrt{2}}\,P)^{-1}
$$

$$
= \frac{1}{\sqrt{2}}\left[I - \frac{h}{\sqrt{2}}\,P\right] + O(h^2) .
\tag{3.13}
$$

(It suffices to retain only the term in h because the whole difference approximation is valid only to this order.) Now write (3.10) as

$$Nh = \begin{bmatrix} 0 & \delta & 0 & \delta* \\ \delta* & 0 & \delta & 0 \\ 0 & \delta* & 0 & \delta \\ \delta & 0 & \delta* & 0 \end{bmatrix}$$

(3.14)

$$+ hQ = \Delta + hQ \quad .$$

Thus

$$M^{-1}N = \frac{1}{\sqrt{2}} (I - \frac{h}{\sqrt{2}} P) (\Delta + hQ) + 0(h^2)$$

$$= \frac{\Delta}{\sqrt{2}} + h \left[\frac{Q}{\sqrt{2}} - \frac{P\Delta}{2} \right] + 0(h^2) \quad .$$

(3.15)

The matrix $M^{-1}N$ is just the transition matrix A_{Σ} (see Sec. 2.2), and the entries are the blocks T, S, R, etc., properly arranged. Observe that if h is neglected, we get simply

$$M^{-1}N \simeq \frac{\Delta}{\sqrt{2}} \quad ,$$

(3.16)

and the block in the (1,1) position is thus zero. A glance at Fig. 12 shows that indeed there can be no transmission to an opposite side without a collision, and (first) collisions are accounted for by the terms in h. Thus $M^{-1}N$ as given by (3.15) is the first collision approximation to the inverse of the transport operator (3.1).

Thus we have the analogue of what was done in [5]. Observe also that the (1,2) block is just δ and that it is an S block. Here no collision is needed for output.

Explicit expressions for the elements of $A_{\Sigma} = M^{-1}N$ can clearly be calculated. It does not seem worth while to list them here.

3.3 Some Computational Experiences

The approaches outlined in this section have been carried out in detail for the example described in the previous section as well as one for which particles were allowed to move in only the compass directions, N, E, S, W. Most of our discussion will be for the first case.

Most striking is the fact that even the initial square Σ involves vectors of length 8 and matrices containing 64 elements. On each doubling, the length of the former increases by a factor of 2, and of the size of the latter increases by a factor of 4. Computer memory restrictions are a clear threat. However, seven doublings were carried out on the CRAY-1 with no strain using only internal memory. The total time required was about 10 seconds. Interestingly enough, at each stage, most of the computer time was required for the last doubling. Hence, the sixth stage took about one second and the seventh took about nine seconds. The fundamental square Σ was chosen with $h = 10^{-2}$. (Recall the unit is a mean free path.) Thus the doublings brought the square to a bit over one and a quarter mean free paths on a side. Direct comparison with other methods proved difficult, but results appeared reasonable.

An additional doubling was carried out in the N, E, S, W case in about the same amount of time. Observe that the first doubling in this case produces a system as large as that with which we <u>start</u> in the previous example.

In all instances standard matrix routines were employed. No efforts were made to optimize programs, and only the readily available machine memory was utilized.

4 REMARKS, SUMMARY, AND CONCLUSIONS–THE FUTURE

We have described in some detail how the conceptual program of Corones can be carried out in a particular 2-D geometry and have described some calculations to demonstrate that the approach is feasible. It may be argued, of course, that a square only one or two mean free paths on a side is not likely to be of great interest.

However, highly efficient programming, plus the use of auxiliary memory, could doubtless have allowed several more doublings without excessive expenditure of time. More important, the advent of the parallel processor probably changes the time and memory considerations greatly. The structure of the equations developed makes parallel processing seem ideal for invariant imbedding.

Assuming all this comes about, of what use might the method be? First, quite irregular regions can be approximated by unions of squares, even if the lengths of their sides are related by multiples of two. Once the basic transition matrices have been calculated and recorded, the joining problems are fairly straightforward.

At a simple level, consider three rectangles of different materials forming a sandwich. Suppose the middle rectangle can be made of any of several materials with the aim of constructing the most effective shield. This problem, even now, is virtually within grasp. Instead of computing each shield separately, one would only have to call up the appropriate A matrices and ''join.''

A geometry that is extremely difficult (or impossible) to handle with all standard computational transport methods except Monte Carlo is illustrated in Fig. 13. Obviously, the imbedding method could be employed quite easily.

What about 3-D? That, at present, does not look too promising. Simple operation counts suggest that invariant imbedding may not be competitive with classical numerical approaches. But that pessimistic view is based on devices similar to the one we have outlined (not on future computing machines). Further research is strongly suggested.

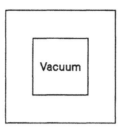

Figure 13. Medium with imbedded vacuum region.

REFERENCES

1.   J. Corones, "A Discrete Model of Transport and Invariant Imbedding in Two or More Dimensions," these proceedings.

2.   Z. Weiss, "Some Basic Properties of the Response Matrix Equations," Nucl. Sci. Eng. **63** (1977), pp. 457-492.

3.   H. Nakata and W. R. Martin, "The Finite Element Response Matrix Method," Nucl. Sci. Eng. **85** (1983), pp. 289-305.

4.   M. Ribarîc, "Functional-Analytic Concepts and Structures of Neutron Transport Theory," Vols. 1 and 2, Slavene Academy of Sciences and Arts, Ljubljana (1973).

5.   G. Hughes, V. Faber, A. B. White, Jr., and G. M. Wing, "On the Advantages of Invariant Imbedding for the Calculation of Deep Penetration Problems," Trans. Amer. Nucl. Soc. **41** (1982), pp. 486-487.

6.   R. M. Redheffer, "On the Relation of Transmission-Line Theory to Scattering and Transfer," J. Math. Phys. **41** (1962), pp. 1-41.

7.   H. C. Van de Hulst, "A New Look at Multiple Scattering," NASA Institute for Space Studies, Goddard Space Flight Center (1963).

8.   J. J. Duderstadt and W. R. Martin, *Transport Theory* (John Wiley and Sons, New York, 1979).

Refined Eigenvalue Estimates in the Case of Lipschitz Continuity

JAMES A. COCHRAN   Department of Mathematics, Washington State University, Pullman, Washington

Recently [3], we have established that the eigenvalues of positive definite kernels $K(x,t)$ whose symmetric derivative $K_{rr}(x,t) \equiv \partial^{2r} K(x,t)/\partial x^r \partial t^r$, as a function of one of its variables, is in Lip α for some α, $0 < \alpha \leq 1$, satisfy $\lambda_n(K) = \mathrm{O}(n^{-2r-1-\alpha})$. In this paper we extend that conclusion to kernels for which $K_{rr}(x,t)$ is in Lip$^*\alpha$ for some α, $0 < \alpha \leq 2$. More importantly, we demonstrate a result which ensures, as a special case, that if $K_{rr}(x,t)$ is in lip$^*\alpha$ for some α, $0 < \alpha < 2$, then $\lambda_n(K) = o(n^{-2r-1-\alpha})$. The method of proof blends ideas from our earlier work with an approach originally suggested by Reade [9].

1 INTRODUCTION

Consider the kernel $K(x,t)$ to be continuous and Hermitian on $-\pi \leq x, t \leq \pi$. If $K(x,t)$ is positive definite as well, then the spectrum of the compact operator

$$Kf(x) \equiv \int_{-\pi}^{\pi} K(x,t) f(t) dt$$

defined on the Hilbert space $\mathcal{L}^2[-\pi, \pi]$ consists of a sequence of positive eigenvalues $\{\lambda_n(K)\}$ which converges to zero. We assume that these eigenvalues are arranged in decreasing order of size and repeated according to their (finite) multiplicities.

Our interest is in asymptotic estimates for $\lambda_n(K)$ when the symmetric derivative

$$K_{rr}(x,t) \equiv \frac{\partial^{2r}}{\partial x^r \partial t^r} K(x,t) \tag{1}$$

of the given kernel has specific smoothness usually allied with the notion of Lipschitz continuity. In [3] we showed that if

$$| K_{rr}(x, t+h) - K_{rr}(x,t) | \leq | h |^\alpha A(x) \tag{2}$$

for some α, $0 < \alpha \leq 1$, where $A(x)$ is integrable, then

$$\lambda_n(K) = O(n^{-2r-1-\alpha}) \tag{3}$$

as $n \to \infty$. Our analysis there, among other things, used a well-known result of Chang [2] and ideas of Reade [7], [8] and of Ha [4]. Moreover, the familiar piecewise linear "hat" basis functions of approximation theory were employed in an essential way.

In our present investigation we replace the single Lipschitz condition (2) by the inequalities

$$\int_{-\pi+\delta}^{\pi-\delta} | K_{rr}(x, x+h) + K_{rr}(x, x-h) - 2K_{rr}(x,x) | \, dx \leq | h |^\alpha A(h)$$

$$\int_{-\pi+\delta}^{\pi-\delta} | K_{rr}(\pm\pi, x+h) + K_{rr}(\pm\pi, x-h) - 2K_{rr}(\pm\pi,x) | \, dx \leq | h |^\alpha B(h) \tag{4}$$

$$\int_{-\pi}^{\pi} \int_{-\pi+\delta}^{\pi-\delta} | K_{rr}(x, t+h) + K_{rr}(x, t-h) - 2K_{rr}(x,t) | \, dt dx \leq | h |^\alpha C(h)$$

where $\delta (0 < \delta < \pi)$ is arbitrary, $| h | \leq \delta$, $0 < \alpha \leq 2$, and A, B, C are nonnegative and bounded. In the absence of further restrictions, the asymptotic eigenvalue estimate (3) is again established. However, if $A(h)$, $B(h)$, and $C(h)$ are all o(1) as $h \to 0$, we can now conclude that

$$\lambda_n(K) = o(n^{-2r-1-\alpha}) \tag{5}$$

This latter result subsumes the important special case wherein $K_{rr}(x,t)$ is such that

$$K_{rr}(x, t+h) + K_{rr}(x, t-h) - 2K_{rr}(x,t) = | h |^\alpha o(1)$$

as $h \to 0$.

Aside from the fact that the inequalities (4) only involve *integrated* Lipschitz conditions for $K_{rr}(x,t)$, the use of second differences here has a number of advantages over first differences as in (2). For example, in the case of the order relation o it asserts a generally weaker hypothesis. Although comparable for $0 < \alpha < 1$ in the case of O, it does provide a distinctly weaker hypothesis again when $\alpha = 1$. Moreover, a unified treatment of K_{rr} and of $\partial K_{rr}/\partial t$ is now possible since if $\alpha > 1$, the condition (4) is equivalent to $\partial K_{rr}/\partial t$ satisfying a standard Lipschitz condition (i.e. with a first difference) of order $\alpha - 1$ (see [1], page 116). The new result is the analogue of a comparable conclusion recently established by Oehring ([6], proof of Theorem 12) for classical exponential Fourier series.

For our demonstration we use an approach originally proposed by Reade [9], appropriately modified and combined with the technique of [3]. In the process, we show how to eliminate the undesirable periodicity assumptions appearing in the Reade hypotheses.

2 PROPERTIES OF JACKSON KERNELS

The estimates needed for our principal results will be based upon finite rank approximations to the kernels in question in which the Jackson kernels

$$J_n(t) \equiv \frac{3}{n(2n^2 + 1)} \frac{\sin^4(nt/2)}{\sin^4(t/2)}$$

play a distinguished role. Since

$$\frac{\sin^2(nt/2)}{\sin^2(t/2)} = \sum_{|k| \leq n-1} (n - |k|) e^{ikt},$$

it follows readily that

PROPERTY 1: $J_n(t)$ is a trigonometric polynomial of degree $2n - 2$

and

PROPERTY 2: $\frac{1}{2\pi} \int_{-\pi}^{\pi} J_n(t) dt = 1.$

Moreover,

LEMMA 1: For $0 < \delta < \pi$ and $-1 < \beta < 3$,

$$\int_0^\delta u^\beta \, J_n(u) du = \mathrm{O}(n^{-\beta})$$

as $n \to \infty$.

Proof: Since $\sin(u/2) > u/\pi$ for all $0 < u < \pi$,

$$\int_0^\delta u^\beta \frac{\sin^4(nu/2)}{\sin^4(u/2)} du \;\leq\; \pi^4 \int_0^\delta u^{\beta-4} \sin^4(nu/2) du$$

$$= n^{3-\beta} \pi^4 \int_0^{n\delta} t^{\beta-4} \sin^4(t/2) dt$$

$$= \mathrm{O}(n^{3-\beta}). \qquad\qquad \square$$

LEMMA 2: Given $0 < \delta < \pi$,

$$\int_\delta^\pi u^\beta J_n(u) du = \mathrm{O}(n^{-3})$$

Proof:

$$\int_\delta^\pi u^\beta J_n(u) du \;=\; \frac{3}{n(2n^2+1)} \int_\delta^\pi u^\beta \frac{\sin^4(nu/2)}{\sin^4(u/2)} du$$

$$\leq\; \frac{3}{n(2n^2+1)} \int_\delta^\pi \frac{u^\beta}{\sin^4(u/2)} du$$

$$=\; \mathrm{O}(n^{-3}). \qquad\qquad \square$$

If

$$J_n(t) = \sum_k a_k e^{ikt},$$

we can define

$$H_n(t) \equiv \sum_k a_{2k} e^{ikt}$$

The remaining necessary results concerning the Jackson kernels then are (see [9] for proofs)

LEMMA 3:

$$\frac{4}{3} J_n(t) - \frac{1}{3} H_n(t) = \sum_k b_k e^{ikt}$$

with $b_k \leq 1$ for all k.

LEMMA 4.   For any continuous kernel $M(x,t)$ which is 2π-periodic in t,

$$\int_{-\pi}^\pi \int_{-\pi}^\pi J_n(u) M(x, x \pm u) du\; dx \quad : \quad \int_{-\pi}^\pi \int_{-\pi}^\pi J_n(x-u) M(x, u) du\; dx$$

$$\int_{-\pi}^\pi \int_{-\pi}^\pi J_n(u) M(x, x \pm 2u) du\; dx \;=\; \int_{-\pi}^\pi \int_{-\pi}^\pi H_n(x-u) M(x, u) du\; dx$$

3 FURTHER PRELIMINARIES

As in [3], we will have need in the sequel for the following eigenvalue results:

LEMMA 5: A Hermitian kernel that differs from a positive definite kernel by a degenerate (separable) kernel of rank N can have at most N negative eigenvalues.

LEMMA 6: If a Hermitian kernel $M(x,t)$ differs from a positive definite kernel $L(x,t)$ by a degenerate kernel of rank N, then for large enough n,

$$\lambda_{n+3N}(L) \leq \lambda_{n+2N}(M) \leq \lambda_n(L).$$

In what follows we will also be concerned with kernels which are 2π-periodic in their second variable. Given an arbitrary Hermitian kernel $L(x,t)$, defined and continuous on $-\pi \leq x, t \leq \pi$, the continuous Hermitian kernel

$$M(x,t) \equiv L(x,t) + \tfrac{1}{2\pi}[(x-\pi)L(-\pi,t) + (t-\pi)L(x,-\pi) - (x+\pi)L(\pi,t) - (t+\pi)L(x,\pi)]$$

$$+ \frac{1}{4\pi^2}[(x-\pi)(t-\pi)L(-\pi,-\pi) - (x-\pi)(t+\pi)L(-\pi,\pi) \tag{6}$$

$$- (x+\pi)(t-\pi)L(\pi,-\pi) + (x+\pi)(t+\pi)L(\pi,\pi)]$$

permits of such an extension. Indeed, since $M(x,-\pi) = 0 = M(x,\pi)$ the new kernel may be readily extended as a periodic function of t with period 2π. We also note that since $M(-\pi,t) = 0 = M(\pi,t)$, the eigenfunctions of $M(x,t)$ vanish for $x = -\pi, \pi$ as well.

If $L(x,t)$ is positive definite, the kernel $M(x,t)$ may or may not share this property. However, by construction $M(x,t)$ differs from $L(x,t)$ by a degenerate kernel and hence, by Lemma 5, has at most a finite number of negative eigenvalues. Excluding these from consideration, it follows that the portion of M associated with the positive eigenvalues, $M^+(x,t)$, is a positive definite and continuous Hermitian kernel which differs itself from $L(x,t)$ by a degenerate kernel and is (may be extended so as to be) 2π-periodic in t. Hence, in view of Lemma 6 we have established

LEMMA 7: Given a positive definite Hermitian kernel $L(x,t)$ which is defined and continuous on $-\pi \leq x, t \leq \pi$, there exists a positive definite and continuous Hermitian kernel $M^+(x,t)$ which

(i) is 2π-periodic in t

(ii) has the same eigenvalue asymptotics as $L(x,t)$, i.e. there exists a positive integer N such that

$$\lambda_{n+3N}(L) \le \lambda_{n+2N}(M^+) \le \lambda_n(L).$$

It remains only to verify that

LEMMA 8: If the positive definite Hermitian kernel $L(x,t)$ of Lemma 7 satisfies the inequalities (4), then the periodic kernel $M^+(x,t)$ of our construction is such that

$$\int_{-\pi+\delta}^{\pi-\delta} |\, M^+(x,x+h) + M^+(x,x-h) - 2M^+(x,x) \,| \, dx \; \le \, |\, h \,|^\alpha \, D(h) \qquad (7)$$

for $|\, h \,| \le \delta$ as well. The nonnegative and bounded function $D(h)$ is $\mathrm{O}(1)$ or $o(1)$ in accordance with the $A(h), B(h)$ and $C(h)$ pertinent to $L(x,t)$.

Proof: By construction $M(x,t)$ satisfies (7) with $D(h) \le A(h)+2B(h)$. But $M(x,t) = M^+(x,t) + M^-(x,t)$ and the smoothness of the degenerate kernel $M^-(x,t)$ is controlled by its eigenfunctions. As a consequence,

$$M^-(x,x+h)+M^-(x,x-h)-2M^-(x,x) = \sum_{k=1}^{N}\lambda_k^- \phi_k^-(x)\overline{[\phi_k^-(x+h) + \phi_k^-(x-h) - 2\phi_k^-(x)]}$$

$$= \sum_{k=1}^{N} \phi_k^-(x) \int_{-\pi}^{\pi} \overline{[M(x+h,t) + M(x-h,t) - 2M(x,t)]\phi_k^-(t)} \, dt$$

and hence

$$\int_{-\pi+\delta}^{\pi-\delta} |\, M^-(x,x+h) + M^-(x,x-h) - 2M^-(x,x) \,| \, dx$$

$$\le \; N \max_{x,k} |\, \phi_k^-(x) \,|^2 \int_{-\pi+\delta}^{\pi-\delta} \int_{-\pi}^{\pi} |\, M(x+h,t) + M(x-h,t) - 2M(x,t) \,| \, dt \, dx$$

$$\le \; N \max_{x,k} |\, \phi_k^-(x) \,|^2 \,|\, h \,|^\alpha \, [C(h) + 2\pi B(h)],$$

again by construction. $\square$

4 THE MAIN RESULTS

THEOREM 1: Let the compact operator L on $\mathcal{L}^2[-\pi,\pi]$ be generated by the continuous, Hermitian, positive definite kernel $L(x,t), -\pi \le x,t \le \pi$. Assume that for some α, $0 < \alpha \le 2$, and arbitrary δ, $0 < \delta < \pi$,

$$\int_{-\pi+\delta}^{\pi-\delta} \mid L(x,x+h) + L(x,x-h) - 2L(x,x) \mid dx \leq \mid h \mid^{\alpha} A(h)$$

$$\int_{-\pi+\delta}^{\pi-\delta} \mid L(\pm\pi,x+h) + L(\pm\pi,x-h) - 2L(\pm\pi,x) \mid dx \leq \mid h \mid^{\alpha} B(h) \qquad (8)$$

$$\int_{-\pi}^{\pi} \int_{-\pi+\delta}^{\pi-\delta} \mid L(x,t+h) + L(x,t-h) - 2L(x,t) \mid dt \, dx \leq \mid h \mid^{\alpha} C(h)$$

where $\mid h \mid \leq \delta$ and A, B, C are nonnegative and bounded. When $\alpha > 1$ we take $\partial L(x,t)/\partial t$ to be continuous. If $A(h), B(h)$ and $C(h)$ are all $o(1)$ as $h \to 0$, then

$$\sum_{k=n}^{\infty} \lambda_k(L) = o(n^{-\alpha})$$

and thus

$$\lambda_n(L) = o(n^{-1-\alpha})$$

as $n \to \infty$. Changing o to O produces equally valid results.

Proof: Our demonstration is based upon the approach appearing in [9]. Instead of $L(x,t)$, however, we consider its positive definite periodic counterpart $M^+(x,t)$ as constructed in the previous section.

Now M^+ is of trace class, and, in view of Lemma 3, $M^{\frac{1}{2}} R M^{\frac{1}{2}} \leq M^+$ where $M^{\frac{1}{2}}$ is the unique positive square root of M^+ and R is the symmetric integral operator of finite rank generated by the separable kernel

$$\frac{4}{3} J_n(x-t) - \frac{1}{3} H_n(x-t)$$

discussed in §2. It follows that $\mathcal{M} \equiv M^+ - M^{\frac{1}{2}} R M^{\frac{1}{2}}$ is Hermitian, with a continuous kernel, and has a trace norm given by

$$\parallel \mathcal{M} \parallel_{tr} = \frac{1}{2\pi} \int_{-\pi}^{\pi} M^+(x,x)dx - \frac{1}{4\pi^2} \int_{-\pi}^{\pi} \int_{-\pi}^{\pi} [\frac{4}{3} J_n(x-t) - \frac{1}{3} H_n(x-t)] M^+(x,t)dx \, dt. (9)$$

Applying the results of Lemma 4, in reverse, to the periodic kernel $M^+(x,t)$, and employing the normalizations of the Jackson kernels J_n, and their evenness, the right-hand side (RHS) of (9) can be rewritten as

$$RHS = \frac{1}{4\pi^2} \int_{-\pi}^{\pi} \int_{-\pi}^{\pi} \Delta_M(x,u) J_n(u)dx \, du$$

$$= \frac{1}{2\pi^2} \int_{0}^{\pi} J_n(u)[\int_{-\pi}^{\pi} \Delta_M(x,u)dx]du \qquad (10)$$

with

$$\Delta_M(x, u) \equiv M^+(x, x) - \tfrac{2}{3}[M^+(x, x + u) + M^+(x, x - u)]$$

$$+ \tfrac{1}{6}[M^+(x, x + 2u) + M^+(x, x - 2u)]$$

$$= \tfrac{1}{6}[M^+(x, x + 2u) + M^+(x, x - 2u) - 2M^+(x, x)]$$

$$- \tfrac{2}{3}[M^+(x, x + u) + M^+(x, x - u) - 2M^+(x, x)].$$

It is this expression that we need to estimate.

We divide the domain of the x-integration in (10) into two parts and begin by recalling that $M^+(x, t)$ is continuous ($\partial M^+/\partial t$ is actually continuous if $\alpha > 1$) and thus $\Delta_M(x, 0) = 0$. Indeed, given $\varepsilon > 0$ we can choose $0 < \delta < \pi$ such that

$$\mid \Delta_M(x, u) \mid < \varepsilon$$

and hence

$$\int_{\pi - 2u \le |x| \le \pi} \mid \Delta_M(x, u) \mid dx < 2u\varepsilon$$

whenever $0 \le u \le \delta$. If $\alpha > 1$, the comparable inequalities are

$$\mid \Delta_M(x, u) \mid < u\varepsilon$$

and

$$\int_{\pi - 2u \le |x| \le \pi} \mid \Delta_M(x, u) \mid dx < 2u^2\varepsilon.$$

In view of Lemmas 1, 2 (with $\beta = 1, 2$), then,

$$\int_0^\pi J_n(u) \int_{\pi - 2u \le |x| \le \pi} \mid \Delta_M(x, u) \mid dx \, du < \begin{cases} \varepsilon O(n^{-1}) & \alpha \le 1 \\ \varepsilon O(n^{-2}) & \alpha > 1. \end{cases}$$

For the remainder of the x-integration we have, by Lemma 8,

$$\int_{-\pi + 2u}^{\pi - 2u} \mid \Delta_M(x, u) \mid dx \le \frac{1}{6}(2u)^\alpha D(2u) + \frac{2}{3}u^\alpha D(u).$$

This time, given $\varepsilon > 0$ we can choose $0 < \delta < \pi$ so that $D(2u) < \varepsilon$ for all $0 \le u \le \delta$. Since $0 < \alpha \le 2$, the application of Lemmas 1, 2 with $\beta = \alpha$ now yields

$$\int_0^\pi J_n(u) \int_{-\pi + 2u}^{\pi - 2u} \mid \Delta_M(x, u) \mid dx \, du < \varepsilon \, O(n^{-\alpha})$$

Combining these results together into (10), and thence (9), we find

$$\parallel M^+ - M^{1/2} R M^{1/2} \parallel_{tr} = o(n^{-\alpha})$$

as $n \to \infty$.

Earlier we noted that $J_n(t)$ is a trigonometric polynomial of degree $2n - 2$. Therefore, $M^{1/2}RM^{1/2}$ has rank at most $4n - 3$. As a consequence, using the Weyl-Courant minimax principle as applied to trace norms,

$$\sum_{4n-2}^{\infty} \lambda_k(M^+) \leq \| M^+ - M^{1/2}RM^{1/2} \|_{tr} = o(n^{-\alpha}).$$

From this, owing to Lemma 7, we immediately obtain the desired result for $\lambda_n(L)$, namely

$$\sum_{k=n}^{\infty} \lambda_k(L) = o(n^{-\alpha}),$$

and hence

$$\lambda_n(L) = o(n^{-1-\alpha})$$

as $n \to \infty$, in view of the monotonicity of the $\lambda_n(L)$. Finally, we note the equal validity of the assertions analogous to the various expressions above which ensue when the order relation o is replaced by O throughout. $\square$

We stated in the introduction that our real interest in this paper concerns asymptotic estimates for the eigenvalues of positive definite kernels whose symmetric derivative satisfies an appropriate condition of Lipschitz type. As outlined in [3], earlier work of Chang [2] makes this all possible.

THEOREM 2: Let the Hermitian, positive definite kernel $K(x,t)$, continuous on $-\pi \leq x, t \leq \pi$, generate a compact operator K on $\mathcal{L}^2[-\pi, \pi]$. Assume that for some positive integer r the symmetric derivative $K_{rr}(x,t)$ given by (1) exists and is likewise continuous on $-\pi \leq x, t \leq \pi$. Moreover, assume that for some α, $0 < \alpha \leq 2$, and arbitrary δ, $0 < \delta < \pi$, $K_{rr}(x,t)$ satisfies the integrated Lipschitz inequalities (4). When $\alpha > 1$ we take $\partial K_{rr}(x,t)/\partial t$ to be continuous. If the nonnegative bounded functions $A(h), B(h)$ and $C(h)$ are all o(1) as $h \to 0$, then

$$\sum_{k=n}^{\infty} k^{2r}\lambda_k(K) = o(n^{-\alpha})$$

and

$$\lambda_n(K) = o(n^{-2r-1-\alpha})$$

as $n \to \infty$. Again changing the order relation o to O leads to equally valid results.

Proof: Given a positive definite Hermitian kernel $K(x,t)$, Kadota [5] has shown that if $K_{rr}(x,t)$ exists and is continuous on the fundamental domain, then K_{rr} is Hermitian and positive definite as well. It follows that we can identify $K_{rr}(x,t)$ with the kernel $L(x,t)$ of Theorem 1, whence

$$\sum_{k=n}^{\infty} \lambda_k(K_{rr}) = o(n^{-\alpha})$$

as $n \to \infty$. Chang [2], however, has established the inequality

$$\lambda_{2n}(K) \leq \text{const.} \ (n-r)^{-2r} \lambda_{n+1}(K_{rr})$$

for $n > r$. The desired relations are thus an immediate consequence of the Chang result and the fact that

$$(n+1)n^{2r}\lambda_{2n}(K) \leq \sum_{k=n}^{2n} k^{2r}\lambda_k(K)$$

Since Theorem 1 remains valid under the change of o to O, so also does the present demonstration. □

5 REMARKS

In Reade's original paper on positive definite Hermitian kernels [8] he established that if $K(x,t)$ was continuously differentiable on its fundamental domain, then

$$\lambda_n(K) = o(n^{-2})$$

as $n \to \infty$. Theorem 1 herein shows that this asymptotic estimate is actually valid if the positive definite Hermitian kernels satisfy much weaker hypotheses which incorporate, as a special case, kernels for which

$$K(x,t+h) + K(x,t-h) - 2K(x,t) = o(h)$$

as $h \to 0$. As Reade himself noted (see also [10]), the eigenvalue estimate obtained cannot be improved even in the class of C^1 kernels.

In Reade's subsequent paper on positive definite kernels [9] he assumed that the kernels $K(x,t)$ under consideration were 2π-periodic in x,t. The construction (6) and Lemmas 7,8 show how this extrinsic periodicity assumption can be removed within the context of the Reade approach. (See also [4].)

For difference kernels $K(x,t) = f(x-t)$, the inequalities (4) are all essentially equivalent and indeed, devolve to the condition that

$$f^{(2r)}(h) + f^{(2r)}(-h) - 2f^{(2r)}(0) = o(h^\alpha).$$

This situation is included in the recent results of Oehring [6], where, in addition, the sharpness of the results is amply explored.

REFERENCES

1. P.L. Butzer and H. Berens, *Semi-groups of operators and approximation*, Springer-Verlag, New York, 1967.

2. S.H. Chang, *A generalization of a theorem of Hille and Tamarkin with applications*, Proc. London Math. Soc. (3), 2(1952), pp. 22-29.

3. J.A. Cochran and M.A. Lukas, *Differentiable positive definite kernels and Lipschitz continuity*, Math. Proc. Camb. Phil. Soc. (to appear).

4. C.W. Ha, *Eigenvalues of differentiable positive definite kernels*, SIAM J. Math. Anal. 17(1986), pp. 415-419.

5. T.T. Kadota, *Term-by-term differentiability of Mercer's expansion*, Proc. Amer. Math. Soc. 18 (1967), pp. 69-72.

6. C. Oehring, *Smooth kernels and Fourier coefficients*, J. Math. Anal. Appl. (to appear).

7. J.B. Reade, *Eigenvalues of Lipschitz kernels*, Math. Proc. Camb. Phil. Soc. 93 (1983), pp. 135-140.

8. J.B. Reade, *Eigenvalues of positive definite kernels*, SIAM J. Math. Anal., 14(1983), pp. 152-157.

9. J.B. Reade, *Eigenvalues of positive definite kernels II*, SIAM J. Math. Anal. 15(1984), pp. 137-142.

10. J.B. Reade, *Positive definite C^p kernels*, SIAM J. Math Anal. 17(1986), pp. 420-421.

Remarks on Some Iterative Methods for an Integral Equation in Fourier Optics

C. W. GROETSCH   Department of Mathematical Sciences, University of Cincinnati, Cincinnati, Ohio

A model problem in Fourier optics involves the solution of a special ill-posed Fredholm integral equation of the first kind. An iterative method for this equation, which is equivalent to Landweber's method, has been studied extensively in recent years. This method suffers from slow convergence as well as sensitivity to errors in the data. Recently some regularized versions of this method have been proposed and numerical investigations have been carried out. However, until now a complete convergence analysis has been lacking. In this note we provide a convergence analysis, with particular attention given to asymptotic relationships between iteration numbers, regularization parameters and data error levels. The general results are applied to the iterative methods of Cesini-Ross and Abbiss et al.

1   INTRODUCTION

In 1873 E. Abbe proposed a theory for the coherent illumination of an object and the reconstruction of an image [11,p.267]. For simplicity, consider a one-dimensional object that is illuminated by a plane wave $e^{i\omega t}$ which is incident normally to the object. In Abbe's theory the formation of an image by an optical instrument consists of two stages. First, each point of the object scatters

the incident wave, and then the instrument collects these

scattered waves and reconstructs the object by bringing the light

back to focus in the image plane.

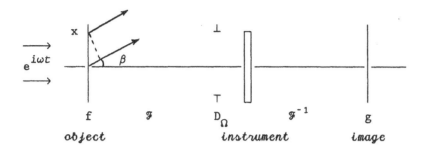

A wave scattered through an angle β by a point in the object at a

distance x from the axis then has an optical path which is

shortened by $\Delta x = x \sin\beta$. This gives rise to a time shift of

$\Delta t = \Delta x/c$ and hence the scattered wave at x with amplitude $f(x)$

has the form

$$f(x)e^{i\omega(t-\Delta x/c)} = f(x)e^{-i2\pi ux}e^{i\omega t}$$

where $u = (\omega \sin\beta)/2\pi c$. Therefore the complete scattered wave in

the direction β is $\hat{f}(u)e^{i\omega t}$ where

$$\hat{f}(u) = \int_{-\infty}^{\infty} f(x)e^{-i2\pi ux}dx$$

is the Fourier transform of the object. A completely faithful

optical instrument for reconstructing the object would then be a

realization of the inverse Fourier transform. However, as a

practical matter, only a limited range of scattered waves can be

collected, say corresponding to $-\Omega \leq u \leq \Omega$. Therefore, if D_{Ω}

represents the domain limiting operator

$$D_\Omega h (u) = \begin{cases} h(u), & |u| \leq \Omega \\ 0, & |u| > \Omega \end{cases}$$

an image $g(y)$ is formed where

$$g = \mathcal{F}^{-1} D_\Omega \mathcal{F} f \qquad (1)$$

and $\mathcal{F}$ represents the Fourier integral operator. We will designate the operator $\mathcal{F}^{-1}D_\Omega\mathcal{F}$ by B_Ω and refer to it as the *band-limiting* operator. Note that by the convolution theorem

$$B_\Omega\varphi (y) = \int_{-\infty}^{\infty} \frac{\sin[\Omega(y-x)]}{\pi(y-x)} \varphi(x) \, dx$$

For an object $f \in L^2(-\infty,\infty)$ of limited extent with support contained in $[-X,X]$ we then have

$$g = B_\Omega D_X f \quad , \quad D_X f = f \qquad (2)$$

or equivalently,

$$g(y) = \int_{-X}^{X} \frac{\sin[\Omega(y-x)]}{\pi(y-x)} f(x) \, dx \qquad (3)$$

We begin by considering some well known properties of the operators B_Ω and D_X acting on the space $L^2(-\infty,\infty)$. First, we note that $R(B_\Omega)$, the range of B_Ω consists of functions whose Fourier transforms have support contained in $[-\Omega,\Omega]$. Therefore, by the Paley-Wiener theorem, functions in $R(B_\Omega)$ are restrictions to the real line of entire functions of exponential type Ω. It follows that no nonzero function in $R(B_\Omega)$ may belong to $N(D_X)$, the nullspace of D_X. Also, if $f \in R(D_X)$, then $\hat{f}$ is analytic and hence if $f \in N(B_\Omega)$, then by (1), $D_\Omega \hat{f} = 0$. Therefore $\hat{f} = 0$ and

hence $f = 0$. We summarize these facts in:

LEMMA 1. $R(B_\Omega) \cap N(D_X) = \{0\} = R(D_X) \cap N(B_\Omega)$.

It follows directly from the lemma that problem (2) can have at

most one solution. In studying (2) it will be convenient to

introduce the compact operator $K = D_X B_\Omega D_X$ given by (3). From (2)

we have $B_\Omega g = g$ and hence

$$Kf = D_X B_\Omega g, \qquad D_X f = f \tag{4}$$

Note that because the image g is band-limited, any solution f of (4)

also satisfies (2) since (4) implies

$$g - B_\Omega D_X f \in N(D_X) \cap R(B_\Omega) = \{0\}.$$

Therefore (2) and (4) are equivalent. Furthermore, since B_Ω and D_X

are self-adjoint projections, it follows that

$$(Kh,h) = \| B_\Omega D_X h \|^2$$

and hence K is positive semi-definite. Moreover, from this

equality we see that if $Kh = 0$, then $D_X h \in R(D_X) \cap N(B_\Omega)$ and

hence $D_X h = 0$. Therefore, $N(K) = N(D_X)$.

It is clear from the definition of K that $\| K \| \le 1$, but if

$Kh = h$ for some h of unit norm, then $D_X h = h$ and hence

$$1 = (Kh,h) = (B_\Omega D_X h, D_X h) = (B_\Omega h, h).$$

Using the fact that $B_\Omega^2 = B_\Omega$, we find

$$\|B_\Omega h - h\|^2 = (B_\Omega h - h, B_\Omega h - h) = 1-2+1 = 0$$

and hence $h \in R(B_\Omega) \cap N(I - D_X) = \{0\}$. Therefore, $\|K\| < 1$. This

proves

LEMMA 2. $K = D_X B_\Omega D_X$ *is a compact positive semi-definite operator with* $N(K) = N(D_X)$ *and* $\|K\| < 1$.

The equation (3) is an example of a Fredholm integral equation of the first kind. Long ago Landweber [10] investigated an iterative method for solving such equations. In the context of (2) his method has been rediscovered a number of times and has come to be known in the physical literature as Gerchberg's method [6]. The method applied to (2) is simply

$$f_0 = 0, \quad f_n = D_X B_\Omega g + (I - K)f_{n-1} \tag{5}$$

There are two difficulties associated with this method. One is attributable to the fact that $\|I - K\| = 1$ and hence the convergence can be exceedingly slow. Another difficulty, the baleful influence of which is familiar to many practitioners, is that equation (4) is *ill-posed* in the sense that small perturbations in the image g can be associated with large changes in the corresponding object f. This ill-posedness is most easily appreciated by expanding the object f in terms of an orthonormal eigensystem $\{v_n; \lambda_n\}$ for K. Using (4) this results in

$$f = \sum_{n=1}^{\infty} \lambda_n^{-1} (v_n, D_X B_\Omega g) v_n$$

Since K is compact, we have $\lambda_n \to 0$ and hence small disturbances in a high order component $(v_n, D_X B_\Omega g) = (B_\Omega v_n, g)$ of the image g can lead to large changes in the reconstructed object f. Such ill-posedness is typical in modeling inverse problems in physics ([5],[9],[13],[16]) and a considerable literature on special approximation methods, called *regularization methods*, for mitigating the instabilities inherent in approximating solutions of such problems has emerged in recent years (see e.g., [2],[7],[8],[12],[16]).

Cesini et al. [4], Ross [14] (see also [15]) and Abbiss et al. [1] have applied special regularized versions of Landweber's method to (4), however, convergence proofs for these methods have not been provided. In the next section we provide a complete convergence analysis for a general regularized iterative method that includes these methods as a special case.

2 A GENERAL REGULARIZED ITERATIVE METHOD

A natural way to accelerate the convergence of the iterative method (5) is to replace it with a method of the type

$$f_0^\alpha = 0, \qquad f_n^\alpha = D_X B_\Omega g + S_\alpha(K) f_{n-1}^\alpha \qquad (6)$$

where $S_\alpha(K)$ is an operator with spectral radius strictly less than that of $I - K$. We accomplish this with a family of continuous functions $S_\alpha(.)$ defined on $[0,1]$ and depending on a parameter $\alpha \in (0,1)$. We assume the following about the functions $S_\alpha(.)$:

$$|S_\alpha(t)| < 1-t \qquad \text{for } t \in [0,1) \text{ and } \alpha \in (0,1) \qquad (7)$$

and

$$S_\alpha(t) \to 1-t \qquad \text{as } \alpha \to 1 \text{ for each } t \in (0,1] \qquad (8)$$

Since the eigenvalues of K lie in $[0,1]$, it follows from (7) that $I - S_\alpha(K)$ is invertible and hence the equation

$$(I - S_\alpha(K))f^\alpha = Kf \qquad (9)$$

has a unique solution f^α. Suppose that $\{\lambda_n; v_n\}$ is an orthonormal eigensystem for K corresponding to the positive eigenvalues λ_n. Since $Kf \in N(K)^\perp$, it follows that Kf can be expanded in terms of the $\{v_n\}$ and we have from (9):

$$f^\alpha = (I - S_\alpha(K))^{-1}Kf = \sum_{n=1}^{\infty} (1 - S_\alpha(\lambda_n))^{-1}\lambda_n(f,v_n)v_n \qquad (10)$$

Since $0 < \lambda/(1 - S_\alpha(\lambda)) < 1$, and since $\lambda/(1 - S_\alpha(\lambda)) \to 0$ as $\lambda \to 0$, we see from (10) that high frequency features of f are damped in f^α. That is, f^α represents an "oversmoothed" version of f.

Conditions (7) and (8) guarantee that

$$m(\alpha) = \max_{t\in[0,1]} \mid S_\alpha(t) \mid < 1$$

Using this it follows that $f_n^\alpha \to f^\alpha$ as $n \to \infty$. In fact,

PROPOSITION 1. $\|f_n^\alpha - f^\alpha\| \le m(\alpha)^n \|f^\alpha\|$.

Proof. First note that $D_X B_\Omega g = D_X B_\Omega^2 D_X f = Kf$, therefore by (6),

$$f_n^\alpha = Kf + S_\alpha(K)f_{n-1}^\alpha .$$

By (9) we then have

$$f_n^\alpha - f^\alpha = -S_\alpha(K)^n f^\alpha$$

and the result follows. $\square$

We now show that the regularized approximations converge to the solution f.

PROPOSITION 2. $f^\alpha \to f$ *as* $\alpha \to 1$.

Proof. Since $D_X f = f$ we have from lemma 2 that $f \in N(D_X)^\perp = N(K)^\perp$. Therefore f has a complete expansion in terms of the eigenfunctions $\{v_n\}$ corresponding to the positive eigenvalues of K. We then have from (10)

$$\|f^{\alpha} - f\|^2 = \sum_{n=1}^{\infty} a_n(\alpha) \ |(f, v_n)|^2 \tag{11}$$

where

$$a_n(\alpha) = \left| \frac{1 - \lambda_n - S_\alpha(\lambda_n)}{1 - S_\alpha(\lambda_n)} \right|^2$$

From (7) and (8) it follows that $0 \leq a_n(\alpha) \leq 1$ and $a_n(\alpha) \to 0$ as $\alpha \to 1$, for each n. From (11) we then find that $\|f^{\alpha} - f\| \to 0$ as $\alpha \to 1$. □

Consider now the case of a noisy image g^δ with $\|g - g^\delta\| \leq \delta$. We will show that the iterative method (5) is a regularizing algorithm in the sense of Tikhonov [16],[7], if the iteration number n and the regularization parameter α are appropriately related to each other and to the noise level δ in the data. The iterative method using the available data is

$$f_n^{\alpha,\delta} = D_X B_\Omega g^\delta + S_\alpha(K) f_{n-1}^{\alpha,\delta} \ .$$

Hence from (6) we have

$$f_n^{\alpha} - f_n^{\alpha,\delta} = [I - S_\alpha(K)]^{-1} [I - S_\alpha(K)^n] D_X B_\Omega (g - g^\delta)$$

and by operator norm estimates we obtain

$$\|f_n^{\alpha} - f_n^{\alpha,\delta}\| \leq \delta/(1 - m(\alpha)).$$

Combining this with the preceding propositions we arrive at the following result.

PROPOSITION 3. *Suppose that $\alpha \to 1$ and $n \to \infty$ as $\delta \to 0$ in such a way that $m(\alpha)^n \to 0$ and $\delta/(1 - m(\alpha)) \to 0$, then $f_n^{\alpha,\delta} \to f$.*

3 TWO SPECIFIC METHODS

Abbiss et al. [1] have performed numerical studies of the regularized iterative method

$$f_n^\alpha = D_X B_\Omega g + \alpha(I - K)f_{n-1}^\alpha \,, \quad f_0^\alpha = 0 \quad (0<\alpha<1).$$

In the general scheme of the preceding section this corresponds to the choice $S_\alpha(t) = \alpha(1 - t)$. For this method we have $m(\alpha) = \alpha$, hence the regularity condition expressed in Proposition 3 calls for the parameter α and the iteration number n to depend on the error level δ in such a way that

$$\alpha \to 1, \; \alpha^n \to 0, \; \text{and } \delta/(1-\alpha) \to 0 \text{ as } \delta \to 0 \qquad (12)$$

Another regularized iterative method has been investigated by Cesini et al.[4] (in Cesini's notation, $D_X B_\Omega = S$ and hence $S^2 = K$) and Ross [14] (see also [15]). The motivation for this method is based on the conversion of the first kind integral equation (4) into an equation of the second kind $(K + \beta I)f_\beta = D_X B_\Omega g$ with $0<\beta<1$ (this is equivalent to simplified regularization). The equation is then converted to

$$f^\alpha = D_X B_\Omega g + [\alpha I - K]f^\alpha$$

where the parameter has been changed to $\alpha = 1-\beta$. This equation is then solved iteratively by the method

$$f_n^\alpha = D_X B_\Omega g + [\alpha I - K]f_{n-1}^\alpha \,, \quad f_0^\alpha = 0 \qquad (13)$$

For this method we have $S_\alpha(t) = \alpha - t$ and again $m(\alpha) = \alpha$. Therefore the regularity condition for the method of Cesini et al. is also given by (12).

Asymptotic order of convergence results in terms of the error level δ may be obtained for each of the methods of this section if the object f satisfies a certain condition. Namely, if $f = Kw$ for some w, then one obtains as in (11)

$$\|f^\alpha - f\|^2 = \sum_{n=1}^{\infty} b_n^2(\alpha) |(w,v_n)|^2$$

where $b_n(\alpha) = O(1-\alpha)$. Therefore the total error $\|f_n^{\alpha,\delta} - f\|$ is of

the order $O(1-\alpha) + O(\alpha^n) + \delta/(1-\alpha)$ and a choice of the parameter
and iteration number of the form $1-\alpha \sim \sqrt{\delta}$, $n \sim \ln(1-\alpha)/\ln \alpha$ gives
a convergence rate for the total error of the order $O(\sqrt{\delta})$.

4   ANOTHER TACK: FREQUENCY LIMITING

Equation (2) can be Fourier transformed to $D_\Omega B_X \hat{f} = \hat{g}$. Since
$B_X \hat{f} = \hat{f}$ and $D_\Omega \hat{g} = \hat{g}$, this may be written

$$\hat{f} = \hat{g} + (I - D_\Omega B_X)\hat{f} = \hat{g} + (I - D_\Omega)B_X \hat{f} .$$

This suggests the iterative method

$$\hat{f}_n = \hat{g} + (I - D_\Omega)B_X \hat{f}_{n-1}, \qquad \hat{f}_0 = 0.$$

Abbiss et al. [1] and Byrne and Wells [3] have noted that in
practice the approximations $\hat{f}_n$ must be truncated to some interval
$[-\Gamma,\Gamma]$ with $\Gamma > \Omega$. This leads to the modified method

$$\hat{f}_n^\Gamma = \hat{g} + (D_\Gamma - D_\Omega)B_X \hat{f}_{n-1}^\Gamma \tag{14}$$

Each of the approximations $\hat{f}_n^\Gamma$ has support contained in $[-\Gamma,\Gamma]$ and
we will show below that this frequency limiting is itself a form
of regularization.

Using the techniques of the first section, it is not hard to
show that $\|(D_\Gamma - D_\Omega)B_X\| = \alpha_\Gamma < 1$ and hence if $\hat{f}^\Gamma$ is defined by

$$\hat{f}^\Gamma = \hat{g} + (D_\Gamma - D_\Omega)B_X \hat{f}^\Gamma \tag{15}$$

then by (14),

$$\|\hat{f}_n^\Gamma - \hat{f}^\Gamma\| = O(\alpha_\Gamma^n) \tag{16}$$

Abbiss et al. have observed that $\hat{f}^\Gamma$ represents an oversmoothed
version of $\hat{f}$ and Byrne and Wells have shown that $\hat{f}^\Gamma$ is optimal in
a sense described in Proposition 4 below. We will take a

simplified and more geometric approach to this characterization of $\hat{f}^{\Gamma}$.

LEMMA 3. *If* $\hat{\varphi} \in R(D_{\Gamma})$ *and* $D_{\Gamma}\hat{\varphi} = \hat{g}$, *then* $(\hat{\varphi} - \hat{f}^{\Gamma}, (I - B_{X})\hat{f}^{\Gamma}) = 0$.

Proof. Since $\hat{\varphi} - \hat{g} = (D_{\Gamma} - D_{\Omega})\hat{\varphi}$ and $\hat{f}^{\Gamma} - \hat{g} = (D_{\Gamma} - D_{\Omega})B_{X}\hat{f}^{\Gamma}$, we have from (15)

$$(\hat{\varphi} - \hat{f}^{\Gamma}, (I - B_{X})\hat{f}^{\Gamma}) = ((D_{\Gamma} - D_{\Omega})(\hat{\varphi} - B_{X}\hat{f}^{\Gamma}), (I - B_{X})\hat{f}^{\Gamma})$$

$$= (\hat{\varphi} - B_{X}\hat{f}^{\Gamma}, (D_{\Gamma} - D_{\Omega})\hat{f}^{\Gamma} - (D_{\Gamma} - D_{\Omega})B_{X}\hat{f}^{\Gamma})$$

$$= (\hat{\varphi} - B_{X}\hat{f}^{\Gamma}, 0) = 0, \quad \text{since } D_{\Omega}\hat{f}^{\Gamma} = \hat{g}. \quad \square$$

The Pythagorean theorem now implies

PROPOSITION 4. [3] *If* $\hat{\varphi} \in R(D_{\Gamma})$ *and* $D_{\Omega}\hat{\varphi} = \hat{g}$, *then* $\|(I - B_{X})\hat{f}^{\Gamma}\| \leq$ $\|(I - B_{X})\hat{\varphi}\|$.

We can interpret this result in the following way: if we denote by $D_{\Omega,\Gamma}$ the operator D_{Ω} acting on the space $R(D_{\Gamma})$ with inner product

$$[\hat{\varphi}, \hat{\psi}] = ((I - B_{X})\hat{\varphi}, \hat{\psi})$$

then $\hat{f}^{\Gamma} = D_{\Omega,\Gamma}^{\dagger}\hat{g}$, where $D_{\Omega,\Gamma}^{\dagger}$ is the Moore-Penrose inverse of $D_{\Omega,\Gamma}$.

Byrne and Wells [3] have shown that $\hat{f}^{\Gamma} \to \hat{f}$ as $\Gamma \to \infty$. If the iteration number n is related to the frequency cut-off Γ in such a way that $\alpha_{\Gamma}^{n} \to 0$, then by (16), $\hat{f}_{n}^{\Gamma} \to \hat{f}$. Finally, if only noisy data g^{δ} is available, then one can show as above that $\hat{f}_{n}^{\Gamma,\delta} \to \hat{f}$ as $\delta \to 0$ if $\delta = o(1 - \alpha_{\Gamma})$.

REFERENCES

[1] J.B. Abbiss, C. De Mol, and H.S. Dhadwal, *Regularized iterative and non-iterative procedures for object restoration from experimental data*, Optica Acta 30(1983), pp.107-124.

[2] J. Baumeister, *Stable Solution of Inverse Problems*, Vieweg & Sohn, Braunschweig/Wiesbaden,1987.

[3] C.L. Byrne and D.M. Wells, *Optimality of certain iterative and non-iterative data extrapolation procedures*, J. Math. Anal. Appl. 111(1985),pp.26-34.

[4] G. Cesini, G. Guattari, and G. Lucarini, *An iterative method for restoring images*, Optica Acta 25(1978),pp.501-508.

[5] H.W. Engl and C.W. Groetsch, eds., *Inverse and Ill-Posed Problems*, Academic Press, Boston, 1987.

[6] R.W. Gerchberg, *Super-resolution through error energy reduction*, Optica Acta 21(1974),pp.709-720.

[7] C.W. Groetsch, *The Theory of Tikhonov Regularization for Fredholm Equations of the First Kind*, Pitman, London, 1984.

[8] B. Hofmann, *Regularization of Applied Inverse and Ill-Posed Problems*, Teubner, Leipzig, 1986.

[9] M.M. Lavrent'ev, V.G. Romanov, and S.P. Shisshatskii, *Ill-posed Problems of Mathematical Physics and Analysis*, American Mathematical Society, Providence, 1986 (translated from the Russian).

[10] L. Landweber, *An iteration formula for Fredholm integral equations of the first kind*, Amer. J. Math. 73(1951), pp.615-624.

[11] S.G. Lipson and H. Lipson, *Optical Physics*, 2nd ed.,Cambridge University Press, Cambridge, 1981.

[12] V.A. Morozov, *Methods of Solving Incorrectly Posed Problems*, Springer-Verlag, New York, 1984 (translated from the Russian).

[13] M.Z. Nashed, *Operator-theoretic and computational approaches to ill-posed problems with applications to antenna theory*, IEEE Trans. Antennas & Prop. AP-29(1981),pp.220-231.

[14] G. Ross, *Iterative methods in information processing for object restoration*, Optica Acta 29(1982),pp.1523-1542.

[15] C.K. Rushforth, *Signal restoration, functional analysis, and Fredholm integral equations of the first kind*, Image Recovery: Theory and Application, H. Stark, ed., Academic Press, Orlando, 1987,pp. 1-27.

[16] A.N. Tikhonov and V.Y. Arsenin, *Solutions of Ill-posed Problems*, Wiley, New York, 1977 (translated from the Russian).

Inversion of a First-Kind Integral Equation as a Plasma Diagnostic

JOSEPH F. MCGRATH and STEPHEN WINEBERG Computational Mathematicians, KMS Fusion, Inc., Ann Arbor, Michigan
GEORGE CHARATIS and ROBERT J. SCHROEDER Experimental Physicists, KMS Fusion, Inc., Ann Arbor, Michigan

1 INTRODUCTION

From an x-ray image recorded on film, it is our task to calculate the radiation profile of a plasma. A description of the diagnostic system for collecting the data is contained in Section 2. In the experiment, the radiation from a three dimensional plasma is projected onto the two dimensional surface of the film. Inverting this projection for the radiation density is a classic poorly posed problem.

A one dimensional slice of the data is the projection of a two dimensional cross section of the plasma. If the plasma is cylindrically symmetric, a slice of the data corresponding to a cross section that is orthogonal to the axis of symmetry is a collection of discrete values of the Abel transform of the radiation profile.

The numerical solution of the Abel equation is a very familiar topic to many of the attendees of this conference. Thus, it is appropriate to discuss at this meeting an important application of the Abel transform. The plasma diagnostic described in Section 2 provides the application. The numerical solution for the radiation profile in the plasma is discussed in Section 3. Using the techniques expounded in reports by Wing (1984) and Faber and Wing (1987), a one dimensional slice of real data is inverted for the radiation profile in a cross section of the plasma based on the assumption of cylindrical symmetry. The solution is computed in terms of the singular value decomposition of the Abel transform.

The original charter for this investigation, however, calls for a three dimensional solution. That is, we would like to invert each projection of a cross section of the plasma for the radiation profile without the assumption of cylindrical symmetry. In this case, the slice of data represents the Radon transform of the cross section of the radiation profile. Some interesting results concerning the solution of the Radon equation are discussed in Section 4. Although the right kind of data has not yet been collected experimentally, the new methods for the Radon equation represent promising results. With data collected from a very limited number of projections of the plasma, the numerical solution will provide some information about the radiation profile without any symmetry assumptions.

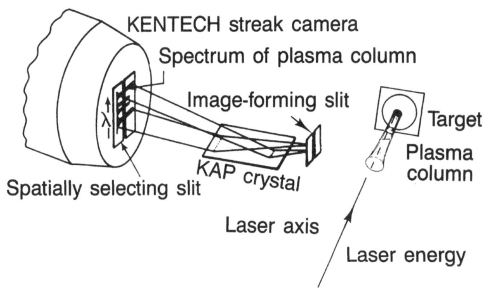

Figure 1. The radiation from a column of plasma is focused by an image-forming x-ray slit onto a KAP (potassium acid phthalate) crystal. A second x-ray slit selects a cross section of the spectrally separated reflections. The KENTECH brand x-ray camera streaks the images across the film. Thus, the film records time-resolved images of the x-ray intensities from a cross section of the plasma.

2 THE PLASMA DIAGNOSTIC

The radiation energy of a plasma contains information about its electron density and temperature. These are important properties in many laser-material interaction experiments. Thus, diagnostic equipment has been designed to take measurements of the radiation energy. A device for this purpose is illustrated in Figure 1.

On the right hand side of the figure, a laser is irradiating the surface of a flat target. The vaporized material is ionized and blown off the surface in a configuration that is approximately cylindrically symmetric about the axis of the laser. The radiation energy from the plasma is focused through an x-ray slit onto a crystal that reflects the energy into a streak camera. The crystal spectrally decomposes the energy into separate inverted images. A second x-ray slit in front of the camera selects a cross section of the images orthogonal to the laser axis. The camera streaks this slice of the image across the film as a function of time. Thus, Figure 2 shows the image, as it is recorded on the film, of a time-resolved cross section of the plasma very close to the target surface (within 100 micrometers).

The two prominent streaks at the top of the figure show the energy of the target at two different wavelengths. Time is increasing from left to right. The electromagnetic energy at 8.459 angstroms turns on very soon after the formation of the plasma by the laser. The 8.021 angstrom x-rays appear about 240 picoseconds later. The cross section of the target is still radiating as its image is swept off the edge of the film on the right of the figure.

The streaks of light across the bottom half of the figure contain images of the target at other frequencies. It is apparent that some sophisticated image processing will have to be done to recover the desired information from these overlapping images of the target that are partially washed out by the continuum radiation. The corresponding image processing research at KMSF is not within the scope of this paper.

Shot number 8099

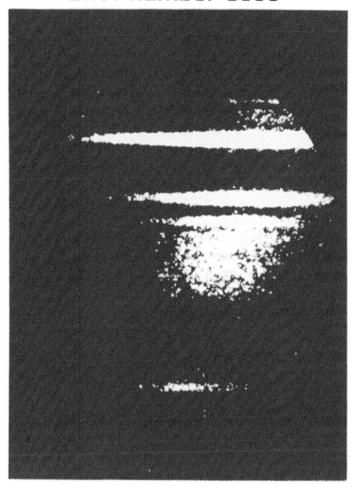

← 8.459 angstroms

← 8.021 angstroms

Figure 2. The spectrum of a sodium target from KMSF shot number 8099 produced this negative of time-resolved x-ray intensities. The streaks on the film are the images at different x-ray frequencies of a cross section of the plasma close to the surface of the target as shown in Figure 1. Thus, while time increases from left to right, spatial and spectral information are both displayed in the vertical direction. Wavelength increases from one horizontal streak to the next. Then, within each streak, the film density readings represent values of the intensity from an edge-on view of the cross section of the plasma.

A microdensitometer is used to measure film densities of the x-ray images of the plasma shown in Figure 2. The digitized readings are converted to intensity values according to the calibration data for the film. The rocking curve of the crystal and the time-of-flight distortion are two other biases that have to be deconvolved from the data. Figure 3 shows a contour plot of the corrected digitized data from the area of the film shown in Figure 2 surrounding the 8.459 angstrom and 8.021 angstrom radiation. Note that the plot in Figure 3 has been stretched in the wavelength direction. Also, the values of the intensity and the spatial dimension have been normalized and scaled to the width of the images. A column of the data, that is, a vertical cut through a section of Figure 3, represents discrete intensity readings from a cross section of the plasma at a fixed point in time.

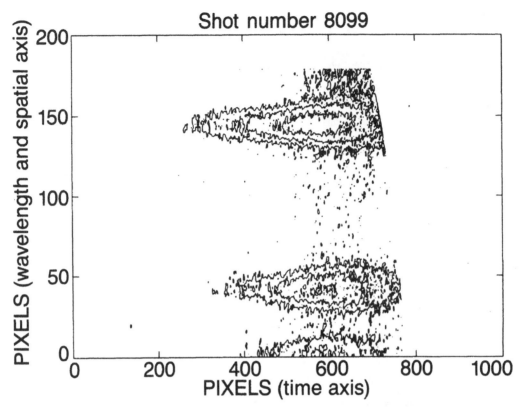

Figure 3. The top two streaks shown in Figure 2 were digitized with a microdensitometer and the data were converted to intensity readings. These contour plots were constructed with an elongated vertical coordinate which is both the wavelength and the spatial dimension.

3 NUMERICAL SOLUTION--CYLINDRICAL SYMMETRY

The column of data corresponding to a vertical line through the 8.459 angstrom radiation at pixel 507 on the horizontal axis of Figure 3 has been selected for demonstrating the analysis discussed in this paper. It is reproduced in Figure 4. Each point on this plot of radiation intensity represents the value of the integral of the power density in normalized units along a line of sight through the plasma. Figure 5 illustrates the mathematical model for the plasma diagnostic. Note that the model is based on the assumption of a refractionless plasma. That is, the electron density is sufficiently low for the x-ray energy to travel along the line of sight.

The value of $\phi(y)$ in Figure 5 is the measured intensity due to the power density $f(x,y)$ along the horizontal line with coordinate y. Thus, $\phi(y)$ is the line integral of $f(x,y)$.

$$\phi(y) = \int_{-\sqrt{R^2-y^2}}^{\sqrt{R^2-y^2}} f(x,y)\ dx \qquad (1)$$

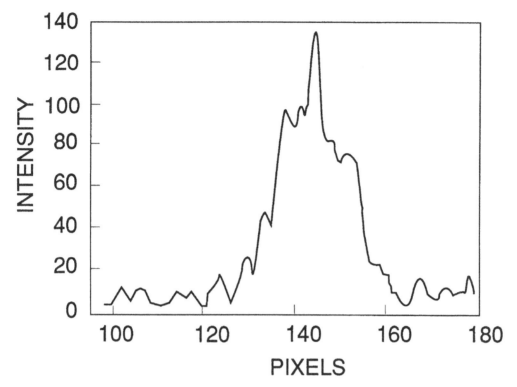

Figure 4. The x-ray intensity at 8.459 angstroms is plotted for a cross section of the plasma close to the surface of the target approximately 850 picoseconds after it turns on. The data shown here corresponds to time pixel 507 between spatial pixels 100 and 180 on Figure 3.

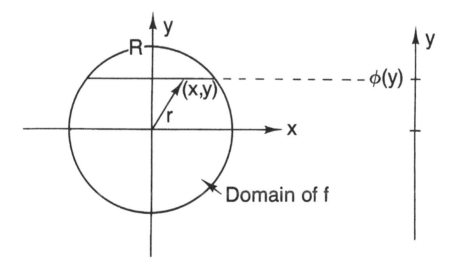

Figure 5. The intensity $\phi(y)$ is the line integral of the radiation density f along the path through the plasma at coordinate y.

If the radiation were symmetric about the laser axis, then the data shown in Figure 4 would have to be symmetric with respect to the peak in the intensity at pixel number 144 to within the noise level of the system. Although it clearly does not possess this property, it is instructive to examine the data based on an assumption of radial symmetry. That is, f depends only on r, the distance from the origin. Hence,

$$f(r) = f(x,y) \tag{2}$$

A change of variables converts the line integral for $\phi(y)$ in equation (1) into

$$2 \int_y^R \frac{r\, f(r)\, dr}{\sqrt{r^2 - y^2}} = \phi(y) \tag{3}$$

which is referred to as the Abel transform of the function f (or of r f(r)). Equation (3) is a first kind Volterra integral equation. In the following paragraphs, we will present our attempts to solve this mathematically poorly posed and numerically ill conditioned problem.

There is more than one way to represent analytically the inverse to the Abel transform. Readable explanations and references for the inversion formulas and for the derivation of the transform itself can be found in Linz (1985) and Vest (1979). Initially, we implemented the inversion formula

$$r\, f(r) = \frac{d}{dr} \int_r^R \frac{y\, \phi(y)}{\sqrt{y^2 - r^2}}\, dy \tag{4}$$

for the solution to equation (3). One alternative to equation (4) involves the integral and differential operators in reverse order. However, Linz advises smoothing the data first and then evaluating the solution with equation (4). A plausibility argument suggests that smoothing the data first and smoothing it further with the integral operator before differentiating produces a more stable numerical scheme. We have implemented this approach with some success.

The key device involved in our initial approach to the numerical evaluation of equation (4) is the midpoint quadrature rule for the numerical integration. Thus,

$$\int_{r_k}^R \frac{y\, \phi(y)}{\sqrt{y^2 - r_k^2}}\, dy \simeq \sum_{i=k}^M \phi(y_i) \int_{r_{i-1}}^{r_i} \frac{y}{\sqrt{r_k^2 - y_i^2}}\, dy \tag{5}$$

where the summation is over the M measurements at the uniformly spaced coordinates y_i and where (r_{i-1}, r_i) is the subinterval of (r, R) centered on y_i. The value of this trick lies with the fact that each of the terms of the summation in formula (5) is integrable. Finally, we employed second order formulas for the numerical differentiation at each r_k to obtain the approximate values of $f(r_k)$.

While we have achieved a certain measure of success with the above method, the discussion in the previous two paragraphs is now purely of academic interest. In a primer on the more general topic of numerical methods for first kind integral equations, Wing (1984) presents the mathematically sound approach applicable to equation (3). While omitting the mathematical foundation, Faber and Wing (1987) have written up a useful guide for implementing the numerical techniques corresponding to the mathematically correct approach. Accordingly, the solution to equation (3) can be evaluated numerically by representing f in terms of the singular vectors of the discrete analog to equation (3).

For simplicity, y is assumed to be positive in equation (3) and in the following analysis. It is easy to generalize the discussion to allow y to be negative. Indeed, our implementation for computing solutions takes measurements at negative values of y into account.

Let

$$K(y,r) = \begin{cases} \dfrac{2\,r}{\sqrt{r^2-y^2}} & \text{for } r > y \\[2ex] 0 & \text{for } r \le y \end{cases} \tag{6}$$

Select an arbitrary partition of the interval [0,R] defined by

$$0 = r_o < r_1 < \ldots < r_N = R \tag{7}$$

Also define

$$r_{j-\frac{1}{2}} = \frac{1}{2}\,(r_{j-1} + r_j) \tag{8}$$

for j = 1, 2, ..., N. Then the generalized midpoint quadrature rule yields the following second order approximation for the Abel transform of f:

$$2 \int_y^R \frac{r\,f(r)\,dr}{\sqrt{r^2 - y^2}} = \int_0^R K(y,r)\,f(r)\,dr$$

$$\simeq \sum_{j=1}^N f(r_{j-\frac{1}{2}}) \int_{r_{j-1}}^{r_j} K(y,r)\,dr \tag{9}$$

If measurements of the Abel transform were taken at the points

$$0 \leq y_1 < y_2 < \ldots < y_m \leq R \qquad (10)$$

then for

$$r_{k-1} \leq y_i < r_k \qquad (11)$$

$$\phi(y_i) = 2 \int_{y_i}^{R} \frac{r \, f(r) \, dr}{\sqrt{r^2 - y_i^2}} = \int_{0}^{R} K(y_i, r) \, f(r) \, dr$$

$$\simeq \sum_{j=1}^{N} f(r_{j-\frac{1}{2}}) \int_{r_{j-1}}^{r_j} K(y_i, r) \, dr$$

$$= f_{k-\frac{1}{2}} \int_{y_i}^{r_k} K(y_i, r) \, dr + \sum_{j=k+1}^{N} f_{j-\frac{1}{2}} \int_{r_{j-1}}^{r_j} K(y_i, r) \, dr$$

$$= 2 \sqrt{r_k^2 - y_i^2} \, f_{k-\frac{1}{2}} + 2 \sum_{j=k+1}^{N} \left(\sqrt{r_j^2 - y_i^2} - \sqrt{r_{j-1}^2 - y_i^2} \right) f_{j-\frac{1}{2}} \qquad (12)$$

where $f_{j-1/2}$ is the approximate value of $f(r_{j-1/2})$.

For a given partition $\{r_j\}_{j=0}^{N}$ and a specified set of nodes $\{y_i\}_{i=1}^{M}$ at which the measurements are to be made, define the matrix K by

$$\frac{1}{2} K_{ij} = \begin{cases} \sqrt{r_j^2 - y_i^2} & \text{if } r_{j-1} < y_i < r_j \\[2mm] \sqrt{r_j^2 - y_i^2} - \sqrt{r_{j-1}^2 - y_i^2} & \text{if } \qquad y_i \leq r_{j-1} \\[2mm] 0 & \text{if } \quad r_j \leq y_i \end{cases} \qquad (13)$$

for $1 \leq j \leq N$ and $1 \leq i \leq M$. Thus, K is an M by N matrix. For example, if the r's and y's are distributed between 0 and R as shown in Figure 6, then K is the following 5 by 3 matrix:

$$\begin{pmatrix} 2\sqrt{r_1^2-y_1^2} & 2\left(\sqrt{r_2^2-y_1^2} - \sqrt{r_1^2-y_1^2}\right) & 2\left(\sqrt{r_3^2-y_1^2} - \sqrt{r_2^2-y_1^2}\right) \\[2ex] 2\sqrt{r_1^2-y_2^2} & 2\left(\sqrt{r_2^2-y_2^2} - \sqrt{r_1^2-y_2^2}\right) & 2\left(\sqrt{r_3^2-y_2^2} - \sqrt{r_2^2-y_2^2}\right) \\[2ex] 0 & 2\sqrt{r_2^2-y_3^2} & 2\left(\sqrt{r_3^2-y_3^2} - \sqrt{r_2^2-y_3^2}\right) \\[2ex] 0 & 0 & 2\sqrt{r_3^2-y_4^2} \\[2ex] 0 & 0 & 2\sqrt{r_3^2-y_5^2} \end{pmatrix} \qquad (14)$$

The solution to the integral equation (3) can be approximated by the solution to the discrete problem

$$Kf = \phi \qquad (15)$$

where

$$f = (f_{1/2},\ f_{3/2},\ \ldots,\ f_{N-1/2})^T \qquad (16)$$

and

$$\phi = (\phi_1,\ \phi_2,\ \ldots,\ \phi_M)^T \qquad (17)$$

When $M > N$ as in the matrix (14), a least squares solution to the overdetermined system (15) supplies the desired results. Even with $M = N$, the singular value decomposition of the matrix K provides the optimal numerical solution technique. This is the approach advocated by Faber and Wing (1987). An explanation of the theory of the singular value decomposition method is contained in the reference by Forsythe, Malcolm, and Moler (1977).

Figure 6. The matrix K in equation (15) is defined by equation (13). K has the entries shown in the expression (14) if the five intensity readings at the coordinates y_1 through y_5 are distributed as shown over the three radial intervals (r_0,r_1), (r_1,r_2), and (r_2,r_3).

Let

$$S = \text{diag } (\sigma_1, \; \sigma_2, \; \ldots, \; \sigma_N) \tag{18}$$

be the N by N diagonal matrix of the singular values of K in decreasing order. Let U and V be the M by M and the N by N orthogonal matrices made up of the left and right singular vectors of K, respectively. Then

$$K = U \; S \; V^T \tag{19}$$

and

$$f = V \; S^{-1} \; U^T \; \phi \tag{20}$$

minimizes the expression

$$||\; Kf \; - \; \phi \; || \tag{21}$$

where U^T in equation (20) is the transpose of the matrix of the first N columns of U and where S is augmented with M - N rows of zeros in equation (19).

Note that N must be large enough to accurately resolve the solution yet small enough so that the number of large singular values, that is, the number of significant left singular vectors, is no greater than the number of measurements M.

We have solved equation (15) twice, once with the data to the right of the peak and again with the data to the left. The results are displayed in Figure 7. While the two solutions show similar qualitative behavior,

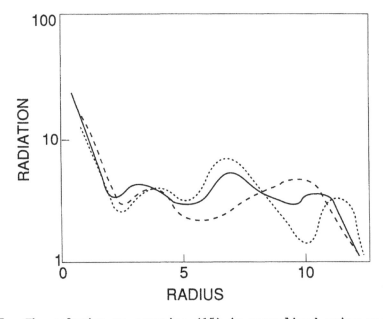

Figure 7. The solution to equation (15) in normalized units was calculated for three sets of data: The dashed line represents the solution for the data to the right of the peak intensity shown in Figure 4, the dotted line for the data to the left, and the solid for the two sets combined.

they are not the same. It remains for us to solve the problem without the assumption of cylindrical symmetry.

In the preceding analysis, only positive values of y were taken into account. However, equation (3), the kernel (6), and the derivation that is composed of the equations and expressions (9) through (14) require only slight modifications to take into account measurements $\phi(y_i)$ for any

$$|y_i| < R \tag{22}$$

Our computer code incorporates this generalization. Consequently, we could solve equation (15) a third time using the data from both sides of the peak in Figure 4. This result is also shown on Figure 7.

The computer program generates the solution in terms of the singular vectors corresponding to the largest singular values. Also, as suggested in Faber and Wing (1987), the least squares solution is computed when M > N, that is, when f is calculated at a fewer number of points than there are measurements. Thus, there are two mechanisms for smoothing the solution: calculating the best value of K $\leq$ N and choosing N < M where K, N, and M are the number of singular vectors used to resolve the solution f, the number of degrees of freedom in the solution (the number of unknowns), and the number of measurements, respectively.

For the solutions shown in Figure 7, there were M = 22 measurements from the right side of the peak in Figure 4 and M = 19 measurements from the left. The dashed curve shows the solution from the data to the right of the peak with K = N = 14. The dotted curve came from the data to the left of the peak with K = N = 12. Finally, the solid curve on Figure 7 is the solution from the M = 40 measurements from both sides of the peak with N = 26 and K = 22. For all three curves, a cubic spline that respects monotonicity was used to interpolate the solutions.

4 NUMERICAL SOLUTION—NO SYMMETRY

If the plasma under investigation were in fact cylindrically symmetric, then the radiation profile from any side view would be symmetric with respect to the center line of the incident laser beam. That is, the data in Figure 4 would be symmetric about a vertical line at the peak in the intensity. However, the results in the previous section do not support the assumption of cylindrical symmetry. In any case, a two-dimensional solution for the radiation from any cross section of the plasma shown in Figure 2 was the objective of the original charter for our investigations.

Two dimensional information concerning the radiation density f(x,y) in a cross section of the plasma can be obtained by collecting data from several directions. In Figure 8, the mathematical model for the diagnostic system is shown at an angle with the horizontal through a cross section of the radiating plasma. The value of the intensity is given by

$$\iint_{RxR} f(x,y)\ \delta[s-(x\cos\theta+y\sin\theta)]\ dxdy = \phi(s,\theta) \tag{23}$$

which is the Radon transform of f. It is equation (23) that we need to solve in place of equation (3).

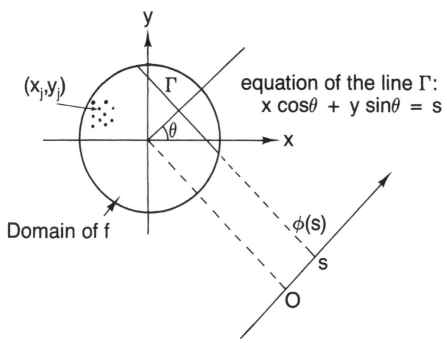

Figure 8. The intensity $\phi(s)$ is the line integral of the radiation density f along the path Γ through the plasma. The nodal points shown in
the second quadrant of the x-y coordinate system illustrate the mesh that
is imposed on the entire domain of f. A basis function h_j is defined at
each node (x_j, y_j) in the mesh.

Equation (23) represents the classical computer assisted tomography
problem. Conventional wisdom has it that many viewing angles are
required for an adequate solution for the unknown f. However, the experimental apparatus for the laser induced plasma in the laboratory at KMSF
is crowded with other diagnostic devices. The expense of the system
shown in Figure 1 and the dynamics of the experiment are additional
limiting factors. Thus, while more than one view is necessary, obtaining
data from more than two or three directions is out of the question.
    In general, solutions to poorly posed problems are found by imposing
restrictions on the solution space. The element of a finite dimensional
subspace that comes closest to satisfying the equations is the accepted
approximate solution to the problem. Thus, the poorly posed problem is
converted into a well posed problem, albeit one that is generally ill
conditioned.
    The tomographic problem is poorly posed when viewed from just one
direction. With two views, it can be technically well posed, but it is
so poorly conditioned that it is not possible to solve. Obtaining data
from additional projections is one way to make the problem sufficiently
well conditioned to solve. It has been our task to find alternatives to
the requirement for multiple projections of the target. This means additional restrictions on the solution space.
    The high contrast solution of a tomographic problem such as a brain
scan requires many viewing angles. Smooth solutions do not require as
many views. While the source of the data shown in Figure 3 may contain

some abrupt behavior, we will settle for the element of a space of smooth functions that is closest to satisfying equation (23).

There is some motivation for the smoothness assumption for solutions to equation (23). Ludwig (1966) established a number of results concerning the Radon transform. With the techniques in Ludwig's paper, it can be shown that equation (23) is equivalent to an expression involving Fourier transforms. The one-dimensional Fourier transform of ϕ with respect to its first variable is

$$
\begin{aligned}
F_1\phi\,(u,\theta) &= \int_R \phi(s,\theta)\,e^{-2\pi ius}\,ds \\[2ex]
&= \int_R \iint_{R\times R} f(x,y)\,\delta(\Gamma)\,dxdy\,e^{-2\pi ius}\,ds \\[2ex]
&= \iint_{R\times R} f(x,y) \int_R \delta(\Gamma)\,e^{-2\pi ius}\,ds\,dxdy \\[2ex]
&= \iint_{R\times R} f(x,y)\,e^{-2\pi iu(x\cos\theta\,+\,y\,\sin\theta)}\,dxdy \\[2ex]
&= F_2 f\,(u\cos\theta,u\sin\theta) \qquad\qquad\qquad (24)
\end{aligned}
$$

which is the two dimensional transform of f. Thus, equation (23) is equivalent to

$$
F_2\,f = F_1\,\phi \qquad\qquad (25)
$$

If f and ϕ are sufficiently smooth in physical space, they can be constructed from their lowest frequency modes in Fourier space.

In a recent paper, Gouldin and Ravichandran (1987) revealed the value of the above interpretation of the Radon transform. They demonstrated the effectiveness of the approach with an example.

The Radon transform (23) was applied to a known function to generate measurement data for each of a number of projections at different viewing angles θ as shown in Figure 8. Then, by imposing assumptions of smoothness, the original function (the solution) was very accurately reconstructed from only three viewing angles by solving the Fourier equation (25). Gouldin and Ravichandran compared the results to a back projection method that utilized the same data to approximate the solution. The latter method which does not incorporate any assumptions of smoothness performed very poorly in comparison. With data from many viewing angles, the back projection method became the better choice. Although both methods produced the correct solution, the back projection method remained more computationally efficient.

The back projection method in computer assisted tomography produces values for the solution at discrete points in the plane. Alternatively, as described in the reference mentioned above, Gouldin and Ravichandran

(1987) impose smoothness conditions by expanding the solution in a fashion similar to the approach used in the finite element method.

In Figure 8, a few of the nodes are shown for a uniform triangular mesh covering the domain of the solution. A basis function is defined at each of the nodes and f is expanded in terms of the basis functions to produce the approximation

$$f(x,y) \simeq \sum_{j=1}^{N} c_j \, h_j(x,y) \qquad (26)$$

Rather than basis functions that are only piecewise smooth such as the ones typically used in finite element analysis, the h_j's are chosen to be very smooth Gaussian-like functions centered at the nodes (x_j, y_j). Let

$$h_j(x,y) = \Psi(x-x_j) \, \Psi(y-y_j) \qquad (27)$$

where Ψ is a globally smooth even function that displays Gaussian-like behavior centered at the origin. That is, $\Psi(r)$ peaks at r=0 and decays exponentially away from the origin.

If the data are also written in terms of the function ψ such that

$$\phi(s,\theta) = \sum_{k=1}^{M} p_k(\theta) \, g_k(s) \qquad (28)$$

where

$$g_k(s) = \Psi(s-s_k) \qquad (29)$$

then both f and ϕ have representations that are consistent with equation (25) and hence also with equation (23).

From the definition of h_j in equation (27), it follows that

$$F_2 \, h_j(u \cos\theta, \, u \sin\theta) = \iint_{R \times R} h_j(x,y) \, e^{-2\pi i u(x \cos\theta, \, y \sin\theta)} \, dxdy$$

$$= \iint_{R \times R} \Psi(x-x_j) \, \Psi(y-y_j) \, e^{-2\pi i u(x \cos\theta, \, y \sin\theta)} \, dxdy \qquad (30)$$

Transform the integration variables with the translation

$$\begin{bmatrix} \xi \\ \zeta \end{bmatrix} = \begin{bmatrix} x-x_j \\ y-y_j \end{bmatrix} \qquad (31)$$

to obtain

$$F_2 \, h_j(u \cos\theta, \, u \sin\theta) = \int\int\limits_{\mathbf{R}\times\mathbf{R}} \Psi(\xi) \, \Psi(\zeta) \, e^{-2\pi i[(\xi+x_j)\cos\theta+(\zeta+y_j)\sin\theta]} \, d\xi d\zeta$$

$$= B_j(u,\theta) \int\int\limits_{\mathbf{R}\times\mathbf{R}} \Psi(\xi) \, \Psi(\zeta) \, e^{-2\pi iu(\xi \cos\theta + \zeta \sin\theta)} d\xi d\zeta$$

$$= B_j(u,\theta) \int\limits_{\mathbf{R}} \Psi(\xi) \, e^{-2\pi iu\xi \cos\theta} d\xi \int\limits_{\mathbf{R}} \Psi(\zeta) \, e^{-2\pi iu\zeta \sin\theta} d\zeta$$

$$= B_j(u,\theta) \, F_1 \, \Psi(u \cos\theta) \, F_1 \, \Psi(u \sin\theta) \tag{32}$$

where

$$B_j(u,\theta) = e^{-2\pi iu(x_j \cos\theta + y_j \sin\theta)} \tag{33}$$

Similarly, the definition of g_k in equation (29) with the translation of the integration variable

$$\xi = s - s_k \tag{34}$$

yields

$$F_1 \, g_k(u) = \int\limits_{\mathbf{R}} \Psi(s-s_k) \, e^{-2\pi ius} \, ds$$

$$= \int\limits_{\mathbf{R}} \Psi(\xi) \, e^{-2\pi iu(\xi+s_k)} \, d\xi$$

$$= E_k(u) \int\limits_{\mathbf{R}} \Psi(\xi) \, e^{-2\pi iu\xi} \, d\xi$$

$$= E_k(u) \, F_1 \, \Psi(u) \tag{35}$$

where

$$E_k(u) = e^{-2\pi i u s_k} \tag{36}$$

Substitute equations (26) and (28) into equation (25) and use equations (32) and (35) to obtain

$$F_2 \sum_{j=1}^{N} c_j h_j = F_1 \sum_{k=1}^{M} p_k g_k \tag{37}$$

or

$$\sum_{j=1}^{N} c_j B_j(u,\theta) \ F_1 \Psi(u \cos\theta) \ F_1 \Psi(u \sin\theta)$$

$$= \sum_{k=1}^{M} p_k(\theta) \ E_k(u) \ F_1 \Psi(u) \tag{38}$$

It remains to choose the function Ψ and to solve equation (38) for the constants c_j given several viewing angles θ and measurements at many points s_k.

As discussed previously, the additional restriction of smoothness for the solution f has been assumed to help solve the problem with a limited number of viewing angles. The desired behavior for f is achieved by making an appropriate selection for Ψ. Since f has compact support, Ψ should be space limited. The smoothness condition indicates that Ψ should be frequency limited. That is, the high frequency Fourier components should be damped out. To achieve these conditions, we will borrow a concept from digital filtering theory.

In the field of signal processing, it is a very tricky problem to define a filter that does a good job of simultaneously limiting the data in space and in the frequency domain. Gouldin and Ravichandran (1987) suggest the Kaiser window function for the Fourier transform of Ψ in the definitions (27) and (29). In a text by Rabiner and Gold (1975), the Kaiser window function is defined to be

$$w(n) = \frac{I_0\left(\beta \sqrt{1-\left(\frac{2n}{N-1}\right)^2}\right)}{I_0(\beta)} \tag{39}$$

for $-\frac{1}{2}(N-1) \le n \le \frac{1}{2}(N-1)$. The parameter β controls the tradeoff between the high frequency modes and the width of the main lobe of the filter. I_0 is the zero order modified Bessel function.

When the Fourier transforms in equation (38) are replaced by their discrete counter parts, it becomes a linear system of the form

$$Ac = b \qquad\qquad (40)$$

to be solved for the unknowns c_j. Once the system (40) is solved, the solution throughout the domain is given by equation (26).

Based on the analysis in this paper, we are currently developing the background information for the design of an experiment. The proposal will contain details concerning the requirements for the diagnostic systems shown in Figure 1.

REFERENCES

V. Faber and G. Milton Wing, The Abel Integral Equation, Los Alamos National Laboratory document LA-11016-MS (1987).

G. E. Forsythe, M. A. Malcolm, and C. B. Moler, Computer Methods for Mathematical Computations, Prentice Hall, Englewood Cliffs, New Jersey (1977).

F. C. Gouldin and M. Ravichandran, "Inversion Scheme for Analysis of Line-of-Sight Spectroscopic Data," Forefronts, vol. 3, no. 8, 3-6, Cornell University (1987).

Peter Linz, Analytical and Numerical Methods for Volterra Equations, SIAM Studies in Applied Mathematics, Philadelphia (1985).

Donald Ludwig, "The Radon Transform on Euclidean Space," Communications on Pure and Applied Mathematics, vol. XIX, 49-81 (1966).

L. R. Rabiner and B. Gold, Theory and Application of Digital Signal Processing, Prentice Hall, Englewood Cliffs, New Jersey (1975).

Charles M. Vest, Holographic Interferometry, John Wiley and Sons, New York (1979).

G. Milton Wing, A Primer on Integral Equations of the First Kind, Los Alamos National Laboratory document LA-UR-84-1234, submitted to the University of California Press (1984)

Application of the Numerical Laplace Transform Inversion to Neutron Transport Theory

B. D. GANAPOL   Department of Nuclear and Energy Engineering, University of Arizona, Tucson, Arizona

A numerical Laplace transform inversion is developed using the Hurwitz-Zweifel method of evaluating the Fourier cosine integral coupled with an Euler-Knopp transformation. The numerical inversion is then applied to problems in linear transport theory concerning slowing down, time-dependence and featuring the determination of the interior scalar flux solution to the one-group stationary transport equation in half-space geometry.

INTRODUCTION

The Laplace transform and inversion have played a prominent role in neutron transport theory ever since the discovery of the neutron. For example, it is well known that the time-dependent neutron transport equation with time-independent parameters can formally be transformed into a stationary transport equation with complex cross sections by application of the Laplace transform. In this way, long and short time behavior can be obtained from the Tauberian theorems. In general, the transformed flux of the time-dependent transport equation is of more theoretical than numerical importance since the inversion can only be performed if the singularities of the transform are known explicitly. These are usually difficult to obtain except in certain cases. In addition, the Laplace transform has found use in generating solutions to the one-group stationary transport equation in a half-space. Again the application has been primarily theoretical yielding representations which are difficult to numerically evaluate.

As indicated above, the difficulty with the Laplace transform method is the inversion to direct space. This problem is inherent in the Laplace transform and, except for the very special cases found in tables, limits the effective use of this method of solution. For this reason, a considerable effort has been made during the past 15 years to develop an accurate and robust numerical inversion that can be applied to a wide variety of transforms. Because of its ill-conditioned nature however, a general all-purpose numerical scheme has yet to be developed. Currently, numerical algorithms based on Pade' approximates, polynomial expansions and the FFT exist to name a few. Unfortunately, the application of these numerical inversions to a general problem is more of an art than a science since there are usually several parameters which must be adjusted to obtain reasonable results. In some cases, these parameters depend sensitively on the result of the inversion and are therefore not known apriori. Another shortcoming of existing inversion schemes is that they are usually designed to give highly accurate inversions for a very specific (and limited) class of functions such as step function and perform poorly for other classes of functions.

In this presentation, I will derive a new numerical inversion which will be applied to obtain solutions to some fundamental problems of neutron transport theory. The method is based on the Hurwitz-Zweifel evaluation of the Fourier cosine integral with subsequent acceleration of the resulting infinite series by an Euler-Knopp transformation. The inversion will be applied to fundamental transport problems in the following areas:

a) one-group time-dependent rod model
b) neutron slowing down
c) one-group time-dependent transport
d) one-group stationary transport.

With the numerical inversion, solutions to basic transport problems are reduced to online computation allowing numerical transport benchmarks to be easily and efficiently generated. These results can then be used to guide code developers but, more importantly, can also be used in the classroom to teach analytical, numerical and computational aspects of transport theory. In addition, this work extends the monumental pioneering effort of Case, de Hoffman and Placzek [1] of the early fifties.

1. THE NUMERICAL LAPLACE TRANSFORM INVERSION

The inversion formula of the familiar Laplace transform pair

$$\bar{f}(s) = \int_0^\infty dt \ e^{-st} \ f(t) \tag{1.1}$$

$$f(t) = \frac{1}{2\pi i} \int_{\gamma-i\infty}^{\gamma+i\infty} ds \ e^{st} \ \bar{f}(s) \tag{1.2}$$

can be reformulated as follows:

$$f(t) = \frac{2}{\pi t} \ e^{\gamma t} \int_0^\infty du \ \text{Re}[\bar{f}(\gamma+iu/t)] \cos u \tag{1.3}$$

where γ is chosen such that the singularities of $\bar{f}(s)$ are to the left of the Bromwich contour $(\gamma-i\infty, \gamma+i\infty)$. Equation (1.3) can then be represented as an infinite series [2]

$$f(t) = \frac{2}{\pi t} \ e^{\gamma t} \left\{ \int_0^{\pi/2} du \ \text{Re}[\bar{f}(\gamma+iu/t)] \cos u \ + \right.$$

$$\left. + \sum_{k=1}^\infty (-1)^k \int_{-\pi/2}^{\pi/2} du \ \text{Re}[\bar{f}(\gamma+i(u+k\pi)/t)] \cos u \right\}. \tag{1.4}$$

The integrals in this representation are evaluated using an accurate Romberg quadrature [3] with error represented by ϵ_R. To accelerate the series convergence, an Euler-Knopp transformation with the Van Wijngaarden modification [3] is applied to the series typically giving remarkable acceleration. The truncation error in the evaluation of Eq. (1.4) is designated as ϵ_T. A contour comparison is also imposed requiring that the function evaluated on two contours defined by γ be within a given error tolerance ϵ_γ

$$\left| f(t,\gamma_j) - f(t,\gamma_{j-1}) \right| / \left| f(t,\gamma_j) \right| < \epsilon_\gamma \quad . \tag{1.5}$$

The contours are chosen such that

$$\gamma_j = (\gamma_{j-1} + \gamma_s)/2 \tag{1.6}$$

where γ_s is a contour greater than or equal to the largest singularity and γ_0 is chosen greater than γ_s. Caution must be exercised since this comparison test may give false convergence, but it will also indicate an inaccurate inversion.

This algorithm has been tested on transforms with simple poles, multiple poles, poles of multiplicity greater than one and branch points and invariably gave results that were within the desired error. The inversions however are sometimes sensitive to the choice of γ_s and γ_0. The result of the inversion algorithmn as applied to

$$\bar{f}(s) = \frac{1}{s+b} \, \frac{1}{\sqrt{s}} \, \frac{1}{\sqrt{s^2+a^2}} \left[\frac{\sqrt{s+2c}-\sqrt{s}}{\sqrt{s}} \right] \tag{1.7}$$

with

$$a = 1/5, \quad b = 5, \quad c = 1/2 \tag{1.8}$$

is given in Fig. 1a. In comparison with the IMSL inversion [4] for an error $\epsilon \equiv \epsilon_R = \epsilon_T = \epsilon_\gamma = 10^{-3}$, three digit agreement is obtained. In addition, the correct asymptotic behavior [$f(t) \rightarrow 1$ as $t \rightarrow \infty$] is observed. As with most inversion schemes, however, inaccurate results are obtained near discontinuities, but in general the qualitative discontinuous behavior is still observable as shown by the inversion for the step function in Fig. 1b.

When any one of the three numerical procedures, integration, summation, and γ-comparison are not convergent, an error flag is set to so indicate to the user. Thus, if the Euler-Knopp acceleration does the opposite, an error diagnostic is given. To accommodate for possible deceleration, the Van Wijngaarden modification forces the original series to be evaluated after the point of divergence. I only address this issue at the request of the conference reviewers, and I must emphasize that in my experience divergence of the Euler-Knopp summation has rarely occurred. When it has, adjustment of γ has always corrected the difficulty. Most likely one could find inversions for which the algorithm fails, but for the transforms considered below I always found convergence of the three numerical procedures.

2. APPLICATIONS OF THE NUMERICAL INVERSION TO TRANSPORT THEORY

2.1 Time-Dependent Rod Model

In this model, neutrons enter a rod of length x and are allowed to scatter only in the forward or backward directions. This is the simplest of transport models and was first considered by G. M. Wing [5] who found the exiting current at the source boundary to be the inversion

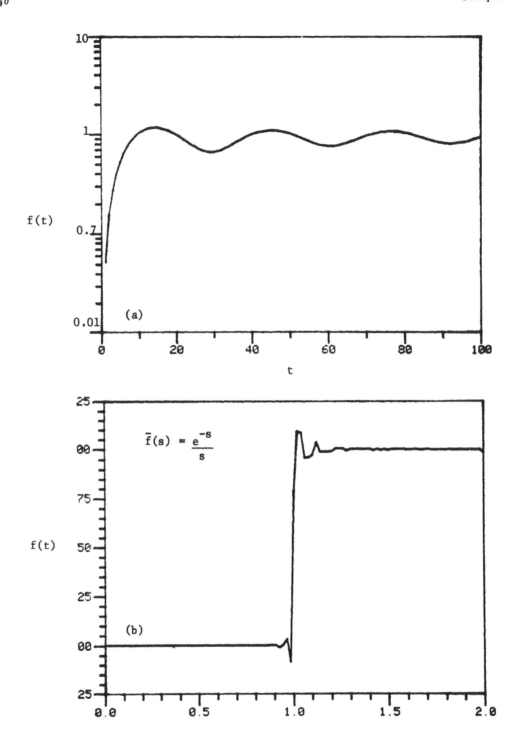

Figure 1   Demonstration of numerical inversion.

$$R(x,t) = \mathcal{L}_t^{-1} \left\{ \frac{1}{s^j} \frac{\tan\left[x\sqrt{1-s^2}\right]}{\sqrt{1-s^2} + s \tan\left[x\sqrt{1-s^2}\right]} \right\} \qquad (2.1)$$

for pulsed (j = 0) and continuously emitting (j = 1) sources. Figures 2a,b give the exiting currents for rods of three lengths. The rod of length x = π/2 is critical with shorter and longer rod lengths being sub- and super-critical respectively. For the continuously emitting source, the three cases have the asymptotically correct behavior as shown in Fig. 2a.

$$R(x,t) \sim \begin{cases} \tan x \;, & \text{subcritical} \\ t - \pi/4 \;, & \text{critical} \\ e^{P_0 t} \;, & \text{supercritical} \;. \end{cases}$$

For the flux from a pulsed source shown in Fig. 2b, the apparent discontinuity at t = 2x is from the reflection off the opposite end of the rod. Criticality is clearly demonstrated in this case for x = π/2.

2.2 Slowing Down with Constant Cross Sections

A standard problem in slowing down theory concerns neutrons undergoing elastic collisions with nuclei of an infinite medium. The appropriate transport equation for J scattering (non-fissioning) species in terms of lethargy u is

$$F(u) = \sum_{j=1}^{J} \frac{c_j}{1-\alpha_j} \int_{u-q_j}^{u} du' \, e^{-(u-u')} \, F(u') + \delta(u) \qquad (2.2)$$

where

$$F(u) \equiv \Sigma \, \phi(u) \;, \quad u \equiv \ln(E/E_0) \qquad (2.3)$$

$$\alpha_j \equiv \left(\frac{A_j-1}{A_j+1}\right)^2 \;, \quad q_j \equiv - \ln \alpha_j \;, \quad c_j \equiv \Sigma_{sj}/\Sigma \;; \qquad (2.4)$$

and the source emits neutrons at u = 0 uniformly in the medium. This equation was first solved by Marschak [6] for J = 1 by application of the Laplace transform to obtain (for J species)

$$\bar{F}(s) = 1/[1 - \bar{Q}(s)] \qquad (2.5)$$

$$\bar{Q}(s) \equiv \sum_{j=1}^{J} \frac{c_j}{1-\alpha_j} \left[\frac{1-e^{q_j(s+1)}}{s+1}\right] \;. \qquad (2.6)$$

Marschak then proceeded to obtain the solution by finding the poles of Eq. (2.5) for J = 1. For J materials, an analytical inversion is no longer practical. With the numerical inversion, however, the number of materials presents no problem. A more efficient inversion of Eq. (2.5) is found by specifying the first few collisions analytically

$$F(u) = F_0(u) + F_1(u) + F_2(u) + \mathcal{L}_u^{-1} [\bar{F}(s)] \qquad (2.7)$$

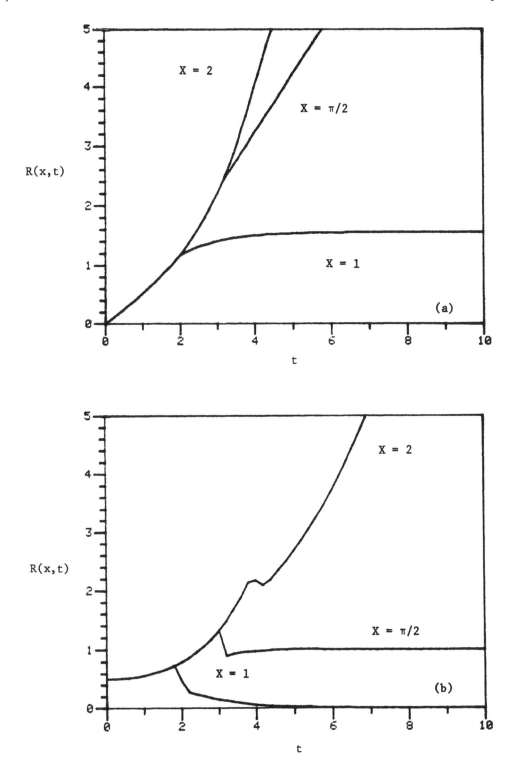

Figure 2   Wing's time-dependent rod transport model.

where

$$F_0(u) = \delta(u) \tag{2.8}$$

$$F_1(u) = e^{-u} \sum_{j=1}^{J} \frac{c_j}{1-\alpha_j} [1 - \theta(u-q_j)] \tag{2.9}$$

$$F_2(u) = \sum_{j=1}^{J} \frac{c_j}{1-\alpha_j} \sum_{k=1}^{J} \frac{c_k}{1-\alpha_k} \left[u - (u-q_j) \theta(u-q_j) - \right.$$

$$\left. - (u-q_k) \theta(u-q_k) + (u-q_k-q_j) \theta(u-q_k-q_j) \right] \tag{2.10}$$

$$\bar{F}(s) = \frac{1}{1-\bar{Q}(s)} - 1 - \bar{Q}(s) - \bar{Q}(s)^2 \quad . \tag{2.11}$$

with θ representing the Heaviside step function. The relative error for the analytical solution with $J = 1$ (for $A = 12$),

$$F(u) = \frac{e^{-u}}{1-\alpha} \sum_{k=0}^{[u/q]} \frac{(-1)^k}{k!} \left[\frac{u-kq}{1-\alpha} \right]^{k-1} \left[\frac{u-kq}{1-\alpha} + k \right] e^{(u-kq)/(1-\alpha)} \quad , \tag{2.12}$$

is compared in Table 1 to the numerical inversion evaluated for several truncation errors ($\epsilon_R = \epsilon_T$). It is observed that the relative error is always within the desired limit except near the Placzek discontinuities (for carbon the primary discontinuities occur at $u = 0.3341$, 0.6682). Figure 3 shows $F(u)$ for a mixture of four materials with varying concentrations. Note that the signature of each material is clearly apparent in the form of a Placzek discontinuity. To this author's knowledge results for composite media with constant cross sections have never before appeared in the literature.

2.3 One-Group Time-Dependent Albedo Problem

Traditionally the Laplace transform has been associated with time-dependent solutions of the transport equation. As mentioned in the introduction, if a stationary result is known, then the corresponding time-dependent result can formally be obtained by including the transform variable s in the cross sections. For example, the scalar flux exiting the free surface of a semi-infinite medium with an isotropic source shining neutrons on the free surface is

$$\psi(0) = \frac{2}{c} \left[1 - \sqrt{1-c} \right] \quad . \tag{2.13}$$

To obtain the flux corresponding to the time-dependent albedo problem with a pulsed source, one need only replace c by $c/(s+1)$ and perform the inversion

$$\psi(0,t) = \frac{2}{c} \mathcal{L}_t^{-1} \left\{ (s+1) \left[1 - \sqrt{1-c/(s+1)} \right] \right\} \quad . \tag{2.14}$$

Table 1 Comparison of Analytical Solution and Inversion

u \ ε_T	10^{-2}	10^{-3}	10^{-4}	10^{-5}	10^{-6}
3.3411E-02	1.0306E-07	1.5666E-07	1.3517E-07	2.7822E-09	1.5604E-09
1.0023E-01	1.3954E-06	1.3954E-06	1.0667E-06	3.3411E-07	7.7788E-08
1.6705E-01	3.1599E-06	3.1918E-06	9.5130E-07	1.7502E-07	7.5134E-09
2.3388E-01	5.6762E-05	3.2200E-05	2.4341E-06	2.5332E-06	8.8477E-08
3.0070E-01	4.3112E-04	3.2846E-05	4.8657E-05	1.1017E-05	1.8696E-06
3.6752E-01	1.2454E-03	6.2939E-05	1.2040E-04	2.0553E-05	2.5557E-06
4.3434E-01	8.2632E-04	1.9848E-04	6.8444E-05	1.4716E-05	1.8293E-06
5.0116E-01	7.1757E-04	1.2001E-04	2.3471E-04	1.0375E-04	3.1883E-05
5.6798E-01	2.9183E-04	2.0237E-04	1.5906E-05	3.4193E-05	3.2926E-06
6.3481E-01	4.6509E-03	9.3369E-04	6.6802E-05	6.2687E-05	1.4354E-05
7.0163E-01	1.9190E-03	2.4579E-03	7.5018E-05	4.8407E-05	1.0666E-05
7.6845E-01	1.0831E-03	3.5655E-04	3.8922E-04	6.9516E-05	3.7514E-06
1.3848E+00	7.4117E-04	1.3526E-04	1.7039E-04	9.0346E-06	8.3573E-06
2.5507E+00	6.4358E-04	1.6578E-05	1.8957E-06	7.5691E-09	1.3055E-08

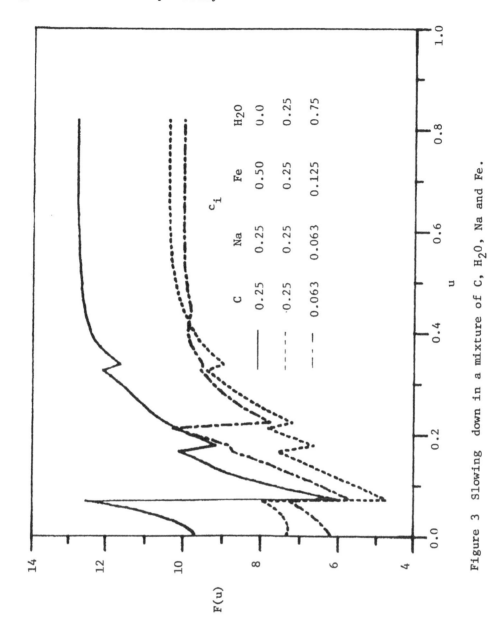

Figure 3 Slowing down in a mixture of C, H_2O, Na and Fe.

If one-group of delayed neutrons with decay constant λ is included, then c in Eq. (2.14) is replaced by

$$c(s) = \tilde{c}[1 + \tilde{c}_f/(s+\lambda)] \tag{2.15}$$

$$\tilde{c} = c_s + (1-\beta)\nu c_f \quad , \quad c_f = \Sigma_f/\Sigma \tag{2.16}$$

$$\tilde{c}_f = c_f \beta/\tilde{c} \quad . \tag{2.17}$$

General sources can also be included by convolution as follows:

$$\psi(0,t) = \frac{2}{c} \mathcal{L}_t^{-1} \left\{ (s+1) \left[1 - \sqrt{1-c(s)/(s+1)} \right] \bar{Q}(s) \right\} \tag{2.18}$$

where $\bar{Q}(s)$ is the Laplace transform of the desired time variation of the source.

Figure 4a shows the variation of the exiting flux for a pulsed source with one delayed neutron group (note that dimensionless parameters are used). The reactor is initially critical ($\tilde{c}=1$) and as time increases the tendency toward super-criticality with increasing strength of the delayed neutron contribution is clearly seen. As a second demonstration, Figure 4b shows the oscillations resulting from the source

$$\Lambda(t) = 1 - J_0(t) \quad . \tag{2.19a}$$

where

$$\bar{\Lambda}(s) = \frac{1}{s} - \frac{1}{\sqrt{s^2+1}} \quad . \tag{2.19b}$$

and

$$\psi(0,t) = \frac{2}{c} \mathcal{L}_t^{-1} \left\{ (s+1) \left[1 - \sqrt{1-c/(s+1)} \right] \bar{\Lambda}(s) \right\} \quad . \tag{2.19c}$$

2.4 Steady State One-Group Problems in the Half-Space

To demonstrate the use of the numerical inversion to find the interior scalar flux for one-group stationary transport problems, we will consider the neutron albedo problem. For this case, neutrons with angular distribution $Q(\mu)$ impinge on the free surface of a semi-infinite medium. The transport equation assuming isotropic scattering is therefore

$$\left[\mu \frac{\partial}{\partial x} + 1 \right] \phi(x,\mu) = \frac{c}{2} \int_{-1}^{1} d\mu' \; \phi(x,\mu') \tag{2.20}$$

$$\phi(0,\mu) = Q(\mu), \quad \mu > 0$$
$$\lim_{x \to \infty} \phi(x,\mu) < \infty, \quad c \leq 1 \quad . \tag{2.21}$$

Then from integral transport theory, the flux exiting the free surface is

$$\phi(0,-\mu) = \frac{c}{2\mu} \int_0^{\infty} dx' \; e^{-x'/\mu} \; \phi(x'), \; \mu > 0 \quad . \tag{2.22}$$

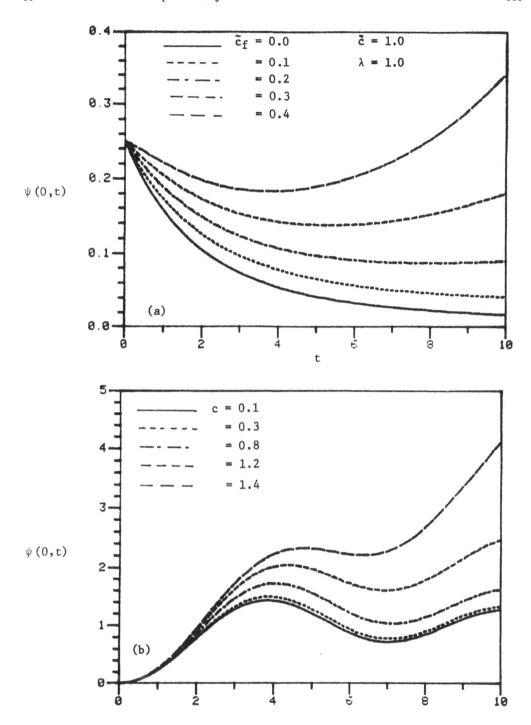

Figure 4   Time variation of flux exiting a half-space: (a) pulsed
source, (b) continuously emitting source.

Note that, if μ is replaced by $1/s$ in Eq. (2.22), $\phi(x)$, the scalar flux, is given by the Laplace transform inversion

$$\phi(x) = \frac{2}{c} \mathcal{L}_x^{-1} \left[\frac{\phi(0,-1/s)}{s} \right] \quad . \tag{2.23}$$

Thus, if we have an explicit expression for the exiting flux $\phi(0,-\mu)$, the inversion can be carried out numerically. Fortunately, in the literature, many such expressions exist.

For a beam $[Q(\mu) = \delta(\mu-\mu_0)]$ and an isotropic source $[Q(\mu) = 1]$, we have

$$\phi(0,-\mu) = \begin{cases} \dfrac{c}{2} \dfrac{\mu_0}{\mu_0+\mu} H(\mu) H(\mu_0) \,, & \text{beam} \\[4mm] 1 - \sqrt{1-c}\, H(\mu) \,, & \text{isotropic} \end{cases} \tag{2.24}$$

where $H(\mu)$ is the Chandrasekhar H- function satisfying the non-linear integral equation

$$H(\mu) = 1 + \frac{c}{2} \mu H(\mu) \int_0^1 d\mu' \frac{H(\mu')}{\mu+\mu'} \quad . \tag{2.25}$$

To obtain $H(1/s)$ as required in the inversion, a shifted Chebyschev quadrature is used in place of the integral in Eq. (2.25) and μ is evaluated at the Lm quadrature points μ_m

$$H_m = \left[1 - \frac{c}{2} \mu_m \sum_{m'=1}^{Lm} \frac{\omega_{m'} H_{m'}}{((\mu_m+\mu_{m'}))} \right]^{-1} \quad . \tag{2.26}$$

This equation is then iteratively solved for H_m. With H_m known, μ is then replaced by $1/s$ in Eq. (2.25) to obtain $H(1/s)$. The scalar flux at $x = 10^{-7}$ is given in Table 2 for various parameters of the algorithm as determined by the numerical inversion and the exact expression for a beam source ($\mu_0=1$) at $x = 0$

$$\phi(0) = H(\mu_0) \quad . \tag{2.27}$$

The underlined digits are those in disagreement with the exact result indicating increasing agreement as the error $\epsilon(\epsilon_R=\epsilon_T=\epsilon)$ is reduced and the quadrature order Lm is increased. The variation of the scalar flux with c is shown in Fig. (5a). It is interesting to note that for $c < 0.5$ the flux always decays with x but for $c > 0.5$ the flux initially increases with x and then decays.

Since the exiting distribution for the Milne problem (source of strength e^{-x/ν_0} at infinity) is

$$\phi(0,-\mu) = \frac{c\,\nu_0}{2} \frac{H(\mu)}{H(\nu_0)} \frac{1}{\nu_0-\mu} \quad , \tag{2.28}$$

Eq. (2.23) can now be applied to obtain $\phi(x)$. ν_0 in this equation is determined from the infinite medium dispersion relation

$$1 - c\nu_0 \tanh^{-1} \frac{1}{\nu_0} = 0 \quad . \tag{2.29}$$

Table 2. Error Estimation of Numerical Laplace Transform Inversion-- Beam Source ($\mu_0 = 1$)

(a) Variation of ϵ ($L_m = 48$)

c/ϵ	10^{-2}	10^{-3}	10^{-4}	exact
0.1	1.03703	1.03681	1.03681	1.03682
0.5	1.25152	1.25125	1.25126	1.25126
0.95	2.07755	2.07711	2.07712	2.07712

(b) Variation of L_m ($\epsilon = 10^{-4}$)

c/L_m	4	16	32	exact
0.1	1.03691	1.03682	1.03682	1.03682
0.5	1.25179	1.25129	1.25127	1.25126
0.95	2.07716	2.07713	2.07713	2.07712

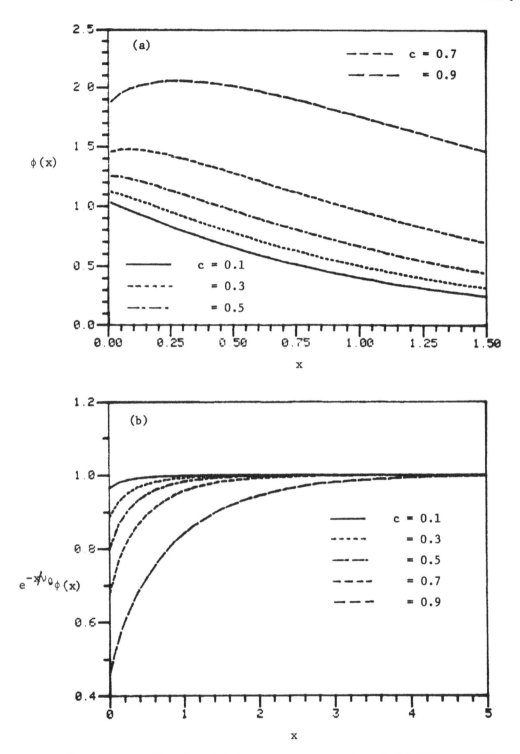

Figure 5 Interior flux for (a) beam source $\mu_0 = 1$ and (b) Milne problem.

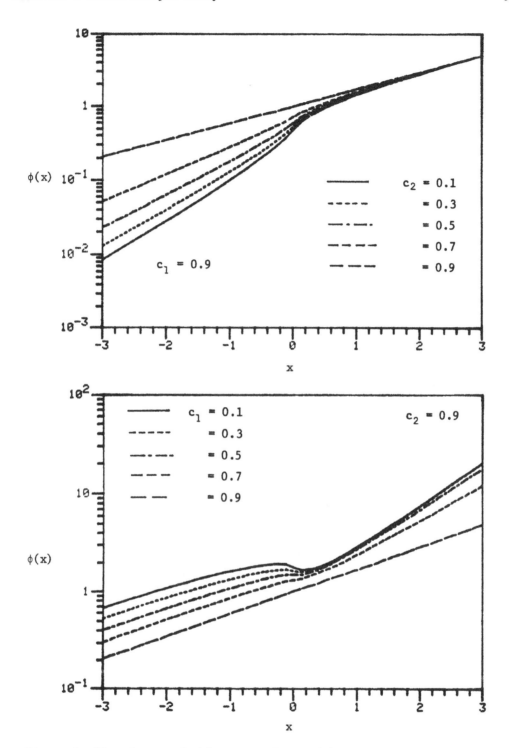

Figure 6   Flux for two half-space Milne problems with source in right
           half-space (1).

Table 3 Half-Space Albedo Problem

Beam Source ($\mu_0 = 1$)

x/c	0.1	0.3	0.5	0.7	0.9
1.0000E-05	1.0368E+00	1.1269E+00	1.2513E+00	1.4448E+00	1.8502E+00
5.0000E-01	6.5564E-01	7.8192E-01	9.6933E-01	1.2852E+00	2.0141E+00
1.0000E+00	4.0547E-01	5.0681E-01	6.6865E-01	9.6854E-01	1.7636E+00
1.5000E+00	2.4936E-01	3.2283E-01	4.4766E-01	6.9964E-01	1.4645E+00
2.0000E+00	1.5291E-01	2.0375E-01	2.9498E-01	4.9425E-01	1.1846E+00
2.5000E+00	9.3594E-02	1.2784E-01	1.9240E-01	3.4427E-01	9.4347E-01
3.0000E+00	5.7214E-02	7.9880E-02	1.2459E-01	2.3747E-01	7.4395E-01
3.5000E+00	3.4942E-02	4.9758E-02	8.0249E-02	1.6262E-01	5.8267E-01
4.0000E+00	2.1324E-02	3.0920E-02	5.1468E-02	1.1074E-01	4.5418E-01
4.5000E+00	1.3006E-02	1.9177E-02	3.2896E-02	7.5090E-02	3.5283E-01
5.0000E+00	7.9287E-03	1.1874E-02	2.0967E-02	5.0734E-02	2.7340E-01

Isotropic Source

x/c	0.1	0.3	0.5	0.7	0.9
1.0000E-05	1.0262E+00	1.0888E+00	1.1715E+00	1.2922E+00	1.5195E+00
5.0000E-01	3.5435E-01	4.2523E-01	5.2910E-01	6.9997E-01	1.0731E+00
1.0000E+00	1.6592E-01	2.1345E-01	2.9031E-01	4.3356E-01	8.0765E-01
1.5000E+00	8.3538E-02	1.1354E-01	1.6625E-01	2.7596E-01	6.1470E-01
2.0000E+00	4.3696E-02	6.2225E-02	9.7257E-02	1.7789E-01	4.6999E-01
2.5000E+00	2.3417E-02	3.4746E-02	5.7632E-02	1.1550E-01	3.6017E-01
3.0000E+00	1.2765E-02	1.9654E-02	3.4448E-02	7.5339E-02	2.7637E-01
3.5000E+00	7.0470E-03	1.1224E-02	2.0718E-02	4.9294E-02	2.1222E-01
4.0000E+00	3.9290E-03	6.4563E-03	1.2518E-02	3.2323E-02	1.6304E-01
4.5000E+00	2.2079E-03	3.7356E-03	7.5910E-03	2.1229E-02	1.2529E-01
5.0000E+00	1.2489E-03	2.1716E-03	4.6163E-03	1.3959E-02	9.6303E-02

Table 4

Single-Space Milne Problem

x/c	0.1	0.3	0.5	0.7	0.9
1.0000E-05	9.6450E-01	8.8737E-01	7.9673E-01	6.7575E-01	4.6065E-01
5.0000E-01	9.9348E-01	9.7640E-01	9.4918E-01	8.9420E-01	7.2301E-01
1.0000E+00	9.9822E-01	9.9305E-01	9.8323E-01	9.5765E-01	8.4155E-01
1.5000E+00	9.9947E-01	9.9780E-01	9.9414E-01	9.8239E-01	9.0775E-01
2.0000E+00	9.9983E-01	9.9928E-01	9.9790E-01	9.9255E-01	9.4591E-01
2.5000E+00	9.9995E-01	9.9976E-01	9.9923E-01	9.9682E-01	9.6817E-01
3.0000E+00	9.9998E-01	9.9992E-01	9.9972E-01	9.9863E-01	9.8124E-01
3.5000E+00	9.9999E-01	9.9997E-01	9.9990E-01	9.9941E-01	9.8893E-01
4.0000E+00	1.0000E+00	9.9999E-01	9.9996E-01	9.9974E-01	9.9346E-01
4.5000E+00	1.0000E+00	1.0000E+00	9.9999E-01	9.9989E-01	9.9614E-01
5.0000E+00	1.0000E+00	1.0000E+00	9.9999E-01	9.9995E-01	9.9772E-01

Two-Space Milne Problem ($c_1 = 0.9$)

x/c_2	0.1	0.3	0.5	0.7	0.9
-5.0000E+00	8.6569E-04	1.4871E-03	3.1424E-03	9.6297E-03	7.2286E-02
-4.0000E+00	2.6747E-03	4.3598E-03	8.4431E-03	2.2197E-02	1.2225E-01
-3.0000E+00	8.4762E-03	1.3013E-02	2.2916E-02	5.1346E-02	2.0674E-01
-2.0000E+00	2.7957E-02	3.9970E-02	6.3266E-02	1.1956E-01	3.4964E-01
-1.0000E+00	9.9461E-02	1.2976E-01	1.8078E-01	2.8256E-01	5.9130E-01
-1.0000E-05	4.8039E-05	5.2993E-01	6.0120E-01	7.1918E-01	1.0000E+00
1.0000E-05	4.8044E-05	5.2998E-01	6.0124E-01	7.1921E-01	1.0000E+00
1.0000E+00	1.4322E+00	1.4550E+00	1.4888E+00	1.5467E+00	1.6912E+00
2.0000E+00	2.7105E+00	2.7237E+00	2.7431E+00	2.7764E+00	2.8601E+00
3.0000E+00	4.7492E+00	4.7569E+00	4.7683E+00	4.7878E+00	4.8370E+00
4.0000E+00	8.1285E+00	8.1730E+00	8.1397E+00	8.1512E+00	8.1302E+00
5.0000E+00	1.3804E+01	1.3806E+01	1.3810E+01	1.3817E+01	1.3834E+01

The interior flux for the Milne problem multiplied by e^{-x/ν_0} is given for several values of c in Fig. (5b).

The inversion method can also be applied to two half-space problems such as the two half-space Milne problem with a source in the right half-space (1) where for $\mu > 0$

$$\phi_1(0,-\mu) = \frac{c_1}{2} \frac{\nu_{01}}{\nu_{01}-\mu} \frac{H_1(\mu)}{H_2(\mu)} \frac{H_2(\nu_{01})}{H_1(\nu_{01})} \tag{2.30}$$

$$\phi_2(0,\mu) = \frac{c_2}{2} \frac{\nu_{01}}{\nu_{01}+\mu} \frac{H_2(\mu)}{H_1(\mu)} \frac{H_2(\nu_{01})}{H_1(\nu_{01})} \tag{2.31}$$

and for a uniform source of strength q_0 in $x > 0$ where

$$\phi_1(0,-\mu) = q_0 \frac{c_1}{c_1-c_2} \left[\left(\frac{1-c_2}{1-c_1}\right)^{1/2} \frac{H_1(\mu)}{H_2(\mu)} - \frac{c_2}{c_1} \right] \tag{2.32}$$

$$\phi_2(0,\mu) = q_0 \frac{c_2}{c_1-c_2} \left[\left(\frac{1-c_2}{1-c_1}\right)^{1/2} \frac{H_2(\mu)}{H_1(\mu)} - 1 \right] \tag{2.33}$$

Figure 6 shows the spatial flux variation with number of secondaries (either c_1 or c_2). When the number of secondaries is equal in both half-spaces, the flux exhibits exponential decay from the source as expected. As c_1 is reduced (Fig. 6a), the flux approaches an exponential ($e^{-x/\nu_{01}}$) for large x but shows more complicated behavior near the interface as a result of the decreased return current from the left half-space. For large negative x the flux again approaches an exponential ($e^{-x/\nu_{02}}$) where ν_{02} depends on c_2. For fixed c_2 (Fig. 6b), the flux now has the same exponential variation for large negative x independent of c_1. Tables 3 and 4, in which flux values are provided, are included as numerical benchmarks for use in method and code development. The tabular values should be correct to one unit in the fourth place.

3. SUMMARY AND CONCLUSIONS

A newly developed numerical Laplace transform inversion based on the Hurwitz-Zweifel expansion has been applied to several fundamental problems in neutron transport theory. This method of numerical evaluation, new in its application to stationary and slowing down problems, provides highly accurate solutions which serve as benchmarks. In addition to what has been presented here, the inversion technique has been applied to stationary half-space problems with anisotropic scattering with equal success. When initially applied to the slab problem, however, the results seem to be limited to only two or three place accuracy at this stage of the development. A weakness of the inversion method is that unfortunately, it is limited to problems with known exiting distributions. For the remaining problems, a new method is currently under development which will be the subject of a future publication.

REFERENCES

1. K. M. Case, F. De Hoffmann, and G. Placzek, Introduction to Neutron Diffusion, LASL Report, 1953.
2. H. Hurwitz, Jr., and P. F. Zweifel, MTAC., 10, 140 (1956).
3. W. H. Press, B. P. Flanney, S. A. Teukolsky and W. T. Vetterling, Numerical Recipes, Cambridge University Press, Cambridge, 1986.
4. IMSL Library Reference Manual, IMSL, Houston, Texas, 1979.
5. G. W. Wing, Introduction to Transport Theory, Wiley, NY 1962.
6. R. E. Marshak, Rev. Mod. Phys., 19, 185 (1947).

A Bayesian Approach to Nonlinear Inversion: Abel Inversion from X-Ray Attenuation Data

KENNETH M. HANSON   Los Alamos National Laboratory, Los Alamos, New Mexico

With a change of variables the Abel transform gives the projection of a 2-D distribution with circular symmetry. Thus the Abel inversion formula allows one to determine the radial dependence of such a distribution from its projection. However, this inversion formula is very sensitive to noise in the projection data. When the projection data are derived from radiographic measurements, further difficulties arise from the necessity to invert the exponential dependence of the measured x-radiation intensity upon material thickness. These difficulties are shown to be overcome by applying a maximum *a posteriori* (MAP) method, which was developed to accomodate nonlinear measurements, to this tomographic reconstruction problem. The MAP method yields a smooth solution in regions where the signal-to-noise ratio is low while maintaining good spatial resolution in regions where it is high.

1   Introduction

In 1826 Abel [1] provided a method for recovering the density distribution of a two-dimensional, circularly symmetric object from its projection. It is possible to extend this 2-D analysis to reconstruct the cross section of a three-dimensional, axially symmetric object from a single radiograph, as depicted in Fig. 1. In industrial radiography, where many objects have nearly circular symmetry, such an approach offers significant benefits as an image analysis tool [2]. These benefits include improved delineation of material boundaries, enhanced display of deviations from axial symmetry (e. g., possibly caused by defects), and estimation of the radial dependence of the attenuation coefficients of the materials. It is often possible to observe in the reconstruction subtle features of the object that are unobservable in the original radiograph. The improvements brought about by the tomographic method are due to the effective removal of overlying material

This work was supported by the United States Department of Energy under contract number W-7405-ENG-36.

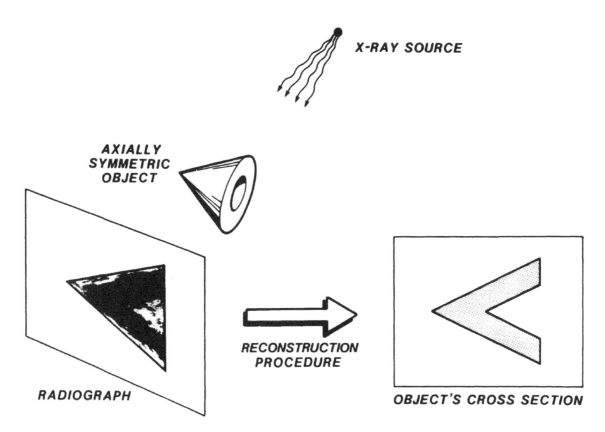

Figure 1: Overview of the tomographic method. An object with axial symmetry is radiographed with the radiographic axis perpendicular to the axis of symmetry. The reconstruction procedure transforms the radiograph into an estimate of the cross section of the object.

from the radiograph and the consequent increase in contrast with which it is possible to display the reconstruction.

Abel inversion has been considered over the years by many authors, most recently in [3,4,5,6]. It has even been scrutinized on a number of occasions by Milt Wing, together with various colleages [7,8,9]. In its discrete form, Abel inversion is nonsingular except at the axis of symmetry. When projection measurements are derived from a radiograph of an object, they are nonlinearly related to the line integral of the linear attenuation values through the object. Near the center of a dense object, the optical density can approach the fog level of the film, resulting in a poor signal-to-noise ratio. This effect, together with the divergence of the Abel inversion at the axis, can result in overwhelming noise in that region of the reconstruction of the object's density. The growth of noise in the reconstruction can be controlled through a nonlinear maximum *a posteriori* (MAP) formulation in which the known uncertainty in the optical density data caused by film noise is balanced with the suspected smoothness in the object's density [10]. In this approach it is possible to include other constraints on the reconstruction, such as non-negativity. The usefulness of this approach is demonstrated by the reconstruction of a solid metal object with axial symmetry from a single radiograph. The MAP approach presented here may be of interest to applied mathematicians because it represents yet another means of obtaining a regularized solution in which prior knowledge about the object function can be incorporated in a consistent way.

2 Statement of the problem

The line integral along a line that is a distance R from the origin of a 2-D function possessing circular symmetry $f(r)$ is given by

$$p(R) = 2 \int_R^\infty \frac{f(r)\,r\,dr}{\sqrt{r^2 - R^2}} \,. \tag{1}$$

By a change of variables, Eq. (1) can be shown to be equivalent to the Abel transform. A solution for $f(r)$, given a known projection $p(R)$, was derived by Abel more than a century and a half ago [1]. Direct evaluation of the Abel inversion formula can lead to mathematical difficulties, which can be overcome by a suitable revision of the formula [5]. An alternative approach based on a discrete model avoids these difficulties. An object with axial symmetry is considered to be composed of a series of nested annuli, as depicted in Fig. 2, whose amplitudes are to be determined. The outermost annulus, whose projection is shown in Fig. 2, is the only annulus that contributes to the ends of the projection interval. Thus, it is possible to determine the amplitude of the outermost annulus from these projection values and consequently its contributions to each inner projection sample. The same analysis is applied to the next annulus and so on until the center is reached. Hence, the solution is easily realized by 'peeling the onion' from the outside to the inside.

A minor variation in this model of nested annuli is useful. Each annulus of the simple model is replaced by two annuli. The density profiles of these two annuli vary as $\frac{1}{2}[1 + cos(\theta)]$ and $\frac{1}{2}[1 - cos(\theta)]$, where θ is the angle between the radiographic axis and the vector between the center of the annulus and its rim. Thus the density of each annulus varies linearly with the transverse distance from the center from zero on one side to full density on the other side. The tapers of the two annuli vary in opposite directions so that, if they are given equal amplitudes, the sum of both densities is the same as that of the original annulus. The result of this decomposition is that the major contribution to the reconstruction on each side of the rotation axis comes predominately from the corresponding side of the projection. This model of annuli with tapered densities is better than one in which the annuli on each side are considered to be entirely distinct because it smoothly bridges both sides. Thus, even when the left and right sides of a projection might otherwise dictate considerably different values at the origin (producing a discontinuity there), this tapered model will tend to produce a smoothly varying reconstruction.

In this discrete model, the relationship between the discretely sampled projections, represented by the vector $\mathbf{p}$, and the amplitudes of the corresponding annuli, represented by the vector $\mathbf{f}$, is given by

$$\mathbf{p} = \mathbf{Hf} \,, \tag{2}$$

where $\mathbf{H}$ is the measurement matrix. The $\mathbf{H}$ matrix is explicitly displayed for a projection of 10 discrete samples in Fig. 3. For this model $\mathbf{H}$ possesses a bow-tie structure so that Eq. (2) is easy to solve by a procedure similar to Gaussian elimination in which the unknowns are determined alternatively from the top and bottom of the set of equations. If the tapered aspect of the model were not employed, $\mathbf{H}$ would have a lower triangular form and straightforward Gaussian elimination could be used.

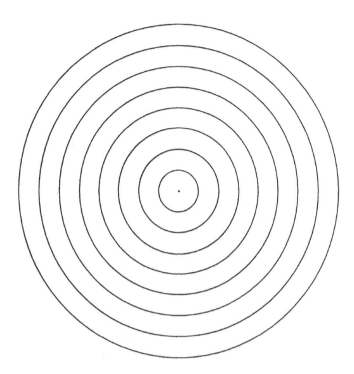

MODEL OF AXISYMMETRIC OBJECT

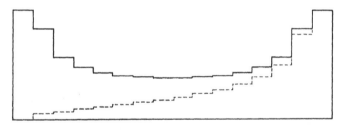

PROJECTION OF OUTER ANNULUS

Figure 2: Model of a 2-D object with circular symmetry composed of a series of annuli whose widths correspond to the width of a pixel in the discretely sampled projection. Below is the projection of the outer annulus. The dashed line shows the projection of the outer annulus when its density is assumed to vary linearly from zero on the left side to full density on the right.

```
4.088  0.000  0.000  0.000  0.000  0.000  0.000  0.000  0.000  0.000
3.083  3.626  0.000  0.000  0.000  0.000  0.000  0.000  0.000  0.385
1.893  2.660  3.097  0.000  0.000  0.000  0.000  0.000  0.443  0.541
1.420  1.598  2.157  2.457  0.000  0.000  0.000  0.539  0.639  0.710
1.121  1.159  1.236  1.504  1.571  0.000  0.752  0.824  0.870  0.896
0.896  0.870  0.824  0.752  0.000  1.571  1.504  1.236  1.159  1.121
0.710  0.639  0.539  0.000  0.000  0.000  2.457  2.157  1.598  1.420
0.541  0.443  0.000  0.000  0.000  0.000  0.000  3.097  2.660  1.893
0.385  0.000  0.000  0.000  0.000  0.000  0.000  0.000  3.626  3.083
0.000  0.000  0.000  0.000  0.000  0.000  0.000  0.000  0.000  4.088
```

Figure 3: The $\mathbf{H}$ matrix, which describes the projection measurements in the model shown in Fig. 2 for a projection consisting of 10 discrete samples.

The projections needed as input for tomographic reconstruction, Eq. (2), may be obtained from radiographic measurements. Suppose that a 3-D object is radiographed with an essentially monoenergetic x-ray source. If the x-rays are detected by a direct-recording film, the optical density of the developed film is proportional to the x-ray intensity. At the intersection of each line segment L that originates at the point source with the film plane, the optical density of the film is given by

$$D(p) = D_0 + D_1 \, exp\{-p\} \,, \tag{3}$$

where the pathlength p is the line integral along line L of the object's linear attenuation coefficient distribution $\mu(x, y, z)$, evaluated at the x-ray energy. In the above equation, D_0 is the background density of the film, which includes the fog level of the film and any contribution from scattered radiation and D_1 is the net density (above D_0) that would be obtained in the absence of the object. Figure 4 illustrates this relationship. The highly nonlinear behavior of Eq. (3) for optical densities near D_0 makes the reconstruction there very sensitive to the choice for D_0. This consideration points out the need to know D_0 very accurately to avoid making serious systematic errors in the reconstruction. Slight variations in D_0 arising from variations in the scattered radiation field may be difficult to take into account.

Putting Eqs. (2) and (3) together, we obtain the overall measurement equation for the measured optical density vector $\mathbf{g}$

$$\mathbf{g} = s\{\mathbf{Hf}\} \,, \tag{4}$$

where, in a change of notation to be consistent with that of Hunt [11], the attenuation law is now represented by the function $s\{\mathbf{p}\}$, which operates on each individual component of the pathlength vector $\mathbf{p} = \mathbf{Hf}$ with the nonlinear scalar function given by Eq. (3). Under certain circumstances, direct inversion of Eq. (4) is possible

$$\hat{f} = \mathbf{H}^{-1} s^{-1}\{\mathbf{g}\} \,. \tag{5}$$

This inverse only exists, however, if $\mathbf{H}$ is nonsingular (possesses no null space). Such is the case for the discrete model of the Abel transformation. However, difficulty can arise in inverting the nonlinear transformation s. When the measured density D is either above $D_0 + D_1$ or below D_0, which is possible because of noise fluctuations, Eq. (3) cannot be meaningfully inverted. The MAP method overcomes this difficulty.

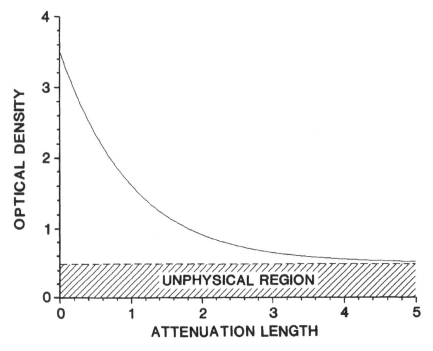

Figure 4: The measurement of x-ray attenuation by film is given by an exponential in the optical density. Values of the optical density below the fog level D_0, 0.5 in this diagram, are theoretically not allowed.

Figure 5: Radiograph of an axially symmetric steel object.

Now consider a 3-D object that possesses axial or rotational symmetry. In any plane that is perpendicular to the symmetry axis, the object's cross section has circular symmetry. If the object is radiographed with a set of parallel x-ray beams all of which lie in such a plane, the pathlengths obtained by applying Eq. (3) will correspond precisely to those in the 2-D situation. Thus, the radial distribution of that cross section can be reconstructed. A line-by-line analysis carried out in this way on each line of the radiograph that is perpendicular to the symmetry axis then yields the complete radial distribution of the object.

Figure 5 displays a radiograph of a test object with axial symmetry. The object is a 70-mm-long right circular steel cylinder, 120 mm in diameter, with a 45° cone removed from one end to a depth of 40 mm. Four 2-mm-square grooves were machined on the flat face as well as on the inside of the cone. The radiograph was taken with a Co-60 source using Kodak AA industrial radiographic film placed in close contact with 0.25-mm-thick lead screens, front and back. The object was placed with its axis of symmetry perpendicular to the radiographic axis in a geometry to closely approximate the parallel beam assumed by the reconstruction procedure. Figure 5 is actually a digital image, 220 by 150 pixels in size with pixel spacing of 0.6 mm. It is displayed here with a contrast comparable to that in the original radiograph. The grooves with the smallest diameters are virtually impossible to see because of their extremely low contrast on the radiograph. The diffuse densities in the radiograph range from about 0.5 to 3.5, the latter corresponding to that produced by the unattenuated x-ray beam.

When the tomographic reconstruction indicated by Eq. (5) is performed on each line of the radiograph, Fig. 5, and the reconstructed radial profiles are combined to form an image, Fig. 6 results. This reconstruction closely resembles the cross section of the object itself. The value of the linear attenuation coefficient reconstructed for the steel is approximately the same as the tabulated value for iron at 1.25 MeV, namely 0.042 mm^{-1}. The grooves, which were very difficult to observe in the original radiograph, are

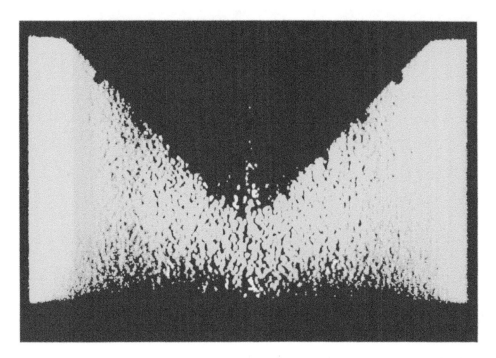

Figure 6: The reconstruction of the cross section of the object derived from Fig. 5. The 2-mm-square grooves machined in the inner cone and on the end face are much more readily visible. These grooves were nearly invisible in the original radiograph.

now easily seen. The tomographic reconstruction procedure has redisplayed the data in such a way that one's ability to derive information from the image is limited by the inherent noise, which was not the case for the original radiograph. The attainment of the ultimate limit in image interpretation dictated by the noise is one of the major goals of image processing!

However, there is severe enhancement of the noise near the axis of symmetry that arises from a nearly singular condition in the reconstruction procedure there. This near singularity is equivalent to the physical statement that as a ring of material with constant thickness decreases in diameter, it produces a smaller and smaller optical-density signal on the radiograph, making it increasingly difficult to observe in the presence of nearly constant film-density noise. The uncontrolled growth of noise in the reconstruction poses difficulty for visual interpretation of the image. The human eye simply cannot perform the necessary averaging over noise of such large amplitude. One possible method to control the noise in the reconstruction is to employ nonlinear maximum *a posteriori* probability (MAP) restoration [2], to be discussed next.

3   The Bayesian Approach

The essence of the Bayesian approach is the assumption that the image to be reconstructed is a random selection from an identifiable ensemble of similar images. In the context of medical imaging, an example of such an ensemble is the collection of all hearts imaged in the same kind of procedure. By using this prior information about the type of image that is expected, one anticipates that the null-space component of the reconstruction might be meaningfully estimated, thereby reducing artifacts. The Bayesian approach provides a way to incorporate prior information about the structure of the reconstructed object into the solution. Of course, other types of prior information can also be incorporated in the Bayesian method of reconstruction. See Refs. [12,13] for a more complete description of the general situation.

Direct inversion of the measurement equations using Eq. (5) tacitly assumes the measurements can be made with infinite accuracy. In reality, all measurements of continuous quantities are subject to random fluctuations called noise. Thus, the measurements should be written as

$$\mathbf{g} = s\{\mathbf{Hf}\} + \mathbf{n} \, , \tag{6}$$

where $\mathbf{n}$ is the noise vector. It must be emphasized that the noise vector is a random variable. It is different for each set of measurements, and its exact value cannot be correctly guessed. Each realization of the vector $\mathbf{n}$ may be regarded as a random selection from an infinitely large ensemble, or collection, of noise vectors. In general, the noise fluctuations may possess an arbitrary probability density distribution. Frequently, the assumption is made that the noise has a multivariate Gaussian distribution with a zero mean

$$P(\mathbf{n}) \sim exp\left\{-\tfrac{1}{2}\mathbf{n}^T \mathbf{R}_n^{-1}\mathbf{n}\right\} \, , \tag{7}$$

where $\mathbf{R}_n$ is the noise covariance matrix, of which the ij element is

$$[\mathbf{R}_n]_{ij} = \langle n_i n_j \rangle \, . \tag{8}$$

The assumption of a normal distribution is often valid. The brackets $\langle \; \rangle$ indicate an average taken over all members of the ensemble of noise vectors. The above expression is general enough to fully characterize noise flucuations that depend upon the strength of the signal being measured or upon the position of the measurement. It can even take into account correlations in the noise. By its definition, $\mathbf{R}_n$ is a positive-definite matrix, and its inverse, needed in Eq. (7), is assured.

Under a wide range of reasonable conditions [14], when averaged over the full ensembles of noise and images, the best estimate for the reconstruction is that particular image $\mathbf{f}$ which maximizes the *a posteriori* conditional probability density of $\mathbf{f}$ given the measurements $\mathbf{g}$. This probability is given by Bayes' formula

$$P(\mathbf{f}|\mathbf{g}) = \frac{P(\mathbf{g}|\mathbf{f})P(\mathbf{f})}{P(\mathbf{g})} \tag{9}$$

in terms of the conditional probability of $\mathbf{g}$ given $\mathbf{f}$, $P(\mathbf{g}|\mathbf{f})$, and the *a priori* probability distributions of $\mathbf{f}$ and $\mathbf{g}$ separately, $P(\mathbf{f})$ and $P(\mathbf{g})$.

Hunt [11] proposed using the Bayesian approach for improved image recovery. It is assumed that $P(\mathbf{g}|\mathbf{f})$ is given by $P(\mathbf{n})$, in which $\mathbf{g}$ is Gaussian distributed about the mean $s\{\mathbf{Hf}\}$, as in Eq. (7),

$$P(\mathbf{g}|\mathbf{f}) \sim exp\left\{-\tfrac{1}{2}(\mathbf{g} - s\{\mathbf{Hf}\})^T \mathbf{R}_n^{-1}(\mathbf{g} - s\{\mathbf{Hf}\})\right\} \, . \tag{10}$$

This conditional probability may be referred to as the probability density distribution of the measurements, since it follows solely from the distribution of the error fluctuations in the measurements. It is also often called the likelihood function [14]. The *a priori* probability density function for the ensemble of images $P(\mathbf{f})$ is assumed to be a multivariate Gaussian distribution with a mean value $\bar{\mathbf{f}}$ and with a covariance matrix $\mathbf{R}_f$:

$$P(\mathbf{f}) \sim exp\left\{-\tfrac{1}{2}(\mathbf{f} - \bar{\mathbf{f}})^T \mathbf{R}_f^{-1}(\mathbf{f} - \bar{\mathbf{f}})\right\}. \tag{11}$$

Under these assumptions, the maximum *a posteriori* (MAP) solution is easily shown to satisfy [11] the MAP equation

$$\mathbf{R}_f^{-1}(\bar{\mathbf{f}} - \mathbf{f}) + \mathbf{H}^T \mathbf{S}_b \mathbf{R}_n^{-1}(\mathbf{g} - s\{\mathbf{H}\mathbf{f}\}) = 0, \tag{12}$$

where $\mathbf{S}_b$ comes from the derivative of s:

$$\mathbf{S}_b = diag\left(\left.\frac{\partial s(u)}{\partial u_i}\right|_{u_i = b_i}\right) \tag{13}$$

evaluated at the arguments of s

$$b_i = [\mathbf{H}\mathbf{f}]_i. \tag{14}$$

The transpose of $\mathbf{H}$ is the familiar backprojection operation in the context of tomographic reconstruction. The first term comes from the derivative of $P(\mathbf{f})$, given by Eq. (11), and the second from $P(\mathbf{g}|\mathbf{f})$, Eq. (10). It can be seen that the MAP solution strikes a balance between its deviation from the ensemble mean $\bar{\mathbf{f}}$ and the solution to the measurement equation, Eq. (4). This balance is controlled by the covariance matrices $\mathbf{R}_f$ and $\mathbf{R}_n$ that specify the confidence with which each deviation is weighted, as well as possible correlations between the deviations.

If s were a linear function, Eq. (12) would be linear in the unknown vector $\mathbf{f}$. This linearity follows from the assumption of normal distributions for the *a priori* and measurement-error probability densities. In this circumstance, the MAP reconstruction method is equivalent to the minimum-variance linear estimator with nonstationary mean and covariance ensemble characterizations [14]. It is also called the minimum mean-square-error method [15]. When the blur function, noise, and ensemble image properties are stationary (do not depend upon position), then $\mathbf{H}$, $\mathbf{R}_f$, and $\mathbf{R}_n$ are Toeplitz matrices, and, in the circulant approximation, Eq. (12) is the same as the well-known Wiener filter [15]. The application of the linear MAP method to tomographic reconstruction was suggested by Herman and Lent [16].

We wish to address situations in which there is a lack of specific information about the object to be reconstructed. It is known that in such cases the use of the MAP method probably cannot provide much benefit in correcting for the null-space deficit in the reconstruction [12,13]. In the present case of Abel inversion, this deficit exists only for the nonlinear transformation s, not for the matrix $\mathbf{H}$. Suppose that we wish to have a smooth solution unless the data legitimately indicate otherwise. Then a common choice for $\mathbf{R}_f$ is

$$\mathbf{R}_f = (\nabla^2 \nabla^2)^{-1}. \tag{15}$$

Since $\nabla^2 \nabla^2$ is a differentiation operator, $\mathbf{R}_f$ is a smoothing operator. Since, in the absence of very accurate data, we would like the solution to approach something smooth, we further choose for $\bar{\mathbf{f}}$ a smooth version of the estimate $\mathbf{f}$

$$\bar{\mathbf{f}} = \mathbf{R}_f \mathbf{f}. \tag{16}$$

In our calculations $\mathbf{R}_f$ is calculated by using a filter that behaves as the inverse fourth power of the spatial frequency. The addition of a small regularizing term avoids blowup at zero frequency. The amount of regularization controls the width of the smoothing function, which in our example is chosen to have a full-width at half-maximum of 11 pixels. To avoid making the analysis any more complicated than necessary, we will assume that the noise in the radiographic measurements is stationary and uncorrelated. Then,

$$\mathbf{R}_n = \sigma_n^2 \, \mathbf{I} \, , \tag{17}$$

where σ_n^2 is the variance of the noise. While this assumption may not be completely valid, it is a sufficiently good approximation in most instances. Since it is desirable to control its strength, $\mathbf{R}_f$ is multiplied by the factor λ, whose value is adjusted to make the rms residuals approximately match the value expected for the noise. The net effect on the MAP equation, Eq. (12), is that $\mathbf{R}_n$ disappears and the coefficient of the $\mathbf{H}^T$ term becomes λ/σ_n^2.

There is a close relationship between the method just described and that of constrained least squares (CLS) [15]. In the latter method, the solution minimizes the norm of some linear operator applied to $\mathbf{f}$, subject to the constraint that the rms residuals match a specified value. In fact the above choice for $\mathbf{R}_f$ amounts to seeking the solution with minimum curvature. However, the usual version of CLS does not make reference to $\bar{\mathbf{f}}$ and, in effect, assumes $\bar{\mathbf{f}} = 0$. Clearly this is not a very good choice for $\bar{\mathbf{f}}$, as it is the default value when the data are very noisy. In noisy regions our choice pushes the solution towards a smoothed value instead of zero, as does the usual CLS algorithm.

Andrews and Hunt [15] suggested that the maximum-entropy algorithm may be viewed as an alternate form of the constrained least-squares approach. This connection is most easily seen from the work of Gull and Daniell [17], in which they merge the maximum-entropy principle with the probabilistic concepts of random noise. They propose that one find the solution that maximizes entropy, subject to the constraint that the calculated rms residuals be equal to a predetermined value. The similarity between the function they wish to minimize and the quadratic form that leads to the MAP equation, (12), prompts one to interpret the maximum-entropy technique in terms of a Bayesian approach [18] in which the *a priori* probability density has a particular form [13]. Independent of whether this Bayesian interpretation is correct or not, the performance of the maximum-entropy algorithm can be understood and interpreted in terms of what one would expect for MAP for that particular prior probability distribution.

4 An Iterative Solution Technique

We have adopted an iterative approach to the solution of Eq. (12) based on the scheme proposed by Hunt [11]. The iteration scheme is given by:

$$\mathbf{f}^0 = \bar{\mathbf{f}} \tag{18}$$

$$\mathbf{f}^{k+1} = \mathbf{f}^k + c^k \mathbf{r}^k \tag{19}$$

$$\mathbf{r}^k = \mathbf{R}_f^{-1}(\bar{\mathbf{f}} - \mathbf{f}^k) + \mathbf{H}^T \mathbf{S}_b \mathbf{R}_n^{-1}(\mathbf{g} - s\{\mathbf{H}\mathbf{f}^k\}) \tag{20}$$

$$c^k = \frac{\mathbf{r}^{kT}\mathbf{q}^k}{\mathbf{q}^{kT}\mathbf{q}^k} \qquad (21)$$

$$\mathbf{q}^k = (\mathbf{R}_f^{-1} + \mathbf{H}^T\mathbf{S}_b\,\mathbf{R}_n^{-1}\mathbf{S}_b\,\mathbf{H})\mathbf{r}^k \qquad (22)$$

where vector $\mathbf{r}^k$ is the residual of Eq. (12), and the scalar c^k is chosen by Eq. (21) to minimize the norm of $\mathbf{r}^k$. When the residual goes to zero, the corresponding $\mathbf{f}^k$ is clearly a solution of the MAP Eq. (12). This iterative scheme is very similar to the one proposed by Herman and Lent [16] for linear MAP image restoration. Their update scheme consisted in incrementing $\mathbf{f}^k$ by Eq. (20) multiplied by $\mathbf{R}_f$, avoiding a potential computational difficulty if $\mathbf{R}_f$ is nontrivial. But it has the disadvantage of providing smooth update vectors. Consequently, it becomes very difficult to achieve good high-frequency response.

We have found that the above scheme does not work well for this problem even though it was adequate for the usual 2-D reconstruction case [12]. To obtain better convergence, we have added a few variations. Instead of using only one update vector, as in Eq. (19), we have used a second vector, alternatively switching between $\mathbf{q}^k$ and the vector

$$\mathbf{H}^{-1}(\mathbf{g} - s\{\mathbf{H}\mathbf{f}^k\}) , \qquad (23)$$

which is similar to the inverse, Eq. 5. We note that the use of the inverse directly did not appear to work as it overcorrected the residuals. The coefficients for each 'of the updating vectors are computed in a manner similar to that given by Eq. (21). This iteration scheme encompasses the modification to Hunt's procedure that was suggested by Trussell and Hunt [19], but with a different choice for the normalization of the update vectors. A scheme similar to that of conjugate gradient was also tried in which each update is constructed to be orthogonal to preceding ones. It appeared to be of little use in improving convergence. We did find it helpful to gradually admit successively highier frequency components to the solution, which is accomplished by low-pass filtering of the update vectors and progressively increasing the cutoff frequency of the filter as the number of iterations increases. For the low-pass filter we used a Gaussian function centered on zero frequency. In our example we used 10 iterations. The half-value frequency of the low-pass filter ranged from 0.210 to 1.0 times the Nyquist frequency, stepping by a factor of 1.189 after each iteration. The computation time on a VAX 8700 was 45 minutes for the MAP reconstruction of 150 lines, each 220 pixels long, compared with one minute for direct inversion.

An advantage to any iterative reconstruction scheme is that constraints may be readily placed upon the reconstructed function $\mathbf{f}^{k+1}$ after each update. Such constraints include upper and lower limits to the reconstruction value, known region of support, etc.

The application of the above MAP technique to the earlier reconstruction problem yields Fig. 7, which shows that the extreme noise fluctuations in Fig. 6 have been controlled well. However, the sharp response to the outer edge of the cylinder is preserved. By reducing extremely large noise fluctuations, the result is rendered more acceptable to the eye [20]. This wide light region near the vertex of the cone probably arises from a slight shift in D_0 caused by variation in the scattered radiation background. Figure 8 shows a comparison between the MAP and direct inversion techniques for a single line taken from the radiograph.

Figure 7: The MAP reconstruction derived from Fig. 5 using $\lambda = 2.0 \times 10^{-5}$ and $\sigma_n = 0.01$. Note that in the noisy regions the amplitude of the noise has been reduced relative to that in Fig. 6, while at the same time maintaining excellent resolution at the outer edge.

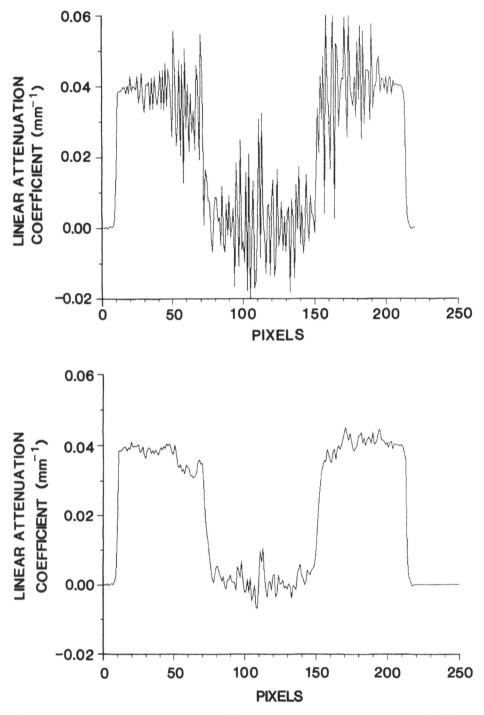

Figure 8: A comparison for one line of data between the reconstruction (top) obtained by direct inversion of the measurement equations and that (bottom) obtained by the MAP technique.

5 Conclusions

It has been shown that the extension of Abel inversion to the reconstruction of a 3-D axially symmetric object from a single radiograph offers significant benefits as an image analysis tool. It has been demonstrated that the noise amplification caused by the nearly singular Abel inversion and the potential difficulty in the inversion of the nonlinear radiographic measurements can be controlled using the MAP method with a simplified choice of a smoothed value of **f** for **f̄** and a smoothing operator for $\mathbf{R}_f$. With these choices the MAP algorithm is similar to the constrained least-squares method [15] corresponding to minimum curvature. We have employed a multifaceted iterative reconstruction algorithm to find the solution with the maximum *a posteriori* probability.

In general the Bayesian approach permits the incorporation of information about the general shape or structure of the object to be reconstructed. Supposing a lack of prior knowledge about the object to be reconstructed, we have only assumed that the reconstructed function should be smooth unless otherwise dictated by the data. If more prior information is available, it may be desirable to incorporate it into the solution. However, one must be careful not to 'load the dice' too heavily or one will obtain only the expected result [13].

Acknowledgments

I wish to express my appreciation to George Brooks for providing the excellent radiographs used in the static example. I have benefited from many fruitful discussions with Rollin Whitman and Allen Mathews.

References

[1] N. H. Abel. Résolution d'un problème de mecanique. *J. Reine u. Angew. Math.*, 1:153–157, 1826.

[2] K. M. Hanson. Tomographic reconstruction of axially symmetric objects from a single radiograph. In *Proc. 16th Inter. Cong. on High Speed Photography and Photonics. (Proc. SPIE, 491:180-187)*, Strasbourg, 1984.

[3] R. W. Porter. Numerical solution for local emission coefficients in axisymmetric self-absorbed sources. *SIAM Rev.*, 6:228–242, 1964.

[4] S. Corti. The problem of Abel inversion in the determination of plasma density from phase measurements. *Let. Nuovo Cim.*, 29:25–32, 1980.

[5] M. Deutsch and I. Beniaminy. Derivative-free inversion of Abel's integral equation. *Appl. Phys. Lett.*, 41:27–28, 1982.

[6] M. Deutsch. Abel inversion with a simple analytic representation for experimental data. *Appl. Phys. Lett.*, 42:237–239, 1983.

[7]   D. S. Carter, G. Pimbley, and G. M. Wing. *On the unique solution for the density function at PHERMEX.* Technical Report T-5-2023, Los Alamos Scientific Laboratory, 1957.

[8]   R. MacRae and G. M. Wing. *Some remarks on problems connected with data analysis in PHERMEX.* Technical Report T-5-2043, Los Alamos Scientific Laboratory, 1958.

[9]   V. Faber and G. M. Wing. *The Abel integral equation.* Technical Report LA-11016-MS, Los Alamos National Laboratory, 1987.

[10] K. M. Hanson. A Bayesian approach to nonlinear inversion: Abel inversion from x-ray attenuation data. Presented at the Fourth Workshop on Maximum Entropy and Bayesian Methods (unpublished), Calgary, August 6-8, 1984.

[11] B. R. Hunt. Bayesian methods in nonlinear digital image restoration. *IEEE Trans. Comp.*, C-26:219–229, 1977.

[12] K. M. Hanson and G. W. Wecksung. Bayesian approach to limited-angle reconstruction in computed tomography. *J. Opt. Soc. Amer.*, 73:1501–1509, 1983.

[13] K. M. Hanson. Bayesian and related methods in image reconstruction from incomplete data. In *Image Recovery: Theory and Application,* Henry Stark, ed., Academic, Orlando, 1987.

[14] A. P. Sage and J. L. Melsa. *Estimation Theory with Applications to Communications and Control.* Robert E. Krieger, Huntington, 1979.

[15] H. C. Andrews and B. R. Hunt. *Digital Image Restoration.* Prentice-Hall, Englewood Cliffs, New Jersey, 1977.

[16] G. T. Herman and A. Lent. A computer implementation of a Bayesian analysis of image reconstruction. *Inform. Contr.*, 31:364–384, 1976.

[17] S. F. Gull and G. J. Daniell. Image reconstruction from incomplete and noisy data. *Nature*, 272:686–690, 1978.

[18] H. J. Trussell. The relationship between image reconstruction by the maximum *a posteriori* method and a maximum entropy method. *IEEE Trans. Acoust., Speech, Signal Processing*, ASSP-28:114–117, 1980.

[19] H. J. Trussell and B. R. Hunt. Improved methods of maximum *a posteriori* restoration. *IEEE Trans. Comput.*, C-27:57–62, 1979.

[20] K. M. Hanson. Image processing: mathematics, engineering, or art? In *Proc. Application of Photo-Optical Instrumentation in Medicine XIII. (Proc. SPIE, 535:70-81),* Newport Beach, 1985.

Numerical Solution of the Magnetic Relief Problem

KATARZYNA JONCA and CURTIS R. VOGEL   Department of Mathematical Sciences, Montana State University, Bozeman, Montana

1   INTRODUCTION

We consider the numerical solution of an inverse problem arising in geophysics. The physical problem is to determine the location of buried magnetized rock from measurements of the magnetic field taken at the surface. Koch and Tarlowski [4] have derived a mathematical model which leads to a system of nonlinear Fredholm first kind integral equations. Preliminary numerical results are presented in [2].

A discussion of the physical problem and the mathematical model appears in section 2. The problem is formulated as a nonlinear operator equation

$$(1.1) \qquad \mathbf{K}(f) = \mathbf{g}.$$

Solutions to this problem are unstable with respect to perturbations in the data $\mathbf{g}$ and the operator $\mathbf{K}$. To achieve stable approximate solutions, Tikhonov Regularization is used, i.e., one solves

$$(1.2) \qquad \min_f \{ \|\mathbf{K}(f) - \mathbf{g}\|^2 + \alpha \|f\|^2 \}$$

for a sequence of values of the regularization parameter α. Ill-posedness, regularization, and the choice of the regularization parameter α are considered in section 3.

Section 4 contains a description of a robust Quasi-Newton/Trust Region algorithm to solve the regularized problem (1.2). This algorithm uses the Singular Value Decomposition of the derivative of the operator $\mathbf{K}$. This ensures numerical stability, it also allows easy computation of an estimate for the optimal value of the regularization parameter, and it gives a characterization of the degree of ill-posedness for this problem.

Numerical results are presented in section 5.

2   THE MATHEMATICAL MODEL

Experimental evidence suggests that in certain situations variations in the magnetic field of the earth depend primarily on the location and shape of the boundary between

This research was partially supported by the NSF under Grant DMS-8602000.

magnetized igneous rock and unmagnetized sediments, which cover this rock. One wishes to determine these attributes from airborne magnetic data. The mathematical formulation of the relationship between the variations in the magnetic field and the location and shape of the boundary leads to a system of nonlinear first kind integral equations. A derivation of this system appears in [4].

Let the z-axis be chosen so that the positive z direction is downward, and the measurements of the magnetic field take place in the plane $z = 0$. Let the boundary surface σ have the parameterization

$$(2.1) \qquad \sigma = \{(x, y, z) : z = h + f(x, y), -\infty < x < \infty, -\infty < y < \infty\}.$$

We refer to $h > 0$ as the *characteristic depth* of the boundary of the surface. f is called the magnetic *relief function*.

Figure 1. Geometry of the Magnetic Relief Problem.

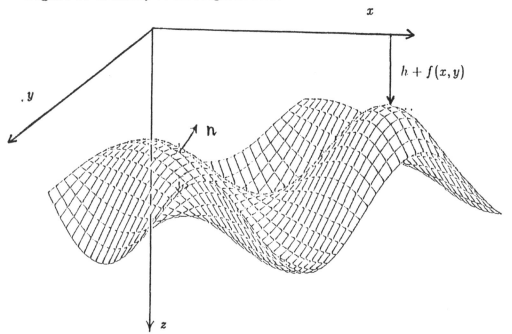

Assuming constant magnetization $\mathbf{M} = [M_x, M_y, M_z]^T$ of the igneous rock and outward unit normal $\mathbf{n}$ to the surface σ, the intensity of the measured magnetic field $\mathbf{H}$ is given by

$$(2.2) \qquad \mathbf{H}(s, t, 0) = -\int \int_\sigma \mathbf{M} \cdot \mathbf{n} \, \nabla \frac{1}{\|(s, t, 0) - (x, y, z)\|} \, dS$$

$$= -\int_{-\infty}^{+\infty} \int_{-\infty}^{+\infty} \frac{(M_x \frac{df}{dx} + M_y \frac{df}{dy} - M_z)[s - x, t - y, -h - f(x, y)]}{\{(s - x)^2 + (t - y)^2 + (h + f(x, y))^2\}^{3/2}} \, dx \, dy.$$

Subscripts "x" and "y" indicate horizontal components, while "z" indicates the vertical component. Letting g_x, g_y, g_z denote the components of $\mathbf{H}$, the following system of nonlinear first kind integral equations must be solved to obtain f:

$$(2.3)\; g_x(s,t) \;=\; K_x(f)(s,t) := \int\int \frac{(M_x\frac{df}{dx} + M_y\frac{df}{dy} - M_z)(s-x)}{\{(s-x)^2 + (t-y)^2 + (h+f(x,y))^2\}^{3/2}} dxdy$$

$$(2.4)\; g_y(s,t) \;=\; K_y(f)(s,t) := \int\int \frac{(M_x\frac{df}{dx} + M_y\frac{df}{dy} - M_z)(t-y)}{\{(s-x)^2 + (t-y)^2 + (h+f(x,y))^2\}^{3/2}} dxdy$$

$$(2.5)\; g_z(s,t) \;=\; K_z(f)(s,t) := \int\int \frac{(M_x\frac{df}{dx} + M_y\frac{df}{dy} - M_z)(-h-f(x,y))}{\{(s-x)^2 + (t-y)^2 + (h+f(x,y))^2\}^{3/2}} dxdy$$

Since measurements are taken within a bounded region, and also for computational convenience, we will assume that the problem is scaled so that f is identically zero outside the domain $\Omega := (0,1) \times (0,1)$, and that measurements $\mathbf{g}(s,t)$ are taken only at points $(s,t) \in \Omega$. The components of $\mathbf{g}$ can be suitably modified so that the integration above takes place over this restricted domain Ω rather than over the entire x-y plane. Thus, the problem can be formulated as a nonlinear operator equation

$$(2.6) \qquad \mathbf{g} = \mathbf{K}(f) := \int\int_\Omega \mathbf{k}(\cdot, x, y, f(x,y), \nabla f(x,y)) dx\, dy.$$

The components of the kernel $\mathbf{k} = [k_x, k_y, k_z]^T$ appear under the integrals on the right hand sides of (2.3)-(2.5). The data $\mathbf{g} = [g_x, g_y, g_z]^T$ is a vector-valued function whose components represent measurements of the magnetic field $\mathbf{H}$. The characteristic depth h gives a measure of the depth of the boundary surface relative to the size of the region Ω.

3 ILL-POSEDNESS AND REGULARIZATION

Since the system (2.3)-(2.5) involves partial derivatives of the solution f, we require f to lie in the Sobolev space $H^1(\Omega)$. We have also assumed that f vanishes outside Ω, so we choose the Hilbert space $X = H^1_0(\Omega)$ consisting of functions $f \in H^1(\Omega)$ satisfying the boundary condition $f(x,y) = 0$, $(x,y) \in \partial\Omega$, with inner product

$$(3.1) \qquad < f, h >_X := \int\int_\Omega \nabla f \cdot \nabla h\, dx\, dy, \quad f, h \in X,$$

and induced norm

$$(3.2) \qquad \|f\|_X := \{\int\int_\Omega \nabla f \cdot \nabla f\, dx\, dy\}^{\frac{1}{2}}, \quad f \in X.$$

On the other hand the components of $\mathbf{g}$ come from measurements taken at discrete points in Ω. One cannot assume that the derivatives of these components are available. Thus we define the space $Y = [L^2(\Omega)]^3 := L^2(\Omega) \times L^2(\Omega) \times L^2(\Omega)$ with the norm

$$(3.3) \qquad \|\mathbf{g}\|_Y := \{\int\int_\Omega [g_x(s,t)^2 + g_y(s,t)^2 + g_z(s,t)^2]\, ds\, dt\}^{\frac{1}{2}}$$

Given arbitrary $\mathbf{g} \in Y$, the problem of finding an $f \in X$ such that $\mathbf{K}(f) = \mathbf{g}$ is *well-posed* provided: (i) for any $\mathbf{g} \in Y$, there exists a solution $f \in X$ such that

$\mathbf{K}(f) = \mathbf{g}$; (ii) the solution f is unique; (iii) the solution f is continuous with respect to perturbations in the data $\mathbf{g}$ and the operator $\mathbf{K}$.

The well-posedness of this particular inverse problem depends on the choice of the spaces X and Y. With the above (physically reasonable) choice of spaces X and Y, problem (1.1) is clearly ill-posed. From (2.3)-(2.5), the components of the kernel $\mathbf{k}$ are smooth in s and t, and hence, for any $f \in X$, the components of $\mathbf{K}(f)$ are smooth. Hence, there exist elements $\mathbf{g} \in Y$ for which (1.1) has no solution. The operator $\mathbf{K}$ can be shown to be compact (see [5]). Consequently, the inverse image of (noncompact) neighborhoods $N_r(g) = \{z \in Y : \|g - y\| \le r\}$, $r > 0$, is unbounded, and we *do not* have continuous dependence of solutions in X on the data in Y. Since $\mathbf{g}$ is obtained from measurements of the magnetic field, one cannot obtain solutions to (1.1) without some sort of stabilization, or regularization.

Using the method of Tikhonov Regularization [8], we replace the ill-posed problem (1.1) by a sequence of regularized problems

$$(3.4) \qquad \min_{f \in X}\{\|\mathbf{K}(f) - \mathbf{g}\|^2 + \alpha\|f\|^2\}$$

For a discussion of existence and stability of solutions f_α to (3.4) and convergence of these solutions as the regularization parameter $\alpha \to 0$ and as the perturbations in the data and the operator tend to zero, see [7].

The operator $\mathbf{K} : X \to Y$ is Frechet differentiable. Given $u \in X = H_0^1(\Omega)$, the z-component of $\mathbf{K}'(f)u$ is given by

$$(3.5) \qquad [K_z'(f)u](s,t) = \int\int_\Omega \frac{\partial k_z}{\partial f}\, u\, dx\, dy + \int\int_\Omega \frac{\partial k_z}{\partial \nabla f} \cdot \nabla u\, dx\, dy$$

$$= \int\int_\Omega \frac{(M_x \frac{df}{dx} + M_y \frac{df}{dy} - M_z) \times (2(h + f(x,y))^2 - (s - x)^2 - (t - y)^2)}{[(s - x)^2 + (t - y)^2 + (h + f(x,y))^2]^3} u(x,y)\, dx\, dy$$

$$+ \int\int_\Omega \frac{(-h - f(x,y))}{[(s - x)^2 + (t - y)^2 + (h + f(x,y))^2]^{3/2}}[M_x, M_y] \cdot [\frac{\partial u}{\partial x}, \frac{\partial u}{\partial y}]\, dx\, dy.$$

Similarly, one can compute the x and y components $K_x'(f)u$ and $K_y'(f)u$ of $\mathbf{K}'(f)u$.

A very important practical consideration is the choice of the regularization parameter α for a given (error contaminated) data set. One would like to choose α so that $\|f_\alpha - f\|$ is minimized, where f is the (unknown) true solution. If the error in the data is discrete and random, then under certain conditions (e.g., see [9]) the method of Generalized Cross Validation (GCV) yields a statistical estimate of the size of $\|\mathbf{K}(f_\alpha) - \mathbf{K}(f)\|^2$, which is related to $\|f_\alpha - f\|$. For Tikhonov Regularization this estimate is given by the GCV functional

$$(3.6) \qquad V(\alpha) := \frac{\|\mathbf{K}(f_\alpha) - \mathbf{g}\|^2}{Trace[I - A(A^TA + \alpha I)^{-1}A]},$$

where $A = \mathbf{K}'(f_\alpha)$ is the derivative operator defined in (3.5). To approximate the minimizer of $\|f_\alpha - f\|$, one attempts to solve

$$\min_{\alpha \ge 0} V(\alpha).$$

For an application of GCV to a nonlinear integral equation arising in the remote sensing of the atmosphere, see [6].

4 NUMERICAL IMPLEMENTATION

In our numerical implementation we took approximate solutions

$$(4.1) \qquad f_n(x,y) = \sum_{j=1}^{n} c_j \Phi_j(x,y), \quad \mathbf{c} = [c_j]_{j=1}^{n} \in R^n.$$

The Φ_j's were taken to be tensor products

$$\Phi_j(x,y) = \phi_{j_x}(x)\phi_{j_y}(y)$$

defined on the unit square Ω. The ϕ_j 's were chosen to be natural cubic spline basis functions (B-splines) defined on a uniform mesh for the unit interval $[0,1]$, each of which satisfies the boundary conditions $\phi_j(0) = \phi_j(1) = 0$. Thus $\{\Phi_j\}_{j=1}^{n}$ forms the basis for an n-dimensional subspace of $X = H_0^1(\Omega)$.

Simulated measurements of the z-component of the magnetic field, $g_{zi} := g_z(s_i, t_i)$, were taken at m points $(s_i, t_i) \in \Omega = (0,1) \times (0,1)$. These points need not lie on a regular grid. We assume $m \geq n$. The square of the L^2 norm, $\|K_z(f_n) - g_z\|^2$, was approximated by the weighted discrete sum

$$(4.2) \qquad \frac{1}{m} \sum_{i=1}^{m} (K_z(f_n)(s_i, t_i) - g_{zi})^2.$$

The nonlinear operator $K_z : X \to L^2(\Omega)$ defined in (2.5) and its derivative $K_z' : X \to L^2(\Omega)$ defined in (3.5) induce discrete operators K_{mn} and $K_{mn}' : R^n$, each mapping R^n into R^m:

$$(4.3) \qquad [K_{mn}(\mathbf{c})]_i = K_z(f_n)(s_i, t_i), \quad 1 \leq i \leq m, \mathbf{c} \in R^n.$$

$$(4.4) \qquad [K_{mn}'(\mathbf{c})]_{ij} = K_z'(f_n)\Phi_j(s_i, t_i), \quad 1 \leq i \leq m, 1 \leq j \leq n.$$

From (3.2), the regularization term $\|f\|^2$ in (3.4) yields the quadratic form

$$\|f_n\|^2 = \mathbf{c}^T B_n \mathbf{c},$$

where B_n is the symmetric positive definite $n \times n$ matrix with components

$$(4.5) \qquad [B_n]_{ij} = \int \int_\Omega \nabla \Phi_i \cdot \nabla \Phi_j \, dx \, dy, \quad 1 \leq i, j \leq n.$$

The finite dimensional analogue of the problem (3.4) is then

$$(4.6) \qquad \min_{\mathbf{c} \in R^n} \{ \frac{1}{m} \sum_{i=1}^{m} ([K_{mn}(\mathbf{c})]_i - g_{zi})^2 + \alpha \mathbf{c}^T B_n \mathbf{c} \}.$$

For notational convenience, we drop the subscripts and multiply the objective function in (4.6) by m. Also, from this point on, "$\|\cdot\|$" indicates the usual Euclidean norm in R^m or R^n. Since B is symmetric positive definite, we can take its Choleski factorization $B = R^T R$, where R is upper triangular. The objective function for problem (4.6) can then be written

$$(4.7) \qquad T_\alpha(\mathbf{c}) := \{\|K(\mathbf{c}) - g\|^2 + m\alpha\|R\mathbf{c}\|^2\}$$

To obtain a minimizer of the nonlinear functional T_α, a Quasi-Newton/Trust Region approach [1] is used. We consider the iteration

$$(4.8) \qquad \qquad c_{k+1} = c_k + s_k,$$

where s_k solves the constrained minimization problem

$$(4.9) \qquad \min_{s \in R^n} \{ \|K(c_k) + K'(c_k)s - g\|^2 + m\alpha \|R(c_k + s)\|^2 \}$$

$$\text{subject to } \|s\| \le \delta_k,$$

and the trust region radius δ_k is chosen to obtain sufficient decrease in $T_\alpha(c)$ at each iteration to guarantee convergence to a (relative) minimizer of T_α.

To solve the constrained quadratic minimization problem (4.9), we use an approach similar to that of Elden [3]. The problem is first diagonalized using the Singular Value Decomposition (SVD). Then a standard dual space technique is used to solve the diagonalized problem. This approach is numerically stable, quite efficient, and it allows easy computation of the GCV function in $V(\alpha)$ given in (3.6). Define

$$b := g - K(c_k),$$

$$A := K'(c_k)R^{-1},$$

let A have the SVD

$$A = UDV^T, \quad U_{m \times m}, V_{n \times n} \text{ orthogonal}, D_{m \times n} = diag\{d_i\},$$

and consider the change of variables

$$(4.10) \qquad \hat{s} = V^T R s, \quad \hat{c} = V^T R c_k, \quad \hat{b} = U^T b.$$

Then (4.9) is equivalent to the diagonalized problem

$$(4.11) \qquad \min_{\hat{s} \in R^n} \{ \|D\hat{s} - \hat{b}\|^2 + m\alpha \|\hat{c} + \hat{s}\|^2 \}$$

$$\text{subject to } \|\hat{s}\|^2 \le \delta_k^2.$$

The unique solution to problem (4.11) has the form

$$(4.12) \qquad \hat{s} = \hat{s}(\mu) = \{ D^T D + (m\alpha + \mu)I \}^{-1} (D\hat{b} - m\alpha\hat{c}),$$

where $\mu \ge 0$ is a Lagrange multiplier. If $\|\hat{s}(0)\| \le \delta_k$, then $s = R^{-1}V\hat{s}(0)$ solves (4.9). Otherwise the constraint is active, and $s = R^{-1}V\hat{s}(\mu)$ solves (4.9), where $\mu \ge 0$ is the unique solution to

$$(4.13) \qquad g(\mu) := \|\hat{s}(\mu)\|^2 - \delta_k^2 = 0.$$

We solve (4.13) using the algorithm given in [1, page 134], which requires $g(\mu)$ and its derivative $g'(\mu)$. In terms of the singular values d_i and the components of $\hat{b}$ and $\hat{c}$,

$$(4.14) \qquad g(\mu) = \sum_{i=1}^{n} [\frac{(d_i\hat{b}_i - m\alpha\hat{c}_i)}{(d_i^2 + m\alpha + \mu)}]^2 - \delta_k^2,$$

so both $g(\mu)$ and $g'(\mu)$ can be obtained easily.

To determine the trust region radius δ_k in (4.9), we used the algorithm in [1, page 143]. The basic idea is to pick δ_k such that the objective function $T_\alpha(c)$ decreases sufficiently, i.e., so that

$$(4.15) \qquad T_\alpha(\mathbf{c}_k + \mathbf{s}) \leq T_\alpha(\mathbf{c}_k) + \epsilon \nabla T_\alpha(\mathbf{c}_k)^T \mathbf{s}$$

where ϵ is a small positive parameter. In addition, δ_k is chosen so that we "trust" the quadratic approximation to $T_\alpha(f)$ within the trust region, i.e.,

$$(4.16) \qquad T_\alpha(c_k + s) \approx \|K(\mathbf{c}_k) + K'(\mathbf{c}_k)s - g\|^2 + m\alpha \|R(\mathbf{c}_k + \mathbf{s})\|^2$$
$$\text{whenever } \|s\| \leq \delta_k.$$

This last requirement allows us to increase the size of the step s under certain conditions, thereby decreasing the number of iterations (4.8)-(4.9) needed to minimize T_α.

The GCV functional in (3.6), which is used to estimate the optimal regularization parameter α, is computed using

$$(4.17) \qquad V(\alpha) = \frac{\frac{1}{m}\|K(c_\alpha) - g\|^2}{[\frac{m-n}{m} + \sum_{i=1}^{n} \frac{\alpha}{(d_i^2 + m\alpha)}]^2},$$

where c_α solves (4.6). The singular values d_i can also be used to obtain (linearized) stability estimates. They are used in the next section to quantify the "degree of ill-posedness" of the magnetic relief problem.

5 NUMERICAL RESULTS

We first conducted a linearized stability analysis to determine how the stability and resolution (i.e., the accuracy of approximation to the solution) varies as the characteristic depth h increases. From physical arguments, one might expect good stability and resolution for small h, with both decreasing as the depth h increases. Since

$$(5.1) \qquad \mathbf{K}(f + \Delta f) - \mathbf{g} \approx \mathbf{K}(f) + \mathbf{K}'(f)\Delta f - \mathbf{g},$$

one might expect to be able to analyze stability and resolution by examining the derivative operator $\mathbf{K}'(f)$.

From (2.5), the nonlinear integral operator K_z corresponding to the vertical component of the magnetic field has as its kernel

$$k_z(s, t, x, y, f, \nabla f) = \frac{(M_x \frac{df}{dx} + M_y \frac{df}{dy} - M_z)(-h - f(x,y))}{\{(s-x)^2 + (t-y)^2 + (h + f(x,y))^2\}^{3/2}}$$

In (3.5), the derivative operator $K'_z(f)$ is expressed as the sum of two linear integral operators. Given $u \in H^1(\Omega)$, the first is applied to $u(x,y)$ and has as its kernel

$$(5.2) \qquad \frac{\partial k_z}{\partial f} = \frac{(M_x \frac{df}{dx} + M_y \frac{df}{dy} - M_z) \times (2(h + f(x,y))^2 - (s-x)^2 - (t-y)^2)}{[(s-x)^2 + (t-y)^2 + (h + f(x,y))^2]^3}$$

The second integral operator is applied to ∇u and has as its kernel

$$(5.3) \qquad \frac{\partial k_z}{\partial \nabla f} = \frac{(-h - f(x,y))}{[(s-x)^2 + (t-y)^2 + (h + f(x,y))^2]^{3/2}}[M_x, M_y]$$

Figure 2 below shows a surface of the first kernel (5.2) evaluated at $f = 0$, $M_x = M_y = M_z = 1$, fixed $s = t = \frac{1}{2}$, and characteristic depth $h = 0.2$.

Figure 2. Derivative Kernel $\frac{\partial k_z}{\partial f}(x,y)$.

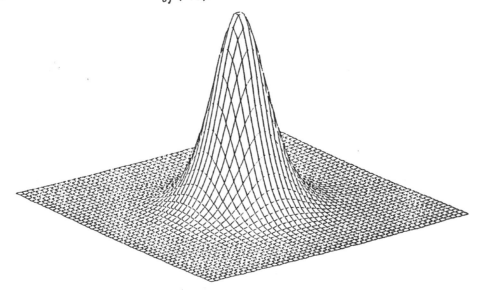

To see how this kernel changes as h increases, we plotted the above kernel on the diagonal $x = y, 0 \leq x \leq 1$ for several different values of h. The solid line in Figure 3 below was obtained for depth $h = 0.1$, while the dotted line represents $h = 0.2$. Similar graphs for the second kernel (5.3) can easily be obtained. One sees from Figure 3 that the kernels become more "flat" and less "delta-like" as the characteristic depth h increases.

A quantitative description of how increasing characteristic depth decreases the resolution is given in Figure 4. Here we plot the singular values of the derivative operator K_z' defined in (4.4) in order of decreasing magnitude for both $h = 0.1$ (stars) and $h = 0.2$ (circles). $n = 8^2 = 64$ tensor product basis functions $\Phi_j(x,y)$ were used in the computations.

Figure 3. Derivative Kernel $\frac{\partial k_1}{\partial f}$ on diagonal $x = y$ for $h = 0.1$ and $h = 0.2$.

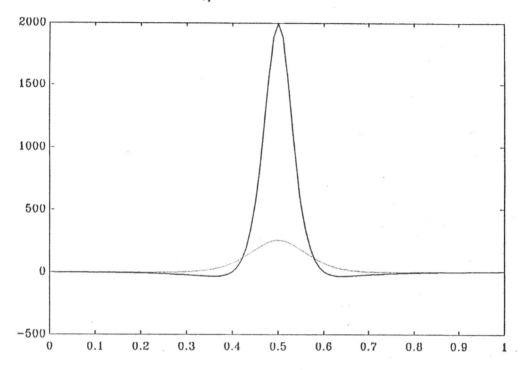

Figure 4. Singular Values of the Derivative.

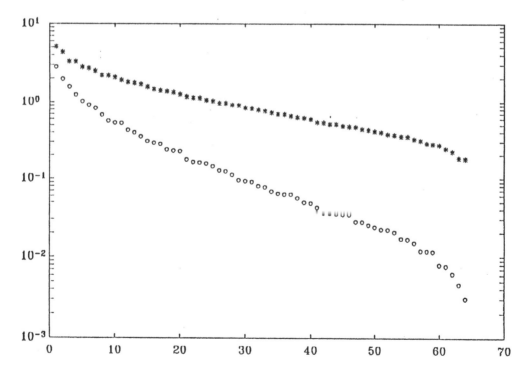

In the above graph, the singular values for $h = 0.2$ decay much more rapidly (note logarithmic scale) than those corresponding to $h = 0.1$. Very small singular values cause instability, so one sees a quantitative decrease in stability with increasing characteristic depth h. Since Tikhonov Regularization has the effect of filtering out singular components associated with small singular values, it should come as no surprise that the amount of resolution for the (regularized) solutions should also decrease with increased depth.

We next generated synthetic data $g_i = K(f)(s_i)$ and obtained approximate solutions using the implementation of Tikhonov Regularization described in section 4. The true magnetic relief function was taken to be a linear combination of Gaussians:

$$-f(x, y) = a_1 \exp[-d_1(x - x_1)^2 - e_1(y - y_1)^2] + a_2 \exp[-d_2(x - x_2)^2 - e_2(y - y_2)^2].$$

The parameters $a_1 = .05, a_2 = .03$ control the magnitude of the solution. $d_1 = d_2 = e_1 = e_2 = 60$ determine the rate of decay of the Gaussians, and $x_1 = y_1 = .4, x_2 = y_2 = .6$ specify the locations of the peaks. For characteristic depth $h = 0.2$, we generated $m = 15^2 = 225$ data points $g_{zi} = g_z(s_{i_x}, t_{i_y})$ on a uniform mesh for Ω. To the data g_{zi} we added pseudo-random error ϵ_i with a Gaussian $N(0, \sigma^2 I)$ distribution. The standard deviation σ was picked so that

$$\frac{\sqrt{E\|\epsilon\|^2}}{\|g_z\|} = .01.$$

To approximate the true solution, we used $n = 8^2 = 64$ tensor product basis functions. We solved the resulting finite dimensional minimization problem (4.6) for a decreasing sequence of regularization parameters $\alpha = 10^{-p}, p = 0, 1, \ldots, 4$. The resulting approximations are shown in Figure 5. Figure 6 shows these approximate solutions restricted to the diagonal $x = y$. The $+$'s represent the true solution, the o's represent the regularized solution for $\alpha = 10^0 = 1$, the solid curve represents the regularized solution for $\alpha = 10^{-2}$, and the dotted curve represents the regularized solution for $\alpha = 10^{-4}$.

Figure 7 shows the norm of the true error $\|e_\alpha\| = \|f_\alpha - f\|$ (indicated by circles) and the GCV functional $V(\alpha)$ (indicated by stars) as functions of α. Note that the true error at first decreases with decreasing α, but then increases noticeably as α becomes small. $V(\alpha)$ follows this behavior somewhat, but it stays very flat for small α and has no well-defined minimizer.

Figure 5. Approximate solutions.

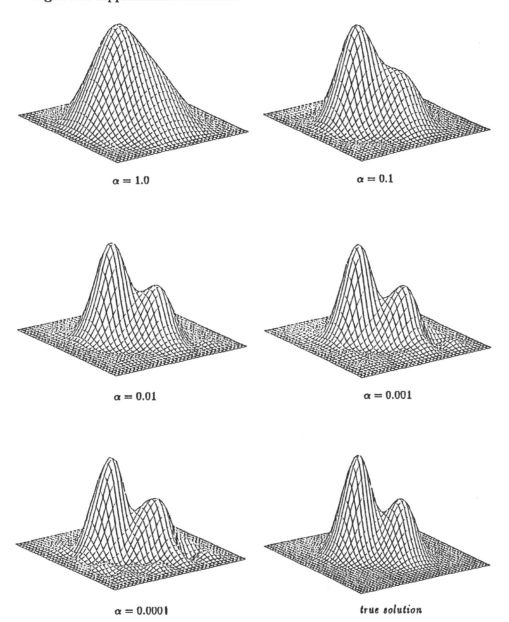

$\alpha = 1.0$

$\alpha = 0.1$

$\alpha = 0.01$

$\alpha = 0.001$

$\alpha = 0.0001$

true solution

Figure 6. Approximate solutions on diagonal $x = y$.

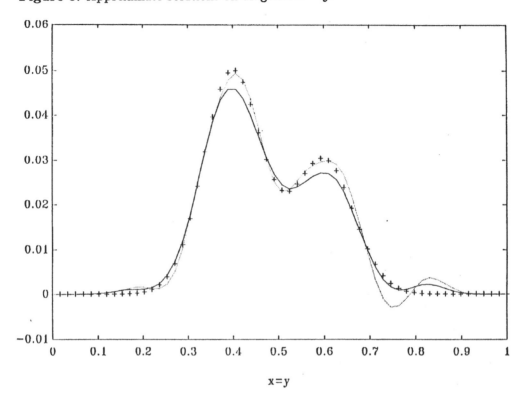

x=y

Figure 7 $\|e(\alpha)\|$ and $V(\alpha)$ vs α.

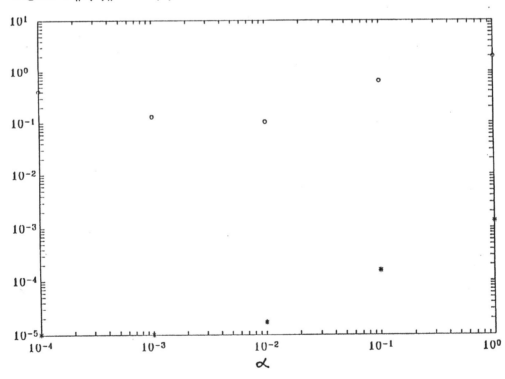

α

ACKNOWLEDGEMENTS. The authors greatly appreciate the assistance of Bob Anderssen and Inga Koch at the CSIRO Division of Mathematics and Statistics and Chris Tarlowski of the BMR Division of Geophysics in Canberra, Australia, in discussions leading to this report. We also acknowledge the support of the San Diego Supercomputer Center, where the computations were performed.

References

[1] J. E. Dennis and R.B. Schnabel (1983). *Numerical Methods for Unconstrained Optimization and Nonlinear Equations*, Prentice Hall, Englewood Cliffs.

[2] K. Jonca and C. R. Vogel (1987). *Numerical solution to the magnetic relief problem–A preliminary report*, Proceedings on Inverse Problems (vol. 17), Centre for Mathematical Analysis, The Australian National University, Canberra ACT 2601.

[3] L. Elden (1977). *Algorithms for the regularization of ill-conditioned least squares problems*, BIT 17, pp. 134-145.

[4] I. Koch and C. Tarlowski, *The magnetic relief problem*, in The 1986 Workshop on Inverse Problems (R.S. Anderssen and G.N. Newsam, Eds.), Centre for Mathematical Analysis, Australian National University, Canberra ACT 2601.

[5] R. H. Martin, Jr., *Nonlinear Operators and Differential Equations in Banach Space*, Wiley, New York, 1976.

[6] F. O'Sullivan and G. Wahba (1985). *A cross validated Bayesian retrieval algorithm for nonlinear remote sensing experiments*, Journal of Computat. Physics 59, pp. 441- 455.

[7] T. Seidman and C.R. Vogel (1987). *Well- posedness and convergence of some regularization methods for nonlinear ill-posed problems*, Tech. Report CMA-R48-86, Centre for Mathematical Analysis, Australian National University, Canberra ACT 2601.

[8] A. N. Tikhonov and V.Y. Arsenin (1977). *Solutions of Ill-Posed Problems*, John Wiley, New York.

[9] G. Wahba (1977). *Practical approximate solutions to linear operator equations when the data are noisy*, SIAM J. Numer. Anal. 14, pp.651-667.

The Weierstrass Transform in X-Ray Analysis

JOHN D. ZAHRT and NATALIE D. CARTER   Diagnostic Physics, Los Alamos National Laboratory, Los Alamos, New Mexico

1   INTRODUCTION

In the past three years much interest and activity has developed in obtaining high spatial resolution ($\sim 30\ \mu$m) x-ray fluorescence analyses [1-3]. The x-ray fluorescence (XRF) technique simply excites a sample with an incident beam of x rays. These may be characteristic (monochromatic) or bremsstrahlung (continuous). The incident rays cause the atoms in the sample to emit (fluoresce) their characteristic x rays. The energy of the fluorescent x rays identifies the atomic species present. The intensity of the fluorescent x rays is related to the concentration. Figure 1 shows an idealized experimental design. The incident beam has finite width (shown as a Gaussian profile) and is centered at x_i on the sample S. The fluorescent x rays could be detected with detectors at D1 or D2. For the discussion in this paper we assume that most of the sample is transparent to both the incident and fluorescent x rays. We do not include absorption or scattering of the x rays.

In experiments, that have been performed to date, the incident bremsstrahlung x-ray beam was delivered to the sample by a collimator and the beam profile was nearly Gaussian. Zahrt has shown that, with Johansson geometry, a diffracted beam of characteristic x rays should also have a Gaussian profile and be narrow enough to be useful [3]. It was also shown in [3] that the sample concentration, c, the x-ray beam intensity, I, and the signal, s, are related by

$$s(x) = \int I(x - y)\ c(y)dy \tag{1}$$

where $I(x)=I_o \exp(-x^2/2\sigma^2)$; $s(x)$ is known by experiment; and $c(y)$ is to be determined. (The limits on this and all other integrals are implied to be from $-\infty$ to ∞ unless specified.)

The integral

$$f(x) = \int \exp[-(x - y)^2]\ \phi(y)/\sqrt{\pi}\ dy \tag{2}$$

is known as the Weierstrass transform of $\phi(y)$ [4]. As an integral equation for $\phi(y)$ this is a Fredholm equation of the first kind and its solution is frought with myriad difficulties. One can build a table of f(x) for given $\phi(y)$. Three such entries were given in [3] and more are given in [4]. An even more extensive table is given in Appendix A.

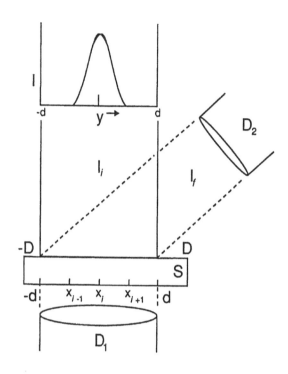

Fig. 1. A schematic of an experimental design. D1 and D2 are detectors, S is the sample, to be stepped along the x_i, and I_o and I_f are the incident and fluorescent intensities respectively.

2   METHODS OF RESOLUTION

We speak here, in the spirit of Milt Wing [5], of resolution of the problem rather than solution because our data are discrete, noisy and may not reflect the properties of the kernel.

One very simple resolution of (1) would require that s(x) be written in terms of the f(x) in the tables. Then c(y) would be written as the corresponding $\phi(y)$. This method, however, lacks the generality that we desire.

Morse and Feshbach [6] supply a classical resolution to the problem. Suppose that $\phi(y)$ can be written as

$$\phi(y) = \sum_m a_m \ H_m(y), \tag{3}$$

where $H_m(y)$ are the Hermite polynomials (the summation here and elsewhere, unless otherwise noted, is assumed to be from 0 to ∞). With the relation

$$\exp(-t^2 + 2xt) = \sum_n (1/n!) \ H_n(x) \ t^n, \tag{4}$$

Eq.(2) becomes

$$f(x) = \sum_n \sum_m (x^n a_m/n!) \int \exp(-y^2) \ H_n(y)H_m(y)dy = \sum_n 2^n \ a_n \ x^n \tag{5}$$

owing to the orthogonality of the Hermite polynomials. By expanding $f(x)$ in a Maclaurin series, we see that

$$a_n = f^{(n)}(0) \ /n!2^n, \tag{6}$$

so $\phi(y)$ is determined from Eq. (3). Taking n derivatives of, most likely, noisy discrete data is a severe problem. It can, on the surface at least, be circumvented by writing

$$f(x) = \sum_j \ b_j \ H_j \ (x) \tag{7}$$

and relating the $H_j(x)$ back to the powers of x. In practice, the sums in Eqs.(3) and (7) must be finite and an error is introduced, but this method could be coded.

An alternative method that is formally correct is indicated by Hirschman and Widder [4]. Define

$$\exp(aD) \ \phi(x) = \sum_n (a^n \ D^n/n!) \ \phi(x) = \phi(x+a). \tag{8}$$

Recognizing that the Laplace transform of a Gaussian is a Gaussian of positive exponent, we take

$$\int \exp(-t^2)\exp(-st)dt/\sqrt{\pi} = \exp(s^2/4). \tag{9}$$

Letting $s = D = d/dx$, we obtain

$$\exp(D^2/4) \ \phi(x) = \int dt \ \exp(-t^2)\exp(-tD) \ \phi(x)/\sqrt{\pi}$$
$$= \int \phi(x-t)\exp(-t^2) \ dt/\sqrt{\pi} = f(x). \tag{10}$$

so finally

$$\phi(x) = \exp(-D^2/4) \ f(x). \tag{11}$$

As stated above, Eq.(11) is formally correct, but again we face the problem of taking many derivatives of noisy data. There are also certain problems of convergence that may arise. For example, applying Eq.(11) works very well for $f(x)=x^2$ but fails for $f(x)=[\sin x \ /x]^2$.

We now introduce some physical considerations relevant to the problem and develop a new algorithm. In the experimental design, the sample will be stepped across the fixed x-ray beam by some fixed amount, say Δ, per step. The data will thus be discrete and have a certain inherent graininess. For us this constitutes the bin structure of the discrete problem, $x_i\text{-}x_{i-1}=\Delta$. We assume that the graininess of the data mirrors itself in the graininess of the determination of the concentration. Furthermore, the full width at half maximum of the Gaussian intensity profile (FWHM = 2.35482σ) is assumed to be of the same order of magnitude as the width of a bin, Δ. If $\sigma \ll \Delta$, then the sample is being discretly analyzed at only relatively few points and there is no deconvolution problem; if $\Delta \ll \sigma$ one is sampling more points than the experimental design justifies. Finally, and not to be overlooked, the signals are obtained at only a finite number of positions $\{x_i\}$ by stepping the sample past the fixed source and detector.

With no scaling, the problem is cast as

$$s(x_i) = A \int_{-d}^{d} \exp[-(x_i - y)^2/2\sigma^2] \ c(y) \ dy. \tag{12}$$

where d is the radius of the field seen by the detector (see Fig. 1). Letting $\frac{x}{\sqrt{2}\sigma} \to x\prime$, $\frac{y}{\sqrt{2}\sigma} \to y\prime$ and $A\sqrt{\pi}\sqrt{2}\sigma \to A\prime$ and then dropping primes, we have

$$s(\sqrt{2}\sigma \ x_i) = \frac{A}{\sqrt{\pi}} \int_{-\infty}^{\infty} \exp[-(x_i - y)^2]c(\sqrt{2}\sigma \ y) \ dy. \tag{13}$$

We have returned to $\pm\infty$ as limits on the integral as d and σ are of the order of 10^{-1} cm and 10^{-3} cm respectively, and the $\frac{d}{\sqrt{2}\sigma} \sim 10^2$. For convenience we also choose $\sigma = \frac{1}{\sqrt{2}}$ to simplify the sequel. Taking cognizance of entries 5 and 6 in Appendix A, we note that $\sin ax$ and $\cos ax$ are eigenfunctions of our integral operator with Gaussian kernel with eigenvalues, e_a, equal to $\exp\left(-\frac{a^2}{4}\right)$ for any a. As the sample is of finite length (-D to D), we see that if we write the signal, $s(x_i)$ as a Fourier series from -D to D [$s(-D) = s(D) = 0$] the concentration will also be given as a Fourier series with modified coefficients. Guided by entries 5 and 6 in Appendix A we write

$$
\begin{aligned}
s(x_i) \ = \ & \frac{A}{\sqrt{\pi}} \int \exp[-(x_i-y)^2] \left\{ a_0 + \sum_{n=0} a_n \exp[(2n+1)^2\frac{\pi^2}{4D^2}]\cos(2n+1)\frac{\pi y}{2D} \right. \\
& + \left. \sum_{n=1} b_n \exp[(2n)^2\frac{\pi^2}{4D^2}]\sin\frac{2n\pi y}{2D} \right\} dy.
\end{aligned}
\tag{14}
$$

where

$$s(x_i) = a_0 + \sum_{n=0} a_n\cos(2n+1)\frac{\pi x_i}{2D} + \sum_{n=1} b_n\sin(2n)\frac{\pi x_i}{2D}. \tag{15}$$

The signal is known at a finite discret (most likely equally spaced) set of points, x_i so that the sums in Eq. (14) would be truncated when x_i - $x_{i-1} \sim \frac{1}{2}\lambda$ where λ is the wavelength of a sine or cosine term in the Fourier series.

The resolution of the problem is thus removed from solving an IFK to computing a Fourier series. The ill-posedness of the problem comes from the exponential decay of the eigenvalues.

Figures 2, 3 and 4 show a model problem to simulate a scan perpendicular to two parallel wires, each 1.2 units in diameter and separated by 0.6 units. Δ was 1 unit, so σ was 0.7 units. The true concentration profile and the intensity profile used to compute the signal matrix are shown in Fig. 2. The signal profile (dashed) and deconvolved concentration profile (solid) are shown in Fig. 3 with no noise. In Fig. 4 Gaussian noise (standard deviation of 0.01) was added to the signal (dashed). The deconvolved concentration profile is also shown (solid). This amount of noise is reasonable as the data come from counting experiments where the signal is usually considered to be $s_i = N_i \pm \sqrt{N}$ and N_i can easily be 10,000. No special care was taken to ensure the positivity of $c(y)$ nor to make the code general or robust. It is obvious from viewing the signal curve that it would be difficult, if not impossible, to discern two rectangular distributions of matter. The deconvolved $c(y)$ makes this quite apparent, although the shape is wrong due to the coarse binning.

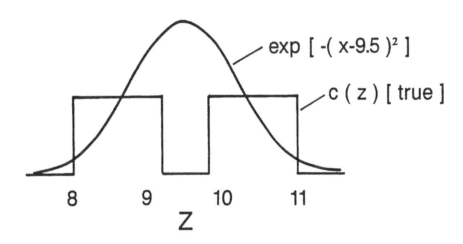

Fig. 2. The beam profile relative to the "true sample", $c(z)$.

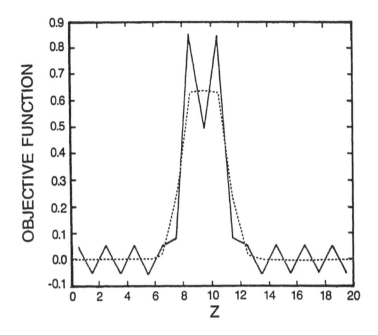

Fig. 3. The signal profile (dash) derived from the curves in Fig. 2 and the deconvolved concentration profile (solid) (by Fourier series with coarse binning and no noise).

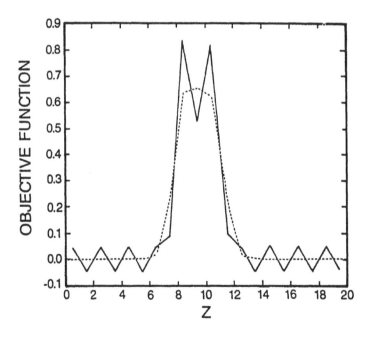

Fig. 4. The same as in Fig. 3 except a 1% Gaussian noise was added to the signal before deconvolution.

Appendix A. Weierstrass Transforms

$$f(x) = \int \pi^{-\frac{1}{2}} \; \exp[-(x-y)^2]\phi(y)dy$$

$\phi(y)$	$f(x)$
y^n	$h_n(x)/2^n \; {*}{*}$
$H_n(y)$	$2^n x^n$
e^{-ay}	$\exp(\frac{a^2}{4})\exp(-ax)$
$e^{-a^2 y^2}$	$\exp[-a^2 x^2/(1+a^2)]/(1+a^2)^{\frac{1}{2}}$
$\sin ay$	$\exp\left(\frac{-a^2}{4}\right)\sin ax$
$\cos ay$	$\exp\left(\frac{-a^2}{4}\right)\cos ax$
$y \sin ay$	$\exp\left(\frac{-a^2}{4}\right)[x\sin ax + \left(\frac{a}{2}\right)\cos ax]$
$y \cos ay$	$\exp\left(\frac{-a^2}{4}\right)[x\cos ax - \left(\frac{a}{2}\right)\sin ax]$
$\sinh ay$	$\exp\left(\frac{a^2}{4}\right)\sinh ax$
$\cosh ay$	$\exp\left(\frac{a^2}{4}\right)\cosh ax$
$\sinh^2 ay$	$\exp(a^2)\sinh^2 ax + \exp(a^2/2)\sinh\left(\frac{a^2}{2}\right)$
$\cosh^2 ay$	$\exp(a^2)\cosh^2 ax - \exp\left(\frac{a^2}{2}\right)\sinh\left(\frac{a^2}{2}\right)$
$y \sinh ay$	$\exp\left(\frac{a^2}{4}\right)[x\sinh ax + \left(\frac{a}{2}\right)\cosh ax]$
$r^2 - y^2, \mid y \mid \le 1$	$(\frac{1}{2})[r^2 - x^2 - \frac{1}{2}][erf(z+r) - erf(z-r)]+$ $\frac{1}{2}erf(-r^2 - x^2)(r\cosh 2rx + x\sinh(2rx)$
$u(a) = \begin{cases} 0 & y < a \\ 1 & y \ge a \end{cases}$	$[1 + erf(x-a)]/2$
$t(y) = \begin{cases} 0 & y < 0 \\ y/a & 0 \le y < a \\ 1 & y \ge a \end{cases}$	$\{\frac{x}{a}[\mathrm{erf}(x) - \mathrm{erf}(x-a)] + \mathrm{erf}(x-a) + 1$ $+ \frac{1}{a\sqrt{\pi}}[\exp(-x^2) - \exp[-(x-a)^2]]\}/2$
$g(y) = \begin{cases} \exp(-a^2 y^2) & -\infty \le y \le 0 \\ 1 & 0 < y \le \infty \end{cases}$	$\frac{1}{2}[1 + \mathrm{erf}(x) + \{\exp[-a^2 x^2/(1+a^2)]/$ $\sqrt{(1+a^2)}\}\{1 - \mathrm{erf}(x/(1+a^2)\}]$

${*}{*}$ The $h_n(x)$ are the same as the $H_n(x)$ (Hermite polynomials) except all terms are positive.

3   CONCLUSIONS

The Weierstrass transform has enjoyed some success in the area of heat conduction [7]. Perhaps some of our entries in Appendix A will be of use in solving such problems. We have also indicated a new area of chemical research that in the near future may make great use of the Weierstrass transform as an integral equation of the first kind. After a review of analytic approaches to the problem, we have sketched a numerical algorithm and given one example of its use.

To the mathematicians, we hope that we have supplied enough motivation that perhaps some new work in the area of the Weierstrass transform will be forthcoming. To the chemists and others using XRF, we hope that we have shown that if the most is to be gained by these experiments, then deconvolution of the data is a necessity.

4   REFERENCES

1. M. Nichols and R. Ryon, An X-Ray Microfluorescence Analysis System With Diffraction Capabilities, Adv. in X-Ray Anal., 29: 423-426 (1986).

2. M. Nichols et. al., Parameters Affecting X-Ray Microfluorescence, Adv. in X-Ray Anal., 30: 45-51 (1987).

3. J. Zahrt, High Spatial Resolution in X-Ray Fluorescence, Adv. in X-Ray Anal., 30: 77-84 (1987).

4. I. Hirschman and D. Widder, *The Convolution Transform*, Princeton University Press, Princeton, 1955; Chap. 8.

5. G. M. Wing, *A Primer on Integral Equations of the First Kind*, Los Alamos National Laboratory report, LA-UR-84-1234, April 1984.

6. P. Morse and H. Feshbach, *Methods of Theoretical Physics*, McGraw-Hill Book Co., New York, 1953; pp. 935-936.

7. D. Widder, *The Heat Equation*, Academic Press, New York, 1975.

Multi-Boundary Alternating Direction Collocation Schemes for Computing Ideal Flows over Bodies of Arbitrary Shape Using Cartesian Coordinates

KEVIN COOPER and P. M. PRENTER   Mathematics Department, Colorado State University, Fort Collins, Colorado

Computation of flows over bodies of arbitrary shape introduces geometric complexities which are reflected in any numerical simulation. If local body fitted curvilinear coordinates are employed together with either finite difference or finite element methods, high computational costs are often involved due to costly mesh generation procedures, point by point storage of Jacobians, and the complexities of the transformed equations.

As an alternative to such methods, we introduce a new procedure, multi-boundary alternating direction collocation. The method takes place entirely in rectangular coordinates, is fully vectorizable for simple flows, involves sparse center banded matrices with no drifting bands, and can be high order accurate. To illustrate the method,we consider it's application to simple ideal flow with rotation and to steady state heat conduction over a variety of geometries.

This research was supported in part by NASA-Ames Research Center under contract #NCA2-181.

1. INTRODUCTION

In this paper we consider the problem of computing fluid flow over one or
more bodies of arbitrary shape in a single code using only rectangular
coordinates together with a fully vectorizable numerical scheme. To this
end we develop a new procedure, <u>multi-boundary alternating direction</u>
<u>collocation</u> (MBADC). The method is comprised of an alternating direction
collocation (ADC) scheme on an embedding rectangle coupled with a correction for interior body boundary conditions each full ADC sweep. For
simplicity, we confine ourselves to ideal two-dimensional flow, although
the methods generalize to more complex flows in both two and three dimensions. An outline of the theory and the numerical procedures in this
setting, together with some numerical output, is presented here. Rigorous
analysis of the methods will occur elsewhere.

To set the problem, let $\Omega = [a,b] \times [c,d]$ and let $\Omega_0 = \cup\{\Omega_k : k \in K\}$ be a
finite union of pairwise disjoint, open, connected subsets of Ω. Let
$\partial\Omega_0 = \cup\{\partial\Omega_k : k \in K\}$ be the boundary of Ω_0 where $\{\partial\Omega_k : k \in K\}$ is pairwise
disjoint and let $\partial\Omega$ be the boundary of Ω. We want to numerically solve the
partial differential equation

$$\begin{aligned}
Lw &= f \quad \text{on } \Omega - \Omega_0 \\
w &= 0 \quad \text{on } \partial\Omega \\
w &= g \quad \text{on } \partial\Omega_0
\end{aligned} \tag{1.1}$$

where

$$\begin{aligned}
L &= X + Y \\
Xu &= -(au_x)_x + cu \\
Yu &= -(bu_y)_y + du
\end{aligned} \tag{1.2}$$

is defined in the Sobolev space $W^2 = (\Omega)$ and where a, b, c, and d are all
positive analytic functions of (x,y) on Ω. When $a=b=1$ and $c=d=0$, $Lu=-\Delta u$
and (1.1) is the equation of ideal fluid flow over one or more bodies.
Typical examples are to compute such flows over a) a disk, b) an airfoil,
c) three disks, d) four bodies of any shape, and e) no body as illustrated
in Figures a-e.

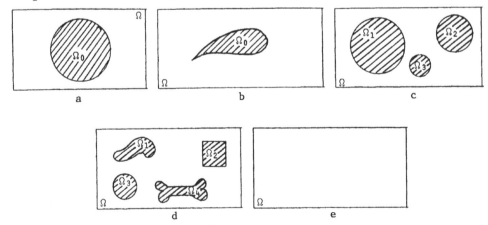

Extant numerical procedures for solving (1.1) include coordinate transform methods, finite element methods, and boundary integral equation methods. Each has its advantages and it's drawbacks. Moreover, they are key methods to which the methods developed in this paper must ultimately be compared.

Multi-Boundary Alternating Direction (MBAD) for the solution of (1.1) with L given by (1.2) is given by the iteration

$$(-X + r_k)u^{k+1/2} = (Y + r_k)u^k + f^k$$

$$(-Y + r_k)u^{k+1} = (X + r_k)u^{k+1/2} + f^k \qquad (1.3)$$

$$f^{k+1} = f^k + N(g - u^{k+1})$$

where the u^k's, $u^{k+1/2}$'s, and f^k's are defined on all of Ω. It is clear that (1.3) is a method of alternating lines which replaces (1.1) by alternating systems of ordinary differential equations. If $a=a(x)$, $c=c(x)$, $b=b(y)$ and $d=d(y)$, then (1.3) is fully vectorizable under any discretization of the associated system of o.d.e.'s. Discretization of the first two equations of (1.3) by a collocation procedure results in MBADC. Convergence of ADC applied to the no body problem, choice of the operator N, and convergence of the iteration (1.3) under collocation to a good approximate solution of (1.1) are the main topics of the ensuing sections.

2. MULTI- AND UNI-BOUNDARY VALUE PROBLEMS

In order to clarify our somewhat curious choice of names, MBAD and MBADC, and to set the stage for analysis of error, we want to distinguish between a _uni-boundary value problem_ and a _multi- boundary value problem_. Given an open, convex region R of R^2, let $W^2(R)$ be the Sobolev space

$$W^2(R) = \{u(x, y): D^a u \in L^2(R) \text{ for } a \leq 2\}$$

where $a = (a_1, a_2)$ is a positive integer pair with $a = a_1 + a_2$ and where $D^a u = \partial^a u/\partial^{a_1} x \, \partial^{a_2} y$ in a weak sense. Letting D(L) denote the domain of L, we say that

$$Lw = f \quad \text{on } \Omega - \Omega_0$$

$$w = 0 \quad \text{on } \partial\Omega$$

$$w = g \quad \text{on } \partial\Omega_0 \qquad (2.1)$$

$$D(L) \subseteq W^2(\Omega - \Omega_0)$$

is a _uni-boundary value problem_ (UBVP) with Ω and Ω_0 as given in Section 1. Problem (2.1) is well-posed and uniquely solvable for all interior bodies Ω_0 whose boundaries $\partial\Omega_0$ are sufficiently smooth. On the other hand, the problem

$$Lu = f_E \quad \text{on } \Omega$$
$$u = 0 \quad \text{on } \partial\Omega$$
$$u = g \quad \text{on } \partial\Omega_0 \qquad\qquad (2.2)$$
$$D(L) \subseteq W^2(\Omega)$$

is a <u>multi-boundary value problem</u> (MBVP) where f_E is some $L^2(\Omega)$ extension of f to all of Ω. Problem (2.2) need not be well-posed or solvable in $W^2(\Omega)$. The terms UBVP and MBVP are two-dimensional extensions of well-known terminology for ordinary differential equations (i.e. two-point versus multi-point boundary value problems). If f_E is an extension of f to Ω which makes (2.2) uniquely solvable, then u restricted to $\Omega - \Omega_0$ must equal w solving (2.1).

The goal of MBADC is to simultaneously compute a well-posed extension f_E of f to all Ω making (2.2) uniquely solvable and to obtain a good numerical approximation to w solving (2.1).

3. SOME COLLOCATION RESULTS

MBADC is based on one-dimensional collocation procedures. In this section we briefly review some cogent results from both one- and two-dimensional collocation. Starting with the ordinary differential operator $\mathcal{X}$ derived from X, let

$$\mathcal{X}u = -(au')'(x) + c(x)u(x)$$

where $a \leq x \leq b$. Let $\mathcal{X}$ carry homogeneous two-point boundary conditions Au(a,b) which make $\mathcal{X}$ strictly positive definite and symmetric on its domain. In particular

$$Au(a, b) = \{a_{11}u(a) + a_{12}u'(a), a_{21}u(b) + a_{22}u'(b)\}$$

is such that

$$(\mathcal{X}u, v) = \int_a^b v(x)\mathcal{X}u(x)dx = (u, \mathcal{X}v)$$

and,

$$(Xu,u) > 0 \qquad \text{for } u \neq 0$$

for all u and v in $D(\mathcal{X})$. Here

$$D(\mathcal{X}) = \{u \in H^2[a, b]: Au(a, b) = (0, 0)\}.$$

Here $H^2[a,b] = \{u: u'' \in L^2[a,b]$ and u' is absolutely continuous on $[a,b]\}$. Let Δ_x be a partition of $[a,b]$ with

$$\Delta_x: a = x_0 < x_1 < x_2 < \ldots < x_n = b$$

and mesh spacing $h_x = \max\{x_1 - x_{i-1}, i=1,2,\ldots,n\}$. Let S_0^x be the piecewise cubic Hermite splines over the partition Δ_x lying in $D(\mathfrak{X})$. In particular,

$$S_0^x = \{u \in D(\mathfrak{X}): u \text{ is a cubic polynomial on each } [x_{i-1}, x_i] \text{ for}$$
$$i = 1, 2, \ldots, n\}.$$

The space S_0^x is 2n-dimensional. Let $x_{11} < x_{12}$ be Gauss points of $[x_{i-1}, x_i]$ for which

$$\int_{x_{i-1}}^{x_i} u(x)dx = w_{11}u(x_{11}) + w_{12}u(x_{12})$$

whenever $u(x)$ is cubic on $[x_{11}, x_{12}]$. Let

$$GX = \{x_{11}, x_{12}: i = 1, 2, \ldots, n\}$$

be the x-Gauss points. Given $f \in L^2[a,b]$, the equation

$$\mathfrak{X}u = f \tag{3.1}$$

is uniquely solvable on $D(\mathfrak{X})$. The Method of Collocation, in this setting, seeks $\underline{u} \in S_0^x$ satisfying

$$\mathfrak{X}\underline{u} \doteq f , \qquad u \in S_0^x \tag{3.2}$$

where

$$f \doteq g \qquad \text{iff} \qquad f(x)=g(x) \text{ for all x in GX.} \tag{3.3}$$

Equation (3.2) is uniquely solvable for h sufficiently small with $\|u^{(j)}-u^{(j)}\|_\infty \leq \gamma\|u^{(4)}\|h^{4-j}$ for f sufficiently smooth (for example, $f \in C^4[a, b]$ and for $j = 0, 1, 2, 3, 4$. Here $\|\ \|_\infty$ denotes the L^∞ and $\|\ \|$ the L^2 norm on $[a,b]$. The papers [6, 13, 21] discuss these matters in detail. Similarily, one can associate the ordinary differential operator $\mathfrak{Y}$ with Y where

$$\mathcal{Y}v(y) = -(bv')'(y) + d(y)v(y)$$

for $c \leq y \leq d$. Attach two-point boundary conditions

$$Bv(c, d) = \{b_{11}v(c) + b_{12}v'(c), \ b_{21}v(d) + b_{22}v'(d)\}$$

to $\mathcal{Y}$ to make it strictly positive definite on $D(\mathcal{Y})$ with

$$D(\mathcal{Y}) = \{v \in H^2[c, d]: Bv(c, d) = (0, 0)\}.$$

Let Δ_y be a partition of $[c, d]$,

$$\Delta_y: c = y_0 < y_1 < y_2 < \ldots < y_m = d,$$

with mesh $h_y = \max\{y_j - y_{j-1}: j = 1, 2, \ldots, m\}$. Let $y_{j1} < y_{j2}$ be the Gauss points of the interval $[y_{j-1}, y_j]$ and let

$$GY = \{y_{j1} \cdot y_{j2}: j = 1, 2, \ldots, m\}$$

be the Gauss points in the y-direction. Let S_0^y be the piecewise cubic Hermite splines in the y-direction with respect to the partition Δ_y with $S_0^y \subseteq D(\mathcal{Y})$. Let

$$\Delta = \Delta_x \times \Delta_y$$

be a partition of Ω with mesh $h = \max\{h_x, h_y\}$. Let

$$S_0 = S_0^x \otimes S_0^y$$

be the tensor product splines on Ω over the mesh Δ and let

$$G = GX \times GY$$

be the Gauss points of Ω. Let $L = X + Y$ with $D(L) = \{u \in W^2(\Omega): Au(\cdot, y) = B(x, \cdot) = (0, 0)$ for all (x, y) in $\Omega\}$. Consider the partial differential equation

$$Lu = f \tag{3.4}$$

where $f \in L^2(\Omega)$. Given (3.4) is uniquely solvable, the method of two-dimensional collocation seeks $\underline{u} \in S_0 \subseteq D(L)$ solving

$$L\underline{u} \stackrel{\cdot}{=} f \tag{3.5}$$

where

$$f = g \quad \text{iff } f(p) = g(p) \text{ for all } p \text{ in } G. \tag{3.6}$$

For f defined on G, the following theorem has been proved by Prenter and Russell [16] when $Lu = -\Delta u + bu$, $b > 0$, and by Percell and Wheeler [14] for more general L.

THEOREM 3.1. *There is a unique* $\underline{u} \in S_0$ *solving* $L\underline{u} = f$.

Error estimates $\|u - \underline{u}\|_\infty$ are also given in both papers, although neither paper presents optimal estimates or seeks superconvergence results at spline knots.

4. EIGENSPLINES AND GENERALIZED EIGENSPLINES

Alternating direction schemes inherently apply to partial differential equations on rectangular domains and their higher-dimensional analogs. Moreover, the convergence of such procedures and their discretizations depends on the underlying finite dimensional approximating subspaces being spanned by tensor product invariant subspaces associated with the splitting operators X and Y. This is certainly the case for ADI as introduced by Douglas and by Peaceman and Rachford using centered difference approximations to the Laplace operator $L = -\Delta$. In this case, the tensor product of eigenvectors associated with centered difference approximations to ∂_x^2 and ∂_y^2 are the spanning set for ADI and are known exactly, as are their eigenvalues. A full theory for alternating direction difference schemes for the general operator $L = X + Y$ as given in Section 1 is not known. However, a good part of the theory we outline in this section for ADC should carry to finite difference schemes as well.

A number of individuals have experimented successfully with ADC applied to partial diffential equations. In particular, we cite the papers of Hayes [10] and of Hayes [11]. However, none of these results includes a theory of convergence of ADC. The recent paper of Dyksen establishes convergence of ADC when $L = -\Delta$ on a uniform mesh. The latter also gave the associated eigenvectors (eigen- splines) and their eigenvalues explicitly for this operator on such meshes. Results of an earlier paper of Cerutti and Parter [2] actually subsume part of the Dyksen results for $L = -\Delta$ on any mesh configuration. However, the latter does not give exact values of associated eigenvectors and eigenvalues. Moreover, Cerutti and Parter were studying heat equation stability results and were not considering alternating direction at all.

In this section we outline a full theory for ADC applied to $L = X + Y$ on a rectangular domain. We establish the convergence of ADC for any mesh configuration. The convergence is in terms of tensor products of complex generalized eigensplines associated with $\mathfrak{X}$ and $\mathfrak{Y}$, respectively. Moreover, even though collocation may carry complex eigensplines and eigenvalues, the real part of the iteration always converges and converges to the full collocation solution of Lu = f as given by Theorem 3.1. It follows that only real calculations are needed, even though the natural embedding may be complex.

With the foregoing in mind, let $\mathfrak{X}$ be the ordinary differential operator associated with X as described in Section 3. We say that $e^X \neq 0$ in S_0^X is an <u>eigenspline</u> of $\mathfrak{X}$ with <u>eigenvalue</u> λ iff

$$\mathfrak{X}e^X \overset{\cdot}{=} \lambda e^X. \tag{4.1}$$

Let

$$\langle u, v \rangle = \sum_{x \in GX} w_x u(x)v(x) \tag{4.2}$$

be two-point, panel Gauss quadrature of uv. The essential features of the following theorem have been proved by both Dyksen and by Cerrutti and Parter.

THEOREM 4.1.   *When $a(x) = 1$, $\mathfrak{X}$ has 2n linearly independent eigensplines* $\{e_i^X: i = 1, 2, \ldots, 2n\}$

$$\mathfrak{X}e_i \overset{\cdot}{=} \lambda_i e_i$$

which have positive real eigenvalues $\{\lambda_i: i = 1, 2, \ldots, 2n\}$. *Moreover, the e_i^X's are mutually orthogonal with respect to the inner product $\langle \, , \, \rangle$ on S_0^X, so that $\langle e_i^X, e_j^X \rangle = 0$ for $i \neq j$ and for $i, j = 1, 2, \ldots, 2n$.*

To carry ADC over to the more general case where $a = a(x,y) = 1$, we need a complex embedding of the entire problem.   To this end, let

$$S_c^X = S_0^X + iS_0^X, \quad S_c^y = S_0^y + iS_0^y, \quad S_c = S_0 + iS_0,$$

and let

$$(u, v)_x = \int_a^b uv \, dx,$$

$$(u, v)_y = \int_c^d uv \, dy,$$

$$(u, v) = \int_\Omega uv.$$

DEFINITION 4.2. $0 \neq e^X \in S_C^X$ *is said to be an* <u>*eigenspline*</u> *of* $\mathfrak{X}$ *with complete* <u>*eigenvalue*</u> $\lambda = \lambda_R + i\lambda_I$ *iff*

$$\mathfrak{X}e^X \doteq \lambda e^X.$$

Convergence of ADC for $L = X + Y$ is based upon the generalized eigensplines of $\mathfrak{X}$ and $\mathcal{Y}$. A natural way of defining such a spline would be to say $0 \neq e^X$ in S_C^X is a generalized eigenspline of $\mathfrak{X}$ with eigenvalue λ in C, C the complex numbers, iff

$$(\lambda I - \mathfrak{X})^j e^X \not\doteq 0, \quad j = 1, 2, \ldots, k - 1$$
$$(\lambda I - \mathfrak{X})^k e^X \doteq 0. \tag{4.3}$$

The number k would be the <u>nullity</u> of $(\lambda I - \mathfrak{X})$ with respect to collocation. Equations (4.3) are defined if one converts them to their matrix equations associated with collocation. However, S_C^X is not globally in the domain of $(\lambda I - \mathfrak{X})^j$ for $j > 1$. To avoid this awkward dichotomy, we turn instead to the recursive properties of gerneralized eigenvectors to define generalized eigensplines.

DEFINITION 4.3. *Let* $0 \neq e_1^X$ *in* S_C^X *be an eigenspline of* $\mathfrak{X}$ *with complex eigenvalue* λ. *The* <u>*generalized eigensplines*</u> *of associated with* $\{e_1^X, \lambda\}$ *are those nonzero splines* $e_2^X, \ldots, e_k^X$ *in* S_C^X *that satisfy*

$$(\lambda I - X)e_i^X \doteq e_{i-1}^X \tag{4.4}$$

for $i=2,3,\ldots,k$. *The largest integer k for which (4.4) obtains is the* <u>*ascent*</u> *of the eigenvalue* λ. *It follows easily that* $\mathscr{P}e_1^X, e_2^X, \ldots, e_k^X\mathscr{P}$ *is linearly independent in* S_C^X.

The following theorem , crucial to convergence of ADC for $L = X + Y$, is proved in [4].

THEOREM 4.4. $\mathfrak{X}$ *has a set of linearly independent generalized eigensplines which span* S_C^X. *Moreover, if* $\mathfrak{X}$ *is strictly positive definite, the eigensplines and generalized eigensplines of* $\mathfrak{X}$ *all have positive real part.*

Now let $\{e_i^x: i = 1, 2, \ldots, N\}$ be a basis of eigensplines and generalized eigensplines associated with $\mathfrak{X}$ and let $\{e_j^y: j = 1, 2, \ldots, M\}$ be a similar set for $\mathfrak{Y}$. Let

$$e_{ij} = e_i^x(x)e_j^y(y). \qquad (4.5)$$

Then e_{ij} is in S_c and

$$\underline{B} = \{e_{ij}: i = 1, 2, \ldots, N \text{ and } j = 1, 2, \ldots, M\}$$

is a basis for S_c. Convergence of the ADC iteration

$$(-X + r_k)u^{k+1/2} \overset{\cdot}{=} (Y + r_k)u^k + f$$
$$(-Y + r_k)u^{k+1} \overset{\cdot}{=} (X + r_k)u^{k+1/2} + f \qquad (4.6)$$

depends on the expansion of the S_c splines u^k, $u^{k+1/2}$, u^{k+1} in the generalized eigensplines of $\underline{B}$, and on the positive definiteness of both $\mathfrak{X}$ and $\mathfrak{Y}$ coupled with Theorem 4.4 and its analog for $\mathfrak{Y}$. The iteration (4.6) leads to a recursion formula on the coefficients of u^k, $u^{k+1/2}$ and u^{k+1} in their $\underline{B}$ expansions . Moreover, this recursion is a contraction mapping which forces convergence of ADC. It is somewhat more complicated than the simple eigenvector expansions that occur in Douglas, Peaceman and Rachford, and Dyksen. In particular, block diagonal coefficient matrices replace simple diagonal coefficient matrices. The details of these observations together with proof of the following theorem are given in [4].

THEOREM 4.5.   *Let $\mathfrak{X}$ and $\mathfrak{Y}$ be strictly positive definite.   For h sufficiently small and for r_k's chosen from any finite set of positive real numbers, ADC always converges.   Moreover, both the real and complex parts of the iteration converge with the real ADC iteration converging to $\underline{u}$ in S_c solving the full collocation equation $L\underline{u} \overset{\cdot}{=} f$ as given in Theorem 3.1.*

5. MULTI-BOUNDARY

We now consider implementing ADC with a correction (constraint) to compute solutions to (1.1) with $L = X + Y$. For simplicity, we consider the model uni-boundary value problem

$$Lw = 0 \quad \text{on } \Omega - \Omega_0$$
$$w = 0 \quad \text{on } \partial\Omega$$
$$w = g \quad \text{on } \partial\Omega_0 \tag{5.1}$$
$$D(L) \subseteq W^2(\Omega - \Omega_0).$$

Continuous multi-boundary alternating direction applied to (5.1) is given by the iteration

$$(-X + r_k)u^{k+1/2} = (Y + r_k)u^k + f^k$$
$$(-Y + r_k)u^{k+1} = (X + r_k)u^{k+1/2} + f^k \tag{5.2}$$
$$f^{k+1} = f^k + N(g - u^{k+1})$$

with u^k, $u^{k+1/2}$ in $W^2(\Omega)$ and f^k defined on Ω. For $F = F(x,y)$ in $L^2(\Omega)$, let $F/\Omega - \Omega_0$ be the restriction of F to $\Omega - \Omega_0$. Ideally, one would like to design the iteration (5.2) in such a way that

1. $f^k/\Omega - \Omega_0 = 0$

2. $f^k \to f$ pointwise on Ω iff u^k and $u^{k+1/2}$ converge to u in $W^2(\Omega)$ with $u/\partial\Omega_0 = g$, and

3.   the sequences (u^k) and $(u^{k+1/2})$ do converge in $W^2(\Omega)$.

If N could be chosen so that 1, 2 and 3 all obtained, then $u = \lim u^k$ would solve the multi-boundary value problem

$$Lu = f \quad \text{on } \Omega$$
$$u = 0 \quad \text{on } \partial\Omega$$
$$u = g \quad \text{on } \partial\Omega_0 \tag{5.3}$$
$$D(L) \subseteq W^2(\Omega).$$

Moreover, since $f/\Omega - \Omega_0 = 0$, $\underline{w} = u/\Omega - \Omega_0$ would satisfy

$$L\underline{w} = 0 \qquad \text{on } \Omega - \Omega_0$$
$$\underline{w} = 0 \qquad \text{on } \partial\Omega$$
$$\underline{w} = g \qquad \text{on } \partial\Omega_0$$
$$D(L) \subseteq W^2(\Omega - \Omega_0).$$

Thus, $\underline{w} = w$ and the iteration (5.2) would solve (5.1).

Discretization of (5.2) by collocating the first two equations leads to the Multi-Boundary ADC (MBADC) iteration

$$(-X + r_k)u^{k+1/2} \doteq (Y + r_k)u^k + f^k$$
$$(-Y + r_k)u^{k+1} \doteq (X + r_k)u^{k+1/2} + f^k \qquad (5.4)$$
$$f^{k+1} = f^k + N(g - u^{k+1})$$

with u^k, $u^{k+1/2}$ and u^{k+1} in S_0 and f^k defined on all of Ω. We would like to choose N so that

$$f^{k+1}/\Omega - \Omega_0 = 0 \qquad (5.5)$$

and so that we obtain convergence of the sequence f^k to f pointwise on G or on Ω. We also want N to force convergence of the sequences (u^k) and (u^{k+1}) to a spline $\underline{u}$ in S_0. If, in addition, $\underline{u}$ is a good approximation to u solving (5.3), then $\underline{u}/\Omega - \Omega_0$ would be a good approximation to w solving (5.1). This outlines the strategy of MBADC.

All of the foregoing discussion is highly heuristic. In the remainder of this section we give a partial outline of the theory of MBADC, show one possible choice of N, and present some numerical examples of MBADC applied to some single and some multi-body problems. MBADC as presented in this paper can diverge without modification of N for interior bodies having narrow x-cross sections. We cite such an example and suggest what can be done to force convergence in such cases. A full theory of the entire MBADC mechanism will appear elsewhere [5].

Starting with the constraint equation

$$f^{k+1} = f^k + N(g - u^{k+1}), \qquad (5.6)$$

we let

$$f^0 = 0$$

and define

$$Nz = -\partial_x^2 Pz. \qquad (5.7)$$

In particular, let

$$I(y) = \{y \in [c, d]: (x, y) \in \overset{\circ}{\Omega}_0 \text{ for some } x \in [a, b]\}$$

where $\overset{\circ}{\Omega}_0$ is the interior of Ω_0. For simplicity, we shall assume Ω_0 is a single convex open subset of Ω so that $K=\{1\}$. Let $z=z(x,y)$ be defined on Ω. Define

$$Pz(x, y) = 0 \quad \text{if } y \in [c, d] - I(y). \tag{5.8}$$

For each $y_0 \in I(y)$, let $e_1(y) < e_2(y)$ be x-coordinates of intersection of $\partial\Omega_0$ with the line $y = y_0$ (see Figure 5.1). Since Ω_0 is open and convex,

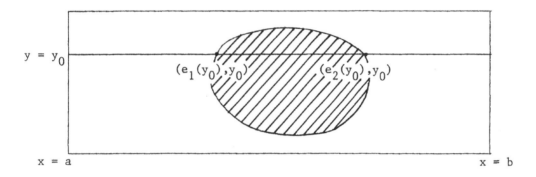

Figure 5.1

there are only two body edges $(e_1(y_0), y_0)$ and $(e_2(y_0), y_0)$ along each line $y = y_0$ with y_0 in $I(y)$. To define $Pz(x, y)$ when $y \in I(y)$ is fixed, let

$$Pz(x, y) = \begin{cases} \text{linear in x on } [a, e_1(y)] \\ \text{cubic in x on } [e_1(y), e_2(y)] \\ \text{linear in x on } [e_2(y), b] \end{cases} \tag{5.9}$$

subject to the constraints

$$\begin{cases} Pz(a,y) = 0 = Pz(b,y) \\ Pz(e_1(y), \ y) = z(e_1(y), \ y) \\ Pz(e_2(y), \ y) = z(e_2(y), \ y) \cdot \\ Pz(\cdot, \ y) \in C^1[a, \ b] \ . \end{cases} \qquad (5.10)$$

Figure 5.2 schematically depicts $Pz(x, \ y)$ versus $z(x,y)$ for fixed $y \in I(y)$.

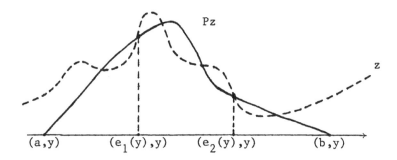

Figure 5.2

Since Pz is linear in x exterior to Ω_0 for each fixed y, it is clear that on $\Omega - \Omega_0$

$$Nz = -\partial_x^2 \ Pz = 0 \ .$$

Moreover, since $f^0 = 0$ and $f^{k+1} = f^k + N(g - u^{k+1})$, it is also clear that

$$f^k/\Omega - \Omega_0 = 0.$$

Thus condition (5.5) is fulfilled. Denote convergence of f^k on the Gauss points G of Ω by

$$f^k \ \overset{\cdot}{\longrightarrow} \ f \ .$$

The following theorem can be proved using the ADC convergence theory of the previous section.

THEOREM 5.1. *Let the sequences* $\{u^k\}$ *and* $\{u^{k+1/2}\}$ *in* S_0 *be defined by the MBADC iteration (5.2) with* $\{f^k\}$ *and N given by (5.6)-(5.10). Then* $f^k \ \overset{\cdot}{\longrightarrow} \ f$

iff $u^k \to \underline{u}$ and $u^{k+1/2} \to \underline{u}$ in S_0, and $\underline{u} = g$ on $\{(e_i(y), y): y \in I(y) \cap GY$ for $i = 1, 2\}$.

The previous theorem neither establishes convergence of MBADC nor provides error estimates for $|w(p) - \underline{w}(p)|$, $p \in \Omega - \Omega_0$. The MBADC algorithm with N given by (5.6)-(5.10) can, in fact, diverge if shape parameters compensating for very narrow body cross-sections are not introduced. In particular, let

$$H(y) = e_2(y) - e_1(y), \quad y \in I(y).$$

Define the <u>x-gauge</u>, GAUGE-X, of Ω_0 by

$$\text{GAUGE-X} = \min\{H(y): y \in I(y)\}.$$

The iteration with N as given diverges whenever GAUGE-X is too small regardless of mesh size. This divergence can be overcome by the introduction of <u>shape-parameters</u>, SHAPE=SHAPE(y) for $y \in I(y)$; into the correction equation. Specifically, let

$$f^{k+1} = f^k + \text{SHAPE} \cdot N(g - u^{k+1}). \tag{5.11}$$

Theorem 5.1 coupled with a number of other intricacies leads to

THEOREM 5.2. *Let N be given by (5.7)-(5.10). There exist shape parameters dependent on Ω_0 but independent of h for which the MBADC iteration (5.4) converges for all meshes sufficiently small when $f^{k+1} = f^k + SHAPE \cdot N(g - u^k)$.*

There are in fact many choices of SHAPE(y) and it is quite simple to make a successful choice. A most important related open question is how to make "near optimal" choices of the linked para-meters $\{r_k\}$ and SHAPE(y), $y \in I(y) \cap Gy$ to accelerate convergence of the iteration. In this context, Wachspress parameters are definitely not the best choice. One also wants to choose SHAPE(y) to conserve as much parallelism as possible.

Error estimates for $|w(p) - w^k(p)|$, $p \in \Omega - \Omega_0$ and $w^k = u^k/\Omega - \Omega_0$, are also important. Being as these results are highly technical, we omit them here. Computationally we experience at least $O(h^2)$ global accuracy on $\Omega - \Omega_0$. Theoretically discrete maximal principle arguments are used and there are open questions on full collocation error estimates on Ω involved. Let GLX = $\{(x, y): y \in I(y) \cap GY$ and $a \leq x \leq b\}$ be the Gauss lines parallel to the x-axis. Letting TOL be a preassigned error tolerance, the same discrete maximal arguments lead to a stopping criteria for the iteration. Specifically, the iteration is stopped whenever $|g(p) - u^k(p)| < \text{TOL}$ for p in GLY $\cap \partial\Omega_0$.

To illustrate the varied geometries to which the algorithm applies, we give six examples in steady state heat conduction and one ideal flow computation. In all cases we used a 32 × 32 grid of Gauss point and took $\Omega = [-2,2] \times [-2,2]$ except examples 4 and 7 where $\Omega = [-2,2] \times [-2,2.5]$ and $\Omega = [-1,1] \times [-1,1]$, respectively. The following list describes Ω_0 for examples 1-6:

1.   A disk of radius .5 centered at (1.25,1.25) with u = 10 on $\partial\Omega_0$.

2.   Two disks of radius .8 and .4 centered at (-.7, 0) and (1, 0), respectively. The boundary of the smaller disk is held at u = 10, while the larger is held at u = -10.

3.   Five disks of radius .5, .4, .5, .4 and .4 centered at (-.75, .75), (1.1, 1.1), (.5, 0), (-1, -1) and (1, -1), respectively. Their boundaries are held at u = 10, -5, 10, 5, and 5, respectively.

4.   A skewed annulus $D_2 - D_1$ embedded in $\Omega = [-2,2] \times [-2,2.5]$. Disk D_2 has radius 2., is centered at (0, 0), and u = 0 on the boundary of D_2. Disk D_1 has center at (0, 1), has radius 1.5, and u = 10 on the boundary of D_1. The region $\Omega_0 = D_2 - D_1$ is multiply connected and we are computing u on $D_2 - D_1$ by correcting f^k on $\Omega - \Omega_0$ rather than Ω_0.

5.   A square S = [.75,1.75] × [.75,1.75] embedded in the rectangle [-2,2] × [-2,2]. The boundary of the square is held at u = 10.

6.   A disk of radius .5 centered at (0, 0) with periodic boundary data u = 10 sin 3πθ on the boundary of the disk. Here θ is the polar angle made with the positive x-axis.

Our final example, example 7, computes the ideal flow of a fluid over a cylindrical wing with rotation. Specifically, we solve

$$\Delta u = 0 \quad \text{in } \Omega - \Omega_0$$
$$u = g_1 \quad \text{on } \partial\Omega$$
$$u = g_2 \quad \text{on } \partial\Omega_0$$

where $\Omega = [-1,1] \times [-1,1]$ and Ω_0 is a disk of radius .5 centered at the origin. Here g_1 and g_2 are boundary data given by the known polar coordinate solution

$$\phi(r, \theta) = u_0\left[r - \frac{a^2}{r}\right]\sin\theta + \frac{\tau}{2\pi}\theta .$$

We let $u_0 = 1$ and $\tau = 2$ where u_0 is the free stream velocity and $\tau = 2$ is the circulation. Figures 5.1 through 5.6 plot the isothermals for examples 1 through 6 using MBADC. Figure 5.7 is the contour plot of the MBADC solution to example 7. The same code was used for all computations with only body edges and boundary data adjusted for each problem. In each case the iteration was stopped when TOL was less than 5% relative error on

$\partial \Omega_0$.   In each case SHAPE(y) $\equiv$ 1. and our 32 $\times$ 32 grid gave 1024 unknowns.
The number of iterations required were 101, 245, 445, 737, 1300, 79 and
35, respectively, with the entire process fully vectorizable.   No attempt
was made at optimizing parameters.   The method should experience a good
reduction in the number of iterations required with a better choice of
both r_k's and SHAPE(y).   This problem is part of our on-going research.

All graphics were done at the NASA-Ames Research Center using their
CONPLOT package.   We are most grateful to Dr. John Murphy at NASA-Ames for
both the graphics and the rotating cylinder data.   An important component
of our on-going research addresses application of MBADC to two-dimensional
viscous flow and is being carried out jointly with John Murphy.

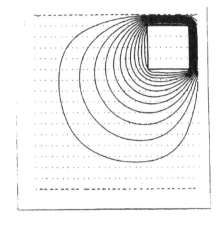 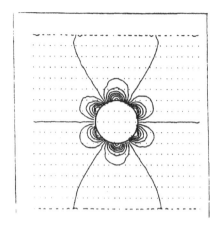

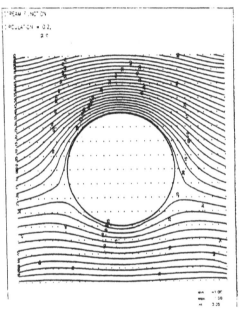

REFERENCES

1. Bellman, R., A note on an inequality of E. Schmidt, *Bull. Amer. Math. Soc.*, 50 (1944), pp. 734--736.

2. Cerutti, John H., and Seymour Parter, Collocation methods for parabolic partial differential equations in one space dimension, *Numer. Math.*, 26 (1976), pp. 227--254.

3. Chan, Tony F. and Faisel Saied, A comparison of elliptic solvers for general two dimensional regions, *SIAM J. Sci. Stat. Comp.*, 6 (1985), pp. 742--60.

4. Cooper, Kevin and P. M. Prenter, Alternating direction collocation for separable P.D.E.: A theory (to be submitted).

5. Cooper, Kevin and P. M. Prenter, Multi-boundary alternating direction collocation schemes for computing flows over arbitrary shapes, to be submitted.

6. deBoor, Carl, and Blair Swartz, Collocation at Gaussian points, *SIAM J. Numer. Anal.*, 10 (1973), pp. 582--606.

7. deBoor, C., and B. Swartz, Collocation approximation to eigenvalues of an ordinary differential equation: The principle of the thing, *Math. Comp.*, 35, no. 151 (1980), pp. 679--94.

8. Douglas, J., and T. Dupont, Collocation methods for parabolic equations in a single space variable, *Lecture Notes in Math.*, Vol. 385, Springer, Berlin and New York, 1966.

9. Dyksen, Wayne R., Tensor product generalized ADI methods for separable elliptic problems, *SIAM J. Numer. Anal.*, 24 (1987), pp. 59--76.

10. Hayes, L., An alternating direction collocation method for finite element approximations on rectangles, *Comp. Math. Appl.* 6 (1980), pp. 45--50.

11. Hayes, L., G. Pinder, and M. Celia, Alternating direction collocation for rectangular regions, *Comp. Meth. Appl. Mech. Eng.* 27 (1981), pp. 265--77.

12. Lapidus, Leon, and G. Pinder, *Numerical Solution of Partial Differential Equations in Science and Engineering*, Wiley, New York, 1982.

13. Locker, J. and P. M. Prenter, Optimal L^2 and L^∞ error estimates for continuous and discrete least squares methods for boundary value problems, *SIAM J. Numer. Anal.* 15, no. 6 (1978), pp. 1151--60.

14. Percell, Peter, and Mary Fanett Wheeler, A C^1 finite element collocation method for elliptic equations, *SIAM J. Numer. Anal.* 17, no. 5 (1980), pp. 605--622.

15. Prenter, P. M., *Splines and Variational Methods*, Wiley, New York, 1975.

16. Prenter, P. M., and R. D. Russell, Orthogonal collocation for elliptic partial differential equations, *SIAM J. Numer. Anal.* 13, no. 6 (1976), pp. 923--39.

17. Russell, R. D., and L. F. Shampine, A collocation method for boundary value problems, *Numer. Math.* 19 (1972), pp. 1--28.

18. Thames, Frank C., Joe F. Thompson, and C. Wayne Mastin, Numerical solution of the Navier Stokes equations for arbitrary two dimensional airfoils, in *Aerodynamic Analyses Requiring Advanced Computers, Part I*, Proceedings of a Conference at Langley Research Center, Hampton, VA, Mar 4-6, 1975.

19. Vainikko, G. M., On the stability and convergence of the collocation method, *Differentsial'nye Uravneniya*, 1 (1965), pp. 244--54.

20. Vainikko, G. M., On convergence of the collocation method for nonlinear differential equations, *USSR Comp. Math. and Math. Phys.*, 6 (1966), pp. 35--42.

21. Wittenbrink, K. A., Higher order projection methods of moment and collocation type for nonlinear boundary value problems, *Computing*, 11 (1973), pp. 255--74.

A Priori L^2 Error Estimates for Galerkin Approximations to Semilinear Parabolic Integro-Differential Equations

YANPING LIN   Department of Pure and Applied Mathematics, Washington State University, Pullman, Washington

In this paper we shall study Galerkin approximations to the solution of the nonlinear integro-differential equation of parabolic type :

$$u_t = \nabla \cdot \{a(u)\nabla u + \int_0^t b(x,t,\tau,u(x,\tau))\nabla u(x,\tau)d\tau\} + f(u)$$

A so-called non-classical H^1 projection will be defined and used in the derivations of the optimal L^2 error estimates for the continuous , Crank-Nicolson and extrapolated Crank-Nicolson approximations.

1. INTRODUCTION

Recently, attention has been given to the numerical approximation of the solution to the equation

$$u_t = \nabla \cdot \{a(u)\nabla u + \int_0^t b(x,t,\tau,u(x,\tau))\nabla u(x,\tau)d\tau\} + f(u), \quad \text{in } Q_T \ (1.1)$$

$$u(x,0) = u_0(x), \quad x \in \Omega \tag{1.2}$$

$$u(x,t) = 0, \quad \text{on} \quad S_T = \partial\Omega \times [0,T], \tag{1.3}$$

where $Q_T = \Omega \times (0,T]$, $T > 0$, Ω is an open bounded subset in R^n with smooth boundary $\partial\Omega$, and the functions a, b, f and u_0 are known .

The problem (1.1)-(1.3) can arise from many physical processes. For example, it can serve as a model in various gas diffusion problems [18] and in heat transfer problems with memory [10, 17]. It is also an example in the general theory of equations of the form

$$u'(t) = Au(t) + \int_0^t B(t,\tau)u(\tau)d\tau + f(u(t)), \quad t > 0, \qquad (1.4)$$

$$u(0) = u_0, \qquad (1.5)$$

in a Banach space E, where A and B are two linear or nonlinear operators [11, 15]. For the formulation of (1.1)-(1.3) and the results on the existence, uniqueness and stability of the solution, we refer the reader to [10, 11, 14, 15, 17, 18].

For approximating the solution u, both finite difference and finite element methods have been considered by several authors. Douglas and Jones [6] and Sloan and Thomée [19] have studied finite difference and time discretization. A similar treatment has been considered in [2] while Galerkin procedures have been formulated by Greenwell and Fairweather [9] and Cannon, Li and Lin [4]. The authors of [9] have obtained optimal L^2 error estimates for a special case of (1.1) and Lin [12] has derived optimal H^1 error estimates for more general equations than (1.1) . An optimal L^2 estimate has been obtained by Lin [13] for a certain linear equation. In order to obtain optimal L^2 error estimates for time-dependent problems, it has been the practice to introduce an auxiliary H^1 projection [22]. In our case equation (1.1) contains two second-order spatial derivatives ; we shall introduce a non-classical H^1 projection in this paper by which optimal L^2 error estimates can be derived. Moreover, this method can be modified and used to obtain optimal L^2 error estimates for other problems in which there are two or more higher order derivatives involved. Equations of Sobolev type provide such examples .

In Section 2 , we shall define some useful notation and formulate the Galerkin procedures. In Section 3 our non-classical H^1 projection will be introduced and its properties will be analyzed. Optimal L^2 error estimates for the continuous and discrete time Galerkin approximations will be demonstrated in Section 4 and Section 5, respectively.

2. NOTATION AND FORMULATION OF GALERKIN APPROXIMATIONS

Let $H^s(\Omega)$ be the Sobolev spaces on Ω with norm

$$||\phi||_s^2 = \sum_{\alpha \le s} ||\frac{\partial^\alpha \phi}{\partial x^\alpha}||^2,$$

where $|| \cdot ||$ denotes the L^2 norm on Ω and $H_0^1(\Omega)$ is the completion of $C_0^\infty(\Omega)$ under the norm $|| \cdot ||_1$. Let

$$(\phi, \psi) = \int_\Omega \phi(x)\psi(x)dx$$

denote the inner product in $L^2(\Omega)$ for scalars. We also use this notation for vectors . Finally, we note that $||\phi||^2 = (\phi, \phi)$.

Let $\{S_h\}_{0<h\leq 1}$ be a family of finite-dimensional subspaces of $H_0^1(\Omega)$, with the following approximation properties:

$$\inf_{\chi\in S_h}\{\|v-\chi\|+h\|v-\chi\|_1\}\leq Ch^s\|v\|_s,\quad v\in H^s\cap H_0^1,\qquad(2.1)$$

where the constant C is independent of h and v.

For the function $a(u)$, we assume that there exist constants C_0, $C_1>0$ such that

$$0<C_0\leq a(u)\leq C_1,\quad u\in R.\qquad(2.2)$$

We also assume that the functions $a(u)$, $b(x,t,\tau,u)$ and $f(u)$ are smooth and bounded together with their derivatives, and $u_0\in H^s(\Omega)\cap H_0^1(\Omega)$.

The continuous Galerkin approximation to the solution u is defined to be a map $U(t):[0,T]\to S_h$ such that

$$(U_t,\chi)+\left(a(U)\nabla U+\int_0^t b(U(\tau))\nabla U(\tau)d\tau,\nabla\chi\right)$$
$$=(f(U),\chi),\ t>0,\ \chi\in S_h,\qquad(2.3)$$
$$(U(\cdot,0)-u_0,\nabla\chi)=0,\ t=0,\ \chi\in S_h.\qquad(2.4)$$

Here and in what follows we shall not write the dependent variable x,t,τ for the function b unless it is necessary.

Let P be a positive integer. Let $\Delta t=T/P$, $t_m=m\Delta t$ and $t_{m+1/2}=(m+1/2)\Delta t$. We also define $g_m=g(x,t_m)$, $g_{m+1/2}=(1/2)(g_{m+1}+g_m)$ and $\partial_t g_m=(\Delta t)^{-1}(g_{m+1}-g_m)$. Thus, the Crank-Nicolson approximation [see 13] is defined to be a sequence $\{U_m\}_{m=0}^P$ such that

$$(\partial_t U_m,\chi)+(a(U_{m+1/2})\nabla U_{m+1/2},\nabla\chi)$$
$$=-\left(\Delta t\sum_{k=0}^m b_{mk}(U_{k+1/2})\nabla U_{k+1/2},\nabla\chi\right)+(f(U_{m+1/2}),\chi),\ \chi\in S_h,\quad(2.5)$$
$$(U_0-u_0,\chi)=0,\ \chi\in S_h.\qquad(2.6)$$

where

$$b_{mk}(v)=\frac{1}{2}\{b(x,t_{m+1},t_{k+1/2},v)+b(x,t_m,t_{k+1/2},v)\},\ k=0,1,\cdots,m-1,$$
$$b_{mm}(v)=\frac{1}{2}b(x,t_{m+1},t_{m+1/2},v),$$

for any function v. We know from [13] that (2.4) will result in the truncation error $O((\Delta t)^2)$ in time. But it will require the solution of a nonlinear algebraic system at each time step. In order to avoid this difficulty, we shall consider the so-called extrapolated Crank-Nicolson Galerkin approximation which is defined by

$$(\partial_t U_m,\chi)+(a(EU_m)\nabla U_{m+1/2},\nabla\chi)$$
$$=-\left(\Delta t\sum_{k=1}^m b_{mk}(EU_k)\nabla U_{k+1/2},\nabla\chi\right)\qquad(2.7)$$
$$-(\Delta t b_{m0}(U_{1/2})\nabla U_{1/2},\nabla\chi)+(f(EU_m),\chi),\quad\chi\in S_h,m=1,2,\cdots,P-1.$$

where $EU_m = (3/2)U_m - (1/2)U_{m-1}$ and where U_0 and U_1 can be obtained by the approximation (2.5)-(2.6) or other similar procedure. We see that (2.7) is also $O((\Delta t)^2)$ order accuracy in time but requires solving only a *linear* algebraic system at each time step.

Let X be a Banach space with norm $||\cdot||_X$ and $\phi : [0,T] \to X$. Define

$$||\phi||^2_{L^2(X)} = \int_0^T ||\phi(t)||^2_X d\tau,$$

$$||\phi||_{L^\infty(X)} = \text{ess} \sup_{0 \le t \le T} ||\phi(t)||_X.$$

Let the space $H^k(H^s)$ be defined by

$$H^k(0,T;H^s) = \{u \in H^s \mid \frac{\partial^j u}{\partial t^j} \in L^2(0,T;H^s), \ j = 0,1,\cdots,k\}$$

and for each $t \in (0,T]$ define

$$||u||^2_{k,s} = \sum_{j=0}^k \{||\frac{\partial^j u}{\partial t^j}||^2_s + \int_0^t ||\frac{\partial^j u}{\partial t^j}||^2_s d\tau\}.$$

Thus we have

$$\int_0^t ||u||^2_{k,s} d\tau \le C||u||^2_{H^k(H^s)}$$

where C is a positive constant independent of u.

REMARK : $||\cdot||_{k,s}$ is *not* the norm on $H^k(H^s)$.

We now give a lemma which will be used throughout this paper.

LEMMA 1.1. If functions $f(t)$ and $g(t)$ are nonnegative and satisfiy

$$f(t) \le Cg(t) + C\int_0^t f(\tau)d\tau, \ 0 \le t \le T, \tag{2.8}$$

then we have

$$f(t) \le C(g(t) + \int_0^t g(\tau)d\tau)\exp\{CT\}, \ 0 \le t \le T. \tag{2.9}$$

Proof: The proof is an elementary application of Gronwall's lemma .

We shall make extensive use of the following two inequalities

$$ab \le \epsilon a^2 + \frac{b^2}{4\epsilon}, \quad \epsilon > 0$$

and

$$||v|| \le C||\nabla v||, \quad v \in H_0^1(\Omega),$$

where C is a positive constant independent of $v \in H_0^1(\Omega)$. We shall denote by C any positive constant which is independent of the parameter h .

3. A NON-CLASSICAL H^1 PROJECTION OF THE SOLUTION

In this section let u be the solution of (1.1)-(1.3). For any function v, we use Rv to denote its weighted H^1 projection into S_h. That is, Rv satisfies

$$(a(u)\nabla(v - Rv), \nabla\chi) = 0, \ \chi \in S_h \tag{3.1}$$

and

$$\|v - Rv\|_p \le C(u)h^{s-p}\|v\|_s, \ 0 \le p \le s, \quad p = 0, 1. \tag{3.2}$$

where $C = C(u)$ is a postive constant. Wheeler [22] utilized Ru in order to obtain optimal L^2 error estimates for Galerkin approximations to parabolic equations. Modifying this idea, we define a map $W(t) : [0, T] \to S_h$ such that

$$\left(a(u)\nabla(u - W) + \int_0^t b(t, \tau, u(\tau))\nabla(u - W)(\tau)d\tau, \ \nabla\chi\right) = 0, \ t > 0, \ \chi \in S_h \tag{3.3}$$

We call this $W(t)$ a non-classical H^1 projection of u into S_h. It is easy to see that (3.3) is actually a system of integral equations of Volterra type. In fact, if $S_h = \text{span}\{\psi_k\}_{k=0}^N$ and the ψ_k, $k = 1, 2, \cdots, N$ are linearly independent, then if $W(t)$ is given by

$$W(x, t) = \sum_{k=1}^N D_k(t)\psi_k(x),$$

(3.3) can be rewritten as

$$A(t)D(t) + \int_0^t B(t, \tau)D(\tau)d\tau = F(t). \tag{3.4}$$

Here $A(t)$, $B(t)$ are matrices, $D(t)$, $F(t)$ are vectors; indeed,

$$D(t) = (D_1(t), \cdots, D_N(t))^T, \ F(t) = (F_1(t), \cdots, F_N(t))^T,$$
$$F_k(t) = (a(u)\nabla u + \int_0^t b(t, \tau, u(\tau))\nabla u(\tau)d\tau, \nabla\psi_k), \ k = 1, 2, \cdots, N,$$
$$A(t) = ((a(u)\nabla\psi_k, \nabla\psi_l)), \text{ and } B(t, \tau) = ((b(t, \tau, u(\tau))\nabla\psi_k, \nabla\psi_l)).$$

From the positivity of A and the linearity of (3.4) we see that the system (3.4) posesses a unique solution $C(t)$. Conseqently, there exist at least one $W(t) \in S_h$ such that (3.3) holds. To establish uniqueness, set $Z = W^1 - W^2$, where W^1 and W^2 are two solutions. Thus, Z satisfies

$$\left(a(u)\nabla Z + \int_0^t b(t, \tau, u(\tau))\nabla Z(\tau)d\tau, \ \nabla\chi\right) = 0, \quad \chi \in S_h.$$

By letting $\chi = Z \in S_h$ it follows that

$$C_0\|\nabla Z\|^2 \le \frac{C_0}{2}\|\nabla Z\|^2 + C\int_0^t \|\nabla Z\|^2 d\tau,$$

from which and Gronwall's inequality we have $Z = 0$. Hence, $W^1 = W^2$.

Now let us study some properties of W. If we set $\eta = u - W$, then we have

LEMMA 3.1. If u, ∇u are bounded and $u \in L^2(H^s)$, then there exists $C = C(u)$, independent of h, such that

$$||\eta|| + h||\eta||_1 \leq Ch^s||u||_{0,s}. \tag{3.5}$$

Proof : Since $W \in S_h$, then we see that

$$\frac{C_0}{2}||\nabla\eta||^2 - \int_0^t C||\nabla\eta||^2 \leq (a(u)\nabla\eta + \int_0^t b(t,\tau,u(\tau))\nabla\eta(\tau)d\tau, \nabla\eta)$$

$$= -(a(u)\nabla\eta + \int_0^t b(t,\tau,u(\tau))\nabla\eta(\tau)d\tau, \nabla(\chi - u)), \quad \chi \in S_h \ .$$

If we take $\chi = Ru$, it follows that

$$\frac{C_0}{2}||\nabla\eta||^2 \leq C\{\int_0^t ||\nabla\eta||^2 d\tau + C(||\nabla\eta|| + \int_0^t ||\nabla\eta||d\tau)\}||\nabla(Ru - u)||$$

$$\leq \frac{C_0}{4}||\nabla\eta||^2 + C\int_0^t ||\nabla\eta||^2 d\tau + C||\nabla(Ru - u)||^2. \tag{3.6}$$

Hence, from the properies of the standard weighted H^1 projection (3.1),

$$||\nabla\eta||^2 \leq Ch^{2s-2}||u||_s^2 + C\int_0^t ||\nabla\eta||^2 d\tau. \tag{3.7}$$

Lemma 1.1 thus implies that

$$||\nabla\eta||^2 \leq Ch^{2s-2}||u||_{0,s}^2. \tag{3.8}$$

We shall now estimate $||\eta||$. Let $\psi \in H^2(\Omega)$ such that

$$(\eta, v) = (a(u)\nabla\psi, \nabla v), \quad v \in H_0^1(\Omega). \tag{3.9}$$

Then, it follows from (3.2) ,(3.3), (3.7) and integration by parts that

$$||\eta||^2 = (a(u)\nabla\psi, \nabla\eta) = (a(u)\nabla\eta, \nabla\psi)$$

$$= (a(u)\nabla\eta + \int_0^t b(u)\nabla\eta d\tau, \nabla\psi) - (\int_0^t b(u)\nabla\eta d\tau, \nabla\psi)$$

$$= (a(u)\nabla\eta + \int_0^t b(u)\nabla\eta, \nabla(\psi - R\psi)) + \int_0^t (\eta, \nabla \cdot b(u)\nabla\psi)d\tau$$

$$\leq Ch^s||u||_{0,s}||\psi||_2 + C(\int_0^t ||\eta||d\tau)||\psi||_2$$

$$\leq C(h^s||u||_{0,s} + \int_0^t ||\eta||d\tau)||\psi||_2. \tag{3.10}$$

Since $||\psi||_2 \leq C||\eta||$ by (3.9), we obtain from (3.7) and Lemma 1.1 that

$$||\eta|| \leq Ch^s||u||_{0,s} \tag{3.11}$$

Hence, Lemma 3.1 has been proved.

LEMMA 3.2. : If u, ∇u, u_t and ∇u_t are bounded, and $u \in H^1(0,T;H^s(\Omega))$, then there exists $C = C(u)$, independent of h, such that

$$||\eta_t|| + h||\nabla \eta_t|| \leq Ch^s||u||_{1,s}. \tag{3.12}$$

If u, u_t, u_{tt}, ∇u, ∇u_t and ∇u_{tt} are bounded and $u \in H^2(0,T;H^s)$, then there exists a $C = C(u)$, independent of h, such that

$$||\eta_{tt}|| + h||\nabla \eta_{tt}|| \leq Ch^s||u||_{2,s} . \tag{3.13}$$

Proof : We differentiate (3.3) to obtain

$$
\begin{aligned}
(a(u)\nabla \eta_t \;\; &+ \;\; (b(t,t,u) + a_t(u))\nabla \eta \\
&+ \;\; \int_0^t b_t(t,\tau,u(\tau))\nabla u(\tau)d\tau, \; \nabla \chi) = 0, \;\; \chi \in S_h.
\end{aligned}
\tag{3.14}
$$

Hence,

$$
\begin{aligned}
\frac{C_0}{2}||\nabla \eta_t||^2 &- C(||\nabla \eta||^2 + \int_0^t ||\nabla \eta||^2 d\tau) \\
\leq \;\; &(a(u)\nabla \eta_t + (b(t,t,u) + a_t(u))\nabla \eta + \int_0^t b_t(u)\nabla \eta d\tau, \; \nabla \eta_t) \\
= \;\; &-(a(u)\nabla \eta_t + (b(t,t,u) + a_t(u))\nabla \eta + \int_0^t b_t(u)\nabla \eta d\tau, \; \nabla(\chi - u_t)).
\end{aligned}
\tag{3.15}
$$

If we now take $\chi = Ru_t$, then it is easy to see from Lemma 3.1 that

$$
\begin{aligned}
&\frac{C_0}{2}||\nabla \eta_t||^2 - Ch^{2s-2}||u||_{0,s}^2 \\
\leq \;\; &\frac{C_0}{4}||\nabla \eta_t||^2 + Ch^{2s-2}||u||_{0,s}^2 + C||\nabla(Ru_t - u_t)||^2 \\
\leq \;\; &\frac{C_0}{4}||\nabla \eta_t||^2 + Ch^{2s-2}||u||_{0,s}^2 + Ch^{2s-2}||u_t||_s^2.
\end{aligned}
\tag{3.16}
$$

Thus, it follows that

$$||\nabla \eta_t||^2 \leq Ch^{2s-2}||u||_{1,s}^2. \tag{3.17}$$

As in (3.9), we define $\psi \in H^2(\Omega)$ such that

$$(\eta_t, v) = (a(u)\nabla \psi, \nabla v), \quad v \in H_0^1(\Omega). \tag{3.18}$$

The remainder of the demonstration is then similar to (3.10).

We now turn to a lemma which will be needed in the L^2 estimation for discrete time Galerkin approximations.

LEMMA 3.3. If $u \in C^3(\bar{Q}_T)$, then there exists $C = C(u)$ such that

$$\|\nabla W\| \leq C\|u\|_{0,1}, \tag{3.19}$$

$$\|\nabla W_t\| + \|W_t\| \leq C\|u\|_{1,1}, \text{ and} \tag{3.20}$$

$$\|\nabla W_{tt}\| \leq C\|u\|_{2,1}. \tag{3.21}$$

Proof : This lemma is an immediate consequence of the above two lemmas.

REMARK : We see from Lemma 3.1 that

$$\|u - W\|_{L^\infty(L^2)} \leq Ch^s, \text{ if } u \in L^\infty(H^s).$$

In the remainder of this paper we assume that the approximation space S_h satisfies sufficient inverse properties [22] to ensure that there exists $C = C(u)$ such that

$$\|W_t\|_\infty + \|\nabla W\|_\infty \leq C(u).$$

(In [22], $S_h = S_h^{2m-1}$ is the span of the tensor product of piecewise Hermite polynomials of degree $2m - 1$, $m \geq 1$ with $u \in L^\infty(0, T; W^{q,\infty}(\bar{\Omega}))$, $q \geq (2+n)/2$ and $2 \leq q \leq 2m$.)

4. L^2 ERROR ESTIMATES FOR THE CONTINUOUS GALERKIN APPROXIMATION

In this section we shall employ the non-classical H^1 projection W defined in Section 3 and derive the optimal L^2 error estimate for the continuous Galerkin approximation.

THEOREM 4.1. Let $U(t)$ be the solution of (2.3)-(2.4). If the solution u of (1.1)-(1.3) is such that u and ∇u are bounded , and $u \in L^\infty(H^s) \cap H^1(H^s)$, then there exists a $C = C(u)$, independent of h , such that

$$\|u - U\|_{L^\infty(L^2)} \leq Ch^s . \tag{4.1}$$

Proof : From the definition of $W(t)$, we see that

$$(W_t, \chi) + (a(u)\nabla W + \int_0^t b(u(\tau))\nabla W(\tau)d\tau, \nabla\chi) = (f(u) - \eta_t, \chi), \quad \chi \in S_h. \tag{4.2}$$

Let $u - U = (u - W) + (W - U) = \eta + \theta$ and subtract (4.2) from (2.4) to obtain

$$(\theta_t, \chi) + (a(U)\nabla\theta + \int_0^t b(U(\tau))\nabla\theta(\tau)d\tau, \nabla\chi)$$

$$= -((a(u) - a(U)\nabla W + \int_0^t (b(u(\tau)) - b(U(\tau)))\nabla W(\tau), \nabla\chi)$$

$$+ ((f(u) - f(U)) - \eta_t, \chi), \quad \chi \in S_h. \tag{4.3}$$

If we take $\chi = \theta$, it follows from Lemma 3.1-Lemma 3.3 that

$$\frac{1}{2}\frac{d}{dt}||\theta||^2 + \frac{C_0}{2}||\nabla\theta||^2 - C\int_0^t||\nabla\theta||^2 d\tau$$

$$\leq C||\nabla W||_\infty(||\theta+\eta|| + C\int_0^t||\theta+\eta||d\tau)\,||\nabla\theta||$$

$$+C(||\theta+\eta|| + ||\eta_t||)\,||\theta||$$

$$\leq \frac{C_0}{4}||\nabla\theta||^2 + C(||\theta||^2 + \int_0^t||\theta||^2 d\tau + ||\eta||^2 + ||\eta_t||^2 + \int_0^t||\eta||^2 d\tau)$$

$$\leq \frac{C_0}{4}||\nabla\theta||^2 + C(||\theta||^2 + \int_0^t||\theta||^2 d\tau) + Ch^{2s}||u||_{1,s}^2 . \tag{4.4}$$

Thus, we see from Lemma 1.1 that

$$||\theta|| \leq C\{h^s||u||_{H^1(H^s)} + ||\theta(0)||\}. \tag{4.5}$$

Notice that

$$||\theta(0)|| \leq ||U(0) - u_0|| + ||W(0) - u_0|| \leq Ch^s||u_0||_s.$$

The desired result now follows from (4.5), Lemma 3.1 and the triangle inequality.

5. L^2 ERROR ESTIMATES FOR DISCRETE GALERKIN APPROXIMATIONS

In this section we shall first give L^2 error estimates for the Crank-Nicolson approximation, then state the result for the extrapolated Crank-Nicolson approximation without proof.

THEOREM 5.1. Assume that $u \in C^3(\bar{Q}_T) \cap H^3(0,T;H^1(\Omega))$, $u \in L^\infty(H^s(\Omega)) \cap H^1(H^s)$, and let $\{U_m\}_{m=0}^P$ be the solution of (2.5)-(2.6). Then there exists C, $r > 0$, independent of h, such that for all $0 < \Delta t \leq r$,

$$\max_{0 \leq m \leq P}||u_m - U_m|| \leq C(h^s + (\Delta t)^2) . \tag{5.1}$$

Proof : Note that

$$\int_0^{t_m+1/2} b(x,t_{m+1/2},\tau,u(\tau))\nabla W(\tau)d\tau$$

$$= \frac{1}{2}\{\int_0^{t_{m+1}} b(x,t_{m+1},\tau,u(\tau))\nabla W(\tau)d\tau + \int_0^{t_m} b(x,t_m,\tau,u(\tau))\nabla W(\tau)d\tau\} + E_m$$

$$= \Delta t\sum_{k=0}^m b_{mk}(u(t_{k+1/2}))\nabla W_{k+1/2} + E_m + E_m',$$

thus if we set $\sigma_m(u,W) = E_m + E_m'$, then we know from Section 2 , Section 3 and [13] that

$$||\sigma_m(u,W)||^2 \leq C(\Delta t)^3\int_{t_m}^{t_{m+1}}(||u||_{2,1}^2 + ||W||_{2,1}^2)d\tau$$

$$+C(\Delta t)^4(||u||_{H^2(H^1)}^2 + ||W||_{H^2(H^1)}^2 + 1)$$

$$\leq C(\Delta t)^3\int_{t_m}^{t_{m+1}}||u||_{2,1}^2 d\tau + C(\Delta t)^4.$$

Moreover, for $\chi \in S_h$, we see by evaluating (3.3) at $t = t_{m+1/2}$ that

$$(a(u(t_{m+1/2}))\nabla u(t_{m+1/2}) + \int_0^{t_{m+1/2}} b(t_{m+1/2}, \tau, u(\tau))\nabla u(\tau)d\tau, \nabla\chi)$$

$$= (a(u(t_{m+1/2}))\nabla W(t_{m+1/2}) + \int_0^{t_{m+1/2}} b(t_{m+1/2}, \tau, u(\tau))\nabla W(\tau)d\tau, \nabla\chi)$$

$$= (a(U_{m+1/2}\nabla W_{m+1/2} + \Delta t \sum_{k=0}^m b_{mk}(U_{k+1/2})\nabla W_{k+1/2}, \nabla\chi)$$

$$+ (\sigma_m(u, W) + \Delta t \sum_{k=0}^m (b_{mk}(u(t_{k+1/2})) - b_{mk}(U_{k+1/2})))\nabla W_{k+1/2}, \nabla\chi)$$

$$+ ((a(U_{m+1/2}) - a(u(t_{m+1/2})))\nabla W_{m+1/2}, \nabla\chi)$$

$$+ (a(u(t_{m+1/2}))\nabla(W(t_{m+1/2}) - W_{m+1/2}), \nabla\chi).$$

Letting $\eta_m = u_m - W_m$ and $\theta_m = W_m - U_m$, we have for any $\chi \in S_h$

$$(\partial_t\theta_m, \chi) + (a(U_{m+1/2})\nabla\theta_{m+1/2} + \Delta t \sum_{k=0}^m b_{mk}(U_{k+1/2})\nabla\theta_{k+1/2}, \nabla\chi)$$

$$= ((f(u(t_{m+1/2})) - f(U_{m+1/2})) + \rho_m + \partial_t\eta_m, \chi)$$

$$+ (\Delta a_m\nabla W_{m+1/2} + \Delta t \sum_{k=0}^m \Delta b_{mk}\nabla W_{k+1/2}, \nabla\chi)$$

$$+ (a(u(t_{m+1/2}))\nabla(W_{m+1/2} - W(t_{m+1/2})) + \sigma_m(u, W), \nabla\chi) \qquad (5.2)$$

where

$$\Delta a_m = a(U_{m+1/2}) - a(u(t_{m+1/2})),$$
$$\Delta b_{mk} = b_{mk}(U_{k+1/2}) - b_{mk}(u(t_{k+1/2})),$$
$$\rho_m = \partial_t u_m - u_t(t_{m+1/2}).$$

Now set $\chi = \Delta t\theta_{m+1/2} \in S_h$ in (5.2). It readily follows that

$$\frac{1}{2}\{||\theta_{m+1}||^2 - ||\theta_m||^2\} + \frac{C_0}{2}\Delta t||\nabla\theta_{m+1/2}||^2 - C(\Delta t)^2 \sum_{k=0}^m ||\nabla\theta_{k+1/2}||^2$$

$$\leq C\Delta t\{||u(t_{m+1/2}) - U_{m+1/2}|| + ||\partial_t\eta_m||\}||\theta_{m+1/2}||$$

$$+ C\Delta t \max_{0 \leq m \leq P-1}||\nabla W_{m+1/2}||_\infty(||u(t_{m+1/2}) - U_{m+1/2}||$$

$$+ \Delta t \sum_{k=0}^m ||u(t_{k+1/2}) - U_{k+1/2}||)||\nabla\theta_{m+1/2}|| \qquad (5.3)$$

$$+ C\Delta t\{||\nabla(W_{m+1/2} - W(t_{m+1/2}))|| + ||\rho_m|| + ||\sigma_m(u, W)||\}||\nabla\theta_{m+1/2}||$$

$$\leq \frac{C_0}{4}\Delta t||\nabla\theta_{m+1/2}||^2 + C\Delta t\{||u(t_{m+1/2}) - U_{m+1/2}||^2 + ||\partial_t\eta_m||^2\}$$

$$+ C\Delta t\{||\nabla(W_{m+1/2} - W(t_{m+1/2}))||^2 + ||\rho_m||^2 + ||\sigma_m(u, W)||^2\}$$

$$+ C(\Delta t)^2 \sum_{k=0}^m ||u(t_{k+1/2}) - U_{k+1/2}||^2.$$

Since

$$\|u(t_{m+1/2}) - U_{m+1/2}\|^2$$
$$\leq 2\|u(t_{m+1/2}) - u_{m+1/2}\|^2 + 2\|u_{m+1/2} - U_{m+1/2}\|^2$$
$$\leq 2\|u(t_{m+1/2}) - u_{m+1/2}\|^2 + 4\|\theta_{m+1/2}\|^2 + 4\|\eta_{m+1/2}\|^2 \quad (5.4)$$

and

$$\|u(t_{m+1/2}) - u_{m+1/2}\|^2 \leq C(\Delta t)^3 \int_{t_m}^{t_{m+1}} \|u_{tt}\|^2 d\tau,$$

$$\|\nabla(W_{m+1/2} - W(t_{m+1/2}))\|^2 \leq C(\Delta t)^3 \int_{t_m}^{t_{m+1}} \|\nabla W_{tt}\|^2 d\tau,$$

$$\|\rho_m\|^2 \leq C(\Delta t)^3 \int_{t_m}^{t_{m+1}} \|u_{ttt}\|^2 d\tau,$$

which may be easily verified, we see that

$$\frac{1}{2}\{\|\theta_{m+1}\|^2 - \|\theta_m\|^2\} + \frac{C_0}{4}\Delta t\|\nabla\theta_{m+1/2}\|^2$$
$$\leq C\Delta t\{\|\eta_{m+1/2}\|^2 + \|\partial_t\eta_m\|^2\} + C(\Delta t)^2 \sum_{k=0}^{m} \|\nabla\theta_{k+1/2}\|^2$$
$$+ C(\Delta t)^4 \int_{t_m}^{t_{m+1}} \{\|u\|_{2,1}^2 + \|u\|_{3,0}^2\} d\tau + C(\Delta t)^5$$
$$+ C(\Delta t)^2 \sum_{k=0}^{m} \{\|\eta_{k+1/2}\|^2 + \|\theta_{k+1/2}\|^2\}. \quad (5.5)$$

Summing on m, we have

$$\frac{1}{2}\|\theta_{m+1}\|^2 + \frac{C_0}{4}\Delta t\sum_{l=0}^{m}\|\nabla\theta_{l+1/2}\|^2$$
$$\leq C(\Delta t)^4 + C\|\theta_0\|^2 + C(\Delta t)^2 \sum_{l=0}^{m}\sum_{k=0}^{l}\|\nabla\theta_{k+1/2}\|^2$$
$$+ C\Delta t\sum_{l=0}^{m}(\|\partial_t\eta_l\|^2 + \|\eta_{l+1/2}\|^2)$$
$$+ C(\Delta t)^2 \sum_{l=0}^{m}\sum_{k=0}^{l}\{\|\eta_{k+1/2}\|^2 + \|\theta_{k+1/2}\|^2\} \quad (5.6)$$
$$\leq C(\Delta t)^4 + C\|\theta_0(0)\|^2 + C\Delta t\|\theta_{m+1}\|^2 + C(\Delta t)^2 \sum_{l=0}^{m}\|\nabla\theta_{l+1/2}\|^2$$
$$+ C\Delta t\sum_{l=0}^{m-1}\{\|\theta_l\|^2 + \Delta t\sum_{k=0}^{l-1}\|\nabla\theta_{k+1/2}\|^2\}$$
$$+ C\Delta t\sum_{l=0}^{m}\{\|\partial_t\eta_l\|^2 + \|\eta_{l+1/2}\|^2 + \Delta t\sum_{k=0}^{l}\|\eta_{k+1/2}\|^2\}$$

and

$$(\frac{1}{2} - C\Delta t)\|\theta_{m+1}\|^2 + (\frac{C_0}{4} - C\Delta t)\Delta t\sum_{l=0}^{m}\|\nabla\theta_{l+1/2}\|^2$$

$$\leq \; C\{||\theta_0||^2 + (\Delta t)^4\} + C\Delta t \sum_{l=0}^{m}\{||\eta_{l+1/2}||^2 + ||\partial_t \eta_l||^2\} \qquad (5.7)$$

$$+ C\Delta t \sum_{l=0}^{m}\{||\theta_l||^2 + \Delta t \sum_{k=0}^{l-1}||\nabla \theta_{k+1/2}||^2\}.$$

If we select a small $r > 0$ such that for all $0 < \Delta t \leq r$

$$\frac{1}{2} - C\Delta t \geq \frac{1}{4}, \quad \text{and} \quad \frac{C_0}{4} - C\Delta t \geq \frac{C_0}{8},$$

then the discrete version of Gronwall's lemma implies that

$$||\theta_{m+1}||^2 \leq C\{||\theta_0||^2 + (\Delta t)^4 + \Delta t \sum_{l=0}^{m}(||\eta_{l+1/2}||^2 + ||\partial_t \eta_l||^2)\}. \qquad (5.8)$$

We know that

$$||\theta_0|| \leq Ch^s ||u_0||_s$$

and

$$\Delta t \sum_{l=0}^{m}\{||\eta_{l+1/2}||^2 + ||\partial_t \eta_l||^2 \;\; \leq \;\; C(||\eta_t||_{L^2(L^2)}^2 + ||\eta||_{L^2(L^2)}^2)$$
$$\leq \;\; Ch^s.$$

Thus, it follows that

$$\max_{0 \leq m \leq P} ||\theta_m|| \leq C(h^s + (\Delta t)^2), \qquad (5.9)$$

and hence

$$\max_{0 \leq m \leq P} ||u_m - U_m|| \;\; \leq \;\; \max_{0 \leq m \leq P} ||\theta_m|| + ||\eta||_{L^\infty(L^2)}$$
$$\leq \;\; C(h^s + (\Delta t)^2). \qquad (5.10)$$

This complete the demonstration of Theorem 5.1 .

For the extrapolated Crank-Nicolson approximation (2.8) , we have

THEOREM 5.2 : Under the same assumptions for u as in Theorem 5.1 , where $\{U_m\}_{m=0}^{P}$ is the extrapolated Crank-Nicolson approximation solution of (2.7) with U_0 and U_1 obtained from (2.5)-(2.6), then there exist C, $r > 0$ such that for all $0 < \Delta t \leq r$,

$$\max_{0 \leq m \leq P} ||u_m - U_m|| \leq C(h^s + (\Delta t)^2) . \qquad (5.11)$$

Proof: The result (5.11) is obtained from an argument similar to that employed in Theorem 5.1 .

REMARK : In this section we have obtained the optimal L^2 error estimate for the Crank-Nicolson and the extrapolated Crank-Nicolson approximations. Thus, one could expect that the same optimal error estimates could be obtained for the predictor-corrector scheme [12] and other similar approximations .

REFERENCES

1. H. Brunner, A survey of recent advances in the numerical treatment of Volterra integral and integral-differential equations, *J. Comp. Appl. Math.,* 8; 76-102 (1982).

2. B. Budak and A. Pavlov, A difference method of solving boundary value problem for a quasi-linear integro-differential equation of parabolic type, *Soviet Math. Dokl.,* 14; 565-569 (1974).

3. J. Cannon , S. Pērez-Esteva and J. Hoek, A Galerkin procedure for the diffusion equation subject to the specification of mass, *SIAM J. Numer. Anal.,* 24; 499-515 (1987).

4. J. R. Cannon, Xinkai Li and Yanping Lin, A Galerkin procedure for an integro-differential equation of parabolic type, submitted

5. J. Douglas, Jr and T. Dupont, Galerkin methods for parabolic equations, *SIAM J. Numer. Anal.,* 7; 575-626 (1970).

6. J. Douglas, Jr and B. Jones, Jr ,Numerical method for integral-differential equations of parabolic and hyperbolic type, *Numer. Math.,* 4; 92-102 (1962).

7. Todd Dupont, L^2-estimates for Galerkin methods for second order hyperbolic equations, *SIAM J. Numer. Anal.,* 10; 880-889 (1973).

8. Todd Dupont, Some L^2 error estimates for parabolic Galerkin methods, The Mathematical Foundations of the Finite Element Method with Applications to Partial Differential equations (ed. A. K. Aziz) , pp. 491-504, Academic Press, New York and London (1973) .

9. E. Greenwell-Yanik and G. Fairweather, Finite element methods for parabolic and hyperbolic partial integro-differential equations, to appear in *Nonlinear Analysis* .

10. M. Gurtin and A. Pipkin, A general theory of heat conduction with finite wave speeds, *Arch. Rat. Mech. Anal.,* 31; 113-126 (1968).

11. M. Heard, An abstract parabolic Volterra integral-differential equation, *SIAM J. Math. Anal.,* 13; 81-105 (1982).

12. Yanping Lin, Numerical methods for nonlinear parabolic integro-differential equations, submitted.

13. Yanping Lin , The optimal L^2 estimate for some linear integro-differential equation of parabolic type, submitted.

14. S. Londen and O. Staffans, Volterra Equations, Lecture Notes in mathematics, **737**; Springer-Verlag, Berlin (1979).

15. A. Lunardi and E. Sinestrai, Fully nonlinear integral-differential equations in general Banach space , *Math. Z.,* 190; 225-248 (1985).

16. H. Meyer, The numerical solution of nonlinear parabolic problems by variational methods, *SIAM J. Numer. Anal.,* 10; 700-722 (1973).

17. R. K. Miller, An integro-differential equation for rigid heat conduction with memory, *J. Math. Anal. Appl.,* 66; 313-332 (1978).

18. M. Raynal, On some nonlinear problems of diffusion, Volterra Equations (ed. Londen and Staffans) , Lecture Notes in Mathematics, 737, pp. 251-266, Spring-Verlag, Berlin (1979).

19. I. Sloan and V. Thomée, Time discretization of an integral-differential equation of parabolic type, *SIAM J. Numer. Anal.,* 23; 1052-1061 (1986).

20. L. Tavernini, Finite difference approximations for a class of semilinear Volterra evolution problems, *SIAM J. Numer. Anal.* 14; 931-949 (1977).

21. V. Volterra, Theory of Functions and Integral and Integral-Differential Equations, Dover, New York (1959).

22. M. F. Wheeler, A priori L_2 error estimates for Galerkin approximations to parabolic partial differential equations, *SIAM J. Numer. Anal.,* 10; 723-759 (1973).

Counting Backward from 10

G. MILTON WING   Computer Research and Applications, Los Alamos National Laboratory, Los Alamos, New Mexico

I wish to start by thanking you all for coming and for being so kind. I hope that this meeting has been instructive and enjoyable, and that the sessions tomorrow will be equally satisfying.

This conference has been made possible through the generous support of the Los Alamos National Laboratory to which I express my thanks. Kudos, to the members of the organizing committee, to Dr. Dan Seth and Jan Hull, and to others who have played a large role in making all this possible.

As for myself, I am having a really good time, and have found this to be one of the most exciting birthdays I have ever had. There was one other which was perhaps even more exciting. Let me explain.

Exactly 18 years ago tonight I was standing on the balcony of my hotel in Mazatlan, Mexico, watching the sun set over the Sea of Cortez (or the Gulf of California, if you prefer). The water was completely calm, the skies cloudless. Just as the upper limb of the sun was disappearing, everything turned green–bright green–for a fraction of a second. At first I thought I was hallucinating, but then I realized that I had just witnessed the "green flash." Now that was exciting! A few minutes later I began to wonder. I had always heard that this rare phenomenon is observed over knife-edge mountains. But over a placid sea? Weeks later I discussed my experience with a physicist friend who was collecting data on the green flash. He assured me that many sightings under just the conditions I described had been reported. My excitement continued.

Now, I regret to say the green flash was not observed this evening. So, I request that to compensate for this deficiency and provide me with the same level of excitement I experienced in 1970 you all help out. Please on the count of three, and for just a moment, everyone in this room TURN GREEN! Please!

One, two, three!

Thank you very much. And Vance, I said just for a moment. It is not necessary to remain green.

And now to the subject of my talk. The title comes from a remark that was made at a Los Alamos colloquium in the early 1950's. In those days old timers from wartime often visited the Laboratory and were invited to give colloquia. Frequently these lectures consisted in part of reminiscences of wartime Lab experiences. Professor Sam Allison of the University of Chicago, who had played a large part in the test near Alamogordo, told of what he considered his most difficult assignment. It was the task of learning to count backwards from ten. The comment stuck in my memory.

A few years ago I learned that in New Mexico one of the preliminary tests for driving while intoxicated consists of having the suspect count backward from a hundred. A HUNDRED! And Sam Allison had admitted to difficulties when starting at ten! Is this a fair test of inebriation or is it a test of the suspect's arithmetic ability?

I began pondering. Just what is the mathematical level of the "ordinary man or woman?" of the "man in the street?" (It is inappropriate here to discuss the "woman in the street.") Where is that level of mathematical understanding acquired? Partly in school, obviously. But after that? My guess is that the average person gets virtually his entire mathematical experience from the media–TV, radio, and the newspapers. Now, I have long felt that the media do a terrible job of handling even the simplest mathematical reporting.

Knowing that this meeting was going to take place, I decided to collect examples of such deplorable reporting. I confined my sources primarily to local ones, because I believe that most people get their information from the daily newspaper or newscast. I also decided to avoid anything that might be the result of a possible typographical error, or the radio or TV equivalent, and to leave out of consideration advertisements deliberately designed to mislead with numbers–aspirin ads are fine examples.

About the time I began this exercise the various professional math societies decided to make the public more "mathematically aware." Indeed there have been a couple of "math awareness" weeks in as many years. During one, one community even printed up Math Awareness bumper stickers. I have nothing against these efforts. But I do believe that they may well be missing the target. Our average person is not the least bit interested in the fact that the Bieberbach conjecture has been settled or even that the four color problem has been solved.

Some of the quotes I shall give in what follows may not be exact. Here and there I have excised some words in the interest of brevity. When I provide a verbatim quote it will be so identified.

Let me begin with a recipe for cooking fish that appeared in the Santa Fe daily paper (The New Mexican) on October 15, 1986. It reads, in part, "Cook fresh fish 10 minutes for each inch of thickness (5 minutes if it is one-half inch, 20 minutes if it is two inches thick, etc.)." That arithmetic of which a third grader should be capable must be carried out in detail for the reader is distressing, to say the least. Although, of course the "etc." suggests an implicit understanding of mathematical induction. It is perhaps better in the short term to provide these explicit arithmetic instructions, than to have the poor cook end up with soggy sole or scorched scrod, but in the long run such detailed explanations can well lead to a feeling on the reader's part that all arithmetic will be done for him or her.

Another example of the same ilk. In June 1987, Ann Landers wrote: "At present, only 20 to 30 percent of those individuals who test positive and feel well will develop the disease

AIDS within five years. (This means 70 to 80 percent will not become ill in this time.)'' This example is perhaps worse than the recipe. The former involves multiplication by fractions. Ann Landers finds it desirable to explain subtraction of integers.

Next is a case in which the arithmetic is carefully carried out for the reader–but incorrectly. This example might be considered a typo, except that the figures were cited over and over again, both in writing and in oral news reports. From the New Mexican of February 27, 1987: ''This winter's snowfall of more than 12 feet (in Los Alamos) broke a 30-year record. Yesterday's snowfall brought the total to 131 inches.'' Remember the old days when everyone knew what a gross was?

In the autumn of 1986 the Public Service Company of New Mexico, in an effort to improve its image (an effort frequently engaged in by utility companies) broadcast the following message. ''Our meter readers are required to make not more than one error in one thousand readings. Actually, they do far, far better than that. They make erroneous readings only one-tenth of one percent of the time.'' According to careful calculations on the Cray, one in a thousand is exactly one-tenth of one percent. Now here is a genuine mistake–not an effort to mislead. The recorded announcement played for about a week over my favorite music station, then stopped. I thought: ''Finally, someone at Public Service has caught the error.'' Not so. Vance Faber heard it played a month later on another station.

The use of alcohol leads to arithmetical and statistical nonsense that can drive any total abstainer to drink. Particularly at holiday time, news articles regularly appear, supposedly explaining how much liquor can be consumed before a person becomes legally unable to drive. Usually reference is made to a number of ''drinks'' with no mention made of alcoholic content, size of the consumer, etc. . . A notable exception appeared in the Santa Fe paper in late December 1986. It was a table giving weight of the consumer, elapsed time since drinking began, total drinks, and probable blood alcohol content. A drink was defined as one and one-quarter ounce of 80-proof liquor, one four-ounce glass of wine, or one twelve-ounce beer. Now, the alcohol content of table wine is about 12 percent. (Dessert wines run about 20 percent, but those are not usually consumed in four-ounce glasses, at least not by people who read charts of this kind.) Thus wine and cocktail figures are compatible.

However, a beer is <u>not</u> a beer. A check with a local distributor reveals that Coors Lite, a fairly popular brew in this region, contains 3.29% alcohol, while some imported stouts are close to 10%. For the beer drinker, the table is totally misleading–off by a factor of two or three, depending on the kind of beer consumed.

In matters of medical research the media are especially bad. It seems quite likely to me that medical statistics are frequently of dubious validity, even in their original form. After the press gets through with them, sheer gibberish often results. This particular problem evidences itself these days on a very regular basis in the almost innumerable fragmentary reports on AIDS. Here is an area in which one would hope for the best possible information. Often the information could hardly be worse.

As a particularly disturbing example, I cite an article in the New Mexican of August 20, 1987. The conspicuous heading reads (and this is an exact quote), "One in five condoms deemed faulty in federal testing." Note the date, just about the time that discussion and advertising of such preventive devices on radio and TV became more or less acceptable. The article itself is somewhat garbled, but the clue comes about one third of the way through the discussion, and again I quote verbatim, ". . .in random testing the FDS inspections over the last four months found that roughly 20% of 204 batches (of condoms) tested failed to meet government standards." Further wading through the journalistic verbiage reveals that a batch fails if at least four out of one thousand devices is faulty. The article goes on to say that although fifty-one thousand condoms were tested, there were no figures on how many failed. An estimate is easy if one believes the heading of the story–about ten thousand, a trivial observation that eluded the reporter and the editor. Incidentally, this was an Associated Press dispatch. At what editorial level the article was rendered totally incorrect cannot be determined. Here is an example of very dangerous reporting based on the total inability to understand trivial statistics. How many people read the headline and nothing more? Frightening!

In a lighter vein, let me digress and discuss something that did not appear in the local media, but did receive very wide distribution. I am indebted to Janet Peterson for this one. It is a "game card" distributed a couple of years ago by one of our fast food vendors, Burger King. Prizes were awarded in a fashion which need not be described here, and ranged from one thousand dollars a month for life down to an order of french fries or a bottle of Pepsi. The probability of winning any particular prize on a single restaurant visit is listed on the back of the card. Also listed is the probability (given in terms of "odds") of winning in 12 visits. This latter number was computed by simply multiplying the one-visit probability by 12 in order to obtain the twelve-visit probability. This, of course, is wrong, but a good approximation when the single probability is very small. Trouble sets in at the free Pepsi level when the odds become one out of one of winning in twelve visits. Undaunted by this result, the preparer of the cards, probably an advertising agency, computed the odds of winning <u>some</u> prize in a single visit. The value given is one in seven. Applying the "multiply by 12" algorithm, the card preparer calculated odds of better than one in one for winning <u>some</u> prize in a dozen visits. This information is printed on the card, millions of which were distributed throughout the country. (For those of you who did not bring calculators, let me mention that the correct answer is about 84 out of 100, making the reasonable assumption of independent events, etc.).

Elementary logic also falls in the domain of mathematics. Therefore the following statement attributed to the Governor of New Mexico by the Santa Fe paper on July 18, 1987, might seem to be grist for our mill: "We really don't know why we (the state) have more money than we really have." Whether this is a misquote or the Governor's actual statement, I do not know. Actually, the statement is really an application of a little-known theorem that Joe Lehner and I proved in the early fifties. Theorem: Two things greater than the same thing are greater than each other. (This result is little-known because referees were stuffy in those days.)

I have one final example that I wish to mention. This is not to say that I found no others. I have simply culled out a few representative goodies. Am I making a mountain out of a molehill? In some cases no immediate harm can be expected (the example about contraceptive testing might, I feel, do a lot of damage). But taken in their entirety the cases discussed do seem to show that the average person receives litle mathematical education through the media. Quite the contrary. Is this distressing to people other than professional mathematicians? It should be. The environment created simply makes life easier for the deceptive advertiser, the automobile salesman who manipulates interest rates, the credit card companies, etc., etc... In a culture that is becoming more and more "numerical" in nature, the media seem to be miseducating the public.

What to do? Letters to the editors or calls to the TV and radio stations may be effective over a short period. A volunteer offer to edit any news that may have arithmetic or statistical content would likely not be received kindly by editors and station managers. Asking that each reporter or newscaster be given an elementary math exam before his appointment would create chaos. Perhaps a start could be made by sponsoring, through the professional societies, a "Quantitative Awareness Week."

And maybe my overall reaction is really too emotional. Think of the horrors a professional grammarian experiences every day!

I promised one more example. It is an article describing the so-called "Monte Carlo" method for the layman. Specifically, how to compute the number pi by throwing darts at a circular dart board. The article describes (with considerable enthusiasm) how this activity can be simulated on a computer. And then a sequence of results is given: After 1,000 throws a pi value of 3.16, after 10,000 a value of 3.18, after 100,000 a value of 3.14. Now I give an exact quote, "The accepted vaue of pi is 3.14159, a figure the program would have reached and further refined with sufficient throws."

This statement is clearly both incorrect and misleading: What does "accepted" mean? It conjures up a picture of "experts" sitting around a conference table discussing and voting on the value of pi, knowing full well that if their program could just run longer they might have to reconvene.

Too technical an example you may say. But I haven't told you the source. This information I regret to say appeared in the February 27, 1987 issue of the Los Alamos Newsbulletin, published weekly by the Los Alamos National Laboratory and distributed to virtually every member of that Laboratory.

Clearly, it is time to stop. And now I shall begin my countdown.

TEN, NINE, EIGHT, SIX. . . I KNEW I would get it wrong!

Thank you.

Index

Printed and bound by CPI Group (UK) Ltd, Croydon, CR0 4YY

22/10/2024

01777634-0019